Electric Circuits
and
Modern Electronics

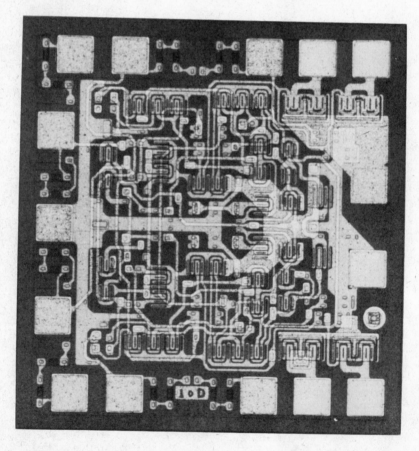

A highly magnified photograph of an integrated D-type, flip-flop circuit. Digital circuits, of which this is an example, are discussed in Chapter 17. Photograph used courtesy of Motorola Inc., Semiconductor Products Division.

Electric Circuits
and
Modern Electronics

L. W. Anderson

Department of Physics

W. W. Beeman

*Department of Physics
and Biophysics Laboratory*

The University of Wisconsin
Madison, Wisconsin

HOLT, RINEHART AND WINSTON, INC.
*New York   Chicago   San Francisco   Atlanta
Dallas   Montreal   Toronto   London   Sydney*

Dedicated to MYA and EMB

Preface

The phenomenal growth of solid state electronics has made available to the scientist and engineer a vast array of highly sophisticated equipment for data acquisition and data processing. No scientist or engineer can ignore these revolutionary developments in instrumentation but many lack the training and background necessary to make the best use of the new instrumentation. The need for this training has been met in numerous ways. In the physics departments of many colleges and universities the first semester of the customary year of undergraduate electricity and magnetism has become a course in circuit theory and solid state electronics. This text is based upon several years experience in teaching such a course at the University of Wisconsin, Madison. Our students have been juniors, seniors, and a few beginning graduate students. About half are physics majors, the rest are other science majors and engineers with the exclusion of electrical engineers. We assume a background of a good general physics course and calculus. Some prior contact with ordinary differential equations is helpful but not essential. No prior knowledge of electric circuits is assumed.

A decision to write another textbook should always be explained. In developing our course we failed to find in any one textbook what we considered to be a proper selection of topics and we failed to find discussions which began at an elementary level but led to a real understanding of the material. If these difficulties are to be avoided in a book of reasonable length one must be selective. Topics we consider of primary importance are discussed in depth. Topics of secondary importance are mentioned only briefly or are developed in problems.

The first seven chapters give a careful and reasonably complete treatment of dc and ac circuit theory. We are convinced that too many students waste a great deal of time in their study of electronics because they do not understand circuit theory. As we shall explain not all of the material of these chapters is needed but in our experience the basics are essential. Chapter 8 introduces nonlinear circuits and the small signal linear approximation using vacuum tubes to develop the ideas. However, the essential ideas, present in Chapter 8 are repeated in later chapters for those who do not wish to include any material on vacuum tubes in their course. Chapter 9 and sections of Chapter 10 discuss some of the solid state physics of metals, insulators, semiconductors, and semiconductor junctions. Since the rest of the book is concerned almost exclusively with solid state electronic devices a major problem is to present enough background in solid state physics for a reasonable understanding of electronic processes in semiconductors. Most of our students have been unhappy when we have attempted to leave out this material and simply present the properties of various devices without a discussion of the physical processes going on inside the device. However an instructor who wishes to replace this material with a somewhat less complete description of solid state physics based perhaps on a chemical approach will find that the material presented in chapters 10 through 18 is still readily useable.

The latter part of chapter 10 and chapter 11 describe the structure and properties of bipolar and field effect transistors and their use in single transistor circuits with incremental signals. We emphasize the use of rapid and approximate methods for the calculation of amplification and of input and output impedances. The concept of feedback is introduced and its central importance in linear electronics is emphasized.

In chapters 16 and 17 we treat nonlinear electronics. Here our The various forms of negative feedback are carefully treated and the essentials of ac and dc amplifiers are presented. The difference amplifier is discussed in detail and an entire chapter (chapter 15) is devoted to the properties and uses of operational amplifiers. In these chapters we gradually deemphasize the single transistor as the unit of circuit analysis and instead lead the student to think in terms of prepackaged or integrated circuits which can be adapted to a wide variety of different functions.

The last three chapters treat nonlinear electronics. Here our main purpose is to introduce the fundamental ideas in the processing of information by digital methods. Binary logic and logic gates and their applications are described. Chapter 18 is a description of a number of common electronic instruments such as the oscilloscope and the electrometer which the student will use in laboratory work. In this chapter we include a fairly complete description of the organization and use of a small digital computer. We believe that the small digital computer is now and will become even more in the future a common laboratory instrument that most scientists and engineers will use on a routine basis.

Our text contains more material than can be presented in a single semester. This is a deliberate choice in the interest of flexibility. The text has been organized so that courses with rather different kinds of emphasis can be taught by the selection of different material. If a minmum of dc and ac circuit theory is needed then chapter 3 on direct current measurements, parts of chapter 6, and all of chapter 7 can be omitted without any serious difficulties in the later discussions. Chapter 8 can be.omitted by those who wish to go directly to solid state electronics without a discussion of vacuum tubes. Chapter 9 and parts of chapter 10 can be replaced by a much shorter discussion of the physical basis of solid state devices if the instructor wishes. The essentials of linear electronics are given in chapters 11 and 12. In addition considerable picking and choosing from the later chapters is possible.

We consider problem solving and laboratory experience to be of the greatest importance. No substitute for, nor simulation of either, is possible. Our students have found much of the material of chapter 18 is helpful in the laboratory. Most of the circuits given in the text can be used in the laboratory. We have included an extensive problem list with each chapter except chapter 18. The problems include some drill exercises but, particularly in the later chapters, there are an increasing number of problems whose solutions illuminate and extend the textual material. The solutions to most of these problems will be found at the end of the text. In two appendices we list the values of the natural constants and also a number of useful references.

Finally it is a pleasure to thank the many sources which have contributed to our own understanding of this subject matter. These include the various companies which have kindly permitted us to use circuit diagrams, component characteristics and illustrations. Their technical publications are usually models of good exposition. A number of individuals have made invaluable criticisms of major parts of the manuscript. These include Professor H. W. Lefevre of the University of Oregon, Professor J. K. Roberge of the Massachusetts Institute of Technology and Professors T. K. Bergstresser, R. R. Borchers, R. N. Dexter, P. R. Moran, and M. A. Thompson, all of the University of Wisconsin. The authors, of course, wish to retain complete responsibility for the inadequacies of the text. This we have assured by not showing our sources the final form in which their suggestions appear.

L. W. Anderson
W. W. Beeman

Madison, Wisconsin
January, 1973

Contents

Electric Circuits
and
Modern Electronics

1/Direct Current Circuits

A direct current (dc) circuit is one in which the potential differences and currents are constant, independent of time. Actually, few circuits of interest meet these requirements but the same physical principles and general methods of circuit analysis apply to all circuits. Such methods are most easily introduced and illustrated using dc circuits. This is the main purpose of the first three chapters of the text.

1.1 Electric Charges and Currents

In the rationalized meter-kilogram-second system (mks),[1] Coulomb's law for the interaction between two point charges at rest in vacuum is written

$$F_{12} = \frac{1}{4\pi\varepsilon_0} \cdot \frac{Q_1 Q_2}{r_{12}{}^2} \qquad (1.1)$$

The magnitude of the force that one charge exerts on the other, F_{12}, is in newtons; the charges, Q_1 and Q_2, are measured in coulombs; and their

---

[1] We shall use the following prefixes to represent the powers of 10:
pico (p), 10^{-12}; nano (n), 10^{-9}; micro (μ), 10^{-6}; milli (m), 10^{-3}; centi (c), 10^{-2}; deci, 10^{-1}; kilo (k), 10^3; and mega (M), 10^6.

separation r_{12} is in meters (m). The force F_{12} is directed along the line connecting Q_1 and Q_2. It is repulsive if Q_1 and Q_2 have the same sign and attractive if Q_1 and Q_2 have opposite signs. The constant ε_0 is called the *permittivity of free space*. The quantity $1/4\pi\varepsilon_0$ is numerically equal to $10^{-7}c^2$ where c is the velocity of light in vacuum. To within a few tenths of 1%,

$$\frac{1}{4\pi\varepsilon_0} = 9 \times 10^9 \frac{\text{newton-meter}^2}{\text{coulomb}^2} \tag{1.2}$$

The total Coulomb force on a charge is the vector sum of the forces due to all other charges. By the introduction of the concept of the electric field the interaction among charges can be broken into two parts. Each charge is the source of an electric field in the space around it and upon each charge a force is exerted by the resultant electric field at its position produced by all other charges. The electric field **E** at a given point is defined by

$$\mathbf{E} = \frac{\mathbf{F}}{Q} \tag{1.3}$$

where **F** is the force on a charge Q at the field point. The direction of E is the direction of the force on a positive charge. Comparing Eqs. 1.1 and 1.3 we see that the magnitude of the electric field produced by a point charge Q is given by

$$E = \frac{1}{4\pi\varepsilon_0} \cdot \frac{Q}{r^2} \tag{1.4}$$

where r is the distance from the source charge Q to the field point. The field is directed radially away from a positive source charge.

Work is done when a charge moves from one point to another in an electric field. The work done by the field on the charge can be written as the line integral of the force along the path of the moving charge

$$W = \int_a^b \mathbf{F} \cdot d\mathbf{l} = Q \int_a^b \mathbf{E} \cdot d\mathbf{l} \tag{1.5}$$

The work per unit charge $V = W/Q$ is by definition the potential difference between the points a and b. Since the electrostatic field is conservative, the work is independent of the path followed and a potential can be assigned to each point in the field. Only potential differences have physical significance, and a zero of potential can be assigned arbitrarily. The sign convention for potential differences is so chosen that if the field does positive work on a positive charge that moves from a to b, then the potential V_a is algebraically greater than V_b. With this sign convention the potentials V_a and V_b satisfy the equation

$$V_b - V_a = -\int_a^b \mathbf{E} \cdot d\mathbf{l} \tag{1.6}$$

The unit of potential difference is the joule per coulomb, which is called a *volt*. The unit of electric field, for which there is no special name, is the newton per coulomb, the equivalent volt per meter or frequently, in practical discussion, the volt per centimeter.

A flow of electric charge is a current. The current through a given surface is simply the net charge passing through the surface per unit time. In the mks system the unit of current is the coulomb per second or ampere. Charges of both signs may move as in an electrolytic solution or the entire current may result from the motion of only one kind of charge. Examples of the latter are the electron beam in a television tube or the drift of the electrons in a metal through a stationary array of positive ions. Almost without exception the circuit effects of positive charges moving in one direction are the same as those of negative charges moving in the opposite direction. In either case the direction of the electric current is by convention the direction of flow of positively charged current carriers. Thus in a metal a current to the right is in fact produced by a flow of electrons to the left.

Except in the phenomenon of superconductivity, which will not concern us, charges moving through a material medium, solid, liquid, or gas, always encounter a resistance to their motion. The flow of charge can be maintained only if an electric field is present. The current is in the direction of the field whether the carriers are positive or negative. Commonly, but not always, the current is proportional to the field and we may write

$$\mathbf{j} = \sigma \mathbf{E} \tag{1.7}$$

In this equation $\mathbf{j}$ is a current density, that is, the current through a unit area at right angles to the direction of flow, and σ is the conductivity of the medium. The resistivity is $\rho = 1/\sigma$.

Equation 1.7 is an expression of Ohm's law. The more familiar form in terms of currents and potential differences is easily developed with the help of Fig. 1.1 in which is shown a conductor of cross-section A and length l.

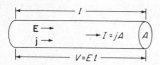

Figure 1.1   The field, current, and potential difference in a long conducting wire of constant cross section. For this wire $V = El = IR = (jA)(\rho l/A)$. The power dissipated in the wire is $P = IV = (E^2/\rho)lA$.

Let us assume that the field and the current density are constant inside the conductor and that the field and therefore the current density are parallel to the length of the conductor. Only the magnitudes of $\mathbf{E}$ and $\mathbf{j}$ are of importance. We can rewrite Eq. 1.7 in the form

$$(El) = (jA)\left(\frac{\rho l}{A}\right) \tag{1.8}$$

The potential difference between the ends of the conductor is $V = El$ and the total current flowing is $I = jA$. This gives the usual form of Ohm's law

$$V = IR \tag{1.9}$$

where $R = \rho l/A$ is the resistance of the conductor. The resistance is measured in volts per ampere or ohms. The unit of resistivity is the ohm-meter or, frequently, the ohm-centimeter.

The current in a conductor always moves from the high potential end to the low potential end. The current carriers, depending on their sign, may move in either direction but that direction is, of necessity, the direction of the force on the carrier from the electric field. Thus work is done by the electric field on the current carriers. In a material medium the carriers very rapidly transmit this energy to the atoms and molecules with which they collide and Joule heating of the medium results. The rate of transfer of energy or the power P dissipated in a conductor of resistance R is given by the charge per unit time that passes through the resistance times the potential difference across the resistance.

$$P = V\frac{dQ}{dt} = IV = I^2R = \frac{V^2}{R} \tag{1.10}$$

In the mks system the power P is in watts.

1.2   Conduction in Metals

Perhaps the simplest electric circuit is a battery between whose terminals is connected a metallic wire. Because chemical processes can provide a continuous supply of electrical energy, an electric field can be maintained within the wire, electrons move in response to the field, and a current flows in the circuit.

We shall discuss in some detail the current flow in metals since it provides a foundation for our later treatment of the behavior of semiconductors. For our purposes a metal is a regular crystalline or polycrystalline array of atoms, each of which contributes one or more of its outer electrons to a sea of negative charge that can move almost freely through the fixed array of positive ions. If the lattice of positive ions were geometrically perfect, the electrons could move without loss of energy, but any lattice imperfections, for instance, impurities, vacancies, dislocations, crystal boundaries, and temperature vibrations (phonons), lead to scattering of the electrons and thus to the transfer of energy from the electric field to the lattice and to electrical resistance.

A simplified model will bring out the essentials of the process. Let us assume that each electron is accelerated freely by the electric field E for a time τ. It then collides with a lattice imperfection and the acceleration starts over again. The acceleration of the electron is Ee/m where e is the charge and m is the mass of an electron. The increment of velocity in the direction of the field during the time τ is $v = Ee\tau/m$. The correct statistical treatment of the motion of the electron cloud in the applied field is complex. In thermal equilibrium in the absence of an electric field the free electrons of a metal move rapidly and randomly, back and forth, as many in one direction as in the opposite. Their speeds are large compared to the increment of speed that can be given in one mean free path by any achievable electric field. The effect of an electric field is to alter slightly this distribution of velocities so that there results a net drift velocity in the direction of the force on the electrons and therefore a net current in the direction of the field. The electrons do not come to rest after each collision but simply scatter in some new direction but with, on the average, some loss of energy. The net work done by the electric field on the electrons is rapidly shared with the lattice. From the correct calculation one obtains for the drift velocity of the free electron cloud just the value $v = Ee\tau/m$. Thus if there are n free electrons per unit volume, the charge density is ne, and the current density $j = nev$ is

$$j = E \frac{ne^2\tau}{m} = \sigma E = \frac{E}{\rho} \qquad (1.11)$$

The constant $\sigma = ne^2\tau/m$ is the conductivity of the material and $\rho = m/ne^2\tau = 1/\sigma$ is the resistivity.

The conductivity is frequently expressed as the product of the charge density ne of carriers and the mobility μ of each carrier. The mobility of a carrier is defined as the drift velocity divided by the field; that is, $\mu = v/E$. From the expression for the drift velocity we have $\mu = v/E = e\tau/m$. Finally then we may write

$$\sigma = \frac{1}{\rho} = \frac{ne^2\tau}{m} = ne\mu \qquad (1.12)$$

It is useful to have some idea of the order of magnitude of these quantities. Let us take copper as an example. If we assume one free electron per copper atom, we find that $n = 8.5 \times 10^{22}$ electrons/cm^3. Multiplying by the charge of the electron, 1.6×10^{-19} coulomb, we find $ne = 1.36 \times 10^4$ coulombs/cm^3. The resistivity of copper at room temperature is 1.7×10^{-6} ohm-cm. From the experimental value of ρ and the calculated ne, we find for the mobility $\mu = 1/ne\rho = 43$ cm^2/volt-sec. Now a large current density in a copper wire is of the order of 10^2 amp/cm^2. The field $E = j\rho$ that produces this current density is 1.7×10^{-4} volt/cm and the drift velocity of the conduction electrons in this field is $v = E\mu = 7.3 \times 10^{-3}$ cm/sec. From $\mu = e\tau/m$ we may also calculate τ, the mean free time between collisions. For copper at room temperature $\tau = 2.4 \times 10^{-14}$ sec (see Prob. 1.1).

To estimate the mean free path of a conduction electron, that is, the distance traveled between collisions, we must know not the drift velocity of the free electron cloud but rather the actual velocities with which electrons dart back and forth in thermal equilibrium. The most energetic of the conduction electrons in copper have speeds of about 1.6×10^8 cm/sec. In Chapter 9 we shall discuss how this comes about; but accepting the result and multiplying by the mean free time $\tau = 2.4 \times 10^{-14}$ sec we find for the mean free path $\lambda = 3.8 \times 10^{-6}$ cm = 380Å (angstroms). This is much greater than the 2.55Å separation of nearest neighbor copper atoms in the metal and supports the statement that the conduction electrons are not scattered by a perfect crystalline array of positive ions.

We now discuss briefly a quite different example of current flow. It will illuminate some of the points we have made and also remind us that Ohm's law is not always valid. Consider the current between two electrodes in air, Fig. 1.2. A few air molecules are ionized by cosmic rays and local radioactivity. The positive and negative ions (or electrons) eventually meet and recombine and a constant charge density is achieved when the rate of recombination equals the rate of ionization.

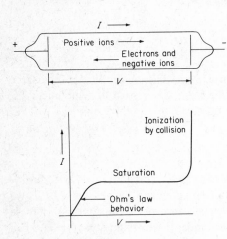

Figure 1.2 The current-voltage characteristic between two electrodes in air. A few ions are always provided by cosmic rays and local radioctivity.

At very low applied voltages the ions behave very much as do the electrons in a metal. There is a drift velocity proportional to the field and an almost constant density of current carriers. Ohm's law is obeyed. However, as the rate at which charge is collected at the electrodes approaches the rate at which it is produced by ionization the current rises more slowly with voltage. Saturation of the current is reached when charges move so rapidly to the electrodes that they do not have a chance to recombine. The charge per second reaching each electrode (that is, the current) is just the charge per second produced by ionization. Raising the voltage now collects each charge more rapidly but it does not increase the current, which is the charge collected each second.

Finally at fields of a few thousand volts per centimeter (the field required varies inversely with the gas density) electrons acquire enough energy in a mean free path to ionize the molecules they strike. Now the density of current carriers rises very rapidly and a spark or a glow discharge results.

Currents limited by the rate of production of carriers also occur in vacuum tubes, and current avalanching from charge multiplication by collision can occur in semiconductors and in the dielectric breakdown of insulators.

Copper and air represent two extremes in the charge density of carriers. In air the density can be less than 10 electron charges per cubic centimeter if sources of ionization other than cosmic rays are excluded. In copper it is 22 orders of magnitude greater. In any medium $jE = E^2/\rho$ is the electrical power per unit volume introduced into the medium. This is easily seen by dividing $P = IV$, the power dissipated in a wire, by Al, the volume of the wire. The ability of the medium to dissipate this power without a destructive increase in temperature places an upper limit on E^2/ρ. If ne is very large and ρ is very small, as in a metal, then E must be small. Drift velocities will be but a small perturbation of the thermal equilibrium velocity distribution and Ohm's law is obeyed. If ρ is very large, as in a gas, then E may be large and drift velocities may become large compared to thermal velocities. In this case nonohmic behavior and ionization by collision are possible.

1.3 Resistors

In ordinary electrical wiring one is interested in carrying a current with the least possible power loss. High conductivity is the prime requirement and copper or aluminium are the customary choices. On the other hand by a resistor we mean an individual circuit component whose purpose is usually to control a current or to provide a needed potential difference by means of the IR drop across the resistor. Resistances from a few ohms to many megohms may be needed. Consideration must be given to accuracy, stability, temperature coefficient, power dissipation, and many other factors.

The resistor most commonly used in electronic circuits is a composite of small carbon particles pressed together in a resinous dielectric binder. Most of the resistance occurs at small areas of contact between adjacent carbon particles. By control of the composition and density of the composite, materials of widely varying resistivities can be formed. A short cylindrical plug can have a very high resistance. It is not easy, however, to mass produce accurately composition resistors to a predetermined value of the resistance. After manufacture the resistors are tested and sorted and then sold as being within $\pm 5\%$ or $\pm 10\%$ of the nominal value. All of a large group of similar resistors will have resistances within the stated limits, but the average may not be the nominal value and the distribution about the average may be far from random. Composition resistors are commonly available with power dissipations of $\frac{1}{4}$, $\frac{1}{2}$, 1, or 2 watts. A sketch of a composition resistor and the color code for the resistance are given in Fig. 1.3.

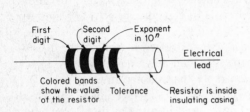

Figure 1.3 A typical composition resistor. The color code for the digits and the exponent is

black	0	green	5
brown	1	blue	6
red	2	violet	7
orange	3	gray	8
yellow	4	white	9

The fourth band is silver for $\pm 10\%$ tolerance and gold for $\pm 5\%$ tolerance. A resistor with successive yellow, violet, green, and gold bands has a resistance $R = 47 \times 10^5$ ohms $\pm 5\%$.

The principal advantages of composition resistors are low cost and small size. Because the resistance element is a short straight plug, the resistors are also quite noninductive. Chief among the disadvantages are limited accuracy and stability. The resistance depends on the temperature and independently on the applied voltage but these effects are mixed in actual use. Most composition resistors have a resistance minimum near room temperature. At 100°C the resistance may be a few percent higher than at room temperature. The effect is greater for high resistances than for low.

In nearly all respects metal wire is much to be preferred in making resistors but there is one great drawback, the resistivity is always very low. A copper wire the diameter of a human hair (about 3×10^{-3} in. or 40 gauge) has a resistance of only 1 ohm/ft. Very long and fine wires must be used if appreciable resistance is needed. Usually the wire is wound as a solenoid. Thus the resistor has appreciable inductance and cannot be used at high frequencies (see Chapter 4 for a discussion of inductance).

In spite of these problems, wire-wound resistors are used whenever accuracy, stability, or high power dissipation are needed. Many alloys have desirable properties. Manganin, an alloy of copper, manganese, and nickel, has a resistivity about 30 times that of copper and near room temperature it has a temperature coefficient of resistivity, $(1/\rho)\ \partial\rho/\partial T$, very close to zero. Manganin is widely used in precision resistors. Near room temperature the temperature coefficient of resistivity of many pure metals including copper and aluminum is about 0.004/°C.

Wire-wound resistors are regularly available in resistances up to 10^5 ohms and power dissipations to 200 watts. If a resistor is to be used as a permanent circuit component, a common accuracy is $\pm 1\%$ or $\pm 0.1\%$. However, laboratory standards can be obtained that are accurate to better than 1 in 10^5 parts.

Still a third and very useful form of resistor is made by depositing a thin film of carbon or metal on a glass or ceramic cylinder. Because the film can be very thin, one has essentially a wire of small cross-sectional area and high resistance per unit length. The length can be maximized by depositing the film as a helical ribbon on the cylindrical support. This is customary for the higher resistances and for high voltage resistors. Thin film resistors are manufactured to accuracies of $\pm 1\%$ or better.

Finally there are available a wide variety of variable resistances and potentiometers (or "pots") using both wire-wound and carbon resistance elements. These provide a third electrical contact that can be moved from one end of the resistor to the other. If only one terminal of the resistor and the movable contact are used in the circuit, we have a variable resistor. If all three terminals are used, the device is a potentiometer. Usually the resistance element is on a circular track and a contact, attached to a dial, rolls or sweeps from one end of the resistance to the other. The most familiar example is the volume control on a TV set. The relationship between angle and resistance, called the *taper*, is frequently linear but other tapers, logarithmic for instance, are widely used. If a very fine division of the total resistance is desired, the resistance element will be deposited as a helix and the tracking contact must both rotate and translate. Ten or more rotations of the dial may be necessary to traverse the entire resistance.

Trimmer resistors are used where small and infrequent adjustments of a resistance are necessary to maximize circuit performance. The adjustment is usually made with a screwdriver.

The conventional symbols for fixed and variable resistors are shown in Fig. 1.4.

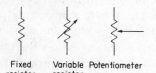

Figure 1.4 The standard circuit symbols for various kinds of resistors.

Fixed   Variable   Potentiometer
resistor   resistor

1.4   Electromotive Forces and IR Drops

Electromotive force (emf) is an unfortunate bit of terminology that history has built into our subject. An emf has the dimensions of a potential difference, an energy per unit charge, not a force. A source of emf is a device, a circuit component, which produces a potential difference and in which electrical energy is given to charges. Positive charge is moved from low to high potential and negative charge from high to low. The ultimate source of the energy may be chemical as in a battery, thermal as in a thermocouple, or mechanical as in a generator and a Van de Graaff machine. A source of emf must be able to supply energy to charges continuously. Thus if a wire is connected externally from the positive to the negative terminal of an emf, a current will flow. The source must be able to recharge the terminals continuously and maintain the current.

The battery is one of the most familiar sources of emf. Let us discuss briefly how a zinc-copper cell operates. If a piece of metallic zinc and a solution of zinc sulfate, originally at the same potential, are brought into contact, a few zinc ions leave the metal and enter the solution. The metal is left negatively charged with respect to the solution. The process continues until the potential difference between the metal and the solution becomes constant at, say, $-V_1$. At this potential difference the chemical free energy released by the solvation of the zinc ion is just balanced by the electrical work that must be done to separate the charges. At equilibrium an electrostatic field exists between an excess of electrons on the surface of the metal and an excess of positive zinc ions in the solution very close to the metal.

At the copper terminal a piece of metallic copper is placed in a copper sulfate solution. Some of the copper ions in solution plate out on the copper metal, which eventually, at equilibrium, assumes a potential of $+V_2$ with respect to the solution. If the solutions are physically separate but in electrical contact, perhaps through a membrane that is porous to H^+ ions, the copper terminal is $V_1 + V_2 = 1.12$ volts positive with respect to the zinc terminal. The separate potentials V_1 and V_2 cannot be directly measured. When the terminals are externally connected, electrons move from the zinc to the copper terminal, additional zinc ions enter the zinc sulfate solution, and additional copper ions plate out of the solution of copper sulfate to maintain the current. This process can continue as long as the supply of reacting chemicals lasts.

Contact potential differences between unlike materials always occur but usually are not sources of an emf. For instance, hard rubber and cat's fur or metallic copper and zinc have slightly different affinities for electrons. When they are in close contact, a few electrons jump from one to the other until a contact potential difference is built up that prevents further transfer of charge. Because no continuing source of chemical energy is available, one does not have a battery.

A cat's fur and hard rubber rod are a source of emf and current, however, when a lecturer repeatedly charges the rod and then discharges it to a grounded object. The input of energy to the system is mechanical. It occurs when the opposite charges are pulled apart, thus greatly increasing the potential difference between them.

A universal property of sources of emf is that the terminal potential difference decreases as the current increases. The governing equation is

$$V = \varepsilon - Ir \tag{1.13}$$

where V is the terminal potential difference; ε is the actual emf, which is available across the terminals only at zero current; and r is the internal resistance of the source. Physically the internal resistance may be the ohmic resistance of the armature of a generator or of the electrodes and electrolyte of a battery. If an external resistance R is placed across the terminals, we have a complete circuit, which we symbolize in Fig. 1.5.

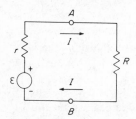

Figure 1.5 A source of emf ε is connected to an external resistance R at the terminals A and B. The internal resistance of the source is r. Note that the circuit symbol for an emf source is an open circle. The positive and negative terminals of the emf source are indicated.

The internal resistance of the source, in this case a battery, is always drawn next to it, in series, and usually A and B, designating the terminals, will be omitted. Using Ohm's law, $V = IR$, Eq. 1.13 may be written

$$\varepsilon = Ir + IR \tag{1.14}$$

The terms of this equation have the dimensions of potential differences, and the equation states that a charge moving around a closed circuit and back to its starting point acquires no net energy. The total increase in potential due to sources of emf is equal to the potential drop across the resistors in going around the loop of the circuit.

1.5 Kirchhoff's Laws

Equation 1.14 gives the relationship between the emf and resistances and the current of the circuit of Fig. 1.5. If the former are known, the current is easily calculated. However, most circuits are much more complicated and one must have systematic procedures for their solution.

Consider the circuit of Fig. 1.6. The two emf's and three resistors are given numerical values. The three currents are the unknowns. Junctions such as A or B where three (or more) currents meet are called *branch points* or *nodes*. Any current path from one node to the next is a branch and any path that includes two or more branches and returns to its starting point is a *loop*. A glance at Fig. 1.6 shows that the circuit contains two nodes, three branches, and three loops. We have counted the loop that goes around the outside enclosing but not traversing the middle branch.

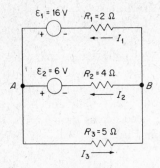

Figure 1.6 A network of three branches and two nodes. The abbreviation V is for volts and Ω for ohms.

A circuit of any complexity can be solved using Kirchhoff's two circuit laws. These are

I. Charge does not accumulate at any junction in a circuit. Thus the sum of the currents entering any node of a circuit must be equal to the sum of the currents leaving that node.

II. The sum of the emf's around any closed loop in a circuit equals the sum of the potential drops around the loop.

Kirchhoff's second or voltage law is simply an extension to any loop of the physical arguments used in support of Eq. 1.14. In writing Eq. 1.14 we also have implicitly used Kirchhoff's first or current law when we assumed that the current through the external resistor was the same as the current through the battery.

Now let us return to Fig. 1.6. Assuming the indicated directions of current flow we may write the three independent equations

$$0 = I_1 + I_2 - I_3$$
$$\varepsilon_1 - \varepsilon_2 = R_1 I_1 - R_2 I_2 \quad \text{or} \quad 16 - 6 = 2I_1 - 4I_2 \qquad (1.15)$$
$$\varepsilon_2 = R_2 I_2 + R_3 I_3 \quad \text{or} \qquad 6 = 4I_2 + 5I_3$$

We have adopted the convention of putting terms involving emf's on the left and terms involving currents on the right. The first equation expresses Kirchhoff's current law applied to node A. Node B gives the negative of the first equation and therefore no new information. The second equation is from Kirchhoff's voltage law applied to a counterclockwise traversal of the upper loop. The third equation is from the lower loop traversed counterclockwise. One could have used the outside loop but it gives nothing new. Traversed counterclockwise the equation is simply the sum of the second and third equations.

Depending on the assumptions for directions of currents and for directions of traversal, the terms in the loop equations may vary in sign but the actual currents determined will always be the same. It is merely necessary to be systematic and consistent. An emf on the left is positive if it traversed from low potential to high potential. An IR drop on the right is positive if it is traversed in the direction of the assumed current, that is, from high potential to low potential. The sign of an emf depends only on the direction of traversal but the sign of an IR drop depends both on the direction of traversal and on the direction of the assumed current.

Solving Eq. 1.15 we find $I_1 = 3$ amp, $I_2 = -1$ amp, and $I_3 = 2$ amp. The negative sign of I_2 means that the current in the middle branch is actually flowing to the right and not, as we had assumed, to the left. Thus the 6-volt battery is being charged. Electrical energy is converted to chemical energy.

We have not indicated internal resistances for the two batteries. They may be considered to be included in the 2- and 4-ohm resistors. However, some caution must be observed. No circuit loop can contain a net emf and no resistance. An infinite current is demanded and both the mathematics and the physics break down. The situation is slightly more subtle if we short out the 4-ohm resistance in our circuit. There is no problem with the equations; three currents can be calculated, but the 6-volt battery is now operating as an emf without internal resistance. We can demand of it whatever current we wish by, for instance, sufficiently reducing the resistance R_3.

Now let us summarize what we have learned from this example. The methods used are entirely general. We are given a network or circuit, that is, an interconnected array of emf's and resistors, and we wish to determine the current in each branch. First we must assume a direction for each branch current; then Kirchhoff's two laws enable us to write down as many independent linear algebraic equations as there are branches. This set of equations is then solved to give the currents in terms of the emf's and the resistances.

The equations must correctly express the physics of Kirchhoff's laws. Thus the signs of the terms in the equations must be chosen consistently. It is for this reason that we must first assume a direction for each branch current. We have chosen to write emf's on the left of each equation and currents and IR drops on the right. In a nodal equation a current entering a node is positive; a current leaving a node is negative. An emf is positive if it is traversed from low potential to high potential. An IR drop is positive if it is traversed in the direction of the assumed current, that is, from high potential to low potential. Other conventions for the writing of the equations will work just as well. Any convention must be applied consistently, and no convention can be used until after current directions have been assumed.

Upon solving the equations a particular branch current may turn out to be a positive or a negative number. If it is positive, the actual branch current is in the same direction as the assumed current. If it is negative, the actual current is in a direction opposite to that of the assumed current.

Kirchhoff's laws have great generality. If the voltages and currents are functions of time, Kirchhoff's laws hold instantaneously although in this case they may lead to differential equations for the currents rather than algebraic equations. We shall begin the discussion of time-varying voltages and currents in Chapter 4.

In Fig. 1.7 we call attention to the two most elementary cases of the reduction of the complexity of a circuit by the combination of circuit elements. From Kirchhoff's laws one easily calculates the values given in Fig. 1.7 of the single equivalent resistances that replace the series and parallel combinations without changing the current out of the battery (see Prob. 1.4).

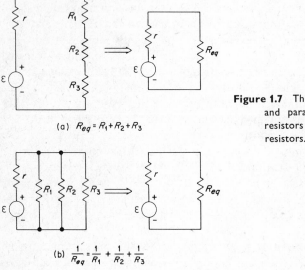

(a) $R_{eq} = R_1 + R_2 + R_3$

(b) $\frac{1}{R_{eq}} = \frac{1}{R_1} + \frac{1}{R_2} + \frac{1}{R_3}$

Figure 1.7 The reduction of series and parallel combinations of resistors to single equivalent resistors.

In succeeding chapters we shall develop and use very general methods of handling complex circuits but it is of the utmost importance that the reader be completely at ease with the elementary circuit considerations of this chapter.

1.6   The Solution of Equations from Kirchhoff's Laws

A set of three linear algebraic equations in three unknowns, such as Eq. 1.15, is usually most easily solved by eliminating variables between equations. If the number of unknowns is greater, a determinant solution may be easier. In any case one must use the determinant form to obtain certain general results, which we shall discuss in the next chapter. Let us use the circuit of Fig. 1.8 as an example. It will also illustrate that a little prior thought sometimes helps.

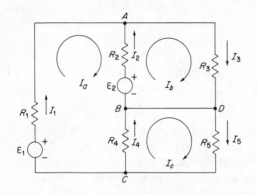

Figure 1.8 A network that leads to five independent Kirchhoff equations.

The circuit shows five branch currents, I_1 through I_5; four nodes, A, B, C, and D; and three simple loops labeled I_a, I_b, and I_c. Later we shall treat I_a, and so on, as currents but for the moment they just identify loops. The circular arrows indicate directions of traversal. The first point to be made is that BD can be considered a single node involving four currents rather than two separate nodes each with three currents. A current I_6 (not shown on Fig. 1.8) actually flows from B to D but the connection is assumed to be of zero resistance; therefore no potential drop is contributed to the circuit. Ignoring I_6 there are five independent equations, which we write

$$\begin{aligned}
&\text{(node } A) & 0 &= I_1 & + I_2 & & - I_3 \\
&\text{(node } BD) & 0 &= & - I_2 & + I_3 & + I_4 & - I_5 \\
&\text{(loop } I_a) & \varepsilon_1 - \varepsilon_2 &= I_1 R_1 & - I_2 R_2 & & & - I_4 R_4 \\
&\text{(loop } I_b) & \varepsilon_2 &= & I_2 R_2 & + I_3 R_3 \\
&\text{(loop } I_c) & 0 &= & & & I_4 R_4 & + I_5 R_5
\end{aligned} \qquad (1.16)$$

Mathematically it is perfectly correct to treat B and D as separate nodes. One obtains the equations $I_4 = I_2 + I_6$ and $I_3 = I_5 - I_6$. The sum of these equations is just the BD nodal equation we have used. If a current appears only in the nodal equations, it means some nodes can be collapsed, thus eliminating the current and reducing the number of unknowns. The solution of Eq. 1.16 for I_1, for example, is $I_1 = \delta_1/\delta$ where δ is the determinant of the coefficients of the currents and δ_1 is the same determinant but with the coefficients of I_1 replaced by the emf's. We have

$$\delta_1 = \begin{vmatrix} 0 & 1 & -1 & 0 & 0 \\ 0 & -1 & 1 & 1 & -1 \\ \varepsilon_1 - \varepsilon_2 & -R_2 & 0 & -R_4 & 0 \\ \varepsilon_2 & R_2 & R_3 & 0 & 0 \\ 0 & 0 & 0 & R_4 & R_5 \end{vmatrix}$$

and   (1.17)

$$\delta = \begin{vmatrix} 1 & 1 & -1 & 0 & 0 \\ 0 & -1 & 1 & 1 & -1 \\ R_1 & -R_2 & 0 & -R_4 & 0 \\ 0 & R_2 & R_3 & 0 & 0 \\ 0 & 0 & 0 & R_4 & R_5 \end{vmatrix}$$

Now the reader may have noticed that our attack on the circuit of Fig. 1.8 is more general than it needs to be. The resistors R_4 and R_5 are really in parallel and their equivalent resistance is in series with R_1. These three resistors can be collapsed to a single resistor and the circuit is then the same as that of Fig. 1.6. The unknowns are just I_1, I_2, and I_3. If I_4 and I_5 are needed, they can be found quickly by calculating $I_1 R_{eq}$ where $1/R_{eq} = 1/R_4 + 1/R_5$ and noting that this same potential drop appears across both R_4 and R_5. One should always be alert for shortcuts. The determinants are a last resort. If there were an emf in series with either R_4 or R_5, the simplification above would not be possible.

The methods of solution we have used thus far assign a current to each branch. One must write and solve as many equations as there are branches in the network. Mathematically there is a different method of solution that uses nonphysical loop currents, the I_a, I_b, and I_c of Fig. 1.8, instead of branch currents. The loop currents go all the way around the loop without change in magnitude. Loop currents automatically satisfy the Kirchhoff's current law nodal equations since any one of the currents that enters a node also leaves it. Thus we need write only as many equations as there are independent loops. There is a slight disadvantage in that the actual physical current in a branch may be a linear combination of different loop currents. Thus, in Fig. 1.8, $I_2 = I_b - I_a$. The equations that result from Kirchhoff's voltage law applied to the loop currents are

(loop I_a) $\varepsilon_1 - \varepsilon_2 = I_a R_1 + (I_a - I_b)R_2 + (I_a - I_c)R_4$

(loop I_b) $\varepsilon_2 = (I_b - I_a)R_2 + I_b R_3$ (1.18)

(loop I_c) $0 = (I_c - I_a)R_4 + I_c R_5$

After collecting the coefficients of the currents the equations are solved either by the elimination of variables or by determinants. In this example the use of loop currents reduces a system of five equations to three.

This reduction in the number of equations is not to be confused with that achieved by the combination of series and parallel resistances. The loop current method treats the circuit as it stands and works even if there are emf's in every branch.

As a numerical example, let us go back to Fig. 1.6 and assign a loop current I_a to the upper loop and I_b to the lower loop, both counterclockwise. We then have

(loop I_a) $16 - 6 = 2I_a + 4(I_a - I_b)$

(loop I_b) $6 = 4(I_b - I_a) + 5I_b$ (1.19)

The solution is very easy and gives as before $I_a = I_1 = 3$ amp, $I_b = I_3 = 2$ amp, and $I_b - I_a = I_2 = -1$ amp.

1.7 Two-Terminal Networks

In the remainder of this chapter we shall discuss some of the properties of two-terminal networks. A two-terminal network is an array of emf's and resistors, perhaps quite complicated, but which is viewed across only two terminals of the network. The quantities of interest are the potential difference between the terminals and the current that can be drawn externally from one terminal to the other. Internal currents and potential differences do not directly concern us. We note that Kirchhoff's current law requires that the same current that leaves the network at one terminal must enter at the other since no net charge can accumulate at any junction within the network.

Consider, for example, the upper left circuit of Fig. 1.9. From our present viewpoint we are interested in just $V(I)$, that is, $V_A - V_B$, the potential difference between A and B, as a function of the current. The current I is assumed to be positive if it leaves the network at terminal A and enters the network at terminal B. Of course in order to actually carry the current I from A to B there must be an external circuit, connected across the terminals A and B. For instance, the external circuit might be the upper right network. For the moment, however, we consider only the upper left network. We may write for the relationship between V and I, which we shall call the current-voltage characteristic of the network

$$V(I) = \varepsilon_{eq} - Ir_{eq}$$

where

$$\varepsilon_{eq} = \frac{\varepsilon_1 r_2}{r_1 + r_2}$$ (1.20)

and

$$r_{eq} = \frac{r_1 r_2}{r_1 + r_2}$$

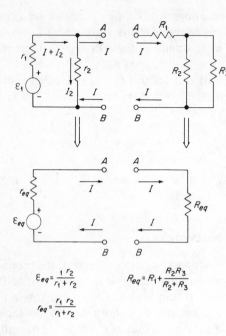

Figure 1.9 Examples of the simplification of two-terminal networks. The networks below have the same current-voltage characteristic across AB as the networks above.

$$\varepsilon_{eq} = \frac{\varepsilon_1 \, r_2}{r_1 + r_2}$$

$$R_{eq} = R_1 + \frac{R_2 R_3}{R_2 + R_3}$$

$$r_{eq} = \frac{r_1 \, r_2}{r_1 + r_2}$$

Equations 1.20 are easily derived. Let a current I_2 flow down through r_2. A current $I + I_2$ then flows up through ε_1 and r_1. Around the internal loop containing r_1 and r_2 we find from Kirchhoff's voltage law $\varepsilon_1 = (I + I_2)r_1 + I_2 r_2$. Remembering that $V(I) = I_2 r_2$, Eq. 1.20 follow at once. The potential difference $V(I)$ will exist between A and B when any external circuit that draws a current I is connected between A and B. It should be noted that r_{eq} is just the parallel combination of r_1 and r_2. It is the resistance one would calculate for the network if ε_1 were replaced by a short circuit.

Now let us consider the upper right network of Fig. 1.9. As before $V(I)$ is the potential difference between A and B, that is, $V_A - V_B$, but we assume that the current I enters at terminal A. The current-voltage characteristic is then $V(I) = I[R_1 + R_2 R_3/(R_2 + R_3)] = IR_{eq}$. The term IR_{eq} appears with a positive sign rather than the negative sign it had in Eq. 1.20 simply because we have reversed the direction of the assumed current, that is, we assumed the current enters at A, whereas in Eq. 1.20 we assumed the current left the network at A.

The direction of the flow of energy is determined as follows. If current leaves a two-terminal network at the positive terminal, the network is a source of electrical energy. If the current enters at the positive terminal, the network absorbs electrical energy.

An active network is one that can provide electrical energy to another circuit. It may also absorb electrical energy from another circuit. In Eq. 1.20, $V = \varepsilon_{eq} - Ir_{eq}$, if $V > \varepsilon_{eq}$, then I is negative; i.e., current enters the network at the positive terminal and the network absorbs energy. A passive network is one that can only absorb but not provide electrical energy. If no emf is present, a network must be passive. If the left and right networks of Fig. 1.9 are connected at A and B, energy flows from the network on the left to the network on the right. The active network on the left is called the *source*; the passive network on the right is called the *load*.

The current-voltage characteristic of a two-terminal network can be determined experimentally using a variable source of voltage which is placed externally across the terminals and which will fix the potential difference at any desired value. Then for each value of the potential difference the current is read. The result is completely determined by the network. The source of voltage has been used only as a probe. In the lower half of Fig. 1.9 we redraw the upper circuits using the equivalent emf's and resistances. We see that there is no way to tell the upper circuits from the lower circuits using only voltage and current measurements at the two terminals A and B. The upper and lower circuits have identical current-voltage characteristics.

Now the results we have just obtained are in fact quite general and of the greatest importance. They are an example of Thevenin's theorem, which states that the current-voltage characteristic of any linear two-terminal network[2] must be of the form $V(I) = \varepsilon - Ir$. Only the parameters ε and r are needed to characterize the network. If ε is finite, the network is active; if $\varepsilon = 0$, the network is passive.

We shall defer the proof of Thevenin's theorem until Chapter 2, but accepting the result we see that we need not probe the network at a large number of different output voltages. The two parameters may be determined experimentally by two measurements. The first might be a measurement of the open circuit $(I = 0)$ output voltage, which gives $V(0) = \varepsilon$. Second, if the output is shorted, $V = 0$, we can measure the short circuit current I_S and obtain $r = \varepsilon/I_S$. The parameters ε and r are referred to as the Thevenin parameters of the network and written ε_{Th} and r_{Th}.

An equivalent statement of Thevenin's theorem is that any linear two-terminal network can be replaced by an emf ε_{Th} and an internal resistance r_{Th} in series.

Let us be quite clear what is implied by the word *replaced*. Some given network is connected to another external network, either active or passive, at two terminals. Our interest is in the currents and potential differences in the external network. The content of Thevenin's theorem is that in solving for the currents in the external network one need not include in the Kirchhoff's equations all the various branches and nodes of the original given network. The given network may be replaced by its Thevenin equivalent emf and internal resistance, that is, by a single branch across the terminals of the external network.

---

[2] The term *linear network* will be defined and discussed in Chapter 2.

In circuit analysis and design one usually does not determine the Thevenin parameters of a network by experimental measurement. Some use of Kirchhoff's laws is necessary but only enough to determine ε_{Th} and r_{Th}. An important result is that the Thevenin resistance r_{Th} of a network is just the equivalent resistance of the passive network that results if all the emf's in the original network are shorted out. This result will be proved in Chapter 2.

We shall now consider an application of Thevenin's theorem. Let us look at the example in Fig. 1.10(a) where to the left of AB we have redrawn the circuit of Fig. 1.6. From the earlier solution we know that $I_3 = 2$ amp if no external current is drawn, and therefore the Thevenin emf, which is just the open circuit voltage, is given by $\varepsilon_{Th} = 5I_3 = 10$ volts. Shorting out the two emf's to the left of AB, we have $1/r_{Th} = \frac{1}{2} + \frac{1}{4} + \frac{1}{5}$ or $r_{Th} = \frac{20}{19}$ ohms.

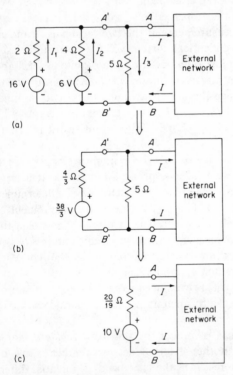

(a)

(b)

(c)

Figure 1.10 The simplification of a network by successive application of Thevenin's theorem, first to the part to the left of $A'B'$, then to the part to the left of AB.

The Thevenin equivalent of the circuit to the left of AB is shown in Fig. 1.10(c). It is only this single branch that is needed in any computation of the currents in the external circuit. The short circuit current provided by the Thevenin equivalent circuit is $I_S = \varepsilon_{Th}/r_{Th} = \frac{19}{2}$ amp. This result can also be calculated directly from Fig. 1.10(a) by shorting the external circuit, that is shorting between A and B. When A and B are shorted, the potential drop across each of the branches to the left of AB is zero and we have $I_1 = 8$

amp, $I_2 = \frac{3}{2}$ amp, and $I_3 = 0$. Thus the short circuit current is $I_S = I_1 + I_2 - I_3 = \frac{19}{2}$ amp.

Successive applications of Thevenin's theorem are often useful as we shall illustrate by calculating once again but in a different way the Thevenin parameters of the circuit in Fig. 1.10(a). We first isolate the part of the circuit to the left of $A'B'$ [Fig. 1.10(a)] and determine the Thevenin parameters as seen from the terminals $A'B'$. If the circuit is open circuited at A' and B', then the potential difference across $A'B'$ is $\frac{38}{3}$ volts. This is the Thevenin emf. The Thevenin resistance is found by shorting the emf's and calculating the resistance of the circuit to the left of $A'B'$. The result is $r_{Th} = \frac{4}{3}$ ohms. This can also be obtained by shorting between $A'B'$. The current that flows is $I_S = \frac{19}{2}$ amp $= \varepsilon_{Th}/r_{Th}$ so that $r_{Th} = \frac{4}{3}$ ohms. The results for the circuit to the left of $A'B'$ are $\varepsilon_{Th} = \frac{38}{3}$ volts and $r_{Th} = \frac{4}{3}$ ohms. These results are shown in Fig. 1.10(b). The circuit to the left of AB is now greatly simplified. The final parameters $\varepsilon_{Th} = 10$ volts and $r_{Th} = \frac{20}{19}$ ohms of Fig. 1.10(c) are calculated by repeating the previous procedure. In the particular example we have used it is probably quicker to calculate directly the final Thevenin parameters. Many cases will arise, however, in which successive applications of Thevenin's theorem are to be preferred.

1.8 Voltage and Current Sources and Power Transfer

One usually finds that efficient operation of a load network demands a particular current-voltage characteristic of the source. For instance, a constant line voltage is needed in a house or laboratory independent of the current drawn. This is the most common case although we shall later discuss situations where a constant current independent of voltage is preferable. Sources that approximate these two extremes are called *voltage* and *current* sources, respectively. Now Thevenin's theorem states that the current-voltage characteristic of any (linear) source is always of the form $V = \varepsilon - Ir$. Let us consider this Thevenin equivalent source to be driving a load resistor R as in Fig. 1.5. The source is a good voltage source if $r \ll R$ where R is any load resistance that may be encountered. The output voltage is $V \simeq \varepsilon$. This follows from $\varepsilon = I(R + r) \simeq IR = V$ if $r \ll R$. For a good current source we must have $r \gg R$. The output current is then almost constant at $I = \varepsilon/r$.

An emf without internal resistance, if it existed, would be a constant voltage source. A constant voltage source is a useful idealization that is often used in circuit theory.

Another useful idealization is the constant current source, that is, a source of electrical energy that provides a constant output current independent of the potential difference across its terminals. In Fig. 1.11 we show an active network containing a constant current source I_0 and an internal

resistance r in parallel with the constant current source. The output voltage across AB is $V = (I_0 - I)r = I_0 r - Ir$. This current-voltage characteristic is indistinguishable from that of an emf and its series resistance r if we make the identification $\varepsilon = I_0 r$ and $r = r$; that is, a constant current source I_0 in parallel with a resistance r is equivalent to an emf $\varepsilon = I_0 r$ in series with a resistance r. Regulated electronic power supplies are available with both very low effective internal resistances (good voltage sources) and very high internal resistances (good current sources).

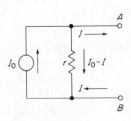

Figure 1.11 A constant current source I_0 and its internal resistance r. The current-voltage characteristic is $V = V_A - V_B = I_0 r - Ir$. Note that the circuit symbol for a constant current source is an open circle with an arrow beside it. The magnitude of the constant current source is indicated beside it. When the numerical value of the current is given, the abbreviation A is used for amperes.

A fundamental problem in circuit design is maximizing the power transfer from a source to a load. Let us consider a battery of emf ε and internal resistance r connected to a variable external resistance R (Fig. 1.5). We wish to maximize $P(R)$, the power dissipation in R. If R is very large, almost no current flows and very little power is dissipated in R. If R is small, large currents flow but most of the power dissipation occurs in the internal resistance of the battery. The maximum $P(R)$ is easily calculated. From $P(R) = I^2 R$ and $I = \varepsilon/(R + r)$ we obtain

$$P(R) = \varepsilon^2 \frac{R}{(R + r)^2} \tag{1.21}$$

Differentiating $P(R)$ and equating the derivative of $P(R)$ to zero, the maximum power transfer is found to occur when $R = r$. The power then dissipated in R is $\varepsilon^2/4R$. The same power is dissipated in the internal resistance r. The current flowing at maximum power transfer is $I = \varepsilon/2R = \varepsilon/2r$. Note that the current is not so large as the short circuit current that can be drawn from the battery, $I_S = \varepsilon/r$, which occurs when $R = 0$. That the maximum power transfer occurs when the internal resistance of the source equals the resistance of the load is a very general and important result.

The maximum in the power transfer function $P(R)$ is quite broad. For $R = 2r$ and $R = r/2$ the power dissipated in R is still 89 % of its maximum value.

Since only the current-voltage characteristics of source and load enter the calculation, the maximum power transfer from a constant current source and a parallel resistance r to a load resistance R requires the same result, $R = r$. If source and load are complex networks as in Fig. 1.9, the condition for maximum power transfer is that the Thevenin resistance of the source is equal to the Thevenin resistance of the load; that is, $R_{Th} = r_{Th}$. If both source and load are active networks, the condition for maximum power transfer will involve both the emf's and the resistances. It is important to note that while the power transfer characteristics of an active network depend only on ε_{Th} and r_{Th}, the internal power dissipation depends on the details of the circuit. If a source is more complex than an emf and a series resistance there may be internal currents and Joule losses even when the output current is zero.

The physical impossibility of sources of truly constant voltage or current can be seen in terms of power transfer. A constant voltage source will transfer an unlimited large power to a sufficiently small external resistance, and a constant current source will transfer an unlimited power to a sufficiently large external resistance.

Problems

1.1   Calculate the mean free time between collisions for the conduction electrons of copper at room temperature. Use the data of Sec. 1.2. Note that a mixture of mks and cgs units is involved.

1.2   In fair weather and flat country there is an electric field near the earth's surface of about 100 volts/meter directed vertically down. What electric charge at the center of the earth would produce the observed field?

1.3   Summing over the surface of the earth the atmospheric field of 100 volts/ meter produces a current to the earth of about 2000 amp. What is the average resistivity of the atmosphere?

1.4   Derive the formulas shown in Fig. 1.7 for the resistances equivalent to the series and parallel combinations. Note that in the series case the same current flows through each resistor while the total voltage drop is the sum of the individual voltage drops. In the parallel case the same voltage is applied across each resistor but the total current carried is the sum of the individual currents.

1.5   For the circuit shown in Fig. 1.12, calculate the equivalent load resistor and the current I in the circuit. What is the voltage drop across the load?

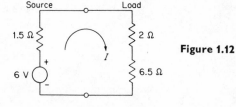

Figure 1.12

1.6   For the circuit shown in Fig. 1.13, calculate the equivalent load resistor and the current I drawn from the emf source. Also, calculate the current in the 6-ohm resistor and the 2-ohm resistor. What is the voltage drop across the load?

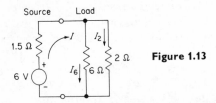

Figure 1.13

1.7   What maximum potential difference can be placed across a 5-ohm, 2-watt resistor without exceeding its rated wattage?

1.8   Calculate the conductance of 30m of #6 copper wire (diameter = 0.41 cm). The conductance is the reciprocal of the resistance and is measured in reciprocal ohms, called *mhos*.

1.9   Find the equivalent resistor between terminals A and B for the circuit shown in Fig. 1.14.

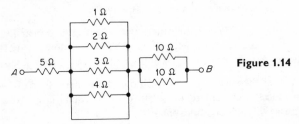

Figure 1.14

1.10   Find the equivalent resistance between terminals A and B for the circuit shown in Fig. 1.15. (*Hint.* Use the symmetry of the problem to simplify your analysis.) What is the equivalent resistance if the wires at the center of the circuit are in electrical contact?

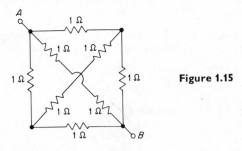

Figure 1.15

1.11 Write the equations corresponding to Kirchhoff's current and voltage laws for the circuit shown in Fig. 1.16. How many independent equations result from Kirchhoff's current law? How many independent equations result from Kirchhoff's voltage law?

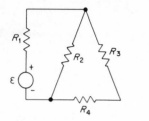

Figure 1.16

1.12 In the circuit shown in Fig. 1.17, find the value of R' such that the complete circuit between terminals A and B has an equivalent resistance R. The three resistors of magnitude R' are called a " T-pad." If an emf ε is applied between terminals A and B, then the T-pad serves to reduce the voltage across R without changing the resistance seen by the emf ε. If an emf ε is applied between A and B, find the voltage drop across R when R' has the value calculated in the first part of this problem.

Figure 1.17

1.13 Find the Thevenin parameters of the circuit shown in Fig. 1.18 when viewed across the ouput.

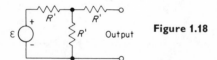

Figure 1.18

1.14 Actually carry out the differentiation of Eq. 1.21 and show that the maximum power is dissipated in a load resistor when the load resistor is equal to the internal resistance of the emf source.

1.15 Calculate the power dissipated in the 10-ohm resistor of the circuit shown in Fig. 1.19.

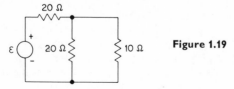

Figure 1.19

1.16   A general principle for dc circuits is that the currents distribute themselves in a network so as to minimize the total power dissipated. In the
example shown in Fig. 1.20, assume that the current I into the resistors
R_1 and R_2 is a constant, and assume Kirchhoff's current law; that is,
$I = I_1 + I_2$. Minimize the total power dissipated, and show that this leads
to the same results as Kirchhoff's voltage law; that is, $I_1 R_1 = I_2 R_2$.

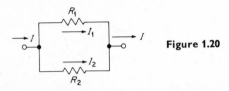

Figure 1.20

1.17   The circuit shown in Fig. 1.21 is excited by a 3-amp current source. Find
the current through the 3-ohm resistor and the voltage across the current
source. Kirchhoff's voltage law should be applied to the loop containing
the two resistors.

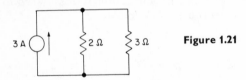

Figure 1.21

1.18   What are the Thevenin parameters ε_{Th}, r_{Th}, and I_S looking into the terminals AB of the current source of Fig. 1.11?

1.19   A voltage source has the Thevenin parameters $\varepsilon_{Th} = 15$ volts and $r_{Th} = 0.5$
ohm. What power will it deliver to a 2-ohm load?

1.20   Design a voltage source with $\varepsilon_{Th} = 5$ volts, $r_{Th} = 10$ ohms, and an internal
$I^2 R$ dissipation of 10 watts when no external current is being drawn. Do
not make use of a battery without internal resistance.

1.21   Two different two-terminal active networks have been connected together
at their positive terminals and at their negative terminals. The polarities
of the terminals are those that were determined before the networks were
connected and when neither was supplying any current. Show that power
always flows from the network with the higher ε_{Th} into the network with
the lower ε_{Th}.

1.22   Two different two-terminal active networks are connected with the zero
load positive terminal of one circuit connected to the zero load negative
terminal of the other. Show that the current always leaves each network
at the terminal that was positive before the connection. Show that the
power transfer can be either from or into the network with the higher ε_{Th}
depending on the Thevenin resistances. Show also that the power transfer

from one network to the other is zero if $\varepsilon_1/r_1 = \varepsilon_2/r_2$ where the emf's and resistances in the equation are the Thevenin parameters of the two networks.

1.23 Calculate ε_{Th} and the maximum r_{Th} for a voltage source that must deliver 100 volts $\pm 1\%$ to all load resistances greater than 50 ohms.

1.24 Sketch the current-voltage characteristics of the following circuits:
a) A 10-volt battery with 1-ohm internal resistance.
b) A 1-amp constant current source with a 10-ohm internal shunt.
c) A 1-amp constant current source with a 10^7-ohm internal shunt.

1.25 A portion of the current-voltage characteristic of a linear two-terminal network is shown in Fig. 1.22. What is the Thevenin equivalent circuit? Given a voltmeter, an ammeter, and assorted resistors, what portion of the current-voltage characteristic can be measured experimentally? How might other parts of the characteristic be measured?

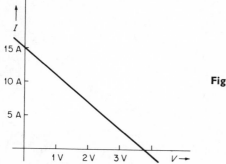

Figure 1.22

1.26 We have shown that by the use of loop currents one can automatically satisfy Kirchhoff's current law and reduce the number of independent equations to those that result from Kirchhoff's voltage law. There is another method, the node voltage method, which enables one to reduce the number of equations that result from Kirchhoff's two laws in a somewhat different manner. In the node voltage method one assigns a voltage to each node. Kirchhoff's voltage law equations are automatically satisfied since the sum of the voltage differences around any loop is zero. Thus one need solve only the Kirchhoff's current equations, writing the branch currents in terms of differences between node voltages. To demonstrate this, consider the circuit of Fig. 1.6. Let us take the voltage at node A as a reference voltage $V_A = 0$ volt and the voltage at node B as V_B. Calculate I_1, I_2, and I_3 in terms of $V = V_B - V_A = V_B$. Write the Kirchhoff's current law equation at one of the nodes expressing the three branch currents in terms of V. Solve first for V and then use V to find I_1, I_2, and I_3. Note that there is only one independent equation to solve for V, whereas if one used the method of loop currents, one would have to solve two independent equations.

2/General Circuit Theorems

There are a number of general circuit theorems that come from a systematic application of Kirchhoff's laws to complex networks. The complete exploitation of these results is usually of interest only to electrical engineers, but some knowledge of them is necessary in ordinary circuit analysis. An example is Thevenin's theorem, which we introduced without proof in the last chapter. Our aim in this chapter is to illustrate and make credible some of these theorems.

2.1 Branches, Nodes, and Loops

An electric circuit or network can be considered as made up of branches, nodes, and loops. Each branch is assumed to consist of a resistance or a resistance and a source of emf in series. A node is the meeting point of three or more branches and a loop is any path along the branches that returns to its starting point without crossing itself.

Our problem is to calculate the branch currents, given the emf's and the resistances. Physical intuition tells us that this is always possible, and we therefore expect as many independent equations from Kirchhoff's laws as there are branches in the circuit. Let us outline the arguments that support our intuition and also tell us how many of the equations are nodal or Kirchhoff's current law equations and how many are loop or Kirchhoff's voltage law equations.

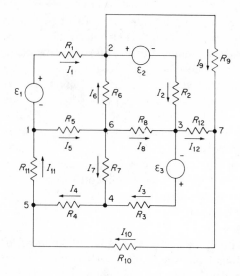

Figure 2.1 A circuit with 12 branches and 7 nodes.

Figure 2.1 shows an electric circuit. The nodes of the circuit are indicated by black dots. There are 7 nodes numbered from 1 to 7. There are 12 branches in the circuit, and the individual branch currents are indicated as I_1 to I_{12}. Loops will be identified by giving the nodes in the sequence that they are met in going around the loop. For instance, the loop 1261 in the upper left of Fig. 2.1 starts at node 1, then goes to node 2, then to node 6, and finally back to node 1. Of course 1261 and 2612 are the same loop. Since there are 12 branches and 12 different branch currents, we must have 12 independent equations in order to solve for the currents in terms of the emf's and the resistances. Let us write down the Kirchhoff's current law nodal equations for the 7 nodes. Adopting the convention that currents entering a node are positive and currents leaving a node are negative, we obtain

(node 1) $I_{11} - I_1 - I_5 = 0$

(node 2) $I_1 + I_6 - I_2 - I_9 = 0$

(node 3) $I_2 + I_8 - I_3 - I_{12} = 0$

(node 4) $I_3 + I_7 - I_4 = 0$ (2.1)

(node 5) $I_4 + I_{10} - I_{11} = 0$

(node 6) $I_5 - I_6 - I_7 - I_8 = 0$

(node 7) $I_9 + I_{12} - I_{10} = 0$

Each branch current occurs in only two node equations, once leaving a node and once entering a node. For example, the current I_1 appears only in equations for nodes 1 and 2. Because I_1 leaves node 1 and enters node 2, it is, according to our convention, negative in the first equation and positive in the second equation. Since each branch current appears in two node equations and with opposite signs, the sum of all seven node equations must be zero. Thus the seven node equations cannot all be linearly independent. Physically this occurs because all the currents flow from one part of the circuit to another; that is, a current that leaves one node must enter another node so that no net current enters or leaves the circuit as a whole. For the particular circuit of Fig. 2.1 the reader can readily convince himself that if any one of the seven nodal equations is eliminated, then the remaining six nodal equations are linearly independent.

For a general circuit with B branches and N nodes it can be shown that there are exactly $N - 1$ linearly independent Kirchhoff's current law nodal equations. This result is of course consistent with our analysis of the circuit shown in Fig. 2.1 where a circuit with $N = 7$ nodes has $N - 1 = 6$ linearly independent nodal equations.

For the general circuit with B branches and N nodes the number of linearly independent loop equations from Kirchhoff's voltage law must be $L = B - (N - 1)$ since B independent equations are required in order to obtain the B different branch currents and since $N - 1$ of these are Kirchhoff's current law nodal equations. Thus for the circuit of Fig. 2.1 there must be $L = B - (N - 1) = 12 - 6 = 6$ linearly independent loop equations.

Six linearly independent loop equations for the circuit of Fig. 2.1 are given below:

Nodal sequence in loop	Loop equation
1261	$\varepsilon_1 = I_1 R_1 - I_6 R_6 - I_5 R_5$
2362	$-\varepsilon_2 = I_2 R_2 - I_8 R_8 + I_6 R_6$
3463	$\varepsilon_3 = I_3 R_3 - I_7 R_7 + I_8 R_8$
16451	$0 = I_4 R_4 + I_{11} R_{11} + I_5 R_5 + I_7 R_7$
3273	$\varepsilon_2 = -I_2 R_2 + I_9 R_9 - I_{12} R_{12}$
37543	$-\varepsilon_3 = -I_3 R_3 - I_4 R_4 + I_{10} R_{10} + I_{12} R_{12}$

$$(2.2)$$

There are of course many other loop equations that can be written; however, the additional loop equations will not be linearly independent of the six we have written. The loop denoted by the nodal sequence 275462 is an example of an additional loop for which we have not written a loop equation. The loop equation resulting from this loop is identically equal to the sum of the loop equations corresponding to the loops 2362, 3463, 37543, and 3273, as given in Eq. 2.2.

We may, of course, as we did in the first chapter, assign loop currents to the L independent loops, solve only the loop equations, and later determine the branch currents as linear combinations of the loop currents.

For many circuits a convenient choice of loops is provided by the elementary loops or meshes. A mesh is a loop which encloses no branch which is not part of the loop. For instance, in Fig. 2.1 the loop defined by the nodal sequence 1261 is an elementary loop or a mesh. However, the loop 275462 is not since it encloses the branches between nodes 2 and 3, nodes 3 and 4, nodes 3 and 6, and nodes 3 and 7. If a circuit is planar, there are just $L = B - (N - 1)$ meshes, and they provide L independent loop equations. It is very easy to identify the meshes, and thus a complete set of independent loop equations can be written quickly. The six independent loop equations given in Eq. 2.2 are a set of mesh equations. The circuit must be planar, however, if one is to use mesh equations; that is, one must be able to draw the circuit so that no branch crosses over or under another branch. The circuit of Fig. 2.1 is planar as is also that of Fig. 2.2 if we consider only the solid lines. When we add the dashed branch to the circuit of Fig. 2.2, it is no longer planar. An obvious and useful property of the mesh equations is that no branch current occurs in more than two mesh equations.

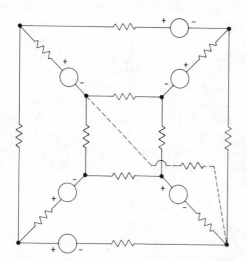

Figure 2.2 A circuit with 12 branches and 8 nodes (ignoring the dashed branch).

We emphasize, however, that our enumeration of the independent node and loop equations holds for any network. Without the dashed branch the circuit of Fig. 2.2 has 8 nodes and 12 branches. There are 7 independent nodal equations and 5 independent loop equations. With the dotted branch there are 8 nodes, 13 branches, 7 node equations, and 6 loop equations.

The reader should experiment with various networks. There is much to be learned. For instance, a necessary condition on a complete set of loop equations is that each branch current appear somewhere in the equations. But this is not sufficient. The four outside meshes of the network of Fig. 2.2 (without the dashed branch) contain all the branch currents but there remains one more independent loop equation, that of the central mesh.

Finally we remark on our requirement that all branches contain a resistance; that is, no ideal constant voltage sources exist. This is physically correct and permits a simpler treatment of network theory, but it may complicate the handling of a practical problem. Voltage sources do exist whose change in output voltage is negligible over the interesting range of currents. We shall not give a general discussion. The proper procedure is usually clear in each particular case.

2.2 The Superposition Theorem

The superposition theorem states that the current in a branch of a linear circuit can be calculated as a sum of partial currents where each partial current is produced by a particular emf in the circuit. A linear circuit is one whose resistances and emf's are independent of the current through them. In order to prove this theorem it is convenient to write Kirchhoff's laws in determinant form. For a circuit with B branches and N nodes we have $N - 1$ Kirchhoff's current law nodal equations each of the form $\sum_j I_j = 0$ where the sum over j includes those currents that enter or leave the particular node being considered. Equations 2.1 are a set of six such nodal equations for the circuit of Fig. 2.1. In a similar fashion we have $L = B - (N - 1)$ Kirchhoff's voltage law loop equations that are of the form $\sum_j \varepsilon_j = \sum_j I_j R_j$ where the sum over the emf's ε_j and the sum over $I_j R_j$ is over those emf's and voltage drops in the loop under consideration. Equations 2.2 are a set of six such equations for the circuit of Fig. 2.1. A complete set of nodal and loop equations can be written in a general abstract form as follows:

$$e_1 = r_{11}I_1 + r_{12}I_2 + \cdots + r_{1B}I_B$$
$$e_2 = r_{21}I_1 + r_{22}I_2 + \cdots + r_{2B}I_B \qquad (2.3)$$
$$\cdots\cdots\cdots$$
$$e_B = r_{B1}I_1 + r_{B2}I_2 + \cdots + r_{BB}I_B$$

The coefficients r_{ij} of the currents I_j have been written in general matrix notation but of course the coefficients of a given current I_j can assume one of only five values, ± 1 when the current I_j occurs in a nodal equation, $\pm R_j$ when the current I_j occurs in a loop equation, or zero if I_j is missing from the equation. R_j is the total resistance of the jth branch including the internal resistance of any emf's that may be in the branch. The e_1, e_2, e_3, and so on, can be different from zero only for loop equations and are linear combinations of the actual emf's ε_1, ε_2, ε_3, and so on, that are encountered in going around the loop.

The solution of these equations for any I_j is the ratio of two determinants.

$$I_j = \frac{\begin{vmatrix} r_{11} & r_{12} & \cdots & e_1 & \cdots & r_{1B} \\ r_{21} & & \cdots & e_2 & \cdots \\ \cdots & & & \cdots \\ r_{B1} & & \cdots & e_B & \cdots & r_{BB} \end{vmatrix}}{\begin{vmatrix} r_{11} & r_{12} & \cdots & r_{1j} & \cdots & r_{1B} \\ r_{21} & & \cdots & r_{2j} & \cdots \\ \cdots & & & \cdots \\ r_{B1} & & \cdots & r_{Bj} & \cdots & r_{BB} \end{vmatrix}} = \frac{\delta_j(r, e)}{\delta(r)} \tag{2.4}$$

The denominator is just the determinant of the coefficients of the currents. In the numerator the column of emf's e_1, e_2, ..., e_B has replaced the jth column occupied by the coefficients of I_j. Let us expand the numerator $\delta_j(r, e)$ in terms of e_1, e_2, and so on, and their codeterminants δ_{1j}, δ_{2j}, and so on. We have

$$I_j = \frac{1}{\delta(r)} [\pm e_1 \delta_{1j}(r) \mp e_2 \delta_{2j}(r) \pm \cdots \pm e_B \delta_{Bj}(r)] \tag{2.5}$$

where the $\pm$ signs come from the rule for the expansion of a determinant in terms of its codeterminants, the $\delta_{kj}(r)$. Now since each e_k is a linear combination of the actual emf's, the ε_k's, we may substitute and express each I_j in terms of the actual emf's. The result is

$$I_j = \frac{1}{\delta(r)} [\varepsilon_1 \beta_{1j}(r) + \varepsilon_2 \beta_{2j}(r) + \cdots + \varepsilon_B \beta_{Bj}(r)] \tag{2.6}$$

where the coefficients $\beta_{kj}(r)$ are linear combinations of the codeterminants δ_{kj} and depend only on the resistances in the circuit.

This is the superposition theorem. It states that the current in any branch of a linear network is a sum or superposition of partial currents each of which is proportional to one and only one of the emf's. To determine the contribution to I_j from the emf ε_k we set all the emf's except ε_k equal to zero. Note that in doing this we must leave in the circuit the internal resistances of all the emf's. Thus each emf in a circuit makes its own contribution to each branch current. The contribution is independent of the other emf's.

A network of resistances and emf's is a linear medium in the same sense that a gas is a linear medium for the transport of sound waves. The response at a given point, if several sources are present, is the sum of the responses that each source would produce if present alone. Figure 2.3(a) will make these points a little clearer. With the help of superposition the current through the 3-ohm resistor can be easily calculated. Let us short out ε_2. The source ε_1 then sees a load of 6 ohms in series with the parallel combination of 6 ohms and 3 ohms. The total resistance is 8 ohms, and since $\varepsilon_1 = 24$ volts, we have a current through ε_1 of 3 amp. Of this, 2 amp flows through the 3-ohm resistor. This is the partial current due to ε_1. From symmetry ε_2 delivers the same partial current, and the total current through the 3 ohms is 4 amp.

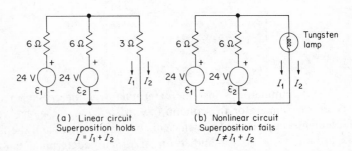

(a) Linear circuit
Superposition holds
$I = I_1 + I_2$

(b) Nonlinear circuit
Superposition fails
$I \neq I_1 + I_2$

Figure 2.3   (a) A linear circuit illustrating the superposition theorem. (b) A nonlinear circuit for which the superposition theorem does not hold.

The superposition theorem does not hold for a nonlinear network. As an example of this, let us now assume that the 3-ohm resistor is a tungsten lamp as in Fig. 2.3(b) and that the resistance of the lamp is 3 ohms when it is carrying 2 amp. Our calculation of the partial currents is the same as before. The resistance of the lamp rises rapidly with increasing temperature, however. If we put 4 amp through the lamp, its resistance might be 4 or 5 ohms. We see in this case that superposition does not hold. The current through the lamp with both emf's in the circuit will be less than the sum of the partial currents. Superposition fails here because the resistance of the lamp is not independent of the current through the lamp so that the Kirchhoff's voltage law relationships between the emf's and the currents are not linear. In general if nonlinear circuit elements are present, the superposition theorem fails. We point out that there are no circuit elements that are truly linear over all values of current and voltage. For example, resistors heat up and the resistance changes. Linearity holds only over a limited range of currents and voltages.

2.3   Thevenin's Theorem

Thevenin's theorem was stated and used without proof in the first chapter. According to Thevenin's theorem, any linear two-terminal network can be replaced by an equivalent network consisting of an emf and a resistance in series. In this section we shall see that Thevenin's theorem is a direct consequence of superposition.

We treat the circuit of Fig. 2.4 where the network of interest is represented by a black box. This is a convenient way of reminding ourselves that we are concerned with the behavior of the network only as it is observed across two of its terminals. Two terminals A and B are brought out of the network in the black box and an external branch containing the emf ε_V is added. It is convenient to treat ε_V as a constant voltage source without internal resistance, but this requires that any loops containing ε_V must also contain a resistance from the part of the loop inside the black box. By different choices of ε_V the potential difference across AB can be fixed at any desired value and the corresponding current I_V can be calculated.

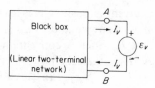

Figure 2.4 A black box containing a linear network that connects to an external circuit through two terminals A and B. According to Thevenin's theorem the I-V characteristic of any linear two-terminal network is $V = \varepsilon_{Th} - I r_{Th}$.

Now by superposition I_V can be expressed as the sum of two partial currents:

$$I_V = I_S - I_\varepsilon = I_S - \frac{\varepsilon_V}{r_{Th}} \tag{2.7}$$

I_S is the partial current produced by all the emf's in the black box. It is obtained when $\varepsilon_V = 0$, that is, when A and B are shorted together. Its magnitude and direction are entirely determined by the contents of the black box. The partial current I_ε is that due to the external emf ε_V. It is obtained when all the internal emf's are set equal to zero, and therefore the black box behaves as a passive network whose equivalent resistance we denote by r_{Th}. Thus we can write $I_\varepsilon = \varepsilon_V / r_{Th}$. The partial current I_ε is proportional to ε_V and can be in either direction. In writing Eq. 2.7 we have assumed that ε_V is connected across AB so that I_S and I_ε are in opposite directions. There is an emf $\varepsilon_V = \varepsilon_V(0)$ for which the two partial currents are equal in magnitude, and therefore $I_V = 0$. We shall denote $\varepsilon_V(0)$ by ε_{Th}. For this emf we have from Eq. 2.7 $\varepsilon_V(0) = I_S r_{Th} = \varepsilon_{Th}$. Since both I_S and r_{Th} depend only on the contents of the black box, so also does $\varepsilon_V(0) = \varepsilon_{Th}$.

Physically it is clear that in the absence of any external circuit between A and B (and therefore $I_V = 0$) the black box must provide a potential difference ε_{Th} between A and B. Therefore when $\varepsilon_V(0) = \varepsilon_{Th}$ is connected across AB, no internal potentials are altered, no internal currents are changed, and I_V must remain zero. We may now rewrite Eq. 2.7 in the form

$$\varepsilon_V = \varepsilon_{Th} - I_V r_{Th}$$

or   (2.8)

$$V = \varepsilon_{Th} - I r_{Th}$$

where V is the potential difference between A and B and I is the current flowing out of A and into B. Thus the I-V characteristic of the black box is the same as that of an emf ε_{Th} and an internal resistance r_{Th} in series. This is Thevenin's theorem.

Our sign convention assumes that the current is positive if it leaves the network at A, the terminal that is positive on open circuit, and enters at B. This is the direction of the short circuit current I_S. Such a convention is appropriate if the network is a source of energy; IV is then the power delivered to a load and is a positive quantity unless $V = \varepsilon_V > \varepsilon_{Th}$. In this case the current reverses and the external circuit, which must be active, sends power into the black box.

Our convention has the disadvantage that the I-V characteristic of a passive network is $V = -I r_{Th}$. Usually for a passive network one reverses the convention and assumes the current to be positive when it enters the network at the positive terminal. Then the form $V = I r_{Th}$ is regained or in general $V = IR$.

For some reason the concept of the I-V characteristic of a two-terminal network seems to cause more trouble than it should. To fix the idea firmly we plot in Fig. 2.5 the I-V characteristics of several circuit elements, a resistance, an emf (constant voltage source), an emf with an internal resistance, a constant current source, and a tungsten lamp whose resistance increases as the current through the lamp increases. The I-V characteristic in each case is simply the plot of the terminal potential difference versus the current flowing between the terminals. Frequently, particularly in the treatment of tubes and transistors, the current is plotted on the vertical axis and the voltage on the horizontal axis.

If the I-V characteristic is a straight line, the network is linear. For nonlinear networks and circuit components there may exist no simple analytic expression for the I-V characteristic, and a graphical plot is then the only convenient way to present the information. The slope of the plot of V versus I for a linear circuit is r_{Th} or $-r_{Th}$ depending on the direction of the assumed current.

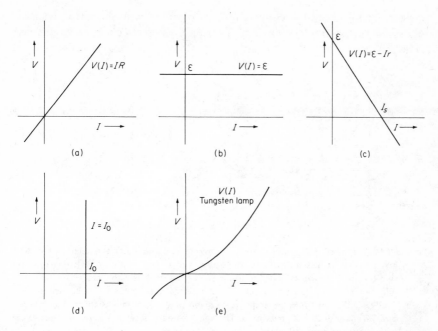

Figure 2.5 The current-voltage characteristic $V(I)$ for several common circuit elements. (a) The I-V characteristic for a resistor. (b) The I-V characteristic of a constant voltage source. (c) The I-V characteristic of an emf and a series resistance. (d) The I-V characteristic of a constant current source. (e) The I-V characteristic of a tungsten lamp.

We may use the I-V characteristic to determine graphically the current and voltage when two different two-terminal networks are connected together. In Fig. 2.6(a) we superimpose the I-V characteristic of an active linear network [Fig. 2.5(c)] and the I-V characteristic of a constant voltage source [Fig. 2.5(b)] for a number of different choices of the constant voltage source ε. The dots at the intersections of two curves give the actual current and voltage. When those two networks are connected, we can look upon the constant voltage sources as a family of two-terminal networks that are used to probe the I-V characteristic of an unknown linear network. This is essentially what was done analytically for the black box of Fig. 2.4 when we deduced Thevenin's theorem from the principle of superposition.

It is important to realize that a linear network may be connected to a nonlinear network without in any way changing the linearity of the first network nor the nonlinearity of the second. The I-V characteristic of a network is completely determined by what's inside it, not by what it's connected to. In Fig. 2.6(b) we show the result of connecting an active linear network to a series of different tungsten lamps. The linearity of the active network is obvious. In Fig. 2.6(c) we connect a series of different linear active networks to a single tungsten lamp. The nonlinearity of the lamp characteristic is now obvious.

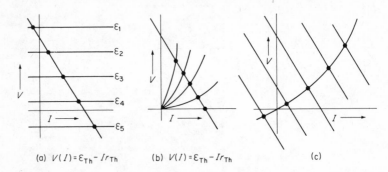

(a) $V(I) = \varepsilon_{Th} - I r_{Th}$ (b) $V(I) = \varepsilon_{Th} - I r_{Th}$ (c)

Figure 2.6 The black dots show the actual current and voltage at the terminals when a given two-terminals network is connected successively to each of a family of other two-terminal networks. (a) An active linear network is connected to a family of different constant voltage sources. (b) An active linear network is connected to a family of different tungsten lamps. (c) A given tungsten lamp is connected to a family of different active linear networks. The various linear networks have different values for ε_{Th} but the same value for r_{Th}.

If a nonlinear element, for instance the tungsten lamp, is part of the active network, for instance the black box of Fig. 2.4, then the I-V characteristic is nonlinear and Thevenin's theorem does not hold.

Because of the great importance of Thevenin's theorem we shall illustrate its use with another example. Let us consider the circuit of Fig. 2.7. We ask for the Thevenin equivalent circuit between terminals A and B. The Thevenin emf is the open circuit voltage between terminals A and B. This can be calculated as follows. Kirchhoff's voltage law applied to the loop currents I_1 and I_2 gives

$$10 = 200I_1 - 100I_2 \quad \text{and} \quad 10 = -100I_1 + 200I_2 \qquad (2.9)$$

The solution of these equations for I_1 and I_2 gives $I_1 = I_2 = 0.1$ amp. The open circuit voltage is $\varepsilon_{Th} = 50I_1 + 50I_2 = 10$ volts. The Thevenin resistance can be found by short-circuiting A and B and calculating the current that flows. The result is $I_S = 0.2$ amp $= \varepsilon_{Th}/r_{Th}$. Combining this with the result that $\varepsilon_{Th} = 10$ volts we find $r_{Th} = 50$ ohms. Let us now ask further what current is drawn from A to B if a load resistor $R_L = 450$ ohms is placed across terminals A and B. The result is immediately obtained and is $I = \varepsilon_{Th}/(r_{Th} + R_L) = 0.02$ amp.

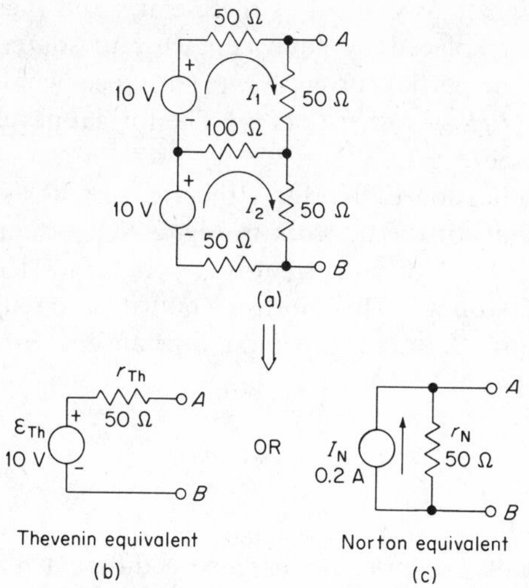

Figure 2.7 (a) A two-terminal network. (b) The Thevenin equivalent circuit for the network of (a). (c) The Norton equivalent circuit for the network of (a).

The reader should note that because of the symmetry of the circuit of Fig. 2.7 no current flows through the 100-ohm resistor. It may be removed without altering any currents or voltages. The parameters ε_{Th} and r_{Th} can then be determined by inspection.

2.4 Norton's Theorem

Norton's theorem states that any linear two-terminal network has an equivalent circuit that consists of a constant current source and a resistance in parallel. This is really an alternate form of Thevenin's theorem. We recall that in Chapter 1 it was pointed out that a constant current source, producing a current I_N, in parallel with a resistance r_N has an I-V characteristic $V(I)$ $I_N r_N - I r_N$ where I is the current drawn out of the two-terminal network. A constant voltage source ε_{Th} in series with a resistance r_{Th} has an I-V characteristic $V(I) = \varepsilon_{Th} - I r_{Th}$. By comparing these two forms for $V(I)$ we see that the two sources are indistinguishable provided $I_N r_N = \varepsilon_{Th}$ and $r_N = r_{Th}$. Now we have shown in the preceding section that any linear two-terminal network can be replaced by its Thevenin equivalent, an emf ε_{Th} and its series resistance r_{Th}. By the arguments just given the network could just as well be replaced by a Norton equivalent, a constant current source I_N and its parallel resistance r_N. This is the content of Norton's theorem. It should be noted that the Norton current I_N is simply the short circuit current I_S that can be drawn from a network. In our development of the superposition theorem from which we derived Thevenin's theorem, we used emf's as the sources of the various partial currents. Any or all of these emf's and their series resistances could have been replaced by equivalent current sources with shunt resistances. In this case the partial current I_{kj}, for instance, would have been proportional to a term $(I_N R_N)_k$ rather than ε_k. The formal development, however, would be the same.

As an example of the application of Norton's theorem, let us ask what is the Norton equivalent circuit for the network in Fig. 2.7. The short circuit current which is the same as the Norton current is $I_N = 0.2$ amp. The Norton resistance is $r_N = r_{Th} = 50$ ohms. The Norton equivalent circuit consisting of a constant current source $I_N = 0.2$ amp and a resistance $r_N = 50$ ohms in parallel is shown in Fig. 2.7(c).

2.5   Reciprocity

In our general discussion of networks we assigned a different real current to each of the B branches and showed that there were B linearly independent equations whose solution gave the B currents. We further have shown that by assigning currents to loops rather than to branches it is possible to satisfy automatically the node equations and solve only the L loop equations.

Let us consider the Wheatstone bridge circuit of Fig. 2.8. There are six branches and three independent loops. We indicate three meshes and the mesh current around each. Kirchhoff's voltage law yields the following three equations.

$$0 = I_1(R_1 + R_2 + r_g) - I_2 r_g - I_3 R_2$$
$$0 = -I_1 r_g + I_2(R_3 + R_4 + r_g) - I_3 R_4 \qquad (2.10)$$
$$\varepsilon = -I_1 R_2 - I_2 R_4 + I_3(R_2 + R_4 + r_\varepsilon)$$

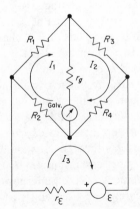

Figure 2.8   A Wheatstone bridge circuit.

The current of greatest interest is $I_g = I_1 - I_2$, the current through the galvanometer. The solution is

$$I_g = (R_1 R_4 - R_2 R_3)\varepsilon/\Sigma$$

where

$$\Sigma = r_g[(R_1 + R_3)(R_2 + R_4) + r_\varepsilon(R_1 + R_2 + R_3 + R_4)] + r_\varepsilon(R_1 + R_2)(R_3 + R_4)$$
$$+ R_1 R_2 (R_3 + R_4) + R_3 R_4(R_1 + R_2) \tag{2.11}$$

Even with only three equations the algebra is rather tedious although the reader may wish to carry it through. The effort will convince one of the very great usefulness of some of the shortcuts furnished, for instance, by Thevenin's theorem.

Our purpose, however, is to point out an interesting feature of these equations. The determinant of the coefficients of the currents

$$\delta(R) = \begin{vmatrix} (R_1 + R_2 + r_g) & -r_g & -R_2 \\ -r_g & (R_3 + R_4 + r_g) & -R_4 \\ -R_2 & -R_4 & (R_2 + R_4 + r_\varepsilon) \end{vmatrix} \tag{2.12}$$

is symmetric about its principal diagonal or, what is equivalent, the jth row and the jth column are identical. In fact, with a proper choice of the loops, the L loop equations of any network can always be written so that the $\delta(R)$ is symmetric.

The symmetry of the determinant leads to an important result that we illustrate using the Wheatstone bridge circuit. Let us write the solution for I_1 in determinant form. We have from Eq. 2.10

$$I_1 = \frac{\varepsilon \begin{vmatrix} -r_g & -R_2 \\ (R_3 + R_4 + r_g) & -R_4 \end{vmatrix}}{\delta(R)} \tag{2.13}$$

Now let us put the emf ε in series with R_1 but without moving r_ε as is shown in Fig. 2.9. The equations for this new circuit are

$$\varepsilon = I_1(R_1 + R_2 + r_g) - I_2 r_g - I_3 R_2$$
$$0 = -I_1 r_g + I_2(R_3 + R_4 + r_g) - I_3 R_4 \tag{2.14}$$
$$0 = -I_1 R_2 - I_2 R_4 + I_3(R_2 + R_4 + r_\varepsilon)$$

Solving for I_3 (which will not have the same value as the I_3 of Fig. 2.8), we obtain

$$I_3 = \frac{\varepsilon \begin{vmatrix} -r_g & R_3 + R_4 + r_g \\ -R_2 & -R_4 \end{vmatrix}}{\delta(R)} \tag{2.15}$$

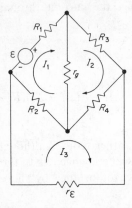

Figure 2.9 The same circuit as in Fig. 2.8 but with the emf moved to the arm containing R_1.

Because the codeterminants in the two numerators (Eq. 2.13 and 2.15) differ only by the interchange of rows and columns, they are equal and we have I_1 (Fig. 2.8) $= I_3$ (Fig. 2.9). This relationship of the codeterminants is a result of the symmetry of $\delta(R)$.

The result we have just illustrated is known as the *reciprocity theorem*. The proof is not difficult; however, we shall merely state the theorem as follows. Suppose a resistanceless emf in branch k of a linear network produces a partial current I_{kj} in branch j. We remove the emf from branch k and place it in branch j and ask what partial current is now produced back in branch k. The answer is that the second partial current is equal to the first. Note that if the emf had an internal resistance, then in relocating the emf its internal resistance must be left behind. We do not change the resistance of any branch.

The theorem can be rephrased in the following less general but very useful form. We have four terminals connecting into an arbitrary passive linear network. We put a constant voltage source (zero internal resistance) across two of them and measure the current between the other two with an ammeter of zero internal resistance. The reciprocity theorem states that if we interchange the voltage source and the ammeter, we shall measure the same current we did before. The network itself has no sense of direction. Actually the theorem holds also for three terminals. One terminal may be shared by the input and the output. In the next section we shall use the reciprocity theorem to reduce the number of variables necessary to describe three- and four-terminal networks.

2.6 Three- and Four-Terminal Networks

Much of our discussion of networks has centered about what are called two-terminal networks. Essentially one postulates a black box that may contain a complex network, and one investigates the voltage and current at two connections made into the box. Thevenin's theorem states that we can always replace a linear two-terminal network with a constant voltage source and a series resistor. If three or more connections are made into the box, the problem becomes more difficult. However, some discussion of three- and four-terminal networks will be useful. The reciprocity theorem is, of course, a statement about four-terminal networks.

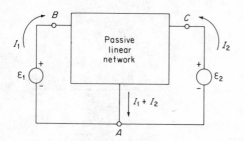

Figure 2.10   A linear passive three-terminal network inside a black box.

As an example, let us consider a three-terminal passive linear network and ask how many parameters are necessary to describe its behavior. In Fig. 2.10 we apply the voltage sources ε_1 and ε_2 between two pairs of terminals; consequently the voltage drop across the other pair is also determined. There are two independent currents I_1 and I_2. From the superposition theorem we may write

$$I_1 = \frac{\varepsilon_1}{R_{11}} + \frac{\varepsilon_2}{R_{12}}$$

$$I_2 = \frac{\varepsilon_1}{R_{21}} + \frac{\varepsilon_2}{R_{22}}$$

(2.16)

The constants R_{11}, and so on, are defined by Eq. 2.16 and have the same dimensions as resistance. For instance, R_{11} is obtained by placing ε_2 at zero, that is, an external short circuit between A and C, and then measuring I_1. The constant R_{11} is ε_1/I_1. There are really only three independent resistance parameters, however, since the reciprocity theorem tells us that $R_{12} = R_{21}$.

This is easily seen from our statement of the reciprocity theorem for a three- or four-terminal passive network. In Fig. 2.10, place an emf ε between A and B. An ammeter of zero resistance placed between A and C (ε_2 is removed) reads the current $I_2 = \varepsilon/R_{21}$. Now interchange the emf and the ammeter. The ammeter now reads $I_1 = \varepsilon/R_{12}$. But the reciprocity theorem tells us that in this experiment $I_1 = I_2$. Therefore we have $R_{12} = R_{21}$. The constant $R_{12} = R_{21}$ must be negative (see Prob. 2.24).

In Fig. 2.11 we illustrate two possible ways of connecting three resistors inside the black box. It is now a simple matter to relate the effective resistances R_{11}, and so on, to the actual internal resistances. One of the voltage sources is put equal to zero and the ratio of the other voltage to the desired current is calculated using the series parallel resistance formulas and Kirchhoff's voltage law. We find

$$(\varepsilon_2 = 0) \qquad \frac{\varepsilon_1}{I_1} = R_{11} = \frac{R_1 R_3}{R_1 + R_3} = R_B + \frac{R_A R_C}{R_A + R_C}$$

$$(\varepsilon_1 = 0) \qquad \frac{\varepsilon_2}{I_2} = R_{22} = \frac{R_1 R_2}{R_1 + R_2} = R_C + \frac{R_A R_B}{R_A + R_B} \qquad (2.17)$$

$$(\varepsilon_2 = 0) \qquad -\frac{\varepsilon_1}{I_2} = -R_{21} = -R_{12} = R_1 = \frac{R_A R_B + R_A R_C + R_B R_C}{R_A}$$

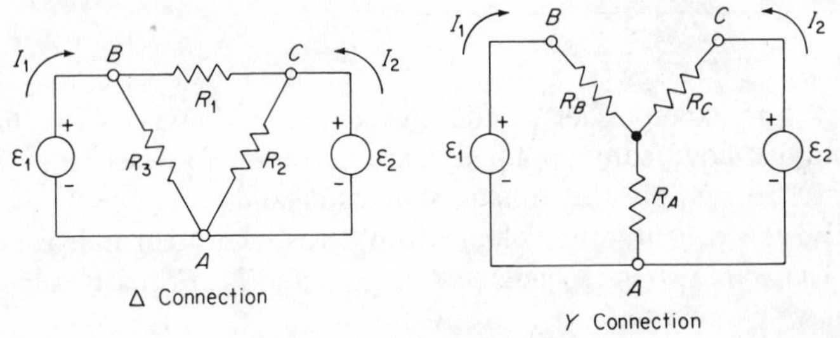

Figure 2. 11 The Δ and Y equivalent circuits for a three-terminal passive network.

Thus either the Δ or the Y connection will provide for the most general current voltage behavior of a three-terminal passive network. Sometimes one is more convenient than the other and we give the relations between the resistances for equivalent Δ and Y networks. In the third equation above, R_1 is already exhibited in terms of R_A, R_B, and R_C. Symmetry demands that

$$R_1 = \frac{R_A R_B + R_A R_C + R_B R_C}{R_A}$$

$$R_2 = \frac{R_A R_B + R_A R_C + R_B R_C}{R_B} \qquad (2.18)$$

$$R_3 = \frac{R_A R_B + R_A R_C + R_B R_C}{R_C}$$

With the help of $R_1 R_A = R_2 R_B = R_3 R_C$ these equations are easily inverted to give

$$R_A = \frac{R_2 R_3}{R_1 + R_2 + R_3}$$

$$R_B = \frac{R_1 R_3}{R_1 + R_2 + R_3} \qquad (2.19)$$

$$R_C = \frac{R_1 R_2}{R_1 + R_2 + R_3}$$

These relations are called the Δ-Y transformation. With this transformation and the series and parallel resistance formulas, any two-terminal passive network can be collapsed to a single resistance and any three-terminal passive network can be collapsed to either the Δ or the Y connection.

In electronics one frequently meets the problem of matching the internal resistance of a source to a load resistance, sometimes with the additional requirement that the output voltage of the source be reduced by an attenuation factor A. Usually this occurs in ac circuits but if, as is often the case, the impedance of the source and load are both resistive, a dc analysis is valid. In Fig. 2.12 we show how this can be accomplished by inserting a Y connection between the source and the load. In this use the Y connection is called a T section network, or T-pad. We make three requirements. Looking to the right through the first dashed line the source must see a load resistance $R_L^* = r_i$; looking to the left through the second dashed line the load must see an internal resistance $r_i^* = R_L$ and an open circuit $(R_L = \infty)$ output voltage $A\varepsilon$. These requirements lead to three equations.

$$R_L^* = r_i = R_B + \frac{R_A(R_C + R_L)}{R_A + R_C + R_L}$$

$$r_i^* = R_L = R_C + \frac{R_A(R_B + r_i)}{R_A + R_B + r_i} \qquad (2.20)$$

$$A = \frac{R_A}{R_A + R_B + r_i}$$

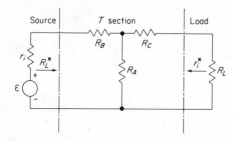

Figure 2.12 A T section used to match a load resistor to a source and to attenuate the voltage seen by the load resistor.

These equations may be solved for R_A, R_B, and R_C in terms of the given parameters r_i, R_L, and A. We could achieve the same results with a Δ connection which in such a use is often called a π section.

We point out that there are limitations on what a passive dc matching network can accomplish. The open circuit output voltage cannot be greater than ε. We may of course increase the internal resistance of the source as seen by the load as much as we wish but we cannot decrease it without simultaneously decreasing the open circuit output voltage. To see this, combine the last two equations to obtain

$$\frac{r_i^*}{r_i} = \frac{R_C}{r_i} + A\left[\left(\frac{R_B}{r_i}\right) + 1\right]$$ (2.21)

Thus $r_i^*/r_i \geq A$, the equality holding when $R_B = R_C = 0$.

We shall now make a few general remarks on some of the different notations used in the treatment of three-terminal networks. We have chosen to write the superposition theorem in the form given in Eq. 2.16 and from this derive the properties of the Δ and Y connections.

More frequently one will see Eq. 2.16 written in terms of conductance parameters

$$I_1 = g_{11}\varepsilon_1 + g_{12}\varepsilon_2$$
$$I_2 = g_{21}\varepsilon_1 + g_{22}\varepsilon_2$$ (2.22)

These are the so-called g parameters and are simply the reciprocals of our R parameters. The reciprocity theorem leads to the result that $g_{12} = g_{21}$. When resistances are used, they are usually the coefficients in the inversion of our equations; that is,

$$\varepsilon_1 = r_{11}I_1 + r_{12}I_2$$
$$\varepsilon_2 = r_{21}I_1 + r_{22}I_2$$ (2.23)

These r parameters are not the same as the R parameters but the relations between them can easily be found by solving Eq. 2.23 for I_1 and I_2 in terms of ε_1 and ε_2 and comparing coefficients with Eq. 2.16. The reciprocity theorem leads to the result that $r_{12} = r_{21}$.

Because they are useful in treating transistor circuits, another set called the *hybrid h parameters* is frequently used. They are defined by the equations

$$\varepsilon_1 = h_{11}I_1 + h_{12}\varepsilon_2$$
$$I_2 = h_{21}I_1 + h_{22}\varepsilon_2$$ (2.24)

Comparison with Eq. 2.16 gives

$$h_{11} = R_{11} \qquad h_{12} = \cdot \frac{R_{11}}{R_{12}} \qquad h_{21} = \frac{R_{11}}{R_{21}} \quad \text{and} \quad h_{22} = \frac{1}{R_{22}} - \frac{R_{11}}{R_{12}R_{21}}$$ (2.25)

Thus for these parameters the reciprocity theorem gives $h_{12} = -h_{21}$. The hybrid h parameters arise naturally if one excites the network of Fig. 2.10 by a constant current I_1 between B and A rather than with the emf ε_1. The emf source ε_2 is still applied between C and A. The dependent variables are then I_2 and ε_1, the potential difference measured between B and A.

As a numerical example of the use of this material, let us find the T circuit that is equivalent to the π network shown in Fig. 2.13. For the equivalent T circuit we obtain, using the Δ-Y transformation,

$$R_A = \frac{R_2 R_3}{R_1 + R_2 + R_3} = \frac{25 \times 75}{200} = 9.375 \text{ ohms}$$

$$R_B = 12.5 \text{ ohms} \tag{2.26}$$

$$R_C = 37.5 \text{ ohms}$$

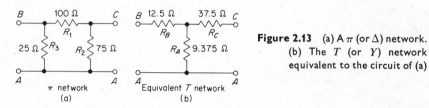

Figure 2.13   (a) A π (or Δ) network.
(b) The T (or Y) network equivalent to the circuit of (a)

The equivalent T circuit is also shown in Fig. 2.13. Let us now ask what are the r parameters (Eq. 2.23) for this circuit? We shall assume that the emf ε_1 is applied between A and B and that the emf ε_2 is applied between A and C. The current I_1 flows in B and out A, and the current I_2 flows in C and out A just as in Fig. 2.11. Under these conditions

$$\varepsilon_1 = 21.875I_1 + 9.375I_2$$
$$\varepsilon_2 = 9.375I_1 + 46.875I_2 \tag{2.24}$$

so that $r_{11} = 21.875$ ohms, $r_{12} = r_{21} = 9.375$ ohms, and $r_{22} = 46.875$ ohms. As predicted by the reciprocity theorem, $r_{12} = r_{21}$.

Problems

2.1   How many nodes and how many branches are there in the circuit illustrated in Fig. 2.14? What is the number of independent equations from Kirchhoff's current law and from Kirchhoff's voltage law? Different points that must be at the same potential should be counted as a single node.

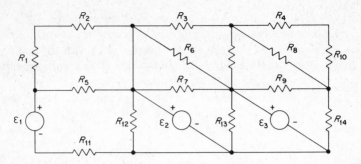

Figure 2.14

2.2   A cube has a 100-ohm resistor along each edge. At each corner three resistors meet and are connected together. A 10-volt source without internal resistance is connected across the body diagonal of the cube. How many independent equations result from Kirchhoff's current law and from Kirchhoff's voltage law? What current flows through the voltage source? Note that the problem is greatly simplified if the symmetry of the array is exploited.

2.3   The 10-volt source is removed from the body diagonal of Prob. 2.2 and placed in one of the edges that connects with the body diagonal. What current now flows through the body diagonal which we assume is a wire with negligible resistance?

2.4   The 10-volt source of Prob. 2.2 is replaced by a 1.0-amp constant current source. Calculate the potential drop across each of the edges.

2.5   In Fig. 2.15, find the current I_1 through the 10-ohm resistor as a function of ε_1 and ε_2. Is the current I_1 a linear function of ε_1 and ε_2? If $\varepsilon_1 = 10$ volts and $\varepsilon_2 = 5$ volts, what is the partial current in the 10-ohm resistor due to each source?

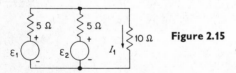

Figure 2.15

2.6   In Fig. 2.16 we show infinite square, triangular, and hexagonal planar arrays of identical resistors of resistance R. Each array covers the entire plane. What is the equivalent resistance of each infinite array measured between adjacent nodes? To solve these problems one must use the superposition principle and exploit the symmetry of the array. Try injecting constant currents into the array at appropriate nodes.

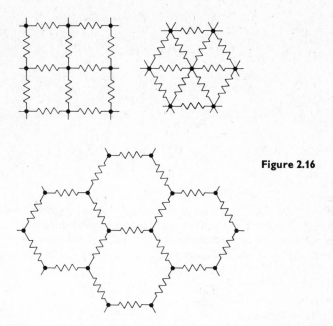

Figure 2.16

2.7 What is the Thevenin equivalent circuit in Fig. 2.17 for the network to the left of terminals A and B? One wishes to dissipate the maximum possible power in a resistance R placed across A and B. What should be the magnitude of R? What power is the 20-volt source then furnishing?

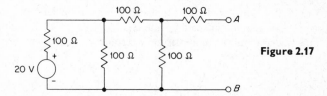

Figure 2.17

2.8 What is the Norton equivalent circuit for the network of Prob. 2.7?

2.9 In Fig. 2.18, what is the Thevenin equivalent circuit between B and ground? This circuit is often used to provide a dc bias to the base of a transistor.

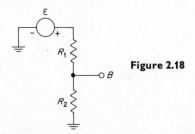

Figure 2.18

2.10   In Fig. 2.19, what is the Thevenin equivalent circuit between C and ground?

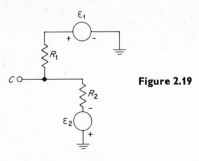

Figure 2.19

2.11   In Fig. 2.20, use Thevenin's theorem to calculate the contribution to I_1 from each emf separately. Then use the superposition theorem to calculate the total current I_1. Next set up the direct Kirchhoff's equations solution for I_1. Which attack seems simpler?

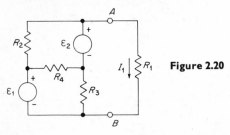

Figure 2.20

2.12   In Fig. 2.21, show by direct calculation using loop currents that the ammeter reading is not affected by interchanging the emf and the ammeter. Assume the emf and the ammeter have zero internal resistance.

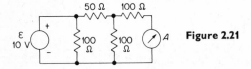

Figure 2.21

2.13   In Fig. 2.22, determine I_1 as quickly and efficiently as possible.

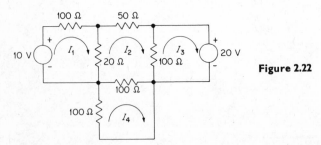

Figure 2.22

2.14 Convert the π (or Δ) network shown in Fig. 2.23 into the equivalent T (or Y) network.

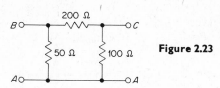

Figure 2.23

2.15 Convert the T (or Y) network shown in Fig. 2.24 into the equivalent π (or Δ) network.

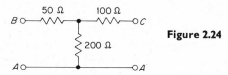

Figure 2.24

2.16 In Fig. 2.25, find the current drawn from the 12-volt battery.

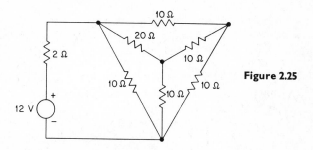

Figure 2.25

2.17 Calculate the R, r, g, and h parameters for the circuit shown in Fig. 2.26.

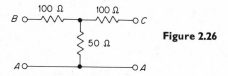

Figure 2.26

2.18 Calculate the R, r, g, and h parameters for the circuit shown in Fig. 2.27.

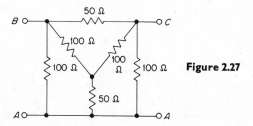

Figure 2.27

2.19   Figure 2.28 shows an infinite chain of identical T sections extending to the right of the terminals AB. What is the equivalent resistance of the chain? If an emf ε is put across AB, what is the voltage at the end of the first T section? at the end of the nth T section? (*Hint*. Note that the resistance of the chain to the right of $A'B'$ is the same as the resistance to the right of AB.)

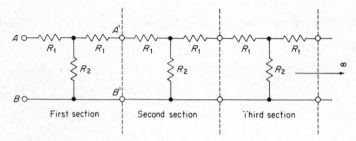

Figure 2.28

2.20   Find the resistances of a unit of the infinite chain of π sections that is equivalent to the chain of T sections of Prob. 2.19. Draw the first few sections of the chain.

2.21   Sketch the I-V characteristic of a constant voltage source ε_0 in series with a tungsten lamp. Does this network have a Thevenin equivalent circuit?

2.22   Figure 2.29 shows the I-V characteristics of a resistor of constant value and of a tungsten lamp. The characteristics intersect at the origin and at a finite value of I and V. This implies, incorrectly, that nonzero currents and voltages may exist when passive networks are connected. Point out the error in the figure, and draw a correct figure.

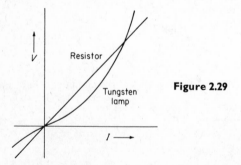

Figure 2.29

2.23   The following measurements indicate the voltage across a resistor that is connected between the terminals of an active two-terminal network. What can you conclude about the network? The voltage measurements were all taken with a VTVM that has an accuracy of $\pm \frac{1}{2}\%$ and an input resistance of 10^{10} ohms.

(R ohms)	Voltage across R (volts)
10^6	299
10^5	299
10^4	299
10^3	296
10^2	271
10	148

2.24 Equation 2.16 gives an expression for the currents shown in Fig. 2.10. Show that the partial currents ε_1/R_{11} and ε_2/R_{12} that contribute to I_1 must have opposite signs. (*Hint.* Consider the sign of the potential difference between B and A when ε_1 is replaced by a small resistor R.)

3/Direct Current Measurements

In this chapter we discuss a few of the common dc measuring instruments, particularly the galvanometer, the volt-ohm-milliammeter, and two null devices—the Wheatstone bridge and the potentiometer. In a null device an unknown resistance or voltage is compared to an accurate standard in a circuit so designed that when a particular current is zero, a simple relation exists between the standard and the unknown. It is usually much easier to determine that a quantity is zero to a certain accuracy than to measure a finite value to the same accuracy. For this reason null devices are very widely used throughout science and engineering.

3.1 The D'Arsonval Galvanometer

A galvanometer is used to measure a current. The current is determined by the force exerted on it by a magnetic field. In the next chapter we shall treat the interaction of currents and magnetic fields in some detail. Here we merely remind the reader that a straight wire of length l which carries a current I and which is situated in a uniform magnetic field B has exerted on it a force $F = BIl \sin \theta$. The force is perpendicular to the plane containing the current and the magnetic field. The quantity θ is the angle between the field and the wire. The standard ampere, the basis of our system of electrical units, is defined in terms of the force between current carrying wires in a very precisely determined geometry. In an ordinary laboratory galvanometer the magnetic field is provided by a permanent magnet, and the current to be measured flows through a coil that can rotate in the magnetic field. Since the magnitude of the field is not known accurately, the instrument must be calibrated against a primary or secondary standard.

The D'Arsonval galvanometer movement shown in Fig. 3.1 is found in nearly all laboratory galvanometers, ammeters, and voltmeters. By shaping the pole pieces as indicated and centering a cylinder of soft iron between them, one has an annular region in which the magnetic field is almost constant in magnitude and radial in direction over an appreciable angular range. When a current flows in the rectangular coil of wire suspended in the annulus, the coil experiences a torque due to the equal and opposite forces produced on the two sides of the current carrying coil by the magnetic field B and given approximately by $\tau = 2r(NIlB) = AINB$ where l is the length of the coil, $2r$ is the width of the coil, $A = 2lr$ is the area of the coil, and N is the number of turns of wire on the coil. The coil may be supported in a variety of ways depending on the sensitivity or ruggedness desired but in all cases a fiber or ribbon suspension or a flat coiled spring provides a restoring torque $\tau = -k\theta$ proportional to the angle of rotation from the equilibrium position at zero current. In the steady state the two torques are equal but opposite in sign.

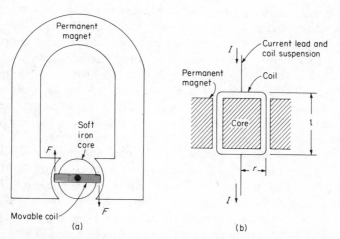

Figure 3.1 A schematic diagram of a D'Arsonval galvanometer movement. (a) Top view. (b) A vertical section through the plane of the coil.

We have then the equation

$$k\theta = AINB \tag{3.1}$$

relating the angular position θ of the coil and the current I. The sensitivity of the galvanometer is defined as $S = \theta/I = ANB/k$. A good laboratory galvanometer using an optical lever of 50 cm to measure the rotation of the coil can detect a current of about 10^{-10} amp. The galvanometer movement of a small panel or box meter frequently has a full-scale deflection of the needle (θ about 90°) for currents in the milliampere range. Full-scale deflections of less than 50 microamp are difficult to achieve if the instrument is to be reasonably portable and shock resistant.

We have outlined the steady state behavior of the D'Arsonval galvanometer, but it must be noted that it is in fact a damped harmonic oscillator. The equation of motion may be obtained as follows. The moment of inertia J times the angular acceleration $d^2\theta/dt^2$ is equal to the restoring torque $-k\theta$ plus the torque due to the current flowing in the galvanometer coil $NABI$. If we include a damping torque proportional to the angular velocity, $\tau = -D\, d\theta/dt$, the angular displacement satisfies the differential equation

$$J\frac{d^2\theta}{dt^2} = -k\theta + NABI - D\frac{d\theta}{dt}$$

or

$$J\frac{d^2\theta}{dt^2} + D\frac{d\theta}{dt} + k\theta = NABI$$

(3.2)

Equations of this form will be treated in detail in later chapters since they also describe the electrical oscillations in ac circuits. For present purposes we need only some very general results. If the damping coefficient D is zero, the solution of Eq. 3.2, $\theta(t)$, is a pure harmonic oscillation about the steady state angular position $\theta = NABI/k$ and with angular frequency $\omega_0 = (k/J)^{1/2}$. If the damping is small, the amplitude of the oscillation dies away exponentially in time and the coil finally comes to rest at the equilibrium position given by $\theta = NABI/k$. For sufficiently large D there is a nonoscillatory exponential approach to the equilibrium position.

Panel and box meters usually have movements whose natural time constant $1/\omega_0$ is a fraction of a second. The time necessary for $\theta(t)$ to reach its equilibrium position is of this same order of magnitude and therefore small compared to the time necessary to read the meter. Quite the opposite is the case with the most sensitive galvanometers. From Eq. 3.1 we see that high sensitivity demands a large area A and a large number of turns N, both of which increase J. High sensitivity also demands a small force constant k. Such galvanometers may have natural periods of several seconds and it becomes important to minimize the time necessary to reach equilibrium. About the best that can be done is the time $1/\omega_0$. This is achieved when the damping, called *critical damping*, is just large enough to eliminate oscillations in $\theta(t)$.

Most of the damping of a sensitive galvanometer is electromagnetic rather than mechanical. Because the coil is moving through a magnetic field, Faraday's law of induction tells us that an emf $\varepsilon(t)$ is induced and a current flows. This current is in addition to the dc current from any external emf. The interaction of the induced current and the magnetic field produces a torque on the moving coil that opposes and damps the oscillations of the coil. The rate at which the mechanical energy of the moving coil is dissipated as Joule heat in resistors is given by $\varepsilon^2(t)/(R + r_g)$. Here r_g is the internal resistance of the galvanometer and R is the resistance of the external circuit through which the induced current flows. Thus by proper choice of R a galvanometer can be critically damped.

In any application of a sensitive galvanometer one must be careful to protect it against overloads. This is particularly true when it is used as a null detector. Near the balance (or null) the ultimate sensitivity is needed. Away from balance very much larger currents are available than are needed for full-scale deflection.

An Ayrton shunt, Fig. 3.2, is frequently used to protect the galvano-meter during the approach to balance. The current source can be applied across a fraction f of the shunt resistance R. If the internal resistance of the current source is sufficiently high, we have for the current through the galvanometer

$$I_g = \frac{IfR}{R + r_g} \tag{3.3}$$

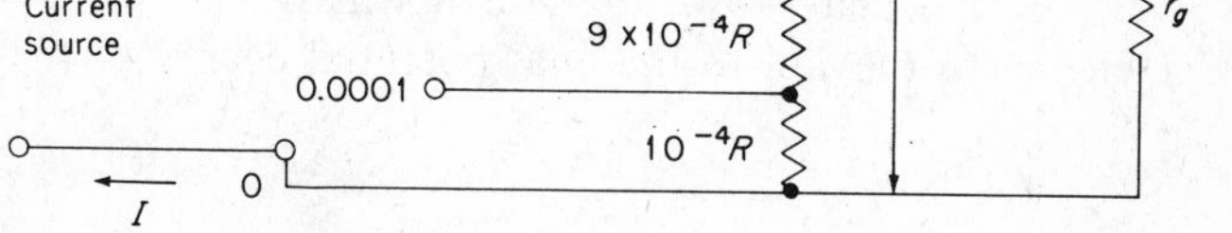

Figure 3.2 An Ayrton shunt of total resistance R connected to a galvanometer whose internal resistance is r_g.

As balance is approached, the switch is moved toward $f = 1$ for maximum sensitivity. Most Ayrton shunts have a setting labeled ∞ which simply connects the current source directly into the galvanometer. This setting can be used if the ultimate in sensitivity is required.

For a current source of high internal resistance, R can be chosen so that $R + r_g$ is the critical damping resistance and the galvanometer will remain critically damped for all settings of the Ayrton shunt. If the internal resistance of the current source is low, consideration must be given to the best match of source and galvanometer characteristics. For instance, a galvanometer of reduced current sensitivity and reduced internal resistance r_g may be better than a more sensitive high resistance galvanometer. A sensitive galvanometer of the kind used with a Wheatstone bridge or a potentiometer might typically have a period between 1 and 10 sec, a coil resistance of a few hundred ohms, and a total resistance for critical damping, $R + r_g$, of a few thousand ohms.

Considering the galvanometer as a current measuring device, it is the steady state rather than the dynamical behavior that interests us. We must understand the latter in order to utilize the former satisfactorily. However, the dynamical properties are directly exploited in a ballistic galvanometer, an instrument that is used to measure total charge rather than current.

Assume we have a galvanometer with a rather long natural period and negligible damping. A short pulse of current is sent through the galvanometer. The pulse lasts for a time τ that must be small compared to the period of the galvanometer. While the current is flowing, it produces a torque $NABI$ that is exerted on the coil. The time integral of the torque on the coil is the increase in angular momentum $J \, d\theta/dt$ of the coil. Let us suppose that before the pulse of current the coil is at rest at its zero current equilibrium position. Just after the pulse the coil has the angular momentum $J \, d\theta/dt$ but because the pulse is very short, the coil has moved scarcely at all and the suspension has exerted no opposing torque. Just after the pulse we have

$$J \frac{d\theta}{dt} = NAB \int_0^\tau I \, dt = NABQ \tag{3.4}$$

where Q is the total charge that the pulse of current transports through the galvanometer.

It is easy to calculate how far the coil will swing by equating the initial kinetic energy of the coil to the final potential energy of the twisted suspension. This gives

$$\frac{1}{2} J \left(\frac{d\theta}{dt}\right)^2 = \frac{(NABQ)^2}{2J} = k\theta_{max}^2$$

or

$$\theta_{max} = \frac{NABQ}{(kJ)^{1/2}}$$

(3.5)

If the natural period of the galvanometer is not too short, θ_{max} is easily observed and Q can be determined. Damping reduces θ_{max} but does not change the direct proportionality of θ_{max} to Q.

The charge on a condenser is easily measured by discharging it through a ballistic galvanometer. A more frequent use is in the measurement of magnetic fields. Assume that an external coil of wire is connected to a ballistic galvanometer. If the magnetic flux through the coil is suddenly changed, it can be shown that the total charge transported through the ballistic galvanometer by the pulse of induced current is directly proportional to the change in magnetic flux $BA \cos \phi$ where B is the magnetic field, A is the area of the external coil of wire, and ϕ is the angle between the field B and the normal to the area A. Thus if the coil of wire is originally perpendicular to a uniform magnetic field and is quickly moved to a region of zero field, the charge transported through the ballistic galvanometer is directly proportional to the magnetic field B.

3.2   Ammeters and Voltmeters

The D'Arsonval galvanometer is the fundamental component of nearly all voltmeters, ammeters, and multimeters. What the galvanometer itself responds to is always the current through its coil. By the choice and arrangement of associated resistors this current can be made proportional to a wide range of currents or voltages applied to the terminals of the complete meter. Usually the sensitivity of the complete meter is limited by the current sensitivity and coil resistance of the galvanometer. Sometimes, however, as in a vacuum-tube voltmeter, the signal may be amplified before it reaches the galvanometer (vacuum-tube voltmeters are discussed in Chapter 18).

In Fig. 3.3 we show the ammeter and voltmeter arrangements of a galvanometer that has a full-scale deflection for a current I_g and whose coil resistance is r_g. The ammeter arrangement has a shunt resistor R in parallel with the galvanometer movement. The full-scale deflection corresponds to a current $I = I_g(R + r_g)/R$ through the galvanometer and the shunt resistor.

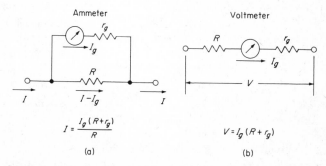

Ammeter

Voltmeter

$$I = \frac{I_g(R+r_g)}{R}$$

$$V = I_g(R + r_g)$$

(a)

(b)

Figure 3.3 A galvanometer movement with external resistors connected for use as an ammeter (a) and a voltmeter (b). The resistance of the ammeter is $R_A = Rr_g/(R + r_g)$. The resistance of the voltmeter is $R_V = R + r_g$.

The voltmeter arrangement has a resistor R in series with the galvanometer movement. The full-scale deflection of the voltmeter corresponds to a voltage of $V = I_g(R + r_g)$. As a voltmeter the maximum full-scale sensitivity of the galvanometer movement is $V = I_g r_g$. The galvanometer movement of a commonly used laboratory multimeter has a full-scale current sensitivity of 50 microamp and a coil resistance of about 5000 ohms. Its maximum voltage sensitivity is therefore 250 millivolts full scale. Frequently a galvanometer movement is characterized by the total resistance needed, $R + r_g$, if 1 volt across the terminals is to produce full-scale deflection. From $V = I_g(R + r_g)$ $= 1$, we see that this is just $1/I_g$, the reciprocal of the full-scale current. The multimeter referred to has a 20,000-ohm/volt movement.

There are advantages, among them greater voltage sensitivity, if r_g is made as small as possible. Low internal resistance is incompatible with current sensitivity, however, since the latter demands a large number of turns of wire on the moving coil.

We now discuss the interaction between the meter and the source furnishing the current or voltage to be measured. There are two aspects: What does the source do to the meter, and what does the meter do to the source? The former is simple enough although pedagogically frustrating. If the source sends a current considerably greater than I_g through the galvanometer, the meter breaks. Sometimes even the source may suffer. The conventional wisdom is almost unanimous. The point can be learned only after the sacrifice of a few meters.

In Fig. 3.4 we turn to the second aspect of the interaction. We show one two-terminal active network acting as a current source (but not a constant current source) and another acting as a voltage source (but not a constant voltage source). We wish to measure I and V by connecting the appropriate meter across the terminals. The meters, we may assume, will accurately measure the applied current or voltage. The difficulty is that the meter perturbs the network. The currents and voltages, after the insertion of the meter, are not the same as they were before. It is easy to see why this must be so. Consider the ammeter case. Originally a current I flows between two terminals at zero potential difference; that is, the terminals are shorted together. We break the short and insert the ammeter. The potential difference between the terminals is no longer zero; it is $I_A r_g R/(R + r_g)$ where I_A is the current through the meter and $r_g R/(R + r_g)$ is the resistance of the meter. The important point is that because $I_A r_g R/(R + r_g) \neq 0$, I_A cannot equal I. Changing one of the voltages or currents in a network will usually change all the others. A similar analysis applies to the voltmeter case. For the ammeter the condition to be met if I_A is to almost equal I can be seen by thinking of the active network as a current source. We ask that the difference in current output between load resistance zero and the load resistance $r_g R/(R + r_g)$ (the ammeter resistance) be very small. This requires that the internal resistance of the source be very large compared to the meter resistance $r_g R/(R + r_g)$. In the voltmeter case we view the active network as a voltage source and demand that loads between $R_L = R + r_g$ (the voltmeter resistance) and $R_L = \infty$ have only a small effect on the output voltage. This requires that the internal resistance of the source be very small compared to the meter resistance $R + r_g$.

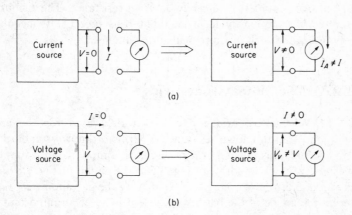

(a)

(b)

Figure 3.4 The diagrams show how the insertion of an ammeter or a voltmeter into a circuit necessarily alters the quantity to be measured. The circle containing the arrow represents the complete meter. Note that in (a) the current source is *not* a constant current source with infinite internal resistance but instead has a finite but large shunt resistance. Similarly, the voltage source is not a constant voltage source with zero internal resistance but instead has a finite series resistance. (a) The quantity to be measured is the current I flowing through a short between the output terminals. The potential difference between the terminals is zero. After the short is broken and the ammeter inserted, the potential difference across the terminals is no longer zero and I_A, the current through the ammeter, is not equal to I. (b) The quantity to be measured is the open circuit ($I = 0$) output voltage V. After the voltmeter is connected, a finite current is drawn and V_V, the potential difference across the volmeter is, not equal to V.

As an example, consider some of the limitations of the multimeter galvanometer for which $I_g = 50$ microamp and $r_g = 5000$ ohms, if we try to use it to measure the output emf of a thermocouple. A thermocouple might in some standard use develop an emf of 10 millivolts and have an internal resistance of 1 ohm. For load resistances considerably less than an ohm, the thermocouple is a good current source and provides 10 milliamp; for loads much greater than an ohm, it is a good voltage source of 10 millivolts magnitude. When the thermocouple is connected to the galvanometer, a current $I = 10^{-2}/5000 = 2.0$ microamp will flow, one twenty-fifth of full scale. The galvanometer, because $r_g \gg 1.0$ ohm, drastically reduces the current available from the thermocouple and it fails as an ammeter. As a voltmeter it is successful in that it does not appreciably change the potential difference it seeks to measure but the minimum full-scale voltage response of the multi-meter galvanometer, $I_g r_g = 250$ millivolts, is too great to permit accurate measurement of the 10-millivolt potential difference put out by the thermo-couple.

A common problem is the measurement of what may be quite large voltages but from sources of very high internal resistance. The vacuum-tube voltmeter, which may easily have an input resistance of 10 megohms, is extremely useful for such measurements (see Chapter 18).

3.3   The Wheatstone Bridge

The Wheatstone bridge circuit is widely used for the accurate determination of an unknown resistance in terms of known standard resist-ances. In the previous chapter it was discussed simply as an example of a three-mesh network. We redraw in Fig. 3.5(a) the circuit of Fig. 2.8 and also show it in Fig. 3.5(b) as a voltage source driving the galvanometer. We omit the internal resistance of the battery because it is never an important factor in the actual use of the circuit.

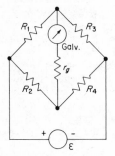

(a) Wheatstone bridge

Figure 3.5 The Wheatstone bridge circuit.

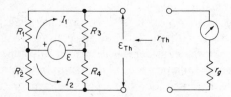

(b) The Wheatstone bridge represented as a voltage source and a galvanometer load

The condition for balance, $I_g = 0$, is $R_1 R_4 = R_2 R_3$ (from Eq. 2.11). The condition for balance can also be obtained from elementary considerations. If $I_g = 0$, the potential difference across the galvanometer branch must be zero. This gives the equations

$$I_1 R_1 = I_2 R_2$$
$$I_1 R_3 = I_2 R_4$$
(3.6)

Division of the first equation by the second leads to the balance condition just stated.

Let us assume that one of the resistances, say R_4, is unknown; that R_1, R_2, and R_3 are known resistances; and that at least one of the known resistances is variable. The known resistances are adjusted until $I_g = 0$. We then have $R_4 = R_2 R_3 / R_1$. The accuracy with which R_4 can be determined is of course no better than the accuracy of the known resistances. The variable resistances usually found in the laboratory are decade resistance boxes that will furnish any resistance up to 10^3 or 10^4 ohms in steps of 1 ohm. Such boxes may be accurate to better than 0.1 %. Very accurate fixed reference resistances are made of manganin wire and are operated in a small oil bath at a known temperature. Accuracies of 1 in 10^5 parts are available. The accuracy of a bridge measurement is also limited by the current sensitivity of the galvanometer. A balance is never better than the smallest detectable off-balance current. But before discussing this, let us consider the bridge from the viewpoint of a source driving a galvanometer load as illustrated in Fig. 3.5(b).

It is a simple matter to calculate ε_{Th}, the open circuit output voltage to the galvanometer, and r_{Th}, the internal resistance of the voltage source. We have $\varepsilon_{Th} = I_1 R_1 - I_2 R_2$. Using $I_1 = \varepsilon/(R_1 + R_3)$ and $I_2 = \varepsilon/(R_2 + R_4)$, we find

$$\varepsilon_{Th} = \varepsilon \left(\frac{R_1}{R_1 + R_3} - \frac{R_2}{R_2 + R_4} \right) = \varepsilon \frac{(R_1 R_4 - R_2 R_3)}{(R_1 + R_3)(R_2 + R_4)}$$
(3.7)

Note that $\varepsilon_{Th} = 0$ gives the balance condition. Now r_{Th} is just the R_1, R_3 parallel combination in series with the R_2, R_4 parallel combination. Therefore

$$r_{Th} = \frac{R_1 R_3}{R_1 + R_3} + \frac{R_2 R_4}{R_2 + R_4}$$
(3.8)

One sees that if there were a resistance r_ε in the battery branch, the calculation of r_{Th} could be accomplished by a Δ-Y transformation. Finally we have

$$I_g = \frac{\varepsilon_{Th}}{r_{Th} + r_g} = \frac{\varepsilon(R_1 R_4 - R_2 R_3)}{R_1 R_3(R_1 + R_4) + R_2 R_4(R_1 + R_3) + r_g(R_1 + R_3)(R_2 + R_4)}$$
(3.9)

This is of course the same expression for I_g that we would obtain by placing $r_\varepsilon = 0$ in the determinant solution of Kirchhoff's equations but it has been obtained much more quickly and in a more convenient form.

Now let us discuss the implications of a small off-balance current. With the assumption that $r_g \ll r_{\text{Th}}$, we combine Eqs. 3.7 and 3.9 and find

$$\frac{(R_1 + R_3)(R_2 + R_4)}{R_1} r_{\text{Th}} \, \Delta I_g = \varepsilon\left(R_4 - \frac{R_2 R_3}{R_1}\right) = \varepsilon \, \Delta R_4 \qquad (3.10)$$

By ΔI_g we mean the smallest off-balance current that the galvanometer can detect and by ΔR_4, the resulting difference between the true value of R_4 and its calculated value of $R_2 R_3/R_1$. Let us assume that R_1, R_2, R_3, and R_4 are approximately equal. We have $r_{\text{Th}} \simeq R_4$ and Eq. 3.10 reduces to

$$4R_4 \, \Delta I_g \simeq \varepsilon \, \frac{\Delta R_4}{R_4} \qquad (3.11)$$

This equation makes clear what maximum resistance can be determined to a given fractional accuracy. For instance if $\varepsilon = 10$ volts, $\Delta I_g = 10^{-8}$ amp and we wish 1 % accuracy ($\Delta R_4/R_4 = 10^{-2}$), then $R_4 \leq 2.5 \times 10^6$ ohms. This limitation is imposed by the galvanometer and the battery. Any inaccuracies of the presumably known resistances are an additional and independent source of error. It is not necessary that the four resistances of the bridge be approximately equal. The reader may wish to examine the equations more carefully and convince himself that no other choice of R_1, R_2, and R_3 gives any appreciable improvement on the predictions of Eq. 3.11.

At the other extreme of very small R_4, measurements become difficult when lead resistances and contact resistances are comparable to the unknown. High accuracy is not possible if R_4 is much less than 1 ohm. Low resistance measurements are usually made with a Kelvin bridge, which measures the ratio of the voltage drop across a known and an unknown low resistance each carrying the same high current. The connections of the voltage comparison circuit are independent of the high current connections and leads and small changes in the high current do not affect the balance.

3.4   The Potentiometer

The potentiometer is probably the most versatile device for the comparison of voltages or for the measurement of any quantity that can be converted to a voltage. The basic circuit is illustrated in Fig. 3.6. The battery ε sends a current I through a slide-wire, which we have taken to be of unit length and total resistance r. It is divided by the slide connection to the galvanometer into sections of length l and $(1 - l)$ and of resistance rl and $r(1 - l)$. The resistance per unit length of the slide-wire must be constant.

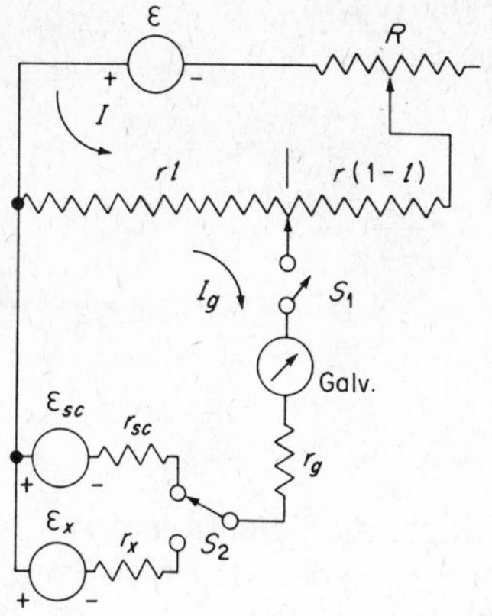

Figure 3.6   A potentiometer circuit.

The variable resistance R is used to control I and to make the instrument direct reading. The current I may be of the order of 10 milliamp. All that is asked of the upper circuit is that the current I remain constant during the operation of the potentiometer. The lower circuit with the galvanometer and emf's ε_x and ε_{sc} involves only very small currents. Readings are taken at $I_g = 0$ to within the sensitivity of the galvanometer and off-balance currents should not exceed a few microamperes. Thus for the lower circuit the branch rl may be considered a voltage source of magnitude Irl.

Let ε_{sc} represent a standard cell of accurately known emf and ε_x, an unknown emf. With the switch S_2 connected to ε_{sc}, the slide-wire is adjusted until $I_g = 0$. Kirchhoff's voltage law then gives us $\varepsilon_{sc} = Irl_{sc}$. The balance is repeated with ε_x connected and we have $\varepsilon_x = Irl_x$. Finally then

$$\varepsilon_x = \frac{l_x}{l_{sc}}\,\varepsilon_{sc} \tag{3.12}$$

All but the crudest potentiometers are provided with numerical dial readings proportional to the fraction l of the total slide-wire resistance appearing in the galvanometer circuit. The potentiometer is made direct reading by setting the dial to the numerical value of ε_{sc} and then with ε_{sc} in place changing R and, therefore, I until balance is achieved. The new balance dial setting with ε_x in the circuit gives ε_x directly.

To determine the sensitivity of a potentiometer we must investigate the currents near balance. As in the case of the Wheatstone bridge this is done most easily using the Thevenin equivalent source seen by the galvanometer. Usually the standard cell has a high resistance r_{sc}. This means that the currents flowing through the galvanometer when the standard cell is in the circuit are small, and the fractional accuracy in determining this current is usually not so good as when the unknown emf is in the circuit. For this reason we investigate the sensitivity of the potentiometer circuit when the standard cell is in the circuit. From inspection we find

$$\varepsilon_{Th} = \varepsilon_{sc} - Irl$$

$$r_{Th} = r_{sc} + \frac{rl[R + r(1 - l)]}{R + r} \tag{3.13}$$

The galvanometer current is

$$I_g = \frac{\varepsilon_{Th}}{r_{Th} + r_g} \tag{3.14}$$

Just as we found for the Wheatstone bridge, the potentiometer is balanced when $\varepsilon_{Th} = 0$. Now let us see how small an ε_{Th} can be detected. For a typical potentiometer and standard cell $r_{sc} \simeq 500$ ohms, $rl[R + r(1 - l)]/(R + r) \simeq 100$ ohms and $r_g \simeq 400$ ohms. The sensitivity of the galvanometer can be 10^{-8} amp or better, thus $\varepsilon_{Th} = I_g(r_{Th} + r_g) = 10^{-8} \times 10^3 = 10^{-5}$ volt. Thus the comparison of two voltages each near 1 volt can easily be made to 1 in 10^5 parts. Of course we have discussed only the limitations imposed by galvanometer sensitivity. A slide-wire 1 meter long, even if it has the required constancy of the resistance per unit length, cannot be read to 1 in 10^5 parts. In a high performance potentiometer, accurate fixed resistors can be switched in at the ends of the wire so that the positioning of the slide becomes less critical.

As we have mentioned, any quantity that can be related to a potential difference can be measured with a potentiometer. In Fig. 3.7 we show some of the possibilities. The potentiometer can be used, for instance, for very accurate measurements of current and therefore for the calibration of ammeters, for the calibration of voltmeters, and for the accurate comparison of resistances. In Fig. 3.7(a) the battery ε sends a current I through a precision resistor R and a box or panel ammeter. The voltage drop IR becomes the ε_x of a potentiometer measurement. The current I is determined to the accuracy with which R is known and ε_x is determined. Figure 3.7(b) illustrates an obvious way of calibrating the scale of a voltmeter. The potentiometer is widely used for the routine calibration of meters. Resistances can be compared as in Fig. 3.7(c), although there is no advantage over a Wheatstone bridge.

Finally, let us compare a potentiometer to a common laboratory voltmeter. Earlier in this chapter we mentioned that the movement in the meter of a particular multimeter had a full-scale sensitivity of 50 microamp. Used as a voltmeter measuring $2\frac{1}{2}$ volts full scale, its resistance is 5×10^4 ohms. The meter can be read to perhaps 1% of full scale. If the error from loading of the voltage source is to be kept less than the scale error, it should not be used to measure a voltage source of internal resistance greater than 500 ohms. By comparison the potentiometer we have described has a just detectable off-balance current of 10^{-8} amp so that the current drawn by the nearly balanced potentiometer is only 5×10^{-3} times the current drawn by the multimeter movement.

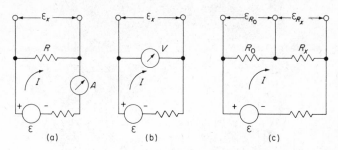

Figure 3.7 Circuits that enable one to use a potentiometer to calibrate (a) an ammeter (b) a voltmeter, and (c) a resistance R. The voltage ε_x is measured with a potentiometer circuit.

3.5   The DC-AC Volt-Ohm-Milliammeter

Accurate determinations of circuit parameters may be made by the bridge methods we have described but such methods are cumbersome and expensive for routine troubleshooting or for checking during the bread-boarding of a new circuit. A small portable multimeter or volt-ohm-milli-ammeter may be used for such purposes. Flexible leads terminating with alligator clips are provided that can be inserted into existing circuits and measurements can be made to an accuracy of 3% to 5%. All measurements are read from a single dc panel meter, which should be as rugged and sensitive as possible. One of the more popular multimeters, the Simpson Model 270, uses a galvanometer movement with a full-scale deflection of 50 micro-amp. Thus the galvanometer movement has a sensitivity of 20,000 ohms/volt. The galvanometer movement has a resistance of about 1800 ohms. In addition the galvanometer movement has a resistor of about 3200 ohms in series so that the effective galvanometer resistance is 5000 ohms rather than the 1800 ohms of the galvanometer movement only.

For use as a dc ammeter, various shunt resistors may be selected using a front panel switch. The Simpson Model 270, which we show in Fig. 3.8, provides full-scale dc current ranges of 50 microamp, 1 milliamp, 10 milliamp, 100 milliamp, 500 milliamp, and 10 amp.

In a similar way, series resistors can be selected so that the meter may be used as a voltmeter with full-scale dc voltage ranges of 2.5 volts, 10 volts, 50 volts, 250 volts, 1000 volts, and 5000 volts.

The meter can also be used for ac voltage measurements (Chapters 4 to 7 discuss ac voltages). A bridge rectifier is used to convert the ac voltage

Figure 3.8   A Simpson Model 270 volt-ohm-milliammeter. (*Photograph courtesy of Simpson Electric Company.*)

into a dc voltage (rectifiers are discussed in Chapter 11). The ac voltage scales can be read in terms of rms volts. If the ac signal is not a pure sine wave, however, then the voltage read from the meter is not a true rms value (the term *rms* is discussed in Chapters 5 and 6). The sensitivity of the meter for ac measurements is 5000 ohms/volt (not 20,000 ohms/volt as is the case for dc). The response of the meter for ac measurements is satisfactory for frequencies up to 50,000 Hz.

The multimeter can also be used for the measurement of a resistance. Let us now consider how the multimeter measures a resistance. We show the basic circuit in Fig. 3.9. From Kirchhoff's laws, one calculates

$$I_g = \frac{\varepsilon}{R_2 + (R_1 + R_2)R/R_1 + (R_1 + R_2)R_x/R_1} \tag{3.15}$$

where R_x is the unknown resistance and ε is an emf source, usually a dry cell. The emf source is an internal part of the multimeter circuit. For measurements of R_x in the range 0–100 ohms one puts $R_1 = 12$ ohms, $R = 0$, and $\varepsilon = 1.5$ volts. The external leads are shorted together making $R_x = 0$ and R_2

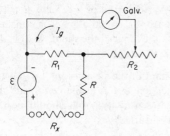

Figure 3.9 The circuit used in a volt-ohm-milliammeter to measure an unknown resistance R_x.

is adjusted until the meter reads full scale; that is, $I_g = 50$ microamp. In this arrangement almost the full 1.5 volts of the battery appears across R_1, and $R_2 = 1.5$ volts/50×10^{-6} amp $= 3 \times 10^4$ ohms. Thus $R_2 \gg R_1$ and the equation for I_g reduces to

$$I_g = \frac{\varepsilon}{R_2(1 + R_x/R_1)} \qquad (3.16)$$

The meter resistance is included in R_2. Since $R_x = 0$ gives an I_g of 50 micro-amp and R_1 is set at 12 ohms, we see that $R_x = 12$ ohms gives a half-scale deflection and $R_x = 120$ ohms gives about 0.09 of full scale. The scale is highly nonlinear with the $R_x \gg R_1$ greatly compressed at the beginning of the scale. Such a scale is provided with the meter and can be made direct reading for a particular choice of R_1. For other ranges of R_x the same meter scale is used but new internal resistances are switched in so that all the readings are simply multiplied by some power of 10. On the highest resistance scale a 7.5-volt battery is used and the circuit is arranged so that most of the battery current flows through the galvanometer. For $R_x > 1$ megohm, however, only rough estimates are possible. With $R_x = 1$ megohm and $\varepsilon = 7.5$ volts, I_g is (at best) 7.5 microamp, only about one-sixth of full scale.

There are three resistance scales on the multimeter. The three scales correspond roughly to 12 ohms, 1.2×10^3 ohms, and 1.2×10^5 ohms for half-scale deflection.

Problems

3.1   The magnetic field, the torque constant of the suspension, and the dimensions of the moveable coil of a galvanometer are given. The coil may be wound from a few turns of heavy wire or many turns of fine wire but the total mass of wire used is fixed. How would you maximize the current sensitivity of the galvanometer? How would you maximize the voltage sensitivity? Comment on the power sensitivity, that is, the deflection of the coil divided by the power input to the coil.

3.2   Assume that the total resistance of the Ayrton shunt of Fig. 3.2 is $R = 10^4$ ohms and that the galvanometer resistance is $r_g = 300$ ohms. If a constant current source of 10^{-7} amp is applied to the input of the shunt, what is the current through the galvanometer for each setting of the shunt?

3.3   The magnetic field and the torque constant of the suspension of a galvanometer are given. A piece of wire of fixed length L is available to wind the rectangular coil. For maximum current sensitivity, what should be the dimensions of the coil and the number of turns?

3.4   If a multimeter has a 20,000-ohm/volt galvanometer movement and an internal resistance of 5000 ohms, what series resistor is required to convert the galvanometer movement into a voltmeter that has full-scale deflection for 2.5 volts? for 10 volts? for 250 volts?

3.5   For the galvanometer movement described in Prob. 3.4, what shunt resistor is required to convert the galvanometer movement into an ammeter with a full-scale deflection for 10^{-2} amp? for 1 amp?

3.6   In the Wheatstone bridge of Fig. 3.5, $R_1 = 1000$ ohms, $R_2 = 1000$ ohms, $R_3 = 1000$ ohms, $R_4 = 1002$ ohms, and $\varepsilon = 10$ volts. What is the Thevenin emf ε_{Th} seen by the galvanometer? What is the Thevenin resistance seen by the galvanometer? If the galvanometer has an internal resistance of 1000 ohms, what is the current through the galvanometer?

3.7   Use the Δ-Y transformation to solve for the current I_g through the galvanometer in a Wheatstone bridge. You may assume the internal resistance r_ε of the emf source is zero.

3.8   A Wheatstone bridge would be balanced if $R_1 = R_2 = R_3 = R_4 = 1000$ ohms. Resistors are not manufactured, however, with absolute precision. Let us suppose that $R_1 = R_2 = R_3 = 1000$ ohms within one in 10^5 parts. R_4 actually has a resistance that is 0.5% larger than its nominal value of 1000 ohms, however. What resistor in parallel with R_4 causes the bridge to balance?

3.9   (a)   As seen from the emf source, can a Wheatstone bridge be reduced to a single resistance by simple series and parallel transformations?
(b)   As seen from the emf source, is there a single equivalent resistance for the Wheatstone bridge circuit?

3.10   Consider the Wheatstone bridge of Fig. 3.5. Suppose that at balance $R_1 = R_2 = R_3 = R_4 = 1000$ ohms. Suppose further that R_4 is slightly larger than 1000 ohms, say about 1001 ohms, so that the bridge is slightly unbalanced and that ε is fixed. If the power transferred to the galvanometer is to be maximized, what should the value of r_g be?

3.11   If the emf source and the galvanometer in a Wheatstone bridge are interchanged, what is the new balance condition? Assume the emf source and galvanometer have zero internal resistance. (*Hint.* Use the Reciprocity theorem.)

3.12   In the bridge circuit shown in Fig. 3.10, no current flows through the galvanometer. Find the value of R_4 and of ε_0.

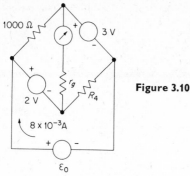

1000 Ω

3 V

+

2 V

r_g

R_4

8×10^{-3}A

ε_0

+   −

Figure 3.10

3.13   In the potentiometer circuit shown in Fig. 3.11, the standard emf ε_{sc} is used to set a standard current $I_s = 10^{-3}$ amp in the potentiometer circuit. The maximum voltage ε_x that can be measured with the potentiometer is 1.5 volts. What is the value of the potentiometer resistor R_P? What is the value of R?

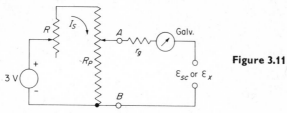

R

I_s

A

Galv.

r_g

$-R_P$

3 V

+

−

B

ε_{sc} or ε_x

Figure 3.11

3.14   Suppose the potentiometer described in Prob. 3.13 is balanced with an emf $\varepsilon_x = 0.25$ volt. For this setting of the circuit, what is the Thevenin equivalent for the circuit to the left of terminals A and B? If the emf ε_x has an internal resistance of 100 ohms, what should the galvanometer resistance be for the power transfer through the terminals AB to be a maximum when the circuit is unbalanced by a small amount?

3.15   Show that the current $I_g = \dfrac{\varepsilon}{R_2 + (R_1 + R_2)(R/R_1) + (R_1 + R_2)(R_x/R_1)}$ for the circuit of Fig. 3.9.

3.16   In using the multimeter to measure resistance it was shown that the "scale" is highly nonlinear. For the meter described where 50 microamp corresponded to full-scale deflection and where $R_1 = 12$ ohms, $R = 0$, $R_2 = 3 \times 10^4$ ohms, and $\varepsilon = 1.5$ volts so that

$$I_g = \frac{1.5}{3 \times 10^4(1 + R_x/12)}$$

plot the galvanometer current versus R_x. This graph shows how nonlinear the resistance "scale" of the meter really is.

3.17   In order to see how the multimeter can be used to measure a large resistance using the same resistance scale as used for small resistances, let us put $R_1 = 2.4 \times 10^4$ ohms, $R_2 = 0.6 \times 10^4$ ohms, $R = 1.2 \times 10^5$ ohms, and $\varepsilon = 7.5$ volts in the circuit of Fig. 3.9. Show that a good approximation for I_g is

$$I_g = \frac{\varepsilon}{(R_1 + R_2)(R/R_1)(1 + R_x/R)}$$

so that the dependence of I_g on R_x has the same functional form as it did for the lower scale but R is now 1.2×10^5 ohms instead of 12 ohms. Is the full-scale current ($R_x = 0$) still 50 microamp?

3.18   Show that Eq. 3.2 for the motion of a galvanometer is satisfied by the function $\theta(t) = \theta_0 \cos(\omega t + \phi) + NABI/k$ where θ_0 and ϕ are arbitrary constants and where $\omega = \sqrt{k/J}$, provided that the damping constant D is zero.

3.19   A meter whose full-scale deflection is 50 milliamp and whose internal resistance is 2 ohms is placed across a 12-volt storage battery whose internal resistance is 0.01 ohm. Discuss what happens to the meter.

4/Capacitors, Inductors, and Transient Currents

For the dc circuits previously considered in Chapters 1 to 3, the emf sources and currents were constant in time. We now begin the study of circuits where the sources and currents may have a complicated time dependence. Such circuits are called *alternating current* (ac) *circuits* although the currents and voltages may not have the simple sinusoidal dependence on time that characterizes the commercial power lines.

In addition to the resistor, two new circuit components, the capacitor and the inductor, are of great importance in ac circuits. In this chapter we study the properties of capacitors and inductors and describe the transient currents that flow when circuits containing capacitors and inductors are driven by suddenly changing voltage sources.

4.1 Gauss's Law

In the first chapter, Coulomb's law, the electric field, and potential difference were introduced. Because the magnitude of the electric field, $E = (1/4\pi\varepsilon_0)(Q/r^2)$, produced by a point source charge Q varies inversely with the square of the distance from the charge, the very useful concept of *lines of force* may be used to visualize the field. From any positive charge Q, lines of force Q/ε_0 in number emanate and upon any negative charge $-Q$ the same number of lines terminate. In electrostatics, lines of force originate and terminate only at charges. No matter what the space distribution of charges, the lines may be so drawn that they are everywhere parallel to the electric field and their density, that is, the number of lines passing through a unit area at right angles to their direction, is numerically equal to the magnitude of the field. The result is easily seen for the case of a point source charge. Because the lines of force are constant in number and proceed radially from the source, their density must vary as $1/r^2$ just as does the electric field.

The general case is summarized in Gauss's law:

$$\int_S \mathbf{E} \cdot d\mathbf{A} = \frac{Q}{\varepsilon_0} \qquad (4.1)$$

where S is any closed surface and Q is the net charge within the surface. The direction of $d\mathbf{A}$ is outward from the surface S. The integral $\int_S \mathbf{E} \cdot d\mathbf{A}$ over any area, closed or not, is called the *electric flux* through that area. It is equal to the number of lines of force through the area. The terms *field* and *flux density* are frequently used interchangeably.

Gauss's law is useful in many general arguments and sometimes provides an easy calculation of the electric field in situations involving symmetry. An important example of such a calculation is the determination of the electric field produced by an infinite uniformly charged plane. Figure 4.1 illustrates the situation. We apply Gauss's law to a cylindrical surface S whose sides are perpendicular to the charged plane and whose ends are parallel to the charged plane. The charged plane cuts through the surface S midway between the ends. By symmetry the electric field must be perpendicular to the charged plane and equal in magnitude but opposite in direction on the two sides of the plane. Gauss's law states that

$$\int_S \mathbf{E} \cdot d\mathbf{A} = 2EA = Q/\varepsilon_0$$

where A is the area of one end of the surface S, E is the magnitude of the electric field at the ends of the surface S, and Q is the total charge inside S. In this expression the factor of 2 comes because there are two ends to the surface S. Since the sides of S and the electric field are parallel, there is no contribution to the Gauss's law integral from the sides of S. It follows that $E = \sigma/2\varepsilon_0$ where $\sigma = Q/A$ is the surface charge density on the plane in coulombs per square meter. It is interesting to note that the field strength E does not depend on the distance from the plane.

Another situation for which the electric field is easily calculated consists of two infinite planes separated by a distance d and carrying equal but opposite uniform surface charge densities σ and $-\sigma$. For this case the electric field is $E = \sigma/\varepsilon_0$ between the two planes and zero outside the two planes. These results are obtained by superimposing the electric fields due to each of the planes.

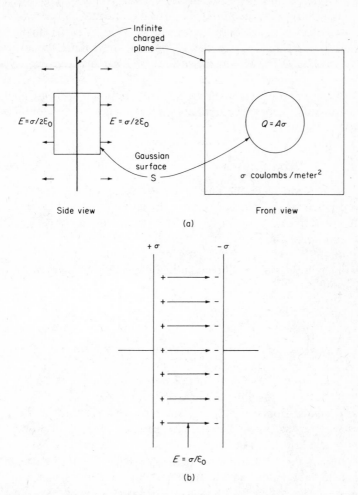

Figure 4.1 (a) The field of a uniformly charged infinite plane. The charge per unit area of the plane is σ. The electric field is $E = \sigma/2\varepsilon_0$ everywhere. The electric field is directed perpendicularly away from the plane if σ is positive. (b) The electric field of a parallel plate capacitor is $E = \sigma/\varepsilon_0$, neglecting fringing fields.

An equipotential surface is, as its name implies, a surface upon which a charge can be moved without work being done. Such a surface must be everywhere perpendicular to the field lines if an electric field is present. The more general importance of equipotential surfaces arises from the existence of good conductors of electricity through which large amounts of charge can move with great freedom. No electrostatic field can exist inside a conductor because internal charges move to the surface of the conductor until their own fields cancel the fields of the external charges. Thus when only electrostatic charges are involved, the surface of a conductor, or any surface within a conductor, is an equipotential surface. The electric field at the surface must be perpendicular to the surface and from Gauss's law we find $E = \sigma/\varepsilon_0$ relating E, the field at the surface, and σ, the charge density on the surface.

Note that it is only in electrostatics that electric fields are eliminated from conductors. If sources of emf are present, persistent fields and currents can exist in conductors.

4.2   Capacitance

Let us introduce the idea of capacitance by discussing the electric field and potential differences produced by an isolated conducting sphere of radius r that carries a total charge Q. The charge distributes itself uniformly on the surface of the sphere and the field outside the sphere is given by $E = (1/4\pi\varepsilon_0)Q/r^2$, the same field that would be produced by a point charge Q at the center of the sphere. The potential difference between the sphere and a point at infinity is

$$V = - \int_{\infty}^{r} \mathbf{E} \cdot d\mathbf{r} = (1/4\pi\varepsilon_0)Q/r.$$

The ratio $C = Q/V$, called the *capacitance* of the system, is $4\pi\varepsilon_0 r$. Capacitance is measured in coulombs per volt or farads. We see that it depends only on geometrical parameters, in this case, the radius of the sphere.

Usually a capacitor, also called a condenser, consists of two conductors insulated from one another, one carrying a charge Q and the other, a charge $-Q$. The potential difference between them is V. For example, assume our sphere of radius r is surrounded concentrically by a spherical shell of radius $r + \Delta r$ where $\Delta r \ll r$. Let the shell have a charge $-Q$. The electric field between the sphere and the shell is the same as if the shell were not present. The potential difference between the sphere and the shell is $V = E\,\Delta r = (1/4\pi\varepsilon_0)Q\,\Delta r/r^2$ and the capacitance of the system is $C = Q/V$ $= 4\pi\varepsilon_0 r^2/\Delta r$. Since the area of the sphere is $A = 4\pi r^2$, the capacitance may be written $C = \varepsilon_0 A/\Delta r$. A segment of this sphere and shell arrangement is essentially the parallel plate geometry used for most commercial capacitors. In the parallel plate geometry two metal plates, each of area A, are separated by a distance d, small compared to the linear dimensions of the plates. The charging process is the transfer of charge from one plate to the other. Thus the plates have equal charges but of opposite sign and all the field lines from one plate terminate on the other.

It is easy to show that (neglecting fringing fields) the capacitance of a parallel plate capacitor is the same as that calculated for the sphere and shell geometry. Let σ be the charge per unit area on one plate as shown in Fig. 4.1. From the preceding section the field between the plates is σ/ε_0, the potential difference between the plates is $\sigma d/\varepsilon_0$, and the total charge on one plate is σA. The capacitance of the parallel plate condenser is therefore

$$C = \frac{Q}{V} = \frac{\varepsilon_0 A}{d} \tag{4.2}$$

Our later discussions of circuit theory will make no great demands on an understanding of the electric polarizability of matter but we remind the reader that in an infinite material medium ε replaces ε_0 in the equations for the electric field and that $\varepsilon_r = \varepsilon/\varepsilon_0$ is the relative permittivity or the dielectric constant of the medium. Thus the capacitance of a parallel plate condenser becomes $C = \varepsilon A/d = \varepsilon_r \varepsilon_0 A/d$ if a material dielectric is between the plates.

The work done in charging a condenser to a final charge Q_0 is

$$W = \int_0^{Q_0} V\, dQ = \int_0^{Q_0} \frac{Q}{C}\, dQ = \frac{1}{2} \frac{Q_0{}^2}{C} \tag{4.3}$$

This can be written in the equivalent forms $W = Q_0 V_0/2 = CV_0{}^2/2$ where V_0 is the final potential difference.

The work done in charging a condenser can be considered stored in the electric field $E = V_0/d$. This very general point of view assigns an energy per unit volume of $CV_0{}^2/2Ad = \varepsilon_r \varepsilon_0 E^2/2$ to the field between the plates and, in fact, to any electric field.

A common form of commercial capacitor consists of two sheets of metal foil separated by a sheet of paper, Mylar, or mica, making essentially a parallel plate capacitor. A lead is attached to each conducting sheet. The sheets are then rolled up and encapsulated. A second sheet of the dielectric is used to insulate the one plate from the other as they are rolled up. Typical values for such capacitors range from 10^{-12} to 10^{-6} farad. The voltage rating for these capacitors is usually ~ 500 volts and is limited by dielectric breakdown, that is, the appearance of conducting channels in the insulating material. Dielectric breakdown may occur suddenly, in which case the mechanism is usually ionization by collision and electron avalanching just as occurs in the breakdown of a gas that we described in Chapter 1. Most dielectrics also have a very small ionic conductivity, which increases rapidly with temperature. Thus if the dielectric is run too hot, I^2R heating from ionic conductivity may lead to a further increase in temperature and current and eventually to breakdown from thermal runaway.

The electric field at which dielectric breakdown occurs is called the *dielectric strength* of the insulator. It varies widely among different kinds of insulating material. For instance, various types of cellulose have dielectric strengths between 250 and 500 volts/mil (1 mil = 10^{-3} in.). Usually the dielectric strength decreases as the thickness of the insulating layer increases. Thus the potential difference that an insulator can support increases more slowly than its thickness.

Metal films can be laid down on a ceramic material to form a capacitor. The advantage of these capacitors is that the dielectric constant ε_r of the ceramic can be very large (more than 100) so that a large capacitance can be obtained in a small volume.

When large capacities are required, as in a power supply filter circuit, electrolytic capacitors are often used. These capacitors are produced by the electrolytic formation of an oxide layer on aluminum or tantalum. The metal forms one electrode and an electrolytic solution forms the other with the oxide layer as the dielectric between the two electrodes. Since the oxide layer can be very thin, extremely large capacities can be obtained in small volumes (up to 10^{-3} farad). The voltage rating of these capacitors is usually low. If the voltage is raised above the limit, either dielectric breakdown occurs or the oxide layer thickens and the capacity is lowered. The electrolytic capacitors are polarized. If the voltage is reversed, the thickness of the oxide layer diminishes and eventually the capacitor shorts out.

Variable parallel plate capacitors are usually made so that either the area A or separation d of the plates can be changed, thus changing C. A common form is a capacitor made up of two sets of interleafing plates. One set of plates is stationary and the other set rotates, varying the area A of overlap between the plates. Another common variable capacitor is a trimmer capacitor where the separation d of the plates is varied. The circuit symbol for a capacitor is shown in Fig. 4.2.

(a)   (b)

Figure 4.2   The circuit symbols for (a) a capacitor (b) a variable capacitor.

4.3   Charging or Discharging a Capacitor

One of the simplest processes involving time-varying currents and potential differences is the charging or discharging of a capacitor through a resistor. By this we mean the flow of charge into one plate and away from the other plate of a capacitor in a series RC circuit. Note that the same charge per second or current that enters one plate of a capacitor must leave the other. Thus we may consider that current flows right through a capacitor without change in magnitude. The total charge in a capacitor, considering the two plates together, is always zero. We shall discuss the charging process with the help of the circuit of Fig. 4.3(a), which shows a capacitor, a resistor, an emf, and a switch S_1 in series. As in the dc case, Kirchhoff's current and voltage laws provide the basic equations from which the circuit behavior is deduced.

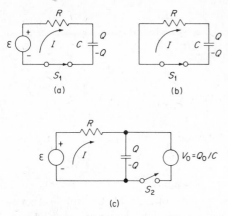

Figure 4.3 Series RC circuits and switching arrangements that can be used to charge or discharge a capacitor through a resistor. In (c) the opening of S_2 initiates a charge or discharge current around the left loop.

Kirchhoff's laws apply at any instant to any circuit whether or not the currents and voltages are functions of time. Applied to Fig. 4.3(a), Kirchhoff's current law simply tells us that at any instant the current is the same in all parts of the circuit. In using the voltage law we must remember that a new kind of potential difference appears, namely Q/C, the potential difference across the capacitor. Before we can write Kirchhoff's voltage law, we must assign the charge Q to one of the plates of the capacitor and $-Q$ to the other. This is equivalent to assuming which plate of the condenser is positive just as assuming a current direction determines which end of a resistor is to be taken as positive. The assumptions on I and Q cannot be made independently of one another, however. The equation $I=dQ/dt$ must hold where dQ/dt is the rate of change of charge on that plate of the capacitor that the current I enters. Thus the assumed current I must enter that plate of the capacitor whose charge is assumed to be Q. Consistent assumptions have been made in Fig. 4.3(a).

Upon solving the equations, I may turn out to be a negative number. This means, as in the dc case, that the real current flows in a direction opposite to the assumed current. If Q turns out to be negative, then the plate of the capacitor given the assumed charge Q in fact carries a negative charge. It is entirely possible that the solution may give different signs to I and Q. In other words the plate that the real current I enters may carry either a positive or a negative charge but dQ/dt for this plate must be positive. For the plate that the real current I leaves, dQ/dt must be negative.

Now let us return to the circuit of Fig. 4.3(a). If the switch S_1 is open, Kirchhoff's voltage law tells us nothing very interesting. The current is zero and the potential difference across the open switch must be $V = \varepsilon - Q/C$ where Q is the charge on the capacitor. Note that a charge Q on the capacitor can persist indefinitely if the switch is open.

If the switch S_1 is closed and we traverse the circuit in the direction of the indicated current, we find

$$\varepsilon = RI + \frac{Q}{C}$$

or   (4.4)

$$\varepsilon = R\frac{dQ}{dt} + \frac{Q}{C}$$

The term on the left is a potential increase due to the emf ε and the terms on the right are the potential drops.

We see that to determine Q and $I = dQ/dt$ as functions of time we must solve a first-order linear differential equation for $Q(t)$. It is characteristic of ac circuits that Kirchhoff's laws give us differential equations rather than the algebraic equations that arise from dc circuits. Fortunately the differential equations we shall meet are easily soluble. Let us briefly discuss the general problem before returning to Eq. 4.4.

By the order of a differential equation is meant the order of the highest derivative of the dependent variable that occurs. Equation 4.4 is a first-order equation. In the next chapter we shall treat second-order equations, that is, equations containing the second derivative of the dependent variable. All the differential equations that we shall need can be put in the form

$$A_2 \frac{d^2f}{dt^2} + A_1 \frac{df}{dt} + A_0 f = A(t) \tag{4.5}$$

where A_2, A_1, and A_0 are constants but $A(t)$ may be a function of the time. The dependent variable $f(t)$ will usually be a charge or a current. The equation is linear because $f(t)$ and its derivatives occur to no more than the first power; that is, terms of the form $(df/dt)^2$ or fd^2f/dt^2, and so on, are absent. If $A(t) = 0$, the equation is said to be *homogeneous*. If $A(t) \neq 0$, the equation is *inhomogeneous*.

The following results, which are easily checked by direct substitution, greatly simplify the solution of equations similar to Eq. 4.5. If $f_1(t)$ and $f_2(t)$ are both solutions of the homogeneous equation, then $F_1 f_1(t) + F_2 f_2(t)$ is also a solution of the homogeneous equation where F_1 and F_2 are constants. This is a result of the linearity of Eq. 4.5. In addition, if $f_1(t)$ is a solution of the homogeneous equation and $f_3(t)$ is a solution of the inhomogeneous equation, then $F_1 f_1(t) + f_3(t)$ is a solution of the inhomogeneous equation. The value of these results is that it is frequently easy to find the general solution of the homogeneous equation. By the general solution we mean a solution containing enough adjustable parameters so that arbitrary initial conditions on the solution can be satisfied. Once the general solution of the homogeneous equation is found, we need find only one particular solution of the inhomogeneous equation. The sum of the general solution of the homogeneous equation and any particular solution of the inhomogeneous equation is a solution of the inhomogeneous equation. It is in fact the general solution of the inhomogeneous equation since it contains the number of adjustable parameters needed to satisfy the initial conditions.

Let us return to Eq. 4.4 and see how the method works. The homogeneous equation always succumbs to an exponential trial solution. Let us try $Q = Ae^{\alpha t}$. Substituting in the homogeneous equation gives

$$Ae^{\alpha t}\left(R\alpha + \frac{1}{C}\right) = 0$$

which requires   (4.6)

$$\alpha = -\frac{1}{RC}$$

Thus we have the solution $Q = Ae^{-t/RC}$. Note that it is the undetermined parameter A, not α, which will permit us to satisfy the appropriate initial condition. The constant $\alpha = -1/RC$ was completely determined by the differential equation itself.

Now a particular solution of the inhomogeneous equation is $Q = \varepsilon C$, and therefore the general solution of the inhomogeneous equation is

$$Q = Ae^{-t/RC} + \varepsilon C \qquad (4.7)$$

To discuss the initial conditions, we must recall the physical situation that our equations describe. We have solved a differential equation that describes the behavior of the circuit [Fig. 4.3(a)] with the switch S_1 closed. We may assume that the switch was closed at $t = 0$ and that at that time $Q(0) = Q_0$ was the charge on the condenser. In principle Q_0 can be set to any value we wish. It is this arbitrary initial condition that we must satisfy by the proper choice of A. If we substitute $t = 0$ in Eq. 4.7, we obtain $Q(0) = Q_0 = A + \varepsilon C$. Solving for A and substituting in Eq. 4.7, we obtain

$$Q = (Q_0 - \varepsilon C)e^{-t/RC} + \varepsilon C$$

or   (4.8)

$$Q = Q_0 e^{-t/RC} + \varepsilon C(1 - e^{-t/RC})$$

We note that we cannot arbitrarily assign an initial value to dQ/dt. Once $Q(0) = Q_0$ is fixed, the differential equation itself gives dQ/dt at $t = 0$. Only one initial condition can be arbitrarily assigned to the general solution of a first-order differential equation. Therefore the general solution will contain only one adjustable parameter.

Because we wished to illustrate a general method of solution, which we shall use many times, our treatment somewhat obscures the simple physical processes of charging and discharging a condenser. These special cases are easily extracted from the general solution, however. The charging to a potential difference ε of an originally uncharged condenser corresponds to $Q_0 = 0$. From the general solution Eq. 4.8 we obtain for the charging of an originally uncharged capacitor in a series RC circuit

$$Q = \varepsilon C(1 - e^{-t/RC})$$

and   (4.9)

$$I = \frac{dQ}{dt} = \frac{\varepsilon}{R} e^{-t/RC}$$

The dependence of both Q and I on t is shown in Fig. 4.4. The discharge through a resistor only of an originally charged condenser corresponds to $\varepsilon = 0$. Thus the equation describing the charge as a function of the time is $R \, dQ/dt + Q/C = 0$. This situation is illustrated in Fig. 4.3(b). From the general solution, Eq. 4.8, we obtain for the discharging of a charged capacitor in a series RC circuit

$$Q = Q_0 e^{-t/RC}$$

and   (4.10)

$$I = \frac{dQ}{dt} = -\frac{Q_0}{RC} e^{-t/RC}$$

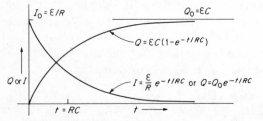

Figure 4.4 The exponential time dependence of the charge and the current in an RC circuit. After three time constants $(t = 3RC)$, the charge or discharge is 95% complete.

The dependence of Q on t is shown in Fig. 4.4. For both charging and discharging the capacitor, the characteristic time constant during which the process reaches e^{-1} of completion is $\tau = RC$. If R is measured in ohms and C in farads, τ is given in seconds.

The initial charge Q_0 can be put on the capacitor and the current started in any of several ways. In Fig. 4.3(a) one would open the switch S_1, put a short across the capacitor to assure its complete discharge, remove the short, and then close S_1. The current immediately jumps to its initial value $I_0 = \varepsilon/R$ and the charging of the capacitor begins. In Fig. 4.3(b) the switch S_1 is opened, a voltage source V_0 is placed across the capacitor that charges it to $Q_0 = V_0 C$. The voltage source is removed, S_1 is closed, and the discharge begins.

Figure 4.3(c) shows another method of establishing the initial conditions. A switch S_2 and a constant voltage source V_0 are shunted across the capacitor. When S_2 is closed, the capacitor charges up to $Q_0 = V_0 C$ and a steady dc current $I_0 = (\varepsilon - V_0)/R$ flows through source ε and the shunt. Kirchhoff's voltage law applied to the outside loop before S_2 is opened gives $\varepsilon = I_0 R + V_0$. Now consider the voltages in the loop containing ε, R, and the capacitor C just after the switch S_2 is opened. Two of the voltages ε and $Q_0/C = V_0$ (because the charge on the capacitor cannot suddenly change) are the same as they were in the outside loop before S_2 opened. Therefore the voltage $I_0 R$ must also be present after S_2 opens if Kirchhoff's voltage law is to hold. Thus the current I_0 simply switches from the shunt to the capacitor when S_2 is opened. Simultaneously the rate of change of the charge on the capacitor jumps from zero to $dQ/dt = I_0$ and Eq. 4.4 controls the subsequent time variation of the charge.

There are two additional points that deserve comment. The first is connected with the closing of the switch S_1 of Fig. 4.3, which in a sense "turns on" the differential equation at $t = 0$. We ask if it is reasonable to suppose that the charge Q_0 on the condenser before $t = 0$ remains unchanged during the closing of the switch S_1. The closing of the switch S_1 is essentially the reduction of the circuit resistance from infinity to R in a very short time Δt. During this time $dQ/dt < Q_0/RC$ because the average circuit resistance is greater than R. We have, therefore, $\Delta Q_0 = (dQ/dt) \Delta t < (Q_0/RC) \Delta t$ or $\Delta Q_0/Q_0 < \Delta t/RC$. The result, which we might have guessed, is that we need not worry about the switch if it is closed in a time very much less than RC.

Secondly, we must qualify our earlier remark that Kirchhoff's laws apply at any instant to any circuit whether or not the currents and voltages are functions of time. In fact, voltage and current signals cannot propagate through a circuit at greater than the speed of light. We may take an instantaneous view of an extended circuit only if the time constants given by Kirchhoff's laws (RC in the present case) are much greater than l/c where l might be the length of a typical current loop of the circuit and c is the speed of light. If l is 1 m, l/c is 3.3×10^{-9} sec and we may already be in some trouble if RC is as small as 10^{-7} sec. We shall not treat such problems but the reader should be aware that they exist. Circuits where $RC \gg l/c$ are called *lumped circuits*, and circuits where $RC \gtrsim l/c$ are called *distributed circuits*.

4.4   Currents and Magnetic Interactions

As the simplest current element creating and interacting with magnetic fields, one may use either the moving charge $Q\mathbf{v}$ (Q is the charge and $\mathbf{v}$ is the velocity of the charge) or a very short linear current element $I\,d\mathbf{l}$ (I is the magnitude of the current and $d\mathbf{l}$ is a line element in the direction of the current). Magnetic interactions are necessarily more complex than electrostatic interactions since the sources are vectors. It is usually not helpful to discuss the magnetic analog of Coulomb's law, that is, the direct interaction between current elements. We introduce the field in vacuum at once with the law of Biot and Savart:

$$\mathbf{B} = \frac{\mu_0}{4\pi} \cdot \frac{Q\mathbf{v} \times \mathbf{r}}{r^3} = \frac{\mu_0}{4\pi} \cdot \frac{I\,d\mathbf{l} \times \mathbf{r}}{r^3} \qquad (4.11)$$

$\mathbf{B}$ is the magnetic field at a distance $\mathbf{r}$ from the moving charge [Fig. 4.5(a)]. $\mathbf{B}$ is perpendicular to both $\mathbf{v}$ (or $d\mathbf{l}$) and $\mathbf{r}$, and the direction of $\mathbf{B}$ is determined by the right-hand screw rule for the direction of the vector product. The magnetic field may also be described by lines of force (also called lines of induction) and the total number of lines through any surface, the flux, is measured in webers. The constant μ_0 is called the *permeability of free space* and has the value $4\pi \times 10^{-7}$ weber/amp meter. $\mathbf{B}$, the magnetic field or flux density, is measured in webers per square meter. A commonly used unit, the gauss, is 10^{-4} weber/meter^2.

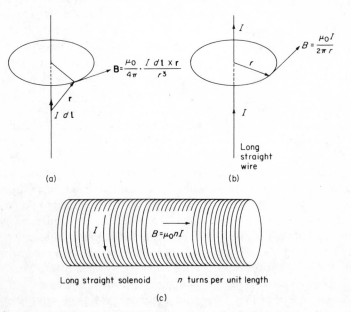

Figure 4.5   The magnetic field B produced by (a) a current element $I\,d\mathbf{l}$, (b) a long straight wire, (c) a long solenoid.

The magnetic fields produced by the extended currents of any real geometry can be obtained by integration of Eq. 4.11. The result of such a calculation for a very long straight wire carrying a current I is $B = \mu_0 I/2\pi r$ where r is the perpendicular distance from the wire to the field point. **B** is perpendicular to the plane containing the wire and the field point. The field inside a very long solenoid can also be calculated. It is parallel to the axis of the solenoid and given by $B = \mu_0 nI$ where n is the number of turns of wire per unit length of solenoid.

There is a fundamental distinction between magnetic and electro-static fields in that isolated magnetic poles, the analog of electric charge, have not been observed. All magnetic fields come from moving charges. There are no magnetic sources from which field lines originate.

From the law of Biot and Savart one can derive an integral form known as Ampere's law, which states that

$$\int_C \mathbf{B} \cdot d\mathbf{l} = \mu_0 I \tag{4.12}$$

In this equation $\mathbf{B} \cdot d\mathbf{l}$ is equal to the component of B tangential to the curve C, times the element of length $d\mathbf{l}$. The integral of the tangential component of **B** is carried out around the closed curve C. The current I is the total current threading through any surface bounded by the curve C. Ampere's law is often useful in calculating the magnetic field in a situation involving sym-metry. It is a magnetic analogy to Gauss's law. Thus the calculation of the **B** field due to a long straight wire such as is shown in Fig. 4.5(b) can be accom-plished as follows. From Ampere's law we know that

$$\int_C \mathbf{B} \cdot d\mathbf{l} = \mu_0 I$$

From symmetry the magnitude of **B** must be the same at any point on a circle of radius r, centered at the wire and in a plane perpendicular to the wire. Thus $2\pi rB = \mu_0 I$ or $B = \mu_0 I/2\pi r$, which is the result we previously quoted. In a similar way one can calculate the field of a solenoid such as is shown in Fig. 4.5(c). Let us select a rectangular path with one side of the rectangle inside the solenoid and parallel to the axis of the solenoid and the other side outside the solenoid. The two ends of the rectangle connect the two sides together. Outside the solenoid the magnetic field is weak. The magnetic field is perpendicular to the ends of the path if the solenoid is very long. Thus the only contribution to the Ampere's law integral is from the side of the rectangle inside the solenoid. Therefore we find

$$\int_C \mathbf{B} \cdot d\mathbf{l} = Bl = \mu_0 NI$$

where l is the length of a side of the path, I is the current in the solenoid, and N is the total number of turns inside the rectangular path. Consequently $B = \mu_0 nI$ where $n = N/l$ is the number of turns per unit length of the solenoid. This result was also given previously.

We have discussed at some length the magnetic fields produced by current elements and extended currents. It remains to give the expression for the force on a moving charge or a current element due to the magnetic field at its position, a magnetic field that must be produced by other current elements in the surrounding space. The fundamental equation is

$$\mathbf{F} = Q(\mathbf{v} \times \mathbf{B}) = I(d\mathbf{l} \times \mathbf{B}) \tag{4.13}$$

Integration of this equation gives the total force on extended currents. It is obvious that a long straight wire of length l at right angles to the field B experiences a total force $F = BIl$ that is at right angles to both B and l. We have used this result in our discussion of the galvanometer in Chapter 3. Equation 4.13 shows that a weber per square meter is equal to a newton per ampere-meter.

4.5   Induced Electromotive Force

In a purely electrostatic field the potential difference between two points is independent of the path between them. It follows that no net work is done on a charge that follows a closed path, or circuit, back to its starting point. Nonelectrostatic forces may do work on a charge moving around a closed circuit and the total work per unit charge is called an *electromotive force* (emf). It has the dimensions of a potential difference. The basic concept of emf was also discussed in Chapter 1.

One of the most important electromotive forces arises from the change with time of the magnetic flux passing through a circuit. Let us assume that the circuit consists of a single loop. The expression for the emf given by Faraday's law of induction is

$$\varepsilon = -\frac{d\phi}{dt} \tag{4.14}$$

where ε, the induced emf, is in volts and the flux

$$\phi = \int_S \mathbf{B} \cdot d\mathbf{A} \tag{4.15}$$

is in webers. The integration is over any surface S bounded by the circuit in question. The minus sign expresses Lenz's law, namely, that the direction of the induced current is such as to try to hold the flux in the circuit constant. This means that the flux, whose change induces the current, plus the flux produced by the induced current tends to remain constant.

The cause of the change in flux linkage may be a relative motion of the magnetic field and the circuit or time-changing currents that produce a time-varying flux linkage in stationary circuits.

In the latter case it is possible to express the induced emf's in terms of time-changing currents and a constant depending only on the geometry of the system.

The flux through a given circuit due to the current in the same circuit is proportional to the current. We have $\phi = LI$ and for the emf induced in a single loop circuit

$$\varepsilon = -\frac{d\phi}{dt} = -L\frac{dI}{dt} \qquad (4.16)$$

The constant L is called the coefficient of self-inductance and is measured in henrys or volt-seconds per ampere. The equations we have written apply only to a circuit of one closed loop.

If the circuit consists of N turns of wire following almost the same path, the self-inductance L is proportional to N^2, it being understood that ε is the emf induced in the entire length of wire. The dependence on N^2 arises because the flux itself is proportional to N, and its time rate of change induces the same emf in each of the N turns. When there are N turns, Eq. 4.16 becomes $\varepsilon = -Nd\phi/dt = -L\,dI/dt$. This gives the useful relation $L = N\phi/I$.

With two separate circuits a time-varying current in one produces a time-varying flux through the other. In this case we have a mutual inductance M also measured in henrys and defined by

$$\varepsilon_1 = -M\frac{dI_2}{dt} \qquad (4.17)$$

The same M also satisfies the reciprocal equation

$$\varepsilon_2 = -M\frac{dI_1}{dt} \qquad (4.18)$$

If each circuit consists of a number of turns of wire, then M is proportional to the product $N_1 N_2$.

In a circuit containing a self-inductance, building up the current from zero to some final value I_0 requires energy. Charge must be moved through the self-inductance against the counter emf. The amount of work is

$$W = \int \varepsilon\, dQ = \int L\frac{dI}{dt}\, dQ = \int_0^{I_o} LI\, dI = \frac{1}{2}LI_0^2 \qquad (4.19)$$

4.6   Magnetic Materials

The essential material on inductance, especially the definition of self-inductance as $L = N\phi/I$, has been covered in Secs. 4.4 and 4.5. The effect of magnetic materials on self-inductance as discussed in Secs. 4.6 and the calculation of the self-inductance of a coil as discussed in Sec. 4.7 may be skipped if time is short. Although Secs. 4.6 and 4.7 contain interesting material, it is not essential for circuit analysis.

The electrons of matter have rapid orbital motions and spins. The resulting currents can be the sources of intense microscopic magnetic fields. Usually no macroscopic effects are observed because within a macroscopic sample of material there are as many electrons revolving or spinning in one direction as in the opposite direction. When an external magnetic field is applied to a material medium, however, the cancellation of the microscopic fields is no longer complete. A macroscopic field from internal currents is observed and it is sometimes very large.

Let us see how the average macroscopic magnetic field within a solid can be measured. Consider the torus or doughnut of Fig. 4.6. It is essentially a solenoid of diameter $r_1 - r_2$ that is bent into a circle so that one end closes against the other. We assume that it is uniformly wound with N turns of wire. The number of turns per unit length is $n = N/2\pi r$ where $r = (r_1 + r_2)/2$ is an average radius of the torus. If r_1 and r_2 are considerably greater than $r_1 - r_2$, the magnetic field within the solenoid is almost constant.

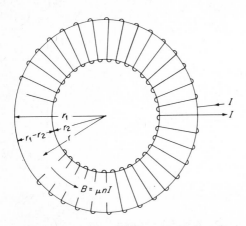

Figure 4.6 The magnetic field produced by a coil with N total turns wound on a magnetic toroid. The number of turns per unit length is $n = N/2\pi r$.

If the interior of the solenoid is empty, the magnitude of the field is $B = \mu_0 nI$, the result we have already derived for a long straight solenoid. This can be experimentally checked using an auxiliary winding of a few additional and independent turns of wire on the solenoid. The auxiliary winding is connected to a ballistic galvanometer, and the current I in the main winding is suddenly switched on or off. The deflection of a ballistic galvanometer (see

Chapter 3) is proportional to the change in the flux that passes through the auxiliary winding. The change in flux is given by $\Delta\phi = BA = \mu_0 nIA$ where A is the cross-sectional area of the solenoid.

When the core of the torus is made of a material medium, the auxiliary coil measures the change in flux and therefore B due to both the solenoid current and the internal currents. To describe the total magnetic field we write $B = \mu nI$ where $\mu = \mu_r \mu_0$. The constant μ_r is called the *relative permeability* of the material. If space is filled with a homogeneous magnetic material, Ampere's current law must be written $\int_C \mathbf{B} \cdot d\mathbf{l} = \mu_r \mu_0 I$. Usually μ_r is slightly less than one (diamagnetic materials) or slightly greater than one (paramagnetic materials) and is a constant independent of the current I. This means that the total magnetic field and therefore the field contributed by internal currents is proportional to the field from the solenoid current.

For a very important class of materials, the ferromagnetic metals and the ferrites, however, the contribution of the internal currents to B can be large. In addition, the contribution usually has a complicated dependence on the solenoid current and on the previous history of the material. The latter effect is called *hysteresis*. The behavior of this class of materials is most easily shown graphically as in Fig. 4.7 where we plot the solenoid current I versus the total magnetic field in the core of the torus. If the material is originally unmagnetized, the dashed curve is followed. For small values of I the contribution to B from internal currents might typically be from a few hundred to a few thousand times greater than the contribution from the solenoid current. At higher values of I the internal currents saturate and the curve of B versus I almost levels off. Further increases in B are independent of the properties of the material inside the solenoid and are proportional to the increases in I, more precisely, $\Delta B = \mu_0 n \Delta I$.

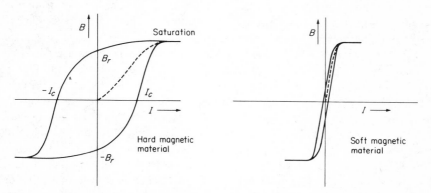

Figure 4.7 A typical hysteresis curve for a ferromagnetic material. The field B_r is called the *retention field* and I_c is called the *coercive current*. For a soft magnetic material, B_r and I_c are small. For a hard magnetic material, B_r and I_c are large. The dashed curve shows how B varies if the material is initially unmagnetized.

When the current in the solenoid is decreased, the initial dashed magnetization curve is not retraced. Instead the solid curve, called the *hysteresis curve*, is followed. At zero current a considerable B field, the remanent magnetization B_r, may remain. The field does not go to zero until a reverse current I_c, called the *coercive current*, is applied. With additional reversals of the current the same hysteresis curve is followed provided that each maximum of the current is large enough to carry the material into the saturation region. In order to return the material to a nearly unmagnetized state when the current is zero, one can apply an alternating current of ever decreasing amplitude. This process is called *demagnetization* or *degaussing*.

When hysteresis is present, some of the magnetic energy of the solenoid is transformed into heat. It can be shown that the area enclosed by the hysteresis curve is proportional to the heat loss per cycle. Thus it is desirable for inductors and transformers for ac circuits to use a soft magnetic material, that is, a material with little remanent magnetization and therefore a narrow hysteresis curve. Examples are silicon iron and Armco iron. For a permanent magnet one needs a large remanent magnetization. These are called *hard* magnetic materials. Alnico is an excellent example.

The most commonly used ferromagnetic metals are iron, nickel, and cobalt. The ferrites are complex oxides containing one or more of the ferromagnetic metals or sometimes only constituents that in the pure state are not ferromagnetic. One of the advantages of the ferrites is their low electrical conductivity. This means that eddy current losses can be almost eliminated. Eddy currents are electrical currents induced within the magnetic core itself by the time-changing flux through the core. If the electrical conductivity is high, as in a ferromagnetic metal, the eddy currents are large and can lead to serious Joule heating losses. The iron used in transformers and ac electrical machinery is usually laminated from thin sheets that have an insulating coating. Thus eddy currents cannot pass from one sheet to the next. On the other hand hysteresis losses can be minimized only by the use of soft magnetic materials. Note that hysteresis losses and eddy current losses arise by two quite different and independent mechanisms.

The magnetic properties of matter are a fascinating field and a field of great technical importance. Let us describe very briefly the mechanisms of the various magnetic responses. Diamagnetism is really Lenz's law applied to atomic and molecular current loops. The change in flux from the solenoid current induces small additional atomic current loops that are in such a direction as to oppose the B field from the solenoid. All materials have a diamagnetic response but it may be overwhelmed if a paramagnetic or ferromagnetic response is also possible. Paramagnetism requires that there be some uncompensated atomic current loops and therefore local magnetic fields, which however average out over a few atomic diameters because thermal agitation randomizes the relative orientation of different atomic currents. When an external B field is present, the atomic current loops tend to orient in the applied field much as does a compass needle. This increases the total macroscopic B field and μ_r is therefore greater than one.

It is a curious fact that the ferromagnetic materials follow an entirely different mechanism. Uncompensated atomic current loops are required as for paramagnetic material but these current loops automatically orient even in the absence of an applied field. The energy that produces the orientation is electrostatic rather than magnetic and was thoroughly understood only after the application of quantum mechanical ideas. The volume in which the orientation is nearly complete is called a *domain*. Domains differ greatly in size but typically might contain 10^{15} atoms. The macroscopic B field inside a domain contributed by the atomic currents can be as high as 2 webers/ meter^2. In an unmagnetized ferromagnetic material the domains are still completely magnetized but the direction of magnetization varies from one domain to another. Two domains with their magnetization in different directions are separated by a region called a *domain wall* where the magnetization varies smoothly from the direction it has in one domain to the direction it has in the other domain. Domain walls may be typically 300 lattice sites or 1000Å across. When an external B field is applied, the domain walls move so that those domains with a component of magnetization in the same direction as B grow at the expense of others and eventually, at saturation, all the domains are lined up with B. The size of the domains and the ease with which they grow and align are sensitive functions of the microcrystalline structure of the ferromagnetic material and of its mechanical hardness. Thus a wide variety of important magnetic properties can be achieved by different metallurgical treatments of the material.

4.7 Inductors

Most inductors are coils wound as straight solenoids [Fig. 4.5(c)] or wound on a closed toroidal core (Fig. 4.6). The latter configuration is more difficult to manufacture but permits more efficient use of the magnetic flux. Near the open ends of a straight solenoid many magnetic field lines pass through the windings, and the field inside is less than $B = \mu_r \mu_0 nI$.

Let us calculate the self-inductance L of a solenoid of length l, cross-sectional area A, and total turns of wire N. We have $\phi = BA = \mu_r \mu_0 nAI = \mu_r \mu_0 NAI/l$. This flux passes through each of the N turns and therefore a change in flux induces in the entire coil an emf

$$\varepsilon = -N\frac{d\phi}{dt} = -\frac{\mu_r \mu_0 N^2 A}{l}\frac{dI}{dt} \tag{4.20}$$

Comparing with the definition of L, namely, $\varepsilon = -L\,(dI/dt)$, we see that

$$L = \frac{\mu_r \mu_0 N^2 A}{l} = \frac{N\phi}{I} \tag{4.21}$$

For example, we may calculate from Eq. 4.21 that an air core ($\mu_r = 1$) solenoid 20 cm long, with $A = 4$ cm^2 and $N = 1000$ turns, has a self-inductance $L = 2.5$ millihenrys. Equation 4.21 overestimates the L of a straight solenoid because of end effects but it is quite accurate for a toroidal solenoid. In this case l is the average circumference of the torus. A ferromagnetic core increases the self-inductance by the factor μ_r that may be large but, as we have discussed, μ_r is not a constant independent of I for ferromagnetic materials. One must consider the possible effects of this nonlinearity on circuit performance.

Variable inductors are usually made with a ferrite core that can be screwed into or out of the coil. Occasionally a variable inductor is made with a movable tap on a coil or with some turns of an inductor coil movable with respect to the others in order to change the flux linkage. The circuit symbol for an inductor is given in Fig. 4.8.

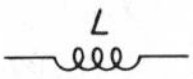

Figure 4.8 The circuit symbol for a self-inductance L.

We have already shown that the energy $LI^2/2$ must be used to build up the current in an inductor from zero to the final value I. This energy is stored in the magnetic field. If we divide the total energy $LI^2/2$ by the volume of the solenoid lA and use $B = \mu_r \mu_0 NI/l$ and Eq. 4.21 to eliminate I and L, we find that the magnetic energy per unit volume is given by $B^2/2\mu_r \mu_0$.

It is important to understand that an inductor is not just a pure inductance. The wire from which the coil is wound has a resistance. In addition to this, because of the proximity of one turn on the inductance to other turns and the fact that the different turns are at different potentials the entire inductor has a distributed capacity. Therefore any real inductor acts like a resistor and an inductor in series, with a capacitor in parallel with this series combination.

In a similar manner a real resistor will always have a distributed capacity from one end of the resistor to the other end, and if the resistor is wire wound, it may have an appreciable inductance. A real capacitor will have some resistance and some inductance in its leads.

The fundamental differences among the three circuit components are really what they do to the electrical energy of the circuit. A resistor can be considered to be a device that dissipates electrical energy as heat. For instance the energy losses in the dielectric material between the plates of a capacitor can, for the purposes of circuit analysis, be represented as I^2R

losses in a fictitious resistor. Thus if a capacitor has a lossy dielectric between its plates, the capacitor will act as though it has a resistor in parallel with the capacitance. The system acts as though the dielectric is conducting a current through the capacitor with I^2R losses. This current is in addition to the current that flows into the capacitor and is stored as charge on the capacitor plates. The total current into the capacitor is the sum of the two currents and so the capacitor and the effective resistance appear to be in parallel.

A capacitor is a device for storing energy in an electric field and an inductor is a device for storing energy in a magnetic field. With the operation of any real circuit component there are always associated heat losses and electric and magnetic fields and therefore a mixture of resistive, capacitive, and inductive behavior.

4.8 The Growth and Decay of Current in an *RL* Circuit

The circuit behavior of inductances will be discussed with the help of the *LR* circuit of Fig. 4.9. The switch *S* can be used to put the emf in or out of the circuit. We shall find that Kirchhoff's voltage law leads to a first-order differential equation that is similar to the one already developed in the discussion of the *RC* circuit. The reader should be certain that he completely understands the earlier example.

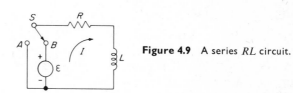

Figure 4.9 A series *RL* circuit.

Let us assume that the emf ε is in the circuit. Thus the switch is in position *B* in Fig. 4.9. We assume the direction of the current to be clockwise. Kirchhoff's voltage law gives $\varepsilon - L\,(dI/dt) = IR$. The *IR* term is a potential drop and the terms on the left are potential increases due to the applied emf ε and the back emf $-L\,dI/dt$. Lenz's law requires the negative sign of the $L\,dI/dt$ term. If the current tries to increase (dI/dt positive), the inductance opposes the increase and therefore behaves as an emf $-L\,(dI/dt)$. We write the differential equation placing $L\,dI/dt$ on the right and obtain

$$\varepsilon = L\frac{dI}{dt} + RI \qquad (4.22)$$

When the term $L\,dI/dt$ is written on the same side of the equation as the *IR* drop, it is called a *potential drop*.

The method of solution for this equation is exactly that used for the RC circuit. The general solution of the homogeneous equation is $I = Ae^{-t/(L/R)}$ where A is a constant, and a particular solution of the inhomogeneous equation is $I = \varepsilon/R$. The desired general solution of Eq. 4.22 is therefore $I = Ae^{-t(L/R)} + \varepsilon/R$.

Let us assume that at $t = 0$ the current in the circuit is $I(0) = I_0$. The equation determining A is then $I_0 = A + \varepsilon/R$. This gives for our final solution

$$I = \left(I_0 - \frac{\varepsilon}{R}\right)e^{-t/(L/R)} + \frac{\varepsilon}{R}$$

or   (4.23)

$$I = I_0 e^{-t/(L/R)} + \frac{\varepsilon}{R}[1 - e^{-t/(L/R)}]$$

We see that the characteristic time constant for the growth or decay of the current in a series LR circuit is L/R. If L is in henrys and R is in ohms, L/R is in seconds.

Now let us return to the circuit of Fig. 4.9 and consider the behavior under switching. If the switch remains at position A for a long time ($t \gg L/R$), the current is zero. Now suppose that we switch suddenly from position A to position B at a time we call $t = 0$. Immediately after the switch is thrown, the current in the circuit must still be zero. The current cannot change instantaneously because this would produce an infinite back emf $-L\,dI/dt$. Therefore the initial condition when one switches from A to B is $I(0) = I_0 = 0$. From Eq. 4.23 we see that the solution for the buildup of a current in a series LR circuit that has initially no current flowing is

$$I = \frac{\varepsilon}{R}[1 - e^{-t/(L/R)}]\qquad\qquad(4.24)$$

After the switch has been in position B for a long time, the current is $I = \varepsilon/R$. Now if we suddenly switch back to position A at a new time, which we call zero, we then have a circuit in which $\varepsilon = 0$ and the differential equation describing the current in this circuit is $L\,dI/dt + RI = 0$. This is exactly the same as the differential equation we solved previously except that $\varepsilon = 0$. The general solution is $I = I_0 e^{-t/(L/R)}$. Applying the initial condition that $I_0 = \varepsilon/R$ we have the solution for the decay of a current in a series LR circuit

$$I = I_0 e^{-t/(L/R)} = \frac{\varepsilon}{R}e^{-t/(L/R)}\qquad\qquad(4.25)$$

It is important to remember that just as the charge on a capacitor cannot change suddenly because an infinite $I = dQ/dt$ would be demanded so also the current through an inductor cannot change suddenly. This would imply an infinite dI/dt and therefore an infinite potential drop across the inductor. Some of the implications of these statements are surprising, and occasionally shocking. Consider the circuit in Fig. 4.10 where with the switch in position A we have $\varepsilon = 1$ volt, $R_1 = 1$ ohm, and $L = 1$ henry in series.

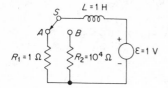

Figure 4.10 A series *RL* circuit in which the resistance can be switched from $R_1 = 1$ ohm to $R_2 = 10^4$ ohms. The abbreviation H is used for henry.

After awhile a steady current of 1 amp will flow. Now we suddenly switch from A to B. The lead to B contains a resistance R_2 of 10^4 ohms. The current of 1 amp cannot suddenly decrease; it must plow through the 10^4 ohms creating a potential difference of 10^4 volts. Simultaneously dI/dt jumps from 0 to 10^4 amp/sec and 10^4 volts (of opposite sign) also appears across the 1-henry inductor. Fortunately this unseemly state of affairs lasts for only 10^{-4} sec. One should open highly inductive circuits with great care. Back voltages of this type will often cause an arc discharge at the contacts of the switch if one tries to rapidly break or disrupt the current in an inductive circuit such as the coils of an electromagnet. The current must be reduced gradually, for instance, by reducing the driving emf or by slowly increasing the resistance in the circuit.

We have emphasized several times what might be qualitatively described as the inertia of a current through an inductor and of a charge on a capacitor. Neither can suddenly change in magnitude. It is from this fact that we have deduced the initial conditions on currents and voltages just after a switching operation, that is, the sudden change from one emf or impedance to another.

Much of modern electronic circuitry has to do with the production and processing of very short voltage and current pulses that, of necessity, involve very rapidly changing voltages and currents. The inductances and capacitances of the circuit will limit how rapidly these changes can occur. Suppose we wish to change the voltage across two terminals between which there is an effective capacitance C. If we differentiate $V = Q/C$, we obtain the important formula $dV/dt = I/C$, which says that if we want a large dV/dt between the terminals, we must provide a large current I to charge the effective capacitance.

In the next section we discuss a number of examples that illustrate the transient behavior of simple circuits.

4.9   The Transient Response of Various AC Circuits

We have treated the transient response of a series RC or RL circuit in some detail but we have not done justice to the very great importance of transient analysis. In this section we discuss the transients excited in several simple circuits by sudden changes in voltage or current sources or in impedances. In Fig. 4.11 we draw the several circuits and show on the left the time dependence of the excitation and on the right the time dependence of the resulting transient output current.

Figure 4.11(a) shows a voltage step of 5 volts driving an RL circuit consisting of 10^2 ohms in series with a 10^{-4}-henry self-inductance. This is just the problem of the buildup of a current in an RL circuit that we have previously discussed. The current is given by $I = 5 \times 10^{-2}(1 - e^{-t/10^{-6}})$ amp. The final current is 50 milliamp and the time constant is $\tau = L/R = 10^{-6}$ sec.

Now let us consider the situation of Fig. 4.11(b). A voltage pulse 5 volts in height and 1 microsec long drives the same series RL circuit as in Fig. 4.11(a). The current starts to build up according to the equation $I = 5 \times 10^{-2}(1 - e^{-t/10^{-6}})$ amp during the first microsec but then the driving voltage vanishes. After this time the current simply dies away exponentially from the value it had when the voltage pulse ended; that is, $I = 5 \times 10^{-2}(1 - e^{-1})$ amp. Thus $I = (5 \times 10^{-2})(1 - e^{-1})e^{-t/10^{-6}}$ amp for times greater than 1 microsec. It can be shown that the solution for times greater than 1 microsec is a linear superposition of two step responses, a positive step at $t = 0$ and a negative step at $t = 1$ microsec.

Figure 4.11(c) shows a voltage step of 5 volts driving a resistor $R = 10^2$ ohms in series with the parallel combination of a resistor $R' = 10^2$ ohms and a self-inductance $L = 10^{-4}$ henrys. We denote the current in the inductor by I_L and the current in the resistor R' in parallel with the inductor by $I_{R'}$. From Kirchhoff's current law the current out of the voltage source is $I = I_{R'} + I_L$. We see that our problem involves three separate currents. Two additional equations can be obtained from Kirchhoff's voltage law and we must solve a set of three simultaneous differential equations for the three transient currents. The equations can be written

$$I = I_{R'} + I_L$$

$$\varepsilon = IR + L\frac{dI_L}{dt} \qquad (4.26)$$

$$\varepsilon = IR + I_{R'}R'$$

Since we are interested in the output current I_L, we eliminate I and $I_{R'}$ from the equations, yielding for I_L the differential equation

$$\frac{R'\varepsilon}{R + R'} = L\frac{dI_L}{dt} + \left(\frac{RR'}{R + R'}\right)I_L \qquad (4.27)$$

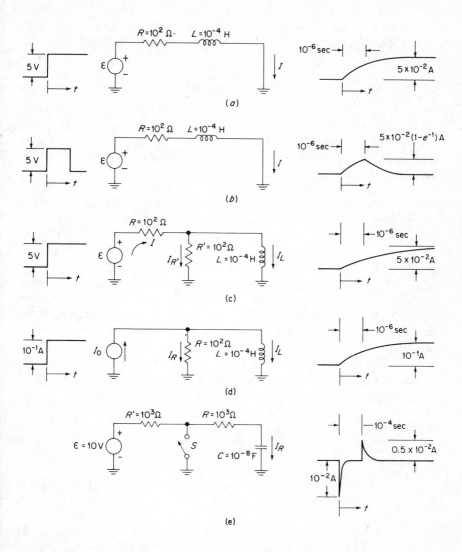

Figure 4.11 The current transients, shown on the right, that are produced in a number of simple circuits by sudden changes in the driving voltages or currents. The time dependence of the sources is shown on the left. In (e) the transient is produced by closing S for 10^{-4} sec and then opening again. The abbreviation F is used for farad.

Now in this particular problem the reader will note that the inductance thinks it is being driven by a Thevenin's equivalent circuit whose parameters are $\varepsilon_{Th} = \varepsilon R'/(R + R') = 2.5$ volts and $r_{Th} = RR'/(R + R') = 50$ ohms. Thus Eq. 4.27 can be written in the same form as for a series RL circuit.

$$\varepsilon_{Th} = I_L r_{Th} + L \frac{dI_L}{dt} \tag{4.28}$$

The solution is $I_L = (\varepsilon_{Th}/r_{Th})(1 - e^{-t/\tau}) = 5 \times 10^{-2}(1 - e^{-t/2 \times 10^{-6}})$ amp. Because the inductance is a dc short, it must eventually draw the short circuit current $I_s = \varepsilon_{Th}/r_{Th} = 5 \times 10^{-2}$ amp. The current builds up to this final value with the time constant $\tau = L/r_{Th} = 2 \times 10^{-6}$ sec.

Actually Thevenin's theorem is only occasionally helpful in transient analysis. It works in this case because the source seen by the inductance is a purely resistive network. In general one must solve a set of simultaneous differential equations equal in number to the branch currents. As in the dc case, loop currents may be used to eliminate the Kirchhoff's current law equations.

It is not necessary that the transient driving the ac circuit be a change in an emf that supplies the circuit. In Fig. 4.11(d) we show a parallel RL circuit driven by a current source that supplies a step function in the current. The current source is assumed to have an infinite impedance both before and after the step in the driving current. Just after the step occurs the current in the inductor must still be zero, and all the current from the current source must go through the resistor R. If the current source provides a constant current I_0 after the step, then we have from Kirchhoff's laws

$$I_0 = I_R + I_L$$
$$L \frac{dI_L}{dt} = RI_R = R(I_0 - I_L) \tag{4.29}$$

The solution to these equations with the initial condition $I_L(0) = 0$ is $I_L = I_0[1 - e^{-t/(L/R)}] = 10^{-1}(1 - e^{-t/10^{-6}})$ amp. This is easily seen by direct substitution.

Sudden changes in the voltage or current sources driving a circuit are usually produced by electronic or mechanical switching. We illustrate this in Fig. 4.11(e), which shows an RC circuit containing a switch S. After S has been open for a long time, it is closed for 100 microsec and then opened again. The capacitor C is charged to $V_C = \varepsilon = 10$ volts before S is closed. After S is closed, the capacitor C discharges to ground through R so that the current in the resistor R is given by $I = (\varepsilon/R)e^{-t/RC} = 10^{-2}e^{-t/10^{-5}}$ amp. After 100 microsec the capacitor C is almost completely discharged since 100 microsec is much longer than the RC time constant $\tau = 10$ microsec.

When the switch S is opened again, the capacitor charges up to 10 volts through the resistors R and R' in series. The time constant for this charging process is $\tau = (R + R')C = 20$ microsec. Thus the current in R is now given by $I = \varepsilon/(R + R')e^{-t/(R+R')C} = 5 \times 10^{-3}e^{-t/2 \times 10^{-5}}$ amp. Of course the currents on charge and discharge are in opposite directions.

The reader should be very careful to understand the examples in this section since this type of analysis for the transient response of simple circuits comes up repeatedly in laboratory work.

Problems

4.1   The capacitance of a parallel plate capacitor is 100 picofarads (1 picofarad $= 10^{-12}$ farad). The area of each plate is 0.5 cm². The material between the plates has a dielectric constant $\varepsilon_r = 10$. What is the separation of the plates?

4.2   Two long aluminum strips each of width w and length l are separated by a dielectric sheet of thickness d and dielectric constant ε_r. Another similar dielectric sheet is laid over the upper aluminum strip and the entire sandwich of four strips is rolled into a tight cylinder. What is the capacitance between the two aluminum sheets? (*Caution.* The answer is not $C = wl\varepsilon_r \varepsilon_0/d$.)

4.3   Assume that the oxide layer in an electrolytic capacitor has a thickness of 10^{-4} cm, a dielectric constant $\varepsilon_r = 8$, and a plate area of 10 cm². What is the capacitance? If the dielectric strength of the oxide layer is 1000 volts/mil, what maximum voltage can be applied to the capacitor?

4.4   A capacitor is assembled from very thin metal sheets and a dielectric material of constant dielectric strength, independent of thickness. The dielectric is available in any thickness. Show that the maximum energy that can be stored in the capacitor is proportional to the volume of the capacitor and is independent of such factors as plate area, plate separation, and capacitance.

4.5   A self-inductance has 1000 turns of wire on a cylindrical form 1 cm in diameter and 3 cm in length. If a ferrite core with $\mu_r = 500$ can be moved into or out of the coil, over what range is the self-inductance variable?

4.6   The coil of Prob. 4.5 is made of 50 gauge copper wire whose diameter is 8×10^{-3} cm. What is the internal resistance of the coil?

4.7   Show that two capacitors in series can be replaced by an equivalent capacitor whose value is $1/C_{eq} = 1/C_1 + 1/C_2$. Show that for two capacitors in parallel the relation is $C_{eq} = C_1 + C_2$.

4.8   Show that two self-inductances in series can be replaced by an equivalent self-inductance whose value is $L_{eq} = L_1 + L_2$. Show that for two self-inductances in parallel the relation is $1/L_{eq} = 1/L_1 + 1/L_2$.

4.9   In the circuit shown in Fig. 4.12 the capacitor is originally uncharged. At $t = 0$ the switch is closed. Calculate the time dependence of the resulting current, of the potential drop across the resistor R, the potential drop across the capacitor, and the charge on the capacitor.

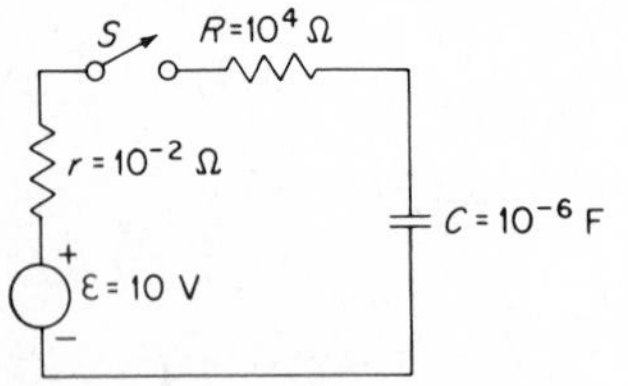

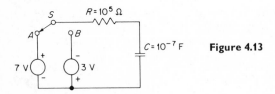

Figure 4.12

4.10   The switch is kept in position A in the circuit shown in Fig. 4.13 until the capacitor is fully charged. Roughly how long need this be? Then at $t = 0$ the switch is moved to position B. Calculate the time dependence of the resulting current.

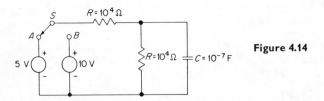

Figure 4.13

4.11   In the accompanying circuit in Fig. 4.14 the switch is kept in position A until the currents are constant. At $t = 0$ it is moved to position B. Calculate the current through the capacitor as a function of time. Review the discussion of the circuit of Fig. 4.11(c).

Figure 4.14

4.12   The switch S in the circuit in Fig. 4.15 is kept in position A for a long time and then at $t = 0$ it is thrown to position B. Calculate as a function of time the current through the inductor L and the current through the 10-volt emf.

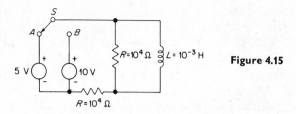

Figure 4.15

4.13 In your own words, describe why an arc is developed across a switch that is used to open the current lead to a large electromagnet. Draw in detail a circuit diagram that will enable one to safely reduce the current in an electromagnet from some large value to zero.

4.14 For the circuit shown in Fig. 4.16, find the currents I_1 and I_2 as functions of the time after the switch S is closed. The capacitors C_1 and C_2 are uncharged initially.

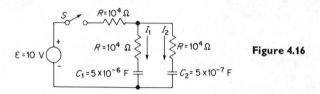

Figure 4.16

4.15 For the circuit shown in Fig. 4.17, find the currents I_1 and I_2 as functions of the time after the switch S is closed.

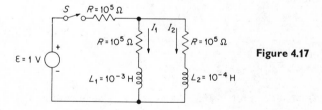

Figure 4.17

4.16 Before $t = 0$ in the circuit shown in Fig. 4.18, the switch is open and the current and the charge on the capacitor are zero. At $t = 0$ the switch is closed. Just after $t = 0$, what is the current in the circuit, and what are the potential drops across R, across L, and across C? What are the current and the potential drops after the switch has been closed a long time?

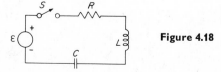

Figure 4.18

4.17   In Fig. 4.19, after having been open a long time the switch S is closed at $t = 0$. Just after $t = 0$, what are the currents through R_1, R_2, L, and C, and what are the potential drops across each? What are these quantities after S has been closed a long time?

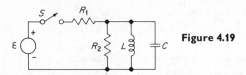

Figure 4.19

4.18   In Fig. 4.20 at regularly spaced time intervals τ, the switch S is suddenly moved from A to B or B to A. The time dependence of the driving emf seen by the RC circuit is shown at the right. Show that after the train of pulses has been on a long time the potential difference across the capacitor rises from V_i to V_f during each period τ that ε_0 is connected and falls from V_f to V_i during each succeeding period when the capacitor is discharged through R. Show that $V_i + V_f = \varepsilon_0$ and that $V_i = V_f e^{-\tau/RC}$. Calculate the dependence of V_i and V_f on τ/RC and ε_0. Give the time dependence of the voltage across the capacitor during the charge and discharge cycles. Sketch this dependence for $\tau \ll RC$, $\tau = RC$, and $\tau \gg RC$.

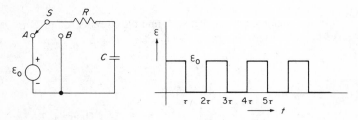

Figure 4.20

4.19   In the circuit of Prob. 4.18, assume that the switching begins at $t = 0$ at which time the charge on the capacitor is zero. For the case $\tau \ll RC$, show how the voltages across the capacitor build up to the steady state values discussed in Prob. 4.18. Show that the average voltage across the capacitor builds up to its steady state value $(\varepsilon_0/2)$ exponentially with a time constant equal to RC.

4.20   A two-terminal network has an effective capacitance between the terminals of 20 picofarads. One wishes to change the voltage between the terminals by 10 volts in 1 nanosec (1 nanosec $= 10^{-9}$ sec). What current must be fed into the network?

5/Sinusoidal AC Currents and Voltages

In this chapter we continue our study of the response of a circuit to an ac current or ac voltage. Of particular importance is the case where the currents and voltages are sinusoidal functions of time, and our main emphasis will be placed upon understanding this case.

5.1 The Description of Sinusoidal Currents and Voltages

A sinusoidal current can be written as $I = I_0 \cos(\omega t + \phi)$ where I_0 is the amplitude or peak value of the current and ω is the angular frequency in radians per second. The frequency in cycles per second is $f = \omega/2\pi$ and the unit of frequency is the hertz (Hz). Nearly all the commercial electrical power in the United States is 60 Hz. The phase of the current is simply the argument of the cosine function. In our expression it is $\omega t + \phi$, a linear function of the time. The quantity ϕ, the phase at $t = 0$, is sometimes called the phase constant. If one deals with a single sinusoidal current or voltage, ϕ is of little interest since it can always be made equal to zero by redefining zero time. More commonly one must simultaneously consider several currents and voltages that have the same frequency but different phases. The phase differences are independent of time and are important circuit parameters. As an example, let us consider the power delivered to a load by an ac source.

In Fig. 5.1 we show an ac source that produces the output voltage $V = V_0 \cos \omega t$. A current $I = I_0 \cos (\omega t + \phi)$ flows into the load. We shall see later in the chapter that phase differences between the voltage and current are usually present if the load contains capacitors and inductors. Now the instantaneous power delivered to the load is

$$P = IV = I_0 V_0 \cos (\omega t + \phi) \cos \omega t \qquad (5.1)$$

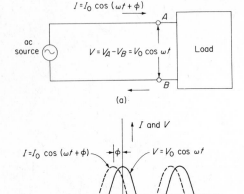

(a)

(b)

Figure 5.1 (a) A sinusoidal ac source drives a load. The current leads the voltage by the phase angle ϕ. (b) The time dependence of the voltage and current. As drawn, the phase $\phi = \pi/3 = 60°$.

To obtain the time average of the power we must average Eq. 5.1 over a complete cycle. We denote the time average of a quantity $F(t)$ over a complete cycle by

$$\langle F \rangle = \frac{1}{T} \int_0^T F(t) \, dt$$

where T is the period of the cycle.

Using the integral

$$\langle \cos (\omega t + \phi) \cos \omega t \rangle = \frac{1}{2\pi} \int_0^{2\pi} \cos (\omega t + \phi) \cos \omega t \, d(\omega t) = \frac{\cos \phi}{2} \qquad (5.2)$$

we obtain for sinusoidal currents and voltages

$$\langle P \rangle = \langle IV \rangle = \tfrac{1}{2} I_0 V_0 \cos \phi$$

$$\langle I^2 \rangle = \frac{I_0{}^2}{2} \qquad\qquad (5.3)$$

$$\langle V^2 \rangle = \frac{V_0{}^2}{2}$$

The quantity $\cos \phi$ is called the *power factor* of the load. Later we shall show that if the load in a circuit is purely resistive, $\phi = 0$ and $\cos \phi = 1$. If the load contains capacitors or inductors, then $-\pi/2 \le \phi \le \pi/2$ and $0 \le \cos \phi \le 1$. The quantities $\langle I^2 \rangle^{1/2} = I_{rms}$ and $\langle V^2 \rangle^{1/2} = V_{rms}$ define the root-mean-square current and voltage. For sinusoidal currents and voltages, Eq. 5.3 gives $I_{rms} = I_0/\sqrt{2}$. If the currents and voltages are not sinusoidal, I_{rms} and V_{rms} must be calculated from the definitions, that is, from

$$I_{rms} = \langle I^2 \rangle^{1/2} = \sqrt{(1/T) \int_0^T I^2(t)\, dt}$$

The root-mean-square quantities are very widely used particularly in the terminology of the commercial 60-Hz power. The advantage of the rms quantities is that for resistive loads the expressions for the average power dissipated have the same form as in the dc case; that is, $\langle P \rangle = I_{rms} V_{rms} = I_{rms}{}^2 R = V_{rms}{}^2/R$. More generally if $\phi \ne 0$, we have $\langle P \rangle = I_{rms} V_{rms} \cos \phi$.

Unless otherwise indicated a numerical value for a sinusoidal variable will always be the rms value, and a numerical value for the power dissipated in a circuit will be the average power.

5.2 Complex Numbers

The use of complex numbers very greatly simplifies the treatment of sinusoidal variables and the differential equations in which they occur. We shall use complex numbers almost exclusively in our development of ac circuit theory. In this section we remind the reader of their fundamental properties.

The unit imaginary number j is defined as $j = \sqrt{-1}$. The various powers of j are $j^2 = -1, j^3 = -j, j^4 = 1, j^5 = j, j^6 = -1$, and so on. A complex number z is written as $z = a + jb$ where a and b are real numbers. Figure 5.2 shows the common graphical representation of a complex number as a point in the complex plane. The imaginary part of the complex number is indicated by the ordinate and the real part of the complex number is indicated by the abcissa. If $z_1 = a_1 + jb_1$ and $z_2 = a_2 + jb_2$, then $z_1 + z_2 = (a_1 + a_2) + j(b_1 + b_2)$ and $z_1 \cdot z_2 = (a_1 a_2 - b_1 b_2) + j(a_1 b_2 + a_2 b_1)$. A complex number may be represented by the polar coordinates r, ϕ instead of the

Cartesian coordinates a, b. We then write $z = r(\cos \phi + j \sin \phi)$ where $r = \sqrt{a^2 + b^2}$ and $\phi = \arctan (b/a)$. The quantity $r = \sqrt{a^2 + b^2}$ is called the *magnitude* or the *modulus* or the *absolute value* of z and is written $|z|$.

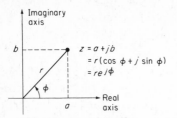

Figure 5.2 The graphical representation of a complex number z as a point in the complex plane.

There exists a most important relationship between the exponential and sinusoidal functions that an examination of the series expansion for $e^{j\phi}$ will reveal. We have

$$e^{j\phi} = 1 + j\phi + \frac{(j\phi)^2}{2!} + \frac{(j\phi)^3}{3!} + \frac{(j\phi)^4}{4!} + \cdots$$

$$= \left(1 - \frac{\phi^2}{2!} + \frac{\phi^4}{4!} - \cdots\right) + j\left(\phi - \frac{\phi^3}{3!} + \frac{\phi^5}{5!} - \cdots\right) \qquad (5.4)$$

$$= \cos \phi + j \sin \phi$$

Thus $z = a + jb = re^{j\phi}$ where $r = \sqrt{a^2 + b^2}$ and $\phi = \arctan (b/a)$. In this form if $z_1 = r_1 e^{j\phi_1}$ and $z_2 = r_2 e^{j\phi_2}$, then $z_1 \cdot z_2 = r_1 r_2 e^{j(\phi_1 + \phi_2)}$ and $z_1/z_2 = (r_1/r_2)e^{j(\phi_1 - \phi_2)}$).

The complex conjugate of a complex number is the complex number with the same real part but the opposite sign for the imaginary part of the number. The symbol for the complex conjugate of $a + jb$ is $(a + jb)^*$. We see that $(a + jb)^* = a - jb$. An important result is that $zz^* = (a^2 + b^2) = r^2 = |z|^2$; that is, a complex number times its complex conjugate is the square of the absolute value of the number.

There is a pitfall in the transformations $r = \sqrt{a^2 + b^2}$ and $\phi = \arctan (b/a)$ that comes about because r is essentially positive and because $\arctan (b/a)$ is a multivalued function of b/a. If the transformations are applied unthinkingly, complex numbers in polar form always end up in the first or fourth quadrant of the complex plane. For example, $z_1 = 1 + j$ and $z_2 = -1 - j$ are obviously different complex numbers. The transformation gives $r = \sqrt{2}$ and $\phi = \arctan 1 = 45°$ for each. What has been forgotten is that $\phi = 225°$ is also a perfectly good solution of $\phi = \arctan 1$ and in fact must be used to represent $z_2 = -1 - j$.

5.3 The Steady State Excitation of an AC Circuit and the Concept of Impedance

In this section we discuss the steady state response of an ac circuit to an emf of the form $\varepsilon = \varepsilon_0 \cos \omega t$ where ε_0 is the peak voltage and ω is the angular frequency of the emf. As a first example, consider the circuit of Fig. 5.3. A resistor and an inductor in series can be connected across an ac source by the switch S. We shall assume that the source has no internal resistance. After the switch is closed, Kirchhoff's voltage law gives the equation

$$L \frac{dI}{dt} + RI = \varepsilon_0 \cos \omega t \qquad (5.5)$$

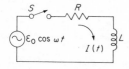

Figure 5.3 A series RL circuit driven by an ac emf $\varepsilon(t) = \varepsilon_0 \cos \omega t$.

In Chapter 4 we treated a similar circuit but there the source was a dc emf ε_0 instead of an ac source. In the dc case we obtained for the current

$$I = Ae^{-t/(L/R)} + \frac{\varepsilon_0}{R}$$

The first term is the general solution of the homogeneous equation and the second term is a particular solution of the inhomogeneous equation. Exponentially damped currents corresponding to the first term are called *transients*. They always occur when a circuit is switched on or off but in most of our discussion of ac circuits the transients will not be our primary concern. We instead are interested in the steady state response which will be a particular solution of the inhomogeneous equation. In the dc case the steady state current is $I = \varepsilon_0/R$. We note that the steady state solution is independent of initial conditions. It tells us nothing of the manner in which the circuit was turned on or of the time at which it was turned on. The steady state solution is simply the current that flows after the circuit has been on for a time long compared to the transient decay time.

Now let us try to find a steady state solution of Eq. 5.5. We shall be greatly assisted by the use of complex functions. Consider the equation

$$L \frac{dI}{dt} + RI = \varepsilon_0 e^{j\omega t} \qquad (5.6)$$

where $I = I_a + jI_b$ is now a complex function. If we make explicit the real and imaginary parts of Eq. 5.6, we find

$$\left(L \frac{dI_a}{dt} + RI_a \right) + j \left(L \frac{dI_b}{dt} + RI_b \right) = \varepsilon_0 \cos \omega t + j\varepsilon_0 \sin \omega t \qquad (5.7)$$

The real parts of the two sides of this equation must be equal, and the imaginary parts must also be equal. Thus we see that any solution I of the complex equation must have real and imaginary parts which separately satisfy the real and imaginary parts of the differential equation. The real part of Eq. 5.7 is identical with Eq. 5.5. The I_a of Eq. 5.7 is the same as the I of Eq. 5.5. Thus we can solve the real Eq. 5.5 by finding a solution to the complex Eq. 5.6 and then taking the real part of that solution.

Let us try the solution $I = I_0 e^{j\omega t}$ where both I and I_0 may·be complex. Substituting in Eq. 5.6, we find

$$(j\omega L + R)I_0 e^{j\omega t} = \varepsilon_0 e^{j\omega t}$$

and   (5.8)

$$I_0 = \frac{\varepsilon_0}{j\omega L + R}$$

The complex number $R + j\omega L$ can be rewritten in polar form,

$$R + j\omega L = [R^2 + (\omega L)^2]^{1/2} e^{j\phi}$$

where $\phi = \arctan(\omega L/R)$. Using this result we have

$$I = \frac{\varepsilon_0}{[R^2 + (\omega L)^2]^{1/2}} e^{j(\omega t - \phi)}$$

and   (5.9)

$$I_a = \frac{\varepsilon_0}{[R^2 + (\omega L)^2]^{1/2}} \cos(\omega t - \phi)$$

where I_a, the real part of I, is the desired steady state solution of Eq. 5.5. We see that there is a phase difference $\phi = \arctan(\omega L/R)$ between the voltage and the current and that the voltage leads the current, that is, reaches a maximum earlier in time than the current. If the load is purely resistive ($L = 0$), then $\phi = 0$. If the load is purely inductive ($R = 0$), then $\phi = \pi/2$.

From the results of the first section (Eq. 5.3) we have $P = (I_{a0}\varepsilon_0 \cos \phi)/2$ for the average power dissipated in the circuit where I_{a0} is the peak value of I_a. All this power is dissipated in the resistor since $\cos(\pi/2)$, the power factor of a pure inductance, is zero. That all the power is dissipated in the resistor is easily verified using $\cos \phi = \cos[\arctan(\omega L/R)] = R/[R^2 + (\omega L)^2]^{1/2}$. With the help of $\varepsilon_0 = I_{a0}[R^2 + (\omega L)^2]^{1/2}$ from Eq. 5.9, we find

$$P = \frac{I_{a0}\varepsilon_0}{2} \cos \phi = \frac{I_{a0}^2 R}{2} = I_{rms}^2 R \qquad (5.10)$$

This result is expected since an inductance stores energy in its magnetic field but dissipates no energy as heat.

The next ac circuit we discuss is the series RC circuit shown in Fig. 5.4. Kirchhoff's voltage law applied to this circuit gives an equation that in complex form can be written

$$R\frac{dQ}{dt} + \frac{Q}{C} = \varepsilon_0\, e^{j\omega t} \tag{5.11}$$

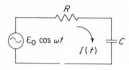

Figure 5.4 A series RC circuit driven by an ac emf $\varepsilon(t) = \varepsilon_0 \cos \omega t$.

We substitute the steady state trial solution $Q = Q_0\, e^{j\omega t}$ and obtain

$$Q\left(j\omega R + \frac{1}{C}\right) = \varepsilon_0\, e^{j\omega t} \quad \text{or} \quad Q = \frac{\varepsilon_0\, e^{j\omega t}}{j\omega R + (1/C)}$$

and upon differentiating Q we find $\tag{5.12}$

$$I = \frac{dQ}{dt} = \frac{j\omega\varepsilon_0\, e^{j\omega t}}{j\omega R + (1/C)} = \frac{\varepsilon_0\, e^{j\omega t}}{R - (j/\omega C)}$$

Putting the complex denominator in polar form gives

$$I = \frac{\varepsilon_0}{[R^2 + (1/\omega C)^2]^{1/2}}\, e^{j(\omega t - \phi)}$$

or $\tag{5.13}$

$$I_a = \frac{\varepsilon_0}{[R^2 + (1/\omega C)^2]^{1/2}}\cos(\omega t - \phi)$$

where $\phi = \arctan(-1/\omega CR)$ and I_a is the real part of I.

Thus in a capacitive circuit the current leads the voltage. In a pure capacitive circuit ($R = 0$) the current leads by $\phi = \pi/2$. As an aid in remembering the phase relations for inductive and capacitive circuits the reader should memorize this phrase: *ELI* the *ICE* man.

In the RC circuit as in the RL circuit all the power is dissipated in the resistor.

From Eq. 5.8 and 5.12 we can write linear relations of the form

$$I(R + j\omega L) = \varepsilon_0\, e^{j\omega t} = V$$

and $\tag{5.14}$

$$I\left(R - \frac{j}{\omega C}\right) = \varepsilon_0\, e^{j\omega t} = V$$

where V is a complex time-dependent potential difference. The coefficient of the current has the dimensions of a resistance and is called the *complex impedance* of the circuit. It is written $Z = R + jX$ where R is the resistance and X is the reactance of the circuit. The relationship $V = IZ$ holds for any linear passive two terminal ac network. V, the voltage across the terminals, must be sinusoidal; that is, when written in complex form, $V = V_0 e^{j\omega t}$. I is then the steady state complex current flowing between the terminals. Thus the problem of calculating the steady state current in an ac circuit is reduced to the problem of finding a complex impedance. We develop these ideas in the next section.

5.4 Impedance and Phasors

We shall see that nearly all the information we need for an understanding of the steady state currents and voltages in ac circuits can be obtained directly from the complex form of these quantities. Because the complex current, the complex voltage, and the complex impedance will be used so frequently, we shall not always distinguish between the complex current and the real current, for instance, but shall simply denote both by I. The reader should keep in mind, however, that physical quantities are real and that the complex notation is just a very convenient method of simultaneously keeping track of amplitudes and phases.

The complex impedance, $Z = R + jX$, which we shall simply call the impedance, is defined by $IZ = V$. It tells us that the ratio of the voltage amplitude to the current amplitude is $V_0/I_0 = (R^2 + X^2)^{1/2}$. The ratio of the rms values is, of course, the same as the ratio of the amplitudes. At the same time the impedance tells us that the voltage leads the current by the phase angle $\phi = \arctan(X/R)$. The impedance is independent of the amplitude and phase of both the current and voltage but does depend on the frequency.

The impedances of our three basic ac circuit elements R, L, and C can easily be determined from the relationships between instantaneous currents and voltages. For a resistor we have $V = IR$. Thus $Z = R$ and no phase difference exists between the voltage and the current. For an inductor, $V = L\, dI/dt$. If $I = I_0 e^{j\omega t}$, then $V = j\omega L I_0 e^{j\omega t} = j\omega L I$. We have $Z = j\omega L$ and the voltage leads the current by $\pi/2$. For a capacitor, $V = Q/C$ or $V = \int (I/C)\, dt$. If $I = I_0 e^{j\omega t}$, then $V = (1/j\omega C)I_0 e^{j\omega t} = (1/j\omega C)I$ and $Z = (1/j\omega C) = -j/\omega C$. Thus the current leads the voltage by $\pi/2$.

All the results above were really obtained in the last section in Eq. 5.14 for the impedance of the series RL and RC circuits. There we also see that the equivalent impedance of two circuit elements in series is the sum of their individual impedances. For the RL circuit, for instance, $Z_{eq} = Z_R + Z_L = R + j\omega L$. This is an extension of the dc result for series resistances.

It is also true that for several circuit elements in parallel the equivalent impedance is $1/Z_{eq} = 1/Z_1 + 1/Z_2 +$ and so on. In addition, as we shall discuss in a moment, the superposition theorem and Thevenin's theorem carry over directly to ac circuits. In fact we can apply almost without change all the rules and theorems of dc circuits to the steady state currents and voltages in an ac circuit driven by a sinusoidal emf. This is possible because in the ac case, Kirchhoff's laws lead to linear algebraic equations that are formally equivalent to the dc equations. The steady state ac currents and voltages and the impedances are complex instead of real but, given sinusoidal emf's, the essential requirement for linearity remains, namely that the voltage across an impedance be proportional to the current through the impedance.

Let us review briefly our results for ac circuits. In the last chapter and earlier in this chapter we used Kirchhoff's voltage law to equate the sum of the instantaneous emf's to the sum of the instantaneous voltage drops around a loop. If a circuit containing a single loop is driven by either a sinusoidal emf or a constant emf, then Kirchhoff's voltage law leads to a differential equation for the current (or the charge) whose solution is the sum of an exponentially decaying transient plus a steady state current. The steady state current depends on the circuit parameters (R, L, C, and so on) and on the voltage source but not on when or how the source is switched on. If the voltage source is sinusoidal and if transients have died away, the potential drop across any circuit element can be written as the current through it times an impedance Z. Kirchhoff's voltage law then leads to a linear algebraic equation relating the complex voltage and the complex current, for instance, Eq. 5.14. We have not yet treated the steady state solutions for circuits containing more than one loop but it is easy to see that if only steady state currents are needed, one need solve only a set of simultaneous linear algebraic equations containing the complex voltages, currents, and impedances.

The general circuit theorems of Chapter 2 all have their ac counterparts. Superposition continues to hold quite generally whether or not transients are present and whether or not the different voltage sources that may be present have the same frequency. All that is needed for superposition is that the differential equations from Kirchhoff's voltage law be linear. It must be remembered, however, that if currents of two different frequencies are simultaneously present in a branch, the potential drop across that branch cannot be written as the total current times an impedance. The potential drop must be written as the current of one frequency times the impedance for that frequency plus the current of the other frequency times the impedance for the other frequency. The ac generalization of Thevenin's theorem is straightforward but it is useful only when the voltage sources present in a circuit all have the same frequency. For this case Thevenin's theorem states that the steady state current-voltage characteristic of any two-terminal black box can be duplicated by an emf $\varepsilon_{Th} = \varepsilon_0\, e^{j\omega t}$ in series with a complex impedance Z_{Th}.

Of course both the Thevenin emf and the Thevenin impedance are functions of the frequency. It is this that limits the usefulness of Thevenin's theorem for ac circuits.

Now let us turn to the circuit of Fig. 5.5 and show how the currents can be calculated for a circuit with more than one loop using the method of complex impedances. For the complex emf we write $\varepsilon = \varepsilon_0 e^{j\omega t}$. The emf drives an impedance Z_{eq} which is the impedance of the resistor and inductor in parallel. We have

$$\frac{1}{Z_{eq}} = \frac{1}{R} + \frac{1}{j\omega L} = \frac{R + j\omega L}{j\omega LR} = \frac{\omega L - jR}{\omega LR} = \frac{[R^2 + (\omega L)^2]^{1/2}}{\omega LR} e^{j\phi}$$

where   (5.15)

$$\phi = \arctan\left(-\frac{R}{\omega L}\right)$$

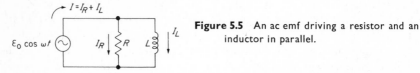

Figure 5.5 An ac emf driving a resistor and an inductor in parallel.

For the complex current I we have

$$I = \frac{\varepsilon}{Z_{eq}} = \frac{\varepsilon_0 [R^2 + (\omega L)^2]^{1/2}}{\omega LR} e^{j(\omega t + \phi)}$$ (5.16)

Remembering from Eq. 5.15 that ϕ is negative, we see that the current lags the voltage and that the phase lag is arctan $(R/\omega L)$. The real current can be obtained by taking the real part of Eq. 5.16 but in fact all the information needed is contained in the complex form.

As one can see, it is important to be able to work with complex numbers in a rapid and sure manner if one is to understand fully ac circuits. A useful aid in working with complex numbers is the graphical representation of the complex number. Consider the complex voltage $\varepsilon = \varepsilon_0 e^{j\omega t}$. The voltage ε is a complex number whose magnitude is ε_0 and whose phase is ωt. The voltage appears in the complex plane as a vector of length ε_0 that rotates around the origin with an angular velocity ω. Thus at time $t = 0$, $\varepsilon = \varepsilon_0$; at $\omega t = \pi/2$, $\varepsilon = j\varepsilon_0$; at $\omega t = \pi$, $\varepsilon = -\varepsilon_0$; at $\omega t = 3\pi/2$, $\varepsilon = -j\varepsilon_0$; and at $\omega t = 2\pi$, $\varepsilon = \varepsilon_0$ once more.

In Fig. 5.6 we use a graphical construction to solve for the currents in the circuit of Fig. 5.5. The current through the resistor $I_R = (1/R)\varepsilon_0 e^{j\omega t}$ has been drawn along the positive real axis, the angular position it would have at $t = 0$. At this same time $I_L = (1/j\omega L)\varepsilon_0 e^{j\omega t}$ must point along the negative imaginary axis since the current through an inductor lags the applied voltage by $\pi/2$. It is the factor $1/j = -j$ which specifies this lag. The total complex current $I = I_R + I_L$ is the vector sum of I_R and I_L. The magnitude of I and its phase with respect to the applied voltage $\varepsilon = \varepsilon_0 e^{j\omega t}$ are immediately obvious from the graphical construction. They are, of course, just the results given in Eq. 5.15.

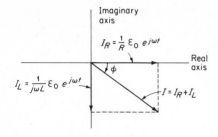

Figure 5.6   The complex current vectors for the circuit of Fig. 5.5. The angular position corresponds to $\omega t = 0$.

Now it is clear that the time factor $e^{j\omega t}$ really contributes nothing to the graphical solution. At some time other than $t = 0$ the whole vector diagram is rotated but the relative positions of the vectors are unchanged and the same magnitude and phase difference information is available. Thus if we divide out $e^{j\omega t}$, the currents and voltages can be considered as stationary vectors in the complex plane. Such vectors are called *phasors*, and their use in the solution of circuit problems is called *phasor analysis*. The phasor representation is what would be seen in a coordinate system in the complex plane that rotates about the origin with the angular frequency ω. The real functions of time that describe the actual currents and voltages are the projections of rotating phasors on the real axis of a stationary coordinate system. Particularly in electrical engineering, circuit analysis using phasors in the rotating coordinate system is referred to as analysis in the frequency domain. Projecting the phasors on the real axis of the stationary coordinate system makes the time dependence explicit and transforms the problem to the time domain. What we have tried to show in this section is that the phasor analysis in the frequency domain provides very great simplifications in the treatment of ac circuits.

Before leaving the subject we emphasize that the vector construction of Fig. 5.6 is really an expression of Kirchhoff's current law applied to either node of the circuit of Fig. 5.5. This is made clearer in Fig. 5.7(a) where we draw the phasor currents I, I_R, and I_L as a closed vector triangle. Their resultant must be zero, and thus the projection on the real axis is always zero as demanded by Kirchhoff's current law.

A similar phasor construction can be obtained from Kirchhoff's voltage law for the voltages summed around any loop. Figure 5.7(b) shows the appropriate construction for the series RL circuit of Fig. 5.3. I_0 is the current which, of course, is in phase with the voltage across the resistor. For convenience $I_0 R$ has been placed along the positive real axis.

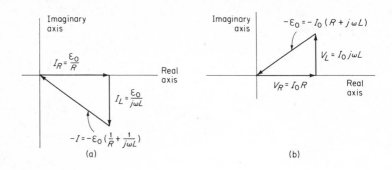

Figure 5.7   (a) The phasor diagram expressing Kirchhoff's current law at one of the nodes of the circuit of Fig. 5.5. (b) The phasor diagram expressing Kirchhoff's voltage law for the voltages around the series RL circuit of Fig. 5.3.

In general a phasor diagram can be constructed for any Kirchhoff's current or voltage law equation.

5.5   The Transient Response of a Series RLC Circuit

In the preceding chapter we discussed the transient behaviour of RL and RC circuits. The extension of the discussion to the RLC circuit has been delayed until after the introduction of complex quantities because they greatly simplify the handling of the problem. We shall see that the RLC transient response is not only important in its own right but is also closely related to the behavior of driven RLC circuits.

Consider the circuit of Fig. 5.8. We assume that the switch has been in position A for a long time and that the circuit has reached the steady state condition where the current is zero and the charge on the capacitor is $Q_0 = \varepsilon C$. At $t = 0$ the switch is moved to B and the capacitor discharges through the resistor and inductor. Our problem is to determine the charge on the capacitor and the current as functions of time. Kirchhoff's voltage law leads to the equation

$$L\frac{dI}{dt} + RI + \frac{Q}{C} = 0 \qquad (5.17)$$

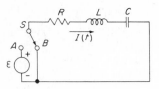

Figure 5.8 A series *RLC* circuit. The switch *S* is used to produce a transient change in the emf.

Let us divide by L and substitute $I = dQ/dt$. We then have the following linear second-order differential equation with constant coefficients:

$$\frac{d^2Q}{dt^2} + \frac{R}{L}\frac{dQ}{dt} + \frac{Q}{LC} = 0 \qquad (5.18)$$

Such an equation can always be solved using trial solutions of the form $Q = Ae^{\alpha t}$ where A and α are constants that may be complex. We substitute the trial solution in Eq. 5.18 and find $(\alpha^2 + \alpha R/L + 1/LC)Q = 0$. Thus if the trial solution is to satisfy the differential equation, α must satisfy the auxiliary equation.

$$\alpha^2 + \alpha\frac{R}{L} + \frac{1}{LC} = 0 \qquad (5-19)$$

Before proceeding we introduce some simplifying notation that also will turn out to have direct physical significance. Let $\gamma = R/2L$, $\omega_0^2 = 1/LC$, and $\omega^2 = \gamma^2 - \omega_0^2$. We may write the two solutions of Eq. 5.19 in the form

$$\begin{aligned}
\alpha_1 &= -(\gamma - \sqrt{\gamma^2 - \omega_0^2}) = -(\gamma - \omega) \\
\alpha_2 &= -(\gamma + \sqrt{\gamma^2 - \omega_0^2}) = -(\gamma + \omega)
\end{aligned} \qquad (5.20)$$

Now $Q = A_1 e^{\alpha_1 t}$ and $Q = A_2 e^{\alpha_2 t}$ are independent solutions of the differential equation. Since the differential equation is linear and homogeneous, the sum $Q = A_1 e^{\alpha_1 t} + A_2 e^{\alpha_2 t}$ is also a solution. It is in fact the general solution. The two constants A_1 and A_2 permit us to satisfy initial conditions ($t = 0$) on both Q and dQ/dt. Because neither the charge on the capacitor nor the current through the inductor and hence the current through the circuit can change instantaneously, the initial conditions are $Q = Q_0 = \varepsilon C$ and $dQ/dt = 0$ at $t = 0$. Thus A_1 and A_2 must satisfy

$$\begin{aligned}
A_1 + A_2 &= Q_0 \\
\alpha_1 A_1 + \alpha_2 A_2 &= 0
\end{aligned} \qquad (5.21)$$

Solving Eq. 5.21 and substituting from Eq. 5.20, we obtain finally for the general solution

$$Q(t) = \frac{Q_0}{2\omega} [(\gamma + \omega)e^{-(\gamma - \omega)t} - (\gamma - \omega)e^{-(\gamma + \omega)t}]$$

or $\qquad (5.22)$

$$Q(t) = \frac{Q_0 e^{-\gamma t}}{2\omega} [(\gamma + \omega)e^{\omega t} - (\gamma - \omega)e^{-\omega t}]$$

From the definitions of our parameters we see that γ and ω_0 are essentially real and positive but that $\omega = \sqrt{\gamma^2 - \omega_0^2}$ may be positive, zero, or imaginary. We do not allow it to be real and negative since that simply interchanges the solution α_1 and α_2 and introduces nothing new. We discuss the three cases separately.

$$\textbf{I} \quad \gamma^2 > \omega_0^2 \qquad \omega = +\sqrt{\gamma^2 - \omega_0^2} < \gamma$$

From the first equation of Eq. 5.22 we see that in this case the charge on the capacitor decays as the sum of two exponentials. One of them, $(\gamma + \omega)e^{-(\gamma - \omega)t}$, is a slowly decaying exponential with a large positive coefficient. It eventually dominates the decay. The other, $-(\gamma - \omega)e^{-(\gamma + \omega)t}$, is a rapidly decaying exponential with a smaller and negative coefficient. Without the second exponential one could not satisfy the boundary condition $dQ/dt = 0$ at $t = 0$. Because the coefficient $-(\gamma - \omega)$ is negative, the second exponential makes a positive contribution to dQ/dt. The first exponential gives a negative contribution to dQ/dt of equal magnitude at $t = 0$. If R is very large or L is very small ($\gamma^2 \gg \omega_0^2$), we expect the decay to approximate that of an RC circuit. It is easy to show that in this limit $(\gamma - \omega) \simeq 1/RC$. Thus the first exponential gives the expected RC decay. The second exponential is small and decays very rapidly. Its reason for existence is to keep $dQ/dt = 0$ at $t = 0$. In this first case that we have discussed the circuit is said to be *overdamped*.

$$\textbf{II} \quad \gamma^2 = \omega_0^2 \qquad \omega = 0$$

This case is called *critical damping*. It is more convenient to work with the second of Eq. 5.22, which for $\omega = 0$ can be put in the form

$$Q(t) = \frac{Q_0 e^{-\gamma t}}{2}\left[2 + \gamma\left(\frac{e^{\omega t} - e^{-\omega t}}{\omega}\right)\right] \tag{5.23}$$

The quantity in parentheses is indeterminate at $\omega = 0$. It can be evaluated by expanding the exponentials as a Taylor's series. We obtain

$$\lim_{\omega \to 0}\left(\frac{e^{\omega t} - e^{-\omega t}}{\omega}\right) = 2t \tag{5.24}$$

Finally we find for the critically damped case

$$Q(t) = Q_0 e^{-\gamma t}(1 + \gamma t) \tag{5.25}$$

We note that the exponential $e^{-\gamma t}$ gives a more rapid decay than does the dominant term for the overdamped case.

$$\textbf{III} \quad \gamma^2 < \omega_0^2 \qquad \omega \text{ is imaginary}$$

In this case the circuit is said to be *underdamped*. Again the second of Eq. 5.22 is easier to work with. We make the substitution $\omega = j\omega^\dagger$ where

$\omega^{\dagger}$ is a positive real number. Remembering that $\sin x = (e^{jx} - e^{-jx})/2j$ and $\cos x = (e^{jx} + e^{-jx})/2$, we can put the equation in the form

$$Q(t) = Q_0 e^{-\gamma t}\left(\cos \omega^{\dagger}t + \frac{\gamma}{\omega^{\dagger}}\sin \omega^{\dagger}t\right) \tag{5.26}$$

This represents an exponentially damped oscillation of angular frequency $\omega^{\dagger} = (\omega_0{}^2 - \gamma^2)^{1/2}$. The sine term is needed to satisfy the initial condition on dQ/dt.

Figure 5.9 shows graphs of the dependence of Q on t for the three cases we have considered.

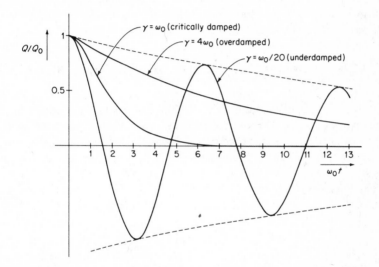

Figure 5.9 Plots of Q/Q_0 versus $\omega_0 t$ for an overdamped, a critically damped, and an underdamped transient in an *RLC* circuit. The dashed lines are $\pm e^{-\varpi_0 t/20}$, the envelopes of the underdamped oscillation. For $\gamma = \omega_0/20$ the sine term in Eq. 5.26 makes a negligible contribution to the transient.

In many applications of oscillating and resonating circuits one wants to minimize the damping, and it is useful to have some criterion of quality. This is usually stated in terms of the fraction of the energy of oscillation that is dissipated per cycle. The electrical energy in the system is given by $E = \frac{1}{2}Q_0{}^2 e^{-2\gamma t}/C$ where we have assumed that the loss in energy per cycle is very small, and we have ignored the oscillations in the rate of loss of energy. The average rate of fractional energy loss is

$$\frac{1}{E}\frac{dE}{dt} = \frac{d\ln E}{dt} = -2\gamma \tag{5.27}$$

If $\gamma^2 \ll \omega_0^2$, the fractional loss per cycle is

$$\frac{\Delta E}{E} = -2\gamma\left(\frac{2\pi}{\omega^\dagger}\right) = -\frac{2\pi R}{L\omega_0} \tag{5.28}$$

The quality factor or Q (not to be confused with the charge on the capacitor) of an oscillating circuit is defined as $Q = -2\pi E/\Delta E = L\omega_0/R$. This is just 2π times the number of oscillations taking place during the decay of the stored energy to e^{-1} of its original value. Thus a 256-vibrations per second tuning fork with a decay time of 3 sec has a $Q = 2\pi \times 256 \times 3 = 4820$. This is much higher than is usually achieved in an RLC circuit. At radio frequencies (10^5 to 10^8 Hz) an RLC circuit might have a Q between 50 and 200. We shall see in the next section that the Q factor also determines the frequency width of the resonant response of a driven oscillator.

5.6   A Series RLC Circuit Driven by a Sinusoidal EMF

Let us now discuss the series RLC circuit driven by a sinusoidal emf as shown in Fig. 5.10. We know from Sec. 5.5 that any transients have damped out if the circuit has been on for a long time. The steady state current in the series RLC circuit is found as follows. First we replace the real emf $\varepsilon_0 \cos \omega t$ by the complex emf $\varepsilon_0 e^{j\omega t}$. The complex impedance for a series RLC circuit is $Z = R + j\omega L - j/\omega C = R + j(\omega L - 1/\omega C)$. Kirchhoff's voltage law in complex notation for the series RLC network is

$$\varepsilon = \varepsilon_0 e^{j\omega t} = ZI = \left[R + j\left(\omega L - \frac{1}{\omega C}\right)\right] I_0 e^{j\omega t}$$

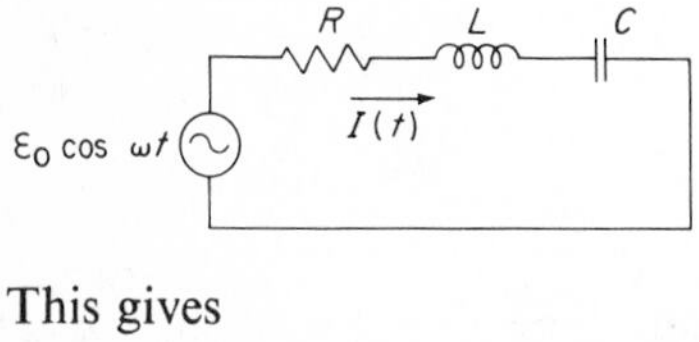

Figure 5.10 A series RLC circuit driven by a sinusoidal emf.

This gives

$$I = \frac{\varepsilon_0 e^{j\omega t}}{R + j[\omega L - (1/\omega C)]} = \frac{\varepsilon_0 e^{j(\omega t - \phi)}}{\{R^2 + [\omega L - (1/\omega C)]^2\}^{1/2}} \tag{5.29}$$

where

$$\phi = \arctan\left[\left(\omega L - \frac{1}{\omega C}\right)\Big/R\right]$$

The current in the series RLC circuit is a maximum when $\omega = \omega_0 = 1/\sqrt{LC}$. When the frequency satisfies this condition, the circuit is said to be driven at the resonant frequency.

The average power dissipated in the series *RLC* circuit is given by $\langle P \rangle = I_{rms}\varepsilon_{rms}\cos\phi = I_{rms}^2 R$. The power as a function of ω is shown in Fig. 5.11. In Fig. 5.11, ω_0 is the resonant frequency, and ω' and ω'' are the angular frequencies where the power dissipated is just half the maximum power dissipated. As a result, ω' and ω'' are called the half-power frequencies.

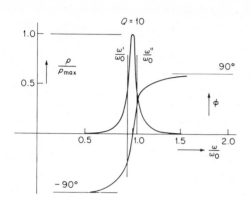

Figure 5.11 Plots of the power dissipation P/P_{max} and phase shift ϕ in a series *RLC* circuit (Fig. 5.10) versus ω/ω_0. The phase angle $\phi = \arctan[(\omega L - 1/\omega C)/R]$ is the angle by which the current lags the voltage. For these curves the quality factor is $Q = 10(\gamma = \omega_0/20)$, the same as for the under-damped oscillation shown in Fig. 5.9. The quantities ω' and ω'' are the half-power frequencies.

At the half-power frequencies, I^2 is reduced to half of its maximum value. Thus ω' and ω'' must satisfy the equation $\omega L - 1/\omega C = \pm R$ or $\omega^2 \pm \omega R/L - 1/LC = 0$. We substitute $\omega_0^2 = 1/LC$ and $Q = L\omega_0/R$ where Q is the quality factor introduced in the last section. Our equation can then be written

$$\left(\frac{\omega}{\omega_0}\right)^2 \mp \left(\frac{\omega}{\omega_0}\right)\frac{1}{Q} - 1 = 0$$

or (5.30)

$$\left(\frac{\omega}{\omega_0}\right) = \mp \frac{1}{2Q} \pm \left[\left(\frac{1}{2Q}\right)^2 + 1\right]^{1/2}$$

Since ω and ω_0 are positive, we must take the positive square root. Also we confine the discussion to high Q circuits; that is, $(1/2Q)^2 \ll 1$. We then obtain $\omega/\omega_0 = 1 \pm 1/2Q$. The two solutions are ω' and ω'' and we write

$$\frac{\omega'' - \omega_0}{\omega_0} = \frac{\omega_0 - \omega'}{\omega_0} = \frac{1}{2Q}$$

or (5.31)

$$\frac{\omega'' - \omega'}{\omega_0} = \frac{1}{Q}$$

Thus the full width at half maximum of the curve of power dissipation versus driving frequency is just the reciprocal of the Q factor. In the previous section $1/Q$ was a measure of the energy dissipated per cycle. In this section it is a measure of the width of a resonance curve. The relationship is, of course, quite general. A sharp resonance always implies a small energy dissipation per cycle. The reader should note that $\omega^\dagger$, the natural frequency of free oscillation from the previous section, is always very much closer to ω_0 than are ω' and ω'' for a high Q circuit. That is, $\omega_0 - \omega^\dagger \ll \omega_0 - \omega'$. From $\omega^\dagger = (\omega_0{}^2 - \gamma^2)$ and $\gamma = R/2L = \omega_0/2Q$ we obtain $(\omega_0 - \omega^\dagger)/\omega_0 = 1/8Q^2$.

An interesting feature of high Q series RLC circuits is that the voltages both across the capacitor and across the inductor can be very much higher than ε_0, the driving voltage, if the circuit is driven at the resonance frequency. This comes about because the capacitor and inductor voltages are $180°$ out of phase at resonance and cancel one another out. Let us calculate V_C, the magnitude of the potential drop across the capacitor. We have, using $\omega_0 L - 1/\omega_0 C = 0$,

$$V_C = \frac{I_0}{\omega_0 C} = \frac{\varepsilon_0}{\omega_0 CR} = \varepsilon_0 \frac{\omega_0 L}{R} = Q\varepsilon_0 \qquad (5.32)$$

In a similar fashion, $V_L = Q\varepsilon_0$ and $V_R = \varepsilon_0$ at resonance. Figure 5.12 shows phasor diagrams for a series RLC circuit driven at resonance and at an arbitrary frequency.

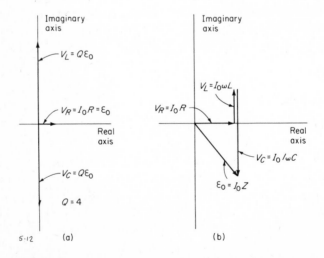

Figure 5.12 (a) A phasor diagram showing the relative magnitudes and phases of the voltages in a series RLC circuit driven at its resonant frequency. The voltage across the resistor V_R is in phase with the current and is equal to the driving voltage ε_0 in magnitude and phase. (b) The voltages in a series RLC circuit driven at a nonresonant frequency. V_L and V_C are drawn from the termination of the previous voltage to show how the phasors add vectorially to give $\varepsilon_0 = I_0 Z$, the driving voltage.

In Eq. 5.29 we saw that the phase lag of the current behind the applied voltage is given by $\phi = \arctan [(\omega L - 1/\omega C)/R]$. At resonance, $\phi = 0$; for $\omega \to \infty$, $\phi \to \pi/2$; and for $\omega \to 0$, $\phi \to -\pi/2$. Most of the phase change takes place in a frequency interval about ω_0 whose width is about the same as the width of the resonance curve. This is easily seen if we remember that ω' and ω'' are the solutions of $\omega L - 1/\omega C = \pm R$. Thus for $\omega = \omega''$ we have $\phi = \arctan (1) = 45°$ and for $\omega = \omega'$ we find $\phi = \arctan (-1) = -45°$. Thus ϕ changes by $90°$ in the frequency interval ω_0/Q. Figure 5.11 shows a graph of ϕ versus ω for a series RLC circuit.

5.7 Mutual Inductance

In the last chapter we briefly introduced the concept of mutual inductance, that is, the induction of an emf in one circuit by the time-varying current in another circuit. Here we shall develop these ideas and apply them to an understanding of the transformer, a very important component in many ac circuits.

Consider the two circuits of Fig. 5.13. Each contains a source of emf and an inductor. The two circuits are close enough together so that some of the magnetic field from each links through the other. Around L_1, for instance, let us express the total magnetic field at any point as $\mathbf{B}_1 = \mathbf{B}_{11} + \mathbf{B}_{12}$ where $\mathbf{B}_{11}$ is the contribution from I_1 and $\mathbf{B}_{12}$ is the contribution from I_2. The total flux through the first circuit is

$$\phi_1 = \phi_{11} + \phi_{12} = \int \mathbf{B}_{11} \cdot d\mathbf{A}_1 + \int \mathbf{B}_{12} \cdot d\mathbf{A}_1$$

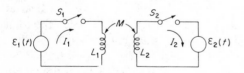

Figure 5.13 Two circuits each containing an emf and a self-inductance. The circuits are coupled by a mutual inductance M.

The fluxes ϕ_{11} and ϕ_{12} are proportional to I_1 and I_2, respectively. We write $\phi_{11} = L_1 I_1$ and $\phi_{12} = M_{12} I_2$. The emf induced in the first circuit when changes in the currents produce changes in the flux is $-d\phi_1/dt = -L_1 dI_1/dt - M_{12} dI_2/dt$. Similarly, the emf induced in the second circuit is $-d\phi_2/dt = -L_2 dI_2/dt - M_{21} dI_1/dt$. The coefficients of mutual inductance M_{12} and M_{21} are independent of the currents but depend on the geometry, that is, the shape of a circuit and the relative positions of the two circuits. It should be noted that the flux $\phi_{12} = M_{12} I_2$ may either add to or subtract from the flux $\phi_{11} = L_1 I_1$ depending on the relative directions of I_1 and I_2. In Chapter 4 it was stated without proof that $M_{12} = M_{21}$. Our first task is to prove this important theorem.

Kirchhoff's voltage law for each circuit gives the equations

$$\varepsilon_1 = L_1 \frac{dI_1}{dt} + M_{12} \frac{dI_2}{dt}$$

$$\varepsilon_2 = L_2 \frac{dI_2}{dt} + M_{21} \frac{dI_1}{dt}$$

(5.33)

where ε_1 and ε_2 are the applied emf's and where the induced emf's have been written on the other side of the equations as voltage drops. We assume that the sources $\varepsilon_1(t)$ and $\varepsilon_2(t)$ may be given any time dependence we wish and that neither circuit contains any resistance. We assume also, for this proof, that the currents and emf's are real. Originally the currents and emf's are zero. We shall calculate the work done by the emf's and stored in the magnetic field when the currents are built up to some final values, say, I_{1f} and I_{2f}. We shall build up the currents in two different ways. First we increase I_1 from zero to I_{1f} while I_2 remains at zero. Then we increase I_2 to I_{2f} while I_1 remains constant at I_{1f}. The second time we reverse the order. I_2 is first increased to I_{2f} and then I_1 is increased from zero to I_{1f}. Let us see how this can be accomplished.

In the first case we give ε_1 some finite value and therefore dI_1/dt becomes finite. When $I_1 = I_{1f}$, we reduce ε_1 to zero and I_1 then remains constant at I_{1f} since there is no resistance in the circuit. While this is going on, the second circuit must be an open circuit so that the mutual inductance term does not lead to an increase in I_2. When I_1 reaches I_{1f}, the source ε_1 will have contributed an energy $L_1 I_{1f}^2/2$ to the magnetic field. The building up of I_2 is slightly more complicated. We make ε_2 finite and I_2 starts to grow. We cannot open the first circuit because we wish to keep I_{1f} flowing. Instead we must program ε_1 so that $\varepsilon_1 = M_{12} \, dI_2/dt$ at each instant. Then the total emf around the first circuit remains zero and I_1 stays constant at I_{1f}. When I_2 reaches I_{2f}, we reduce both ε_2 and ε_1 to zero and the currents I_{1f} and I_{2f} remain constant. During the buildup of I_2, the source ε_2 contributes an energy $L_2 I_{2f}^2/2$ to the magnetic field. Because I_1 remains constant, the only flux changes seen by the second circuit are from dI_2/dt.

There remains one more energy term, however. While I_2 is increasing, the emf $M_{12} \, dI_2/dt$ appears in the first circuit. The constant current I_{1f} flows through this emf and through the emf ε_1 that has been programmed to equal $M_{12} \, dI_2/dt$. Thus energy is given to the magnetic field at the rate $I_{1f} M_{12} \, dI_2/dt$ and extracted from ε_1 at the same rate, $I_{1f}\varepsilon_1 = I_{1f} M_{12} \, dI_2/dt$. We integrate and obtain for the energy stored in the magnetic field

$$I_{1f} M_{12} \int (dI_2/dt) \, dt = I_{1f} M_{12} \int_0^{I_{2f}} dI_2 = M_{12} I_{1f} I_{2f}$$

This energy is provided by the source ε_1. To summarize, we have found that in building up to the final currents I_{1f} and I_{2f} a total energy

$$U_1 = \tfrac{1}{2}L_1 I_{1f}^2 + \tfrac{1}{2}L_2 I_{2f}^2 + M_{12} I_{1f} I_{2f} \tag{5.34}$$

is transferred from the sources to the magnetic field. Of this energy the first and last terms are contributed by ε_1 and the middle term is contributed by ε_2. Now it is easy to see that if we build up the currents in the second way, that is, first I_2 and then I_1, we find the result

$$U_2 = \tfrac{1}{2}L_1 I_{1f}^2 + \tfrac{1}{2}L_2 I_{2f}^2 + M_{21} I_{1f} I_{2f} \tag{5.35}$$

In this case ε_1 contributes only the first energy term while ε_2 contributes the last two.

The argument is now essentially completed. Because the total energies U_1 and U_2 of Eq. 5.34 and 5.35 must be the same, it follows that $M_{12} = M_{21}$. That $U_1 = U_2$ is obvious since each is the integral over all space of the magnetic energy density $B^2/2\mu_r\mu_0$. We have

$$U_1 = U_2 = \int \frac{B^2}{2\mu_r\mu_0} \, dV$$

with B depending only on the final currents and not on how they were built up.

Dropping the final subscript on the currents and writing $M_{12} = M_{21} = M$, we obtain

$$U = \tfrac{1}{2}L_1 I_1^2 + \tfrac{1}{2}L_2 I_2^2 + M I_1 I_2 \tag{5.36}$$

If there is any magnetic field any place, the magnetic energy U must be positive. It can be zero only if all currents are zero. Of the terms on the right those involving the self-inductances are necessarily positive or zero but $M I_1 I_2$ may be negative. Thus we see that for a given L_1 and L_2 the mutual inductance M cannot be indefinitely large in magnitude without contradicting the fact that $U \geq 0$. The actual condition on M is obtained as follows. Assume that $U = 0$ and that $I_2 \neq 0$, and let $\alpha = I_1/I_2$. Equation 5.36 can then be written, after dividing by I_2^2,

$$L_1\alpha^2 + 2M\alpha + L_2 = 0$$

and solving $\hspace{4cm}$ (5.37)

$$\alpha = \frac{-2M \pm \sqrt{4M^2 - 4L_1 L_2}}{2L_1}$$

Now this equation can have no real solutions for α. If it did, we would have $\alpha = I_1/I_2 = C$ where C is a real number and thus $U = 0$ when I_1 and I_2 are not zero. The condition that no real (but only complex) solutions exist for α

is $L_1 L_2 > M^2$. This is an important result as we shall see in the discussion of transformers.

It is interesting to discuss the limiting case $L_1 L_2 = M^2$. Our solutions are then $\alpha = I_1/I_2 = -M/L_1$ or $I_1/I_2 = -\sqrt{L_2}/\sqrt{L_1}$. Direct substitution of these currents in Eq. 5.36 verifies that the magnetic field energy U is in fact zero. Because $U = 0$ forbids the existence of any real currents, I_1 and I_2 must, in some sense, cancel one another. The solution must be the mathematical fiction of interpreting a zero current as two equal but opposite currents in superposition. But there remains a difficulty. Our equations give $I_1\sqrt{L_1} = -I_2\sqrt{L_2}$, not $I_1 = -I_2$. To see the meaning of this we recall from the previous chapter that, for different coils of the same size and shape but differing in the total number of turns, the self-inductance of each coil is proportional to the square of the number of turns on that coil. Since I_1 and I_2 must follow the same path, the windings carrying I_1 and I_2 must have the same size and shape so that we may substitute $N \propto \sqrt{L}$ and write $I_1 N_1 = -I_2 N_2$. For any coil the product IN is the total current through a surface that cuts all the windings. The magnetic field in the coil depends only on the product IN. For instance, consider a long solenoid either straight or bent into a torus. Suppose there is one uniform winding of N_1 turns and right on top of it (or interlaced), an independent uniform winding of N_2 turns. The total magnetic field is by superposition proportional to $I_1 N_1 + I_2 N_2$. Therefore, oppositely directed currents satisfying $I_1 N_1 = -I_2 N_2$ and, therefore, $I_1\sqrt{L_1} = -I_2\sqrt{L_2}$ will produce very little magnetic field anyplace. Thus for this system $L_1 L_2$ is only slightly greater than M^2. We shall see that the use of iron cores permits structures in which $L_1 L_2$ is only slightly greater than M^2 and also, because nearly all the magnetic field is in the iron, the requirement that the windings for L_1 and L_2 be very close together is relaxed.

A two-coil device, usually designed to maximise M, is called a *mutual inductor* or a *transformer*. At low frequencies, 60 Hz or audio frequencies, transformers are commonly made with two coils wound around a soft iron core that provides excellent flux linkage between the coils. The core will be laminated to reduce eddy current losses. For high-frequency use the coils are wound on a ferrite core of high resistivity, or sometimes an air core may be used. The circuit symbols for transformers are shown in Fig. 5.14.

(a)   (b)

Figure 5.14 The symbols for (a) a mutual inductance and (b) a mutual inductance or transformer wound on a ferromagetic coil.

In the next section we shall discuss the steady state behavior of transformers when driven by an ac source.

5.8 Transformers

Figure 5.15 illustrates schematically the construction of a typical transformer. There are two independent windings around the core: one containing N_1 turns and the other, N_2. The winding containing the source is called the *primary* (N_1 in our discussion) and the winding containing the load (N_2) is called the *secondary*. L_1, L_2, and M have their usual meanings.

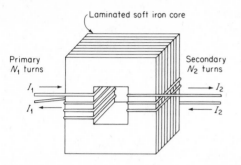

Figure 5.15 Schematic of a transformer with a laminated soft iron core.

The reader should review the discussion of the solenoidal iron core inductor given in Chapter 4 since those results are immediately applicable to the transformer. We found in Chapter 4 that if N turns are wound on a toroidal core of cross-section area A and circumference l a current I will produce a total flux in the core of $\phi = \mu_r \mu_0 NAI/l$ where μ_r is the relative permeability of the material of the core. The self-inductance of such a winding is $L = N\phi/I = \mu_r \mu_0 N^2 A/l$.

In the discussion of the previous chapter it was assumed that the N turns were uniformly distributed around the circumference of the torus. For such a winding nearly all the flux is inside the torus, no matter what the torus is made of. If the torus is made of a ferromagnetic material such as iron so that $\mu_r \gg 1$, the atomic currents that provide most of the flux are equivalent to a current sheet on the surface of the torus. This current sheet also distributes itself almost uniformly along the circumference of the torus. Actually if the cross-sectional area A varies somewhat from one place to another, the atomic currents adjust themselves so that the same total flux threads through the core independent of variations in A. Thus the flux is confined to the region where μ_r is large; that is, the flux is confined to the inside of the ferromagnetic core of the torus.

If $\mu_r \gg 1$, it is observed that the magnitude and distribution of the atomic currents, and therefore of the flux, are largely independent of how the N turns of the coil are distributed. The N turns may all be wound on one side of the iron core. If there are two independent windings, they can be localized as shown in Fig. 5.15 and the magnetic flux that flows through one coil is almost the same as the flux through the other coil. This flux is $\phi = \mu_r \mu_0 NAI/l$ just as when the windings are uniform.

We can now write down the self- and mutual inductances for the iron core transformer. We have $L_1 = \mu_r \mu_0 N_1{}^2 A/l$ and $L_2 = \mu_r \mu_0 N_2{}^2 A/l$. The mutual inductance follows from the assumption that all the flux is confined to the iron core. The primary then sends a flux $\phi_1 = \mu_r \mu_0 N_1 I_1 A/l$ through the secondary. This gives $M_{21} = N_2 \phi_1/I_1 = \mu_r \mu_0 N_1 N_2 A/l$. The secondary sends a flux $\phi_2 = \mu_r \mu_0 N_2 I_2 A/l$ through the primary and $M_{12} = N_1 \phi_2/I_2 = \mu_r \mu_0 N_1 N_2 A/l = M_{21} = M$. Thus we have rederived for this case the general result $M_{12} = M_{21} = M$. More important, however, a comparison of the formulas shows that $L_1 L_2 = M^2$. Referring to the discussion of the previous section we see that this cannot be entirely true. Some flux does leak out of the core and therefore not all the flux through one winding flows through the other. It is customary to write $M^2 = k^2 L_1 L_2$ where k is called the *coupling constant* between the coils. In general, $0 < k^2 < 1$. For an iron core transformer, k^2 is very close to unity. We make the assumption $k^2 = 1$ in what follows.

To illustrate the steady state behavior of a transformer, let us study the circuit shown in Fig. 5.16. Kirchhoff's second law enables us to write the equations for this circuit as

$$\varepsilon = \varepsilon_0 e^{j\omega t} = I_1 R + j\omega L_1 I_1 + j\omega M I_2$$

and   (5.38)

$$0 = j\omega M I_1 + j\omega L_2 I_2 + R_L I_2$$

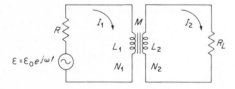

Figure 5.16 An emf source $\varepsilon_0 e^{j\omega t}$ and an internal resistance R coupled by a transformer to a load resistor R_L.

It follows that $I_2 = -j\omega M I_1/(R_L + j\omega L_2)$. Substituting this expression for I_2 in the first equation, we find

$$\varepsilon = \varepsilon_0 e^{j\omega t} = \left(R + j\omega L_1 + \frac{\omega^2 M^2}{R_L + j\omega L_2} \right) I_1$$

or   (5.39)

$$\varepsilon = \varepsilon_0 e^{j\omega t} = \left[\frac{R R_L + j\omega(R_L L_1 + R L_2) + \omega^2(M^2 - L_1 L_2)}{R_L + j\omega L_2} \right] I_1$$

Now R is simply the internal resistance of the voltage source plus the resistance of the primary windings and leads. In a good transformer driven by a good voltage source, $R \ll \omega L_1$. In addition, the load resistance usually satisfies the condition $R_L \ll \omega L_2$. With these approximations and using $M^2 = L_1 L_2$, we obtain the equation $\varepsilon = (R + L_1 R_L / L_2)I_1$. From the expressions for L_1 and L_2 we have $L_1/L_2 = N_1{}^2/N_2{}^2$ and thus finally

$$\varepsilon = \left(R + \frac{N_1{}^2}{N_2{}^2} R_L\right) I_1 \tag{5.40}$$

In other words the transformer causes the load resistance R_L as seen from the primary circuit to appear as if transformed to $(N_1{}^2/N_2{}^2)R_L$. The total impedance seen by the emf in the primary circuit is $R + (N_1/N_2)^2 R_L$. This impedance is real and hence the current is drawn in phase with the voltage. The power flows through the transformer to the load resistor where it is dissipated. In addition to enabling one to feed power to a load in an ac circuit, transformers are often used in electronic circuits for impedance matching. Thus a load resistor R_L of 8 ohms can be matched to a source impedance of 800 ohms if $N_1/N_2 = 10$.

If we had solved our original coupled equations (Eq. 5.38) for ε vs I_2 instead of ε vs I_1, we would have found that

$$-\frac{N_2}{N_1} \varepsilon = \left(R_L + \frac{N_2{}^2}{N_1{}^2} R\right) I_2 \tag{5.41}$$

In other words as viewed from the secondary circuit the emf ε is transformed to $\varepsilon_2 = -(N_2/N_1)\varepsilon$ and the resistance R of the source is transformed to $(N_2{}^2/N_1{}^2)R$.

Transformers are often used to feed a load with an emf different than the input emf. Thus a 6.3-volt filament transformer in a television set has a 110-volt primary voltage and the output voltage is N_2/N_1 times 110 volts = 6.3 volts. This implies that $N_1/N_2 = 17.5$.

Figure 5.17 shows the circuits equivalent to the circuit of Fig. 5.16 when the circuit is viewed from either the primary or the secondary. As seen from the primary circuit [Fig. 5.17(a)], the impedance R_L is transformed to $(N_1/N_2)^2 R_L$. The physical significance of this is easily understood. Since $I_2 = (N_1/N_2)I_1$ and since $\varepsilon_2 = -(N_2/N_1)\varepsilon$, we see that the power dissipated in the secondary $P = |I_2||\varepsilon_2|/2 = |I_1||\varepsilon|/2$ is just equal to the power to the primary of the transformer; that is, the transformer is lossless and does not absorb power. The impedance seen by the primary is $|\varepsilon|/|I_1| = (N_1/N_2)^2 |\varepsilon_2|/|I_2| = (N_1/N_2)^2 R_L$. The impedance transformation is required if the power dissipated in the secondary circuit is to be equal to the power to the primary of the transformer. In our discussion of the physical significance of the impedance transformation we have assumed for simplicity that $R = 0$.

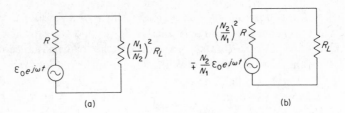

Figure 5.17 Circuits equivalent to the circuit of Fig. 5.16. (a) As seen by the source. (b) As seen by the load resistor.

Problems

5.1 Calculate the complex impedance of the circuit shown in Fig. 5.18 when driven by an ac source whose frequency is $f = 10^4/2\pi$ Hz. Is the circuit capacitive or inductive? Repeat the calculations for $f = 10^6/2\pi$ Hz.

5.2 Calculate the complex impedance of the circuit shown in Fig. 5.19 when driven by an ac source whose frequency is $f = 10^8/2\pi$ Hz. Is the circuit capacitive or inductive?

5.3 In Prob. 5.2 an ac source of 10 volts, frequency $f = 10^8/2\pi$ Hz, and zero internal impedance is placed between terminals A and B. What current flows in the 10^{-6}-henry inductor? What current flows in the 100-ohm resistor? What is the phase of the current in the 100-ohm resistor relative to the phase of the voltage applied between terminals A and B?

5.4 An emf $\varepsilon_0 \cos \omega t$ is shown driving a series RC circuit in Fig. 5.20. When the switch is open, the current in the circuit and the charge on the capacitor are zero. At $t = 0$ the switch is closed. Find the current and the charge as functions of time for all $t > 0$. Identify the transient and the steady state parts of the solutions.

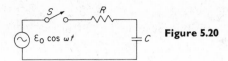

Figure 5.20

5.5   For the circuit shown in Fig. 5.21, calculate the current that flows from the emf source, the current in the resistor, and the current in the capacitor. Draw a phasor vector diagram showing the applied voltage of the emf source, the total current drawn from the emf, the current in the resistor, and the current in the capacitor.

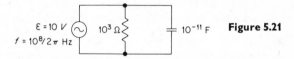

Figure 5.21

5.6   What is the resonant frequency of the circuit shown in Fig. 5.22? What is the Q factor of the circuit? What is the phase difference between the voltage across the capacitor and the voltage across the inductor? For what range of driving frequencies is the rms voltage across the inductor greater than the rms voltage across the capacitor? What is the impedance of the circuit at resonance?

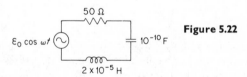

Figure 5.22

5.7   In Prob. 5.6, let $\varepsilon_0 = 10$ volts and $\omega = \omega_0$ be the resonant frequency of the circuit. What current flows through the circuit, and what is the voltage across each circuit element? Make clear whether your answers are rms or peak values.

5.8   Repeat Prob. 5.7 when the source frequency is $\omega = \omega_0/2$ and when it is $\omega = 2\omega_0$.

5.9   Calculate the impedance of the circuit shown in Fig. 5.23 if $f = 10^8/2\pi$ Hz.

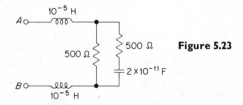

Figure 5.23

5.10   Let the circuit of Prob. 5.9 be driven by an emf of 3 volts and frequency $f = 10^8/2\pi$ Hz. Draw a phasor diagram illustrating Kirchhoff's current law at one of the nodes.

5.11   Derive the following relation for a series RLC circuit driven at the frequency ω:

$$|Z|^2 = ZZ^* = R^2 \left[1 + Q^2 \left(\frac{\omega}{\omega_0} - \frac{\omega_0}{\omega} \right)^2 \right]$$

where Z is the complex impedance, Q is the quality factor, and ω_0 is the resonant frequency.

5.12   In the approximation that Q is large and $\omega' \simeq \omega'' \simeq \omega_0$, use the result of Prob. 5.11 to derive the half-power frequencies ω' and ω''. See Eq. 5.31.

5.13   Figure 5.24 shows in one loop a source $\varepsilon_0 e^{j\omega t}$ driving an inductor L_1 and a second loop containing an ammeter (without resistance) and an inductor L_2. The two circuits are weakly coupled and an induced emf $M_{12}\, dI_2/dt$ appears in the first circuit and an induced emf $M_{21}\, dI_1/dt$ appears in the second circuit. Now interchange the voltage source and the ammeter, and calculate I_1' the current through the ammeter in the new position. From reciprocity, one has $I_2 = I_1'$. Show that this implies $M_{12} = M_{21}$.

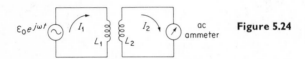

Figure 5.24

5.14   An iron core transformer (unity coupling constant) has a turns ratio of $N_1/N_2 = 3$. If an emf source of 6 volts and 900 ohms internal resistance is applied to the primary of the transformer, what current will flow in the secondary that has a resistive load of 10^2 ohms?

5.15   A well-made audio transformer is used to match an 8-ohm secondary load to a primary source impedance of 800 ohms. The 8-ohm load is to dissipate 32 watts of power. What is the rms current in the primary and in the secondary? What is the ratio of primary to secondary turns? What power is dissipated in the primary? What is the phase of the primary current with respect to the voltage driving the primary current?

5.16   In the series RL circuit shown in Fig. 5.25, calculate the impedance, the current, the phase angle between voltage and current, and the power dissipated in the resistor. Do the calculations for $f = 10^4/2\pi$ Hz and for $f = 10^6/2\pi$ Hz.

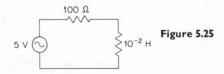

Figure 5.25

5.17   Calculate the primary and the secondary current for the circuit shown in Fig. 5.26. The transformer is an ideal transformer. The turns ratio is $N_1/N_2 = 1/3$.

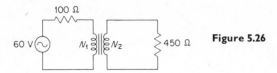

Figure 5.26

5.18   Calculate the currents I_1 and I_2 of Fig. 5.16 (Eq. 5.38) but without the assumption that the resistances are small compared to inductive reactances. Show that for very small ω we obtain $I_1 \simeq (\varepsilon_0/R)e^{j\omega t}$ and $I_2 \simeq 0$. We see that knowing only the turns ratio of a transformer one cannot predict its performance in a circuit. Comment on Prob. 5.14, 5.15, and 5.17.

5.19   A primary and a secondary winding are made on a long solenoid. The two windings occupy almost the same region of space. The primary has a resistance of 40 ohms and a self-inductance of 10^{-2} henry. The secondary has a resistance of 400 ohms and a self-inductance of 9×10^{-2} henry. What is the coupling constant k between the two coils? What is the turns ratio? Plot the open circuit output voltage of the secondary as a function of the frequency ω if the primary is driven by a source $\varepsilon_0 e^{j\omega t}$ where $\varepsilon_0 = 60$ volts.

5.20   A series RLC circuit is driven by a 10-volt ac source. $R = 100$ ohms, $L = 10^{-1}$ henry and $C = 10^{-7}$ farad. At resonance, what energy is stored in the circuit? At what average rate does the source feed energy into the circuit?

5.21   Show that in addition to the inequality $L_1 L_2 \geq M^2$ any mutual inductor will also satisfy the inequality $L_1 + L_2 \geq 2M$.

5.22   In Fig. 5.27, show that if $L_1 L_2 = M^2$ and if $R \ll \omega L_1$ and $1/\omega C \ll \omega L_2$, then the capacitor in the secondary circuit appears as if transformed to $(N_2/N_1)^2 C$ as seen from the primary circuit. Find the primary current.

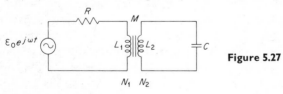

Figure 5.27

6/Alternating Current Circuits

In Chapter 5 we developed methods for treating circuits where the driving emf is sinusoidal. In this chapter we shall apply the ideas developed in the previous chapter to a variety of useful circuits.

6.1 Power Transfer in an AC Circuit

In Chapter 1 we discussed the power transfer to a load resistor from a dc emf source. It was shown that the power transfer was a maximum when the load resistor was equal to the internal resistance of the dc source. We discuss here the power transfer to a passive load impedance from an ac source that consists of an ac emf in series with a source impedance. This circuit is shown in Fig. 6.1. The frequency is assumed to be fixed.

It is easily shown that for maximum power transfer the source and load resistances (the real parts of the impedances) must be equal and the source and load reactances (the imaginary parts of the impedances) must be equal in magnitude but opposite in sign. Suppose we let R_S be the source resistance and X_S be the source reactance. Then $Z_S = R_S + jX_S$. Similarly, $Z_L = R_L + jX_L$. Now

$$I_0 = \frac{\varepsilon_0}{(R_S + R_L) + j(X_S + X_L)} \tag{6.1}$$

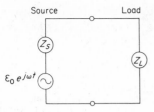

Source   Load

Figure 6.1   A source, consisting of an ac emf in series with a source impedance Z_S, driving a load impedance Z_L.

The average power dissipated in the load is $P = I_{rms}^2 R_L$ where $I_{rms} = |I_0|/\sqrt{2}$. For any given R_S and R_L this power is a maximum when $|I_0|$ is a maximum. Thus we must have $X_L = -X_S$. The power transfer can now be expressed as $P = R_L \varepsilon_{rms}^2/(R_S + R_L)^2$. But this is just the expression for the dc case where we know that maximum power transfer requires $R_L = R_S$. This completes the proof. We note that at maximum power transfer the current and voltage are in phase. With $X_L = -X_S$ the circuit is at series resonance for ω, the given driving frequency.

As an example, suppose an ac source emf has an angular frequency $\omega = 10^5$ rad/sec and a source impedance that consists of a 100-ohm resistor in series with a 10^{-3}-henry inductor. The source impedance therefore is $Z_S = (100 + j100)$ ohms. We ask, What should the load impedance be to maximize the power dissipated in the load? From our previous discussion we know that $X_L = -100$ ohms and $R_L = 100$ ohms so that $Z_L = (100 - j100)$ ohms. This load impedance can be provided by a 100-ohm resistor in series with a 10^{-7}-farad capacitor.

6.2   The Parallel Resonant Circuit

In Chapter 5 we discussed the series resonant circuit. In this section we wish to discuss the parallel resonant circuit of Fig. 6.2. In Fig. 6.2 we have assumed that the inductive arm contains all the resistance and that the capacitive arm has no resistance. This is often an excellent approximation since inductances are usually wound using a long, small diameter wire so that the resistance is not negligible, whereas the leads to the capacitor may be short and hence low resistance. Either dielectric material in the capacitor or magnetic material in the inductor will absorb some power and this loss may be represented as a resistance in parallel with that arm of the circuit. We ignore such effects in the circuit of Fig. 6.2.

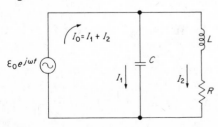

Figure 6.2   A parallel resonant circuit. The emf ε_0 and the currents I_0, I_1, and I_2 are phasors. The factor $e^{j\omega t}$ is omitted in the analysis.

For the circuit of Fig. 6.2 we calculate the current in each arm with the result $I_1 = \varepsilon_0/(-j/\omega C) = j\omega C\varepsilon_0$ and $I_2 = \varepsilon_0/(R + j\omega L)$. The total current drawn from the emf source is

$$I_0 = I_1 + I_2 = \frac{\varepsilon_0}{Z} = \varepsilon_0 \left(\frac{1}{R + j\omega L} + j\omega C \right) \tag{6.2}$$

where $1/Z = 1/(R + j\omega L) + j\omega C$ is the reciprocal of the equivalent complex impedance for the parallel resonant circuit. The expression for $1/Z$ can be rewritten as

$$Z^{-1} = \frac{1 - \omega^2 LC + j\omega RC}{R + j\omega L}$$

or

$$Z^{-1} = \frac{R + j\omega L[(\omega^2 LC - 1) + (R^2 C/L)]}{R^2 + (\omega L)^2} \tag{6.3}$$

and finally

$$Z^{-1} = \frac{R + j\omega L[(\omega^2/\omega^2_0 - 1) + 1/Q^2]}{R^2 + (\omega L)^2}$$

where $\omega_0{}^2 = 1/LC$ and $Q^2 = L/R^2 C$ are the same parameters used in the discussion of the series resonant circuit. Parallel resonance is said to occur when I_0 and ε_0 are in phase, that is, when $[(\omega^2/\omega_0{}^2 - 1) + 1/Q^2] = 0$ or $\omega^2 = \omega_0{}^2(1 - 1/Q^2)$. For a high Q circuit, $\omega \simeq \omega_0$ at resonance. Assuming Q is large and ω near ω_0, we have $I_0 = \varepsilon_0/Q^2 R$. For very low frequencies, $I_0 \simeq I_2 \simeq \varepsilon_0/R$ and, for very high frequencies, $I_0 \simeq I_1 \simeq j\omega C\varepsilon_0$. At parallel resonance the current I_0 is near a minimum. The frequency ω which minimizes I_0 is not quite the same as the frequency which puts I_0 and ε_0 in phase but both rapidly approach ω_0 as Q becomes large.

Neither I_1 nor I_2 show any resonance behavior but for $\omega \simeq \omega_0$ the currents I_1 and I_2 are almost $180°$ out of phase and are almost the same amplitude. Thus their sum I_0 is small.

Sometimes the parallel resonant circuit is shown as a resistor, an inductor, and a capacitor all in parallel. In this case the current $I_0 = I_R + I_L + I_C$ shows a parallel resonance, both of phase and amplitude, at $\omega = \omega_0$. As we have stated, however, the circuit we treat is often a better approximation to real parallel resonant circuits.

6.3   A Network for Shifting the Phase of an AC Voltage

It is useful for some applications to have a circuit that will take an input ac voltage and put out an ac voltage that can be shifted in phase from the input voltage by a variable amount. The output voltage should not change in amplitude over the entire range of phase. A circuit that can perform this function is shown in Fig. 6.3. An input voltage V_A is applied between the input terminal A and ground. A resistive divider is used to set the voltage between terminal B and ground at $V_A/2$. The phasor diagram of Fig. 6.4 shows that the voltage difference between points B and D is constant in magnitude but has a variable phase. This is understood as follows. The voltage across the resistor R plus the voltage across the capacitor C must sum to V_A.

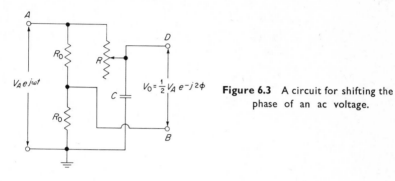

Figure 6.3 A circuit for shifting the phase of an ac voltage.

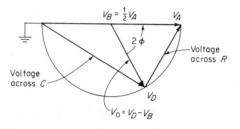

Figure 6.4 A phasor diagram for the voltages in the circuit of Fig. 6.3.

The phasor voltage across the resistor R must be at right angles to the phasor voltage across the capacitor C. Therefore the voltage V_D must lie on a circle with its center at $V_A/2$. Therefore the output voltage is

$$V_0 = V_D - V_B = (V_A/2)e^{-j2\phi}$$

To show the same result analytically is a straightforward exercise in the manipulation of complex numbers. We have

$$V_D = \frac{1}{j\omega C}\frac{V_A}{[R + (1/j\omega C)]} \quad \text{and} \quad V_B = \frac{1}{2}V_A$$

Therefore

$$V_0 = V_D - V_B = \left[\frac{1}{(j\omega CR + 1)} - \frac{1}{2}\right]V_A \tag{6.4}$$

$$V_0 = \frac{1}{2}V_A\left(\frac{1 - j\omega CR}{1 + j\omega CR}\right) = \frac{1}{2}V_A e^{-j2\phi}$$

where $\phi = \arctan(\omega CR)$. Thus as $R \to 0$, $2\phi \to 0$, as $R \to \infty$, $2\phi \to \pi$. The possible variation in phase is somewhat less than 180°. The circuit provides a constant amplitude and the calculated phase shift only when the current drawn between B and D is small compared to the currents flowing in the other branches of the circuit.

6.4 A Constant AC Current Source

The circuit shown in Fig. 6.5 will provide a constant amplitude ac current I_2, through the resistor R independent of the value of R if $\omega = 1/\sqrt{LC}$ and provided that R is much greater than the resistance of the inductor L (note that the resistance of the inductor L is not shown). In order to see this, let us calculate I_2 as follows. Treating I_1 and I_2 as loop currents, Kirchhoff's voltage law gives two equations

$$V_0 = \left(\frac{-j}{\omega C} + j\omega L \right) I_1 - j\omega L I_2$$

$$0 = -j\omega L I_1 + (R + j\omega L) I_2$$

(6.5)

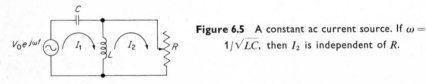

Figure 6.5   A constant ac current source. If $\omega = 1/\sqrt{LC}$, then I_2 is independent of R.

Thus the assumption $\omega = 1/\sqrt{LC}$ gives $-j/\omega C + j\omega L = 0$, and $I_2 = jV_0/\omega L = j\sqrt{C/L} \, V_0$ follows from the first equation. We see that both the amplitude and phase of I_2 are independent of R and that I_2 leads V_0 by $\pi/2$. If we look back into the source from the load resistor, we see that the source has a Thevenin equivalent impedance consisting of an inductor and a capacitor in parallel. These are driven at their resonant frequency and have therefore an infinite impedance, a result that follows from our earlier discussion of parallel resonance. A constant current source must, of course, have an infinite internal impedance.

If R is increased indefinitely, $I_2 R$ and therefore also the potential across the inductor increases without limit. Since I_2 remains constant, it follows that I_1 must increase indefinitely. We can calculate I_1 by substituting for I_2 in the second of Eq. 6.5. We obtain

$$I_1 = \frac{(R + j\omega L)}{(\omega L)^2} V_0$$

(6.6)

which verifies the unlimited increase of I_1 with R. Just as in the dc case this unphysical result stems from our neglect of internal resistance that will occur primarily in the voltage source V and in the windings of the inductor L. Such internal resistance limits all impedances, currents, and voltages to finite values and also limits the usefulness of the constant current source to values of $R \ll |Z_{int}|$, where Z_{int} is the source impedance seen by the load resistor R. For $R \ll \omega L$, Eq. 6.6 gives $I_1 \simeq I_2$. Thus very little current flows through the inductor.

6.5 Filter Circuits

It is often useful to have circuits that have a very high impedance for certain frequencies and a low impedance for other frequencies. These circuits are called *filters* because when they are inserted between an emf source and a load, they filter out the source frequencies for which they have a high impedance. In order to make these ideas clear, let us consider some specific examples.

A high-pass filter is shown in Fig. 6.6. Suppose an input voltage V_{in} is applied between the terminals A and B. The output voltage between terminals C and B is

$$V_{out} = \frac{R}{[R - (j/\omega C)]} V_{in} = \frac{V_{in}\,e^{j\phi}}{[1 + (1/\omega RC)^2]^{1/2}} \tag{6.7}$$

where $\phi = \arctan(1/\omega RC)$. The derivation assumes that the current drawn between C and B by the load is small compared to the current through R.

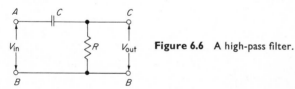

Figure 6.6 A high-pass filter.

We see that V_{out}/V_{in} approaches unity (very little attenuation) if $\omega \gg 1/RC$ but that V_{out}/V_{in} is small (large attentution) if $\omega \ll 1/RC$. In Fig. 6.7 we interchange the R and C of Fig. 6.6, producing a low-pass filter. Low frequencies are passed with little attenuation but high frequencies are shorted out by the condenser.

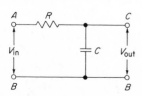

Figure 6.7 A low-pass filter.

Band-pass filters have low attenuation for a limited range of frequencies but high attenuation for frequencies on either side of the pass band. A band reject filter has a high impedance for a limited range of frequencies. From the discussion of series and parallel resonant circuits it can be seen that if an inductor L is placed in series with C in Fig. 6.6, the circuit will pass a band of frequencies centered at $\omega = 1/\sqrt{LC}$. If L is in parallel with C in Fig. 6.6, the circuit will reject or attenuate frequencies near $\omega = 1/\sqrt{LC}$.

It is common to measure the attentuation of an ac voltage in decibels (dB). The attenuation of a voltage in decibels is defined as $\alpha_{dB} = -20 \log |V_{out}/V_{in}|$ where the logarithm is to the base 10.

As an aside we note that in other fields, for instance acoustics, the attenuation of a sound wave is defined as $\alpha_{dB} = -10 \log (P_{out}/P_{in})$ where P_{out} is the output power and P_{in} is the input power. Our present definition is equivalent to $\alpha_{dB} = -10 \log |V_{out}^2/V_{in}^2|$ but V_{out}^2 and V_{in}^2 bear no necessary relation to the actual input and output powers. In the circuits of Fig. 6.6 and Fig. 6.7, for instance, the input power is finite since the attenuator network dissipates power in the resistor R. The output power may be very small if the output load has a high impedance. In fact in the derivation we assumed the output power to be zero. The definition $\alpha_{dB} = -20 \log |V_{out}/V_{in}|$ is very widely used in circuit theory. It is best to accept it at face value.

For the high-pass filter of Fig. 6.6 we find using Eq. 6.7 and the definition of α_{dB}

$$\alpha_{dB} = -20 \log \left| \frac{V_{out}}{V_{in}} \right| = -20 \log \frac{1}{[1 + (\omega_0/\omega)^2]^{1/2}} \qquad (6.8)$$

where $\omega_0 = 1/RC$. Note that this is not the definition of ω_0 used in RLC circuits. A plot of α_{dB} for the high-pass filter is shown in Fig. 6.8. It is clear from the plot and from Eq. 6.8 that for frequencies well below ω_0 the attenuation increases by 6 dB/octave (that is, α_{dB} increases by 6 dB when ω changes by a factor of 2) or 20 dB/decade (that is, α_{dB} increases by 20 dB when ω changes by a factor of 10) as the frequency decreases.

In Fig. 6.7 we show the low-pass filter that results if the R and C of Fig. 6.6 are interchanged. The phasor equation is

$$V_{out} = \frac{V_{in}\, e^{j\phi}}{(1 + (\omega/\omega_0)^2]^{1/2}} \qquad (6.9)$$

where $\phi = \arctan \omega RC$ and $\omega_0 = 1/RC$. This expression for V_{out} assumes, as before, that the output of the filter is used to drive a very high resistance, that is, that the output is not severely loaded. For frequencies well below ω_0 we see that $V_{out} = V_{in}$ and for frequencies well above ω_0 we see that V_{out} is attenuated by 6 dB/octave or 20 dB/decade. These results are shown in Fig. 6.9.

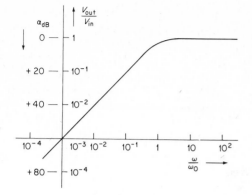

Figure 6.8 Plots of α_{dB} and V_{out}/V_{in} versus ω/ω_0 for the high-pass filter of Fig. 6.6. Note that for $\omega \ll \omega_0$ the attenuation α_{dB} changes by 20 dB/decade change in ω. The quantities V_{out}/V_{in} and ω/ω_0 are plotted logarithmically.

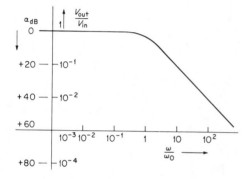

Figure 6.9 Plots of α_{dB} and V_{out}/V_{in} versus ω/ω_0 for the low-pass filter of Fig. 6.7.

Another low-pass filter using an LR circuit is shown in Fig. 6.10. If the input voltage to the circuit of Fig. 6.10 is V_{in}, then the output voltage without a load is given by Eq. 6.9 but where $\phi = \arctan(-\omega L/R)$ and where $\omega_0 = R/L$. Therefore this circuit will pass without attenuation frequencies much lower than ω_0 and will attenuate frequencies much higher than ω_0.

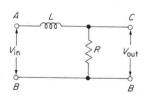

Figure 6.10 A low-pass filter.

A typical application of a low-pass filter is the reduction of ripple in the output of a rectifier of a dc power supply as illustrated in Fig. 6.11. We shall show later that the output of a full-wave rectifier driven from a 60-Hz line has a large dc component and in addition ac ripple frequencies of 120 Hz, 240 Hz, etc. Let us calculate the reduction in ripple amplitude provided by the filter. We assume that the internal resistance of the power

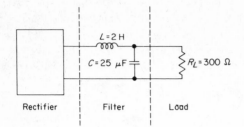

Figure 6.11 A low-pass filter used to help eliminate all frequencies except 0 Hz (dc) from the output of a power supply rectifier.

supply and the resistance of the inductor are small compared to the 300-ohm load resistor. The entire dc voltage of the rectifier therefore appears across the load. The ripple frequencies however are strongly attenuated by the 2-henry inductor (called a *choke*) and the 25-microfarad capacitor. For the 120-Hz ripple we have

$$Z_L = j\omega L = j2\pi \times 120 \times 2 = j1500 \text{ ohms}$$

$$Z_C = \frac{-j}{\omega C} = \frac{-j}{2\pi \times 120 \times 25 \times 10^{-6}} = -j50 \text{ ohms} \qquad (6.10)$$

Because $Z_C \ll R_L$, we may ignore the latter in calculating the output ripple. We obtain for the ratio of the 120-Hz amplitude between output and input

$$\left| \frac{V_{\text{out}}}{V_{\text{in}}} \right| = \left| \frac{Z_C}{Z_L + Z_C} \right| = \left| \frac{-j50}{j(1500 - 50)} \right| \simeq \frac{1}{30} \qquad (6.11)$$

In decibel language this is an attenuation of $\alpha_{\text{dB}} = 20 \log 30 \simeq 30$ dB. For this filter the attenuation expressed as an amplitude ratio is proportional to $1/\omega^2$. Thus 240 Hz is reduced by approximately $\frac{1}{120}$.

A useful band reject filter, the twin T circuit, is shown in Fig. 6.12. Its characteristics can be calculated in several different ways but one of the most informative is to make use of two Y-Δ transformations. The circuit then appears as two Δ networks in parallel among the points A, B, and D.

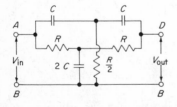

Figure 6.12 A twin T circuit used as a band reject filter.

Figure 6.13 shows the steps in the transformation. The equations relating the Y and Δ parameters are given in Chapter 2. In the present application the resistances of Eq. 2.14 are replaced by the appropriate complex impedances from Fig. 6.13. The impedances of the Δ circuits are given below where $\omega_0 = 1/RC$.

$$Z_1 = 2R\left(1 + \frac{j\omega}{\omega_0}\right)$$

$$Z_2 = Z_3 = R\left(1 - \frac{j\omega}{\omega_0}\right)$$

$$Z_2' = -2R\frac{\omega_0{}^2}{\omega^2}\left(1 + \frac{j\omega}{\omega_0}\right) \tag{6.12}$$

$$Z_1' = Z_3' = R\left(1 - \frac{j\omega}{\omega_0}\right) = Z_2 = Z_3$$

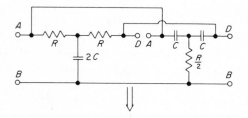

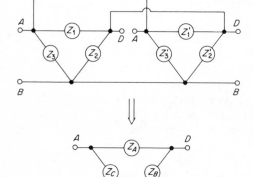

Figure 6.13 The transformation of the twin T network of Fig. 6.12 into an equivalent Δ network.

Finally Z_A, Z_B, and Z_C are just the parallel combinations of the corresponding Z_1 and Z_1', and so on; that is, $Z_A = Z_1 Z_1'/(Z_1 + Z_1')$. Carrying out the calculations we obtain

$$Z_A = \frac{2R[1 + (j\omega/\omega_0)]}{1 - (\omega^2/\omega_0{}^2)}$$

$$Z_B = Z_C = \frac{R}{2}\left(1 - \frac{j\omega}{\omega_0}\right) \tag{6.13}$$

With the help of these impedances we can calculate the output voltage and current between B and D for any assumed load. The important characteristic of the twin T filter, however, is obvious from the expression for Z_A. We see that $Z_A \to \infty$ as $\omega \to \omega_0$. Thus the output voltage drops to zero at $\omega = \omega_0$. If the load draws very little current (that is, $Z_{load} \gg Z_B$), the output voltage can be written $V_{out} = V_{in}[Z_B/(Z_A + Z_B)]$. The short circuit output current is V_{in}/Z_A. Thus the Thevenin impedance of the filter is $Z_{Th} = Z_A Z_B/ (Z_A + Z_B)$. Note that the Thevenin impedance remains finite at $\omega = \omega_0$. The output goes to zero because $\varepsilon_{Th} = V_{out}$ goes to zero.

The twin T is an example of a circuit that provides a resonant behavior without the use of inductors. This can be important. Inductors are bulky and expensive and the frequency range over which resistive and capacitive effects can be ignored is usually limited.

We remind the reader that although we calculate the behavior of the twin T by transforming to an equivalent Δ circuit, this does not mean that we can actually construct a real Δ circuit that provides the calculated impedances. This is an important difference between reactive and purely resistive circuits. When only resistances are involved, one can construct the Δ circuit that is equivalent to a given Y circuit or one can physically replace a complicated two-terminal resistive network by a single resistance. On the other hand so simple an impedance as a resistance and capacitance in parallel cannot be replaced for all frequencies by any series combination of resistors, capacitors, and inductors.

A problem that arises in the design of filters is the matching of the input and output of the filter to the source impedance and to the load impedance with satisfactory attenuation and phase shift. In order to to this one must analyze the filter circuit in a manner somewhat analogous to our treatment of the T pad. Of course the analysis is very much more complicated. An excellent discussion of the design of filters including impedance matching can be found in *Radio Engineers' Handbook*, F. Terman (McGraw-Hill Co., New York, 1943).

6.6 The Coupling of Resonant Circuits by a Mutual Inductance

Figure 6.14 shows two resonant LC circuits coupled by a mutual inductance. Circuits of this type are important in providing the coupling from one stage of a tuned amplifier to the next, as discussed in Chapter 13. The Kirchhoff voltage law equations for determining I_1 and I_2 are the following:

$$\varepsilon = j\left(\omega L_1 - \frac{1}{\omega C_1}\right) I_1 + j\omega M I_2$$

$$0 = j\omega M I_1 + j\left(\omega L_2 - \frac{1}{\omega C_2}\right) I_2$$

(6.14)

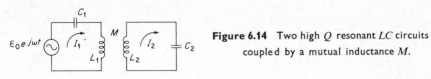

Figure 6.14 Two high Q resonant LC circuits coupled by a mutual inductance M.

The solutions are

$$I_1 = \frac{-j\varepsilon[\omega L_2 - (1/\omega C_2)]}{\{[\omega L_1 - (1/\omega C_1)][\omega L_2 - (1/\omega C_2)] - \omega^2 M^2\}}$$

$$(6.15)$$

$$I_2 = \frac{-j\omega M\varepsilon}{\{[\omega L_1 - (1/\omega C_1)][\omega L_2 - (1/\omega C_2)] - \omega^2 M^2\}}$$

The currents are infinite (because we have omitted the resistances) when the denominators are zero. Using $\omega_1^2 = 1/L_1 C_1$ and $\omega_2^2 = 1/L_2 C_2$ and introducing the coupling constant $k^2 = M^2/L_1 L_2$, the condition that the denominator be zero can be written

$$\left(\frac{\omega^2}{\omega_1^2} - 1\right)\left(\frac{\omega^2}{\omega_2^2} - 1\right) - k^2\left(\frac{\omega^2}{\omega_1^2}\right)\left(\frac{\omega^2}{\omega_2^2}\right) = 0 \qquad (6.16)$$

Nearly always the two coupled circuits will have the same individual resonant frequencies and we write $\omega_1 = \omega_2 = \omega_0$. Equation 6.16 then gives

$$\left(\frac{\omega^2}{\omega_0^2} - 1\right) = \pm k\left(\frac{\omega^2}{\omega_0^2}\right)$$

or $$(6.17)$$

$$\omega = \frac{\omega_0}{(1 \mp k)^{1/2}}$$

Thus the coupling of two identical resonant circuits causes the natural frequency ω_0 of the system to split into two frequencies, $\omega_0/(1 + k)^{1/2}$ and $\omega_0/(1 - k)^{1/2}$.

In any real circuit, resistance is present and the infinite currents at resonance are avoided. The analysis is somewhat more complicated and we shall merely give the results in graphical form. Figure 6.15 shows two identical circuits each containing a resistance and coupled by the mutual inductance M. In Figs. 6.16 and 6.17 we plot $|I_1 R/\varepsilon_0|$ and $|I_2 R/\varepsilon_0|$ as functions of ω/ω_0 for a number of values of the parameter kQ where Q, the quality factor, has been

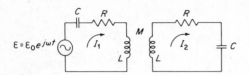

Figure 6.15 Two identical resonant circuits coupled by a mutual inductance M. Plots of the currents as functions of frequency are shown in Figs. 6.16 and 6.17.

taken to be 100. For small values of kQ the natural width of the resonances produced by the finite Q obscures the splitting due to the coupling and a single maximum is observed at $\omega = \omega_0$. The resonance is, of course, wider than it would be in the absence of coupling. For sufficiently large values of kQ two separate resonances are observed near $\omega = \omega_0/(1 \mp k)^{1/2}$. If $M = k = 0$, the circuits are independent. The current $I_2 = 0$ and I_1 behaves as in a simple series resonance circuit. This is obvious if one places $M = 0$ in Eq. 6.15.

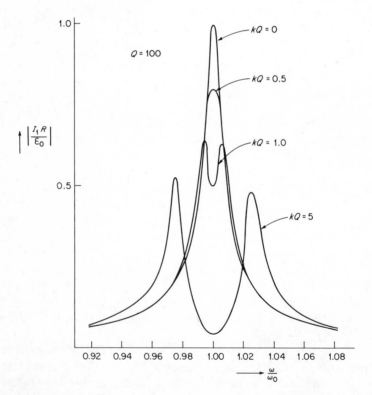

Figure 6.16 Plots of $|I_1 R/\varepsilon_0|$ versus ω/ω_0 for the coupled resonant circuits of Fig. 6.15.

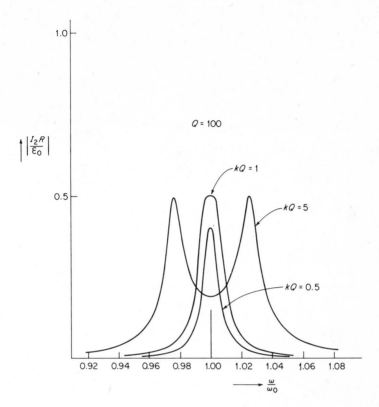

Figure 6.17 Plots of $|I_2 R/\varepsilon_0|$ versus ω/ω_0 for the coupled resonant circuits of Fig. 6.15. For $kQ = 0$, $I_2 = 0$ since this corresponds to zero coupling between the circuits.

6.7 AC Bridges

A common method for determining the resistance of a resistor is to use a Wheatstone bridge. This technique was described in Chapter 3. One advantage of the use of a bridge for determining a resistance is that the basic measurement involves the observation of a null voltage or current so that the method has good sensitivity. In this section we discuss methods for measuring self-inductance, mutual inductance, and capacitance with ac bridges. In addition, ac bridges are sometimes used to measure the frequency of the ac voltage driving the bridge.

Figure 6.18 shows an ac bridge. The use of an ac bridge is analogous to the use of a Wheatstone bridge. Three of the impedances are usually known and one of these three impedances is a variable impedance (both the real and imaginary parts of the impedance must be variable). The fourth impedance is unknown. The purpose of the bridge is to measure both the real and

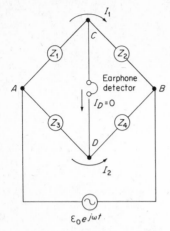

Figure 6.18 An ac bridge at balance, $I_D = 0$. The currents are drawn I_1 flows through Z_1 and Z_2, and I_2 flows through Z_3 and Z_4.

imaginary parts of the unknown impedance. At balance the ac current flowing in the detector from C to D is zero. Therefore the same current I_1 flows through the branch from A to C and through the branch from C to B. In a similar manner the same current I_2 flows through the two branches from A to D and D to B. Since the detector current is zero at balance, we have $V_{CD} = 0$ and $I_1 Z_1 = I_2 Z_3$ and $I_1 Z_2 = I_2 Z_4$. It follows that

$$\frac{Z_1}{Z_2} = \frac{Z_3}{Z_4}$$

or   (6.18)

$$Z_1 Z_4 = Z_2 Z_3$$

Because the real and imaginary parts of the two sides of Eq. 6.18 must separately be equal, we actually have two equations relating the various real quantities. Two unknown bridge parameters can therefore be determined by the adjustment of two or more of the known parameters. The detector for the null measurement with the ac bridge may be a pair of earphones as indicated in the drawing although other devices such as an oscilloscope are often used if earphones are not convenient or if ω is not in the audio range.

The current in the detector when the bridge is unbalanced can be calculated in a manner similar to that used for calculating the detector current in an unbalanced Wheatstone bridge. If we assume that Z_{source} is very small, then the Thevenin equivalent voltage source seen by the detector is

$$\varepsilon_{\text{Th}} = \varepsilon_0\, e^{j\omega t} \left(\frac{Z_1}{Z_1 + Z_2} - \frac{Z_3}{Z_3 + Z_4} \right) = \varepsilon_0\, e^{j\omega t}\, \frac{(Z_1 Z_4 - Z_2 Z_3)}{(Z_1 + Z_2)(Z_3 + Z_4)} \quad (6.19)$$

This is equivalent to Eq. 3.6 for the Wheatstone bridge. Equations 6.19 and 3.6 are not identical because we have numbered the arms of the bridges in a

different manner. Compared to Chapter 3, arms 2 and 3 have been interchanged. The balance condition is $\varepsilon_{\text{Th}} = 0$. The Thevenin impedance of the
bridge is (see Eq. 3.7)

$$Z_{\text{Th}} = \frac{Z_1 Z_2}{Z_1 + Z_2} + \frac{Z_3 Z_4}{Z_3 + Z_4} \qquad (6.20)$$

and the detector current can be written $I_D = \varepsilon_{\text{Th}}/(Z_{\text{Th}} + Z_D)$ where Z_D is the
impedance of the detector. Often with a Wheatstone bridge the Thevenin
resistance is greater than the galvanometer resistance and the bridge acts
as a current source for the galvanometer. In this case the greater the current
sensitivity of the detector, the more accurate is the balance that can be
achieved. With an ac bridge, particularly if an oscilloscope is used as detector, one may have $Z_{\text{Th}} < Z_D$. In this case the bridge is a voltage source,
and a detector of maximum voltage sensitivity is needed.

Now let us consider a few examples of actual ac bridges. The Owen
bridge, shown in Fig. 6.19, is a simple and very useful bridge. Let us derive
the balance condition for this bridge. At balance

$$I_1\left(-\frac{j}{\omega C_1}\right) = I_2 R_3 \quad \text{and} \quad I_1\left(-\frac{j}{\omega C_2} + R_2\right) = I_2(R_4 + j\omega L_4)$$

It follows that $\qquad\qquad\qquad\qquad\qquad\qquad\qquad\qquad\qquad (6.21)$

$$\left(-\frac{j}{\omega C_1}\right)(R_4 + j\omega L4) = R_3\left(-\frac{j}{\omega C_2} + R_2\right)$$

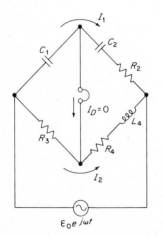

Figure 6.19 An Owen bridge. This ac bridge is
used to measure an unknown inductance L_4
and its series resistance R_4.

We can obtain $L_4/C_1 = R_2 R_3$ by equating the real parts of both sides of the
equation above and $R_4/C_1 = R_3/C_2$ from equating the imaginary parts. These
two equations must both be satisfied in order to balance the Owen bridge.

Usually this bridge is used to measure an unknown self-inductance L_4 and its internal resistance R_4 in terms of known capacitors C_1 and C_2 and known resistors R_2 and R_3.

An important result to note is that the balance conditions are independent of ω as long as the components do not vary with ω. This is a very useful property since one usually wishes to measure L_4 and R_4 without measuring ω with great accuracy.

A problem with ac bridges is that in achieving one balance condition the other one may be upset. For example, in order to balance the Owen bridge we must approach the conditions

$$L_4 = R_2 R_3 C_1$$

and

$$R_4 = \left(\frac{C_1}{C_2}\right) R_3$$

(6.22)

Suppose we were to vary R_3 until $L_4 = R_2 R_3 C_1$ and then attempt to vary C_1 until $R_4 = (C_1/C_2)R_3$. Unfortunately C_1 also occurs in the balance equation for L_4. We destroy the first balance condition in achieving the second. The problem is avoided if we vary R_2 to obtain the L_4 balance and C_2 to obtain the R_4 balance. The convergence to balance of an Owen bridge is relatively rapid.

Let us now consider the Schering bridge, which is useful for the measurement of an unknown capacitor and the series resistance of the unknown capacitor. A Schering bridge is shown in Fig. 6.20. The balance condition for the Schering bridge can be found as follows. Since $I_D = 0$, we find that

$$I_1 \frac{R_1(-j/\omega C_1)}{R_1 - (j/\omega C_1)} = I_2 R_3$$

$$I_1\left(-\frac{j}{\omega C_2}\right) = I_2\left(R_4 - \frac{j}{\omega C_4}\right)$$

so that

(6.23)

$$\frac{R_1(-j/\omega C_1)}{R_1 - (j/\omega C_1)}\left(R_4 - \frac{j}{\omega C_4}\right) = R_3\left(-\frac{j}{\omega C_2}\right)$$

or

$$R_1\left(-\frac{j}{\omega C_1}\right)\left(R_4 - \frac{j}{\omega C_4}\right) = R_3\left(-\frac{j}{\omega C_2}\right)\left(R_1 - \frac{j}{\omega C_1}\right)$$

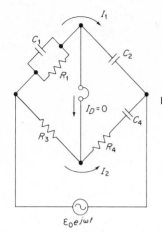

Figure 6.20 A Schering bridge. This ac bridge is used to measure an unknown capacitance C_4 and its series resistance R_4.

Separating the real and imaginary parts of Eq. 6.23 yields

$$R_4 = \frac{C_1}{C_2} R_3$$

and   (6.24)

$$C_4 = \frac{R_1}{R_3} C_2$$

In this bridge it is common to vary R_1 and C_1 in searching for impedance values such that $I_D = 0$.

As another example, let us consider a Carey Foster bridge for measuring an unknown mutual inductance. The Carey Foster bridge is shown in Fig. 6.21. The balance conditions for the Carey Foster bridge are obtained as follows. At balance the current in the detector is $I_D = 0$. This occurs when the voltage across the detector is zero. Note that the voltage from C to D is not zero since the mutual inductance produces a voltage drop in the detector branch from C to D. At balance the voltage from A to C is not equal to the voltage from A to D. When the voltage across the detector is zero, then $I_1(R_1 - j/\omega C_1) = I_2(R_3 + j\omega L_3 - j\omega M)$ and $I_1 R_2 = I_2 j\omega M$. From these equations it follows that $(R_1 - j/\omega C_1)(j\omega M) = R_2(R_3 + j\omega L_3 - j\omega M)$. Separating the real and imaginary parts of this equation yields the results

$$M = R_2 R_3 C_1$$

and   (6.25)

$$L_3 = (R_1 + R_2) \frac{M}{R_2}$$

Normally R_1 and C_1 are varied until conditions are found such that $I_D = 0$.

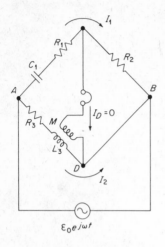

Figure 6.21 A Carey Foster bridge. This ac bridge is used to measure an unknown mutual inductance M.

Sometimes an ac bridge circuit is used for the determination of a frequency. For a bridge of this type it is obvious that one wants the balance conditions to depend strongly on ω. An example of a bridge used for frequency determination is the Wien bridge shown in Fig. 6.22. The derivation of the balance conditions for the Wien bridge will be left as a problem for the student. The balance conditions are

$$\omega^2 = \frac{1}{R_3 C_3 R_4 C_4}$$

and   (6.26)

$$\frac{R_1 R_4}{C_4} = \frac{R_2 R_4}{C_3} + \frac{R_2 R_3}{C_4}$$

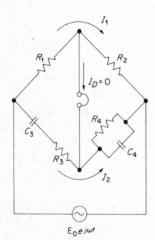

Figure 6.22 A Wien bridge. This bridge is used to determine the angular frequency ω of the driving voltage.

The parameters are often chosen so that $R_3 = R_4$ and $C_3 = C_4$ so that the balance conditions become $\omega = 1/R_3 C_3$ and $R_1 = 2R_2$. It is clear that if one knows R_1, R_2, R_3, R_4, C_3, and C_4, then from the balance conditions one readily obtains ω, the angular frequency of the driving voltage, when the bridge is balanced.

The Wien bridge has similarities to the twin T filter. Both circuits use only resistors and capacitors and provide an output voltage that goes to zero at a certain resonant frequency.

There are a great many varieties of ac bridges. A careful study of the literature will usually enable one to select a bridge that is suitable for a particular application. An excellent reference for a person who is interested in learning more about ac bridges is *Basic Electrical Measurements*, 2nd ed., M. B. Stout (Prentice-Hall, Inc., Englewood Cliffs, N.J., 1960).

6.8 Attenuation, Differentiation, and Integration

In many applications a circuit is useful only if its properties are independent of frequency over some range of frequencies. A resistive network is, of course, such a circuit but many important circuit functions cannot be performed by resistive networks. In addition, stray capacitances always occur in circuits. For these and other reasons, capacitive and inductive reactances are usually quite unavoidable. A common frequency-independent circuit using resistors and capacitors is the compensated attenuator. It is used in the following way.

Frequently one wishes to sample and display the time variation of the voltage at some point in a circuit. An oscilloscope (see the discussion in Chapter 18) whose input impedance is a resistance and a capacitance (including stray capacitance) in parallel is often used for such a measurement. The voltage to be sampled may be larger than the oscilloscope can handle or the input impedance of the oscilloscope may be so low that it overloads the circuit. In either case an attenuator probe is used that presents to the oscilloscope only some fraction of the voltage sampled. Figure 6.23 shows both an uncompensated and a compensated attenuator. The difficulty with the uncompensated attenuator is easily seen. At low frequencies the attenuation ratio is $V_{\text{out}}/V_{\text{in}} = R_2/(R_1 + R_2)$. At sufficiently high frequencies the capacitance C_2 shorts out the output and the attenuation ratio approaches zero. Thus the circuit properties are not independent of frequency.

This is corrected in a compensated attenuator by placing the capacitance C_1 in parallel with R_1. The attenuation ratio is now

$$\frac{V_{\text{out}}}{V_{\text{in}}} = \frac{Z_2}{Z_1 + Z_2}$$

where $\hspace{11cm}$ (6.27)

$$Z_1 = \frac{R_1(-j/\omega C_1)}{[R_1 - (j/\omega C_1)]} \quad \text{and} \quad Z_2 = \frac{R_2(-j/\omega C_2)}{[R_2 - (j/\omega C_2)]}$$

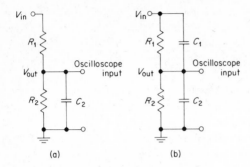

Figure 6.23 (a) An uncompensated attenuator. (b) A compensated attenuator. When used with an oscilloscope the external components of a compensated attenuator are just R_1 and C_1. R_2 and C_2 are the internal input resistance and capacitance of the oscilloscope itself.

Let us put $R_1 = \alpha R_2$ and $1/C_1 = \alpha(1/C_2)$. We then have $Z_1 = \alpha Z_2$ and

$$\frac{V_{\text{out}}}{V_{\text{in}}} = \frac{1}{\alpha + 1} \qquad (6.28)$$

Thus the attenuation ratio is independent of frequency although the total impedance of the attenuator is not. The resistor chain draws a constant current independent of ω when the magnitude of V_{in} is constant but the current through the capacitor chain increases as ω. The relation $V_{\text{out}}/V_{\text{in}} = 1/(1 + \alpha)$ holds for each chain separately, however. Thus no current flows from one chain to the other at the output point.

A typical value for the impedance at the input to an oscilloscope is a 10^6-ohm resistor in parallel with a (47×10^{-12})-farad capacitor. The attenuation in the probe is usually chosen to be 10 so that the probe will consist of a (9×10^6)-ohm resistor in parallel with a capacitor that is variable over a range near 5×10^{-12} farad. The variable capacitor is adjusted until the probe is properly compensated.

The properties demanded of an attenuator are just the opposite of those needed in a filter. A filter takes a signal, for instance, the output of a rectifier, that may consist of a mixture of many different frequencies and eliminates all but a few. An attenuator on the other hand must attenuate all frequencies by the same amount if its output signal is to have the same shape as the input signal.

This is a convenient point to mention a few ideas that will be treated more fully in the next chapter. Many time-dependent voltages and currents are not sinusoidal; in fact, they may not even repeat in any regular manner. For instance, Fig. 6.24(a) shows a rectangular voltage pulse. Pulses such as those in Fig. 6.24 are very common in all kinds of timing and counting circuits. A pulse or waveform of any shape can be constructed by adding together a great many pure sinusoidal waves of different frequencies. The relative amplitudes of the different frequencies characterize the pulse. A useful general rule is that a pulse that lasts a time τ requires a range of frequencies $\Delta f = \Delta\omega/2\pi = 1/\tau$ to give a reasonable reproduction of the pulse.

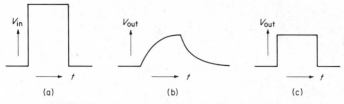

Figure 6.24   (a) A rectangular voltage input signal to an attenuator. (b) The resulting output signal from an uncompensated attenuator such as the one in Fig. 6.23(a). (c) The output signal from a properly compensated attenuator such as the one in Fig. 6.23(b). The dc attenuation ratio $R_2/(R_1 + R_2)$ is one-half.

When a pulse is fed into a circuit, each of its component waves (Fourier components) is changed in phase and amplitude by the frequency-dependent impedance of the circuit. If the impedance is a strong function of ω in the range $\Delta\omega$, the output pulse is badly distorted in shape. If the impedance is almost independent of ω in the range $\Delta\omega$, the pulse shape is little changed.

Figure 6.24(b) shows the output pulse for an uncompensated attenuator. Because the high-frequency response is poor, it fails to reproduce the steep edges of the pulse. In Fig. 6.25 we show the output pulse shapes that can result from improper adjustment of C_1 in a compensated attenuator.

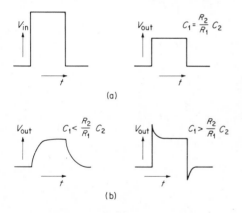

Figure 6.25   (a) The input and output voltages of a properly adjusted compensated attenuator. The dc attenuation ratio is $R_2/(R_1 + R_2) = 1/2$. (b) Output voltage waveforms when the compensated attenuator is improperly adjusted.

Now let us discuss two useful RC circuits; one can differentiate and the other can integrate a time-varying electrical signal. We shall find that our circuit requirements are not an output independent of ω but rather an output directly proportional to ω (differentiation) or inversely proportional to ω (integration). An RC circuit for differentiating is shown in Fig. 6.26.

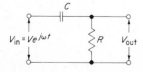

Figure 6.26   An RC network for differentiating a voltage signal. If $\omega \ll 1/RC$, then $V_{\text{out}} = RC\,dV_{\text{in}}/dt$.

If the input voltage is $V_{in} = Ve^{j\omega t}$, the output voltage is

$$V_{out} = \frac{R}{R - (j/\omega C)} Ve^{j\omega t}$$

If $\omega RC \ll 1$,

$$V_{out} = j\omega RC Ve^{j\omega t} \qquad (6.29)$$

or

$$V_{out} = RC \frac{dV_{in}}{dt}$$

The requirement $\omega \ll 1/RC$ means that only those signals can be handled that are made up of sinusoidal components whose frequency is much less than $1/RC$. Qualitatively speaking, the input signal shouldn't change very much in the time RC. A pulse will be adequately differentiated only if it is long compared to RC. This can be seen in still another way. If RC is small compared to the length τ of the pulse, the capacitor will charge up to $V_C = Q/C \simeq V_{in}$. We then have $V_{out} = IR = R\, dQ/dt \simeq RC\, dV_{in}/dt$. One of the disadvantages of a good differentiator is that on the average the output signal is small compared to the input signal.

An integrating circuit is shown in Fig. 6.27. If $V_{in} = Ve^{j\omega t}$, we have

$$V_{out} = \frac{-j/\omega C}{R - (j/\omega C)} Ve^{j\omega t}$$

If $\omega RC \gg 1$, then $\qquad\qquad\qquad\qquad\qquad\qquad\qquad (6.30)$

$$V_{out} = \frac{1}{j\omega RC} Ve^{j\omega t}$$

$$V_{out} = \frac{1}{RC} \int V_{in}\, dt$$

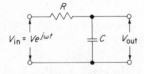

Figure 6.27   An RC network for integrating a voltage signal. If $\omega \gg 1/RC$, then $V_{out} = (1/RC) \int V_{in}\, dt$.

The result can also be seen by noting that if RC is large, most of the input voltage appears across the resistor. Thus $V_{in} \simeq IR$ and

$$V_{out} = \frac{Q}{C} = \frac{1}{C} \int I\, dt = \frac{1}{RC} \int V_{in}\, dt$$

The requirement $\omega \gg 1/RC$ means that integration fails for very low-frequency signals. We see that if the input has a dc component, the condenser steadily charges up. The charge remains proportional to the integral of the dc component of V_{in} only for a time $\ll RC$. Voltage signals whose average value is zero are easily integrated but as in the case of differentiation the broader the frequency range over which the integrator operates, the smaller is the average voltage output. Figures 6.28 and 6.29 show numerical examples of the operation of RC circuits for integration and differentiation.

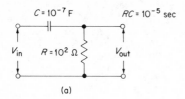

(a)

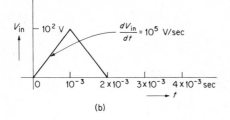

(b)

Figure 6.28 (a) A differentiating RC network with a time constant $RC = 10^{-5}$ sec. (b) An input voltage waveform to the differentiating network. (c) The output voltage $V_{out} = RC\, dV_{in}/dt$.

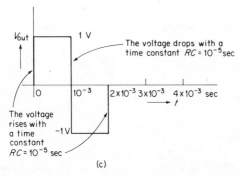

(c)

Particularly in the literature dealing with short pulses the term *differentiation* is sometimes used rather loosely. Suppose an approximately rectangular pulse, which is 1 volt in magnitude and 1 μsec long, is fed into an RC circuit with a time constant of 10^{-7} sec as shown in Fig. 6.30. The output is a double positive and negative pulse that resembles, superficially, the derivative of the input pulse; however, the condition $\omega RC \ll 1$ is not satisfied. Suppose the rise time and fall time of the input pulse is 10^{-8} sec. Frequency components $\omega \simeq 10^8$ rad/sec are involved and we have $\omega RC = 10$.

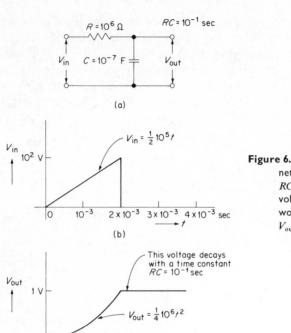

(a)

(b)

Figure 6.29 (a) An integrating RC network with a time constant $RC = 10^{-1}$ sec. (b) An input voltage waveform to the network. (c) The output voltage $V_{\text{out}} = 1/RC \int V_{\text{in}}\, dt$.

(c)

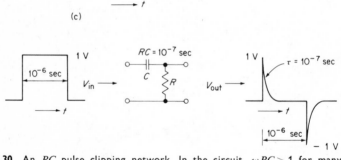

Figure 6.30 An RC pulse clipping network. In the circuit, $\omega RC > 1$ for many of the frequencies of the input pulse shown.

The pulse is not being differentiated at all as shown by the fact that the peak amplitude of the double pulse is almost the same as the amplitude of the input pulse. The high-frequency components of the leading edge are going right through the condenser and appearing across the resistor. Such circuits are very useful but their function is properly described as RC pulse clipping rather than differentiation.

Now suppose the input pulse is as shown in Fig. 6.31. The input pulse rises linearly to its maximum value in 10^{-8} sec, remains constant for 10^{-6} sec, and then falls linearly to zero in 10^{-7} sec. The RC time of the network is reduced to 10^{-10} sec. In this case the output pulse is an accurate derivative of the input. However, the actual amplitude of the output is $V_{out} = RC\,(dV/dt) = 10^8 \times 10^{-10} = 10^{-2}$ volt on the rising edge and 10^{-3} volt on the falling edge. Thus the output is small compared to the input, as it must be.

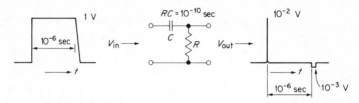

Figure 6.31   An RC differentiating network satisfying $\omega RC \ll 1$ and therefore providing an accurate derivative of the input pulse. The leading edge of the pulse rises at a constant rate from 0 to 1 volt in 10^{-8} sec so that during the rise of the pulse $RC\,dV_{in}/dt = 10^{-2}$ volt. The trailing edge of the pulse falls at a constant rate from 1 to 0 volt in 10^{-7} sec so that during the fall of the pulse $RC\,dV_{in}/dt = -10^{-3}$ volt.

Problems

6.1   An ac source has a frequency $10^6/2\pi$ Hz, an open circuit emf of 10 volts rms, and a source impedance that consists of a resistance of 10^3 ohms in series with an inductance of 10^{-3} henry. If one wishes to transfer the maximum power to a load, what should the impedance of the load be? What actual circuit components will give this impedance? What is the maximum power that can be transferred to the load?

6.2   Is a parallel LC circuit capacitive or inductive for frequencies below the resonant frequency? Above the resonant frequency, is the circuit capacitive or inductive?

6.3   For the circuit shown in Fig. 6.32, calculate the resonant frequency, the impedance at resonance, the quality factor Q, the average power dissipated in the resistor at resonance, and the maximum instantaneous power supplied by the emf at resonance.

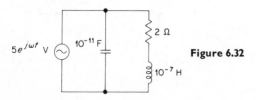

Figure 6.32

6.4   For the parallel resonant circuit shown in Fig. 6.33, what is the load impedance seen by the emf, the source impedance seen by the resistor, the current supplied by the emf, and the average power dissipated in the resistor? Give answers for very low frequencies, very high frequencies, and the resonant frequency.

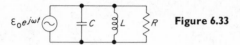

$\varepsilon_0 e^{j\omega t}$ C L R **Figure 6.33**

6.5   A circuit consists of a 10^{-6}-henry inductor, a 10^{-6}-farad capacitor, and a 10^5-ohm resistor all in parallel. What is the impedance of this combination? What is the resonant frequency? What is the impedance of the circuit at resonance? If the circuit is driven by a 1-volt rms source at resonance, what current flows, and what is the average power dissipated in the circuit?

6.6   For the low-pass filter shown in Fig. 6.34, graph the magnitude of the output voltage divided by the input voltage versus the angular frequency ω. Plot the frequency on a logarithmic scale. Also plot the phase angle of the output voltage versus the frequency.

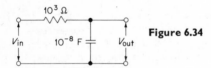

$10^3 \, \Omega$

V_{in} $10^{-8} \, F$ V_{out} **Figure 6.34**

6.7   For the high-pass filter shown in Fig. 6.35, graph the magnitude of the output voltage divided by the input voltage versus the angular frequency ω. Also, plot the phase angle of the output voltage versus the frequency.

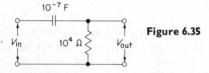

$10^{-7} \, F$

V_{in} $10^4 \, \Omega$ V_{out} **Figure 6.35**

6.8   Sketch an RL filter that has the same frequency response as the low-pass RC filter in Prob. 6.6. Sketch an RL filter that has the same frequency response as the high-pass RC filter in Prob. 6.7.

6.9   For the twin T circuit of Fig. 6.12, suppose $R = 10^3$ ohms and $C = 1.6$

$\times 10^{-6}$ farad. What is the frequency for complete rejection? Plot the magnitude of V_{out}/V_{in} as a function of the frequency. Plot the relative phase between the input and the output as a function of the frequency.

6.10 Derive the balance conditions for the Wien bridge of Fig. 6.22.

6.11 Derive the balance conditions for the ac bridge shown in Fig. 6.36. If the circuit is balanced when $R_1 = 10^5$ ohms, $C_1 = 10^{-10}$ farad, $R_2 = 10^3$ ohms, and $R_3 = 10^3$ ohms, what is the value of R_4 and L_4?

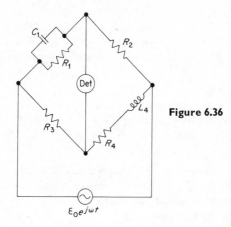

Figure 6.36

6.12 Derive the balance conditions for the ac bridge shown in Fig. 6.37. (*Hint:* Use the Δ-Y transformation to simplify the problem.)

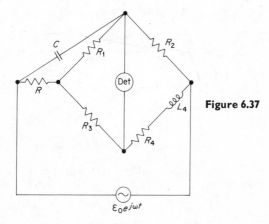

Figure 6.37

6.13 Derive the balance conditions for the ac bridge shown in Fig. 6.38.

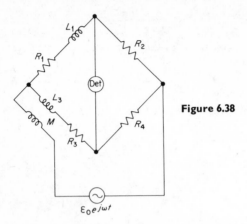

Figure 6.38

6.14 In the discussion of the compensated attenuator we indicated in Fig. 6.25 how a square pulse would be altered if $R_1 C_1 \neq R_2 C_2$. Discuss carefully why the distorted output pulse appears as shown when $C_1 > (R_2/R_1)C_2$ and when $C_1 < (R_2/R_1)C_2$.

6.15 Suppose that the input impedance of a oscilloscope is 470×10^3 ohms in parallel with 10^{-10} farad. What network of resistors and capacitors can produce a compensated attenuator that attenuates a signal by a factor of 5 at the input of the oscilloscope?

6.16 Given the circuit and input voltage shown in Fig. 6.39, plot the output voltage as a function of time. What is the current through the resistor at $t = 4 \times 10^{-5}$ sec? What is the voltage across the capacitor at $t = 10^{-1}$ sec?

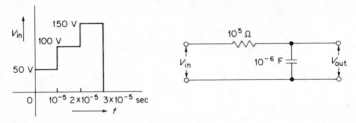

Figure 6.39

6.17 For the circuit shown in Fig. 6.40, what is the magnitude of V_{out}/V_{in}? At very high frequencies, what is the rate of increase of the attenuation for

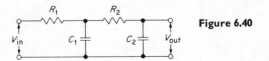

Figure 6.40

this circuit? Express your answer in decibels per decade. If $\omega R_1 C_1 \gg 1$ and if $\omega R_2 C_2 \gg 1$, what is V_{out} in terms of derivatives or integrals of V_{in}?

6.18 A rectangular pulse is 10^{-3} sec in duration. Sketch a circuit that will "differentiate" or RC clip this pulse. The "derivative" should not have a width at half height greater than 10^{-7} sec.

6.19 A square pulse is 10^{-3} sec in duration. Sketch a circuit that will integrate this pulse. The integral should be maintained for 10^{-1} sec without a loss of more than 5%.

6.20 Plot the magnitude of $V_{\text{out}}/V_{\text{in}}$ as a function of frequency for the circuit shown in Fig. 6.41. Given that $R_1 C_1 \gg R_2 C_2$, identify three regions where the change in attentuation per decade change in frequency is constant. What is the rate of change of attentuation at very high frequency? If $\omega R_1 C_1 \ll 1$ and $\omega R_2 C_2 \ll 1$, express V_{out} in terms of derivatives or integrals of V_{in}.

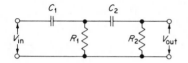

Figure 6.41

6.21 In the circuit of Fig. 6.15, show that for Q large $|I_2/I_1|_{\max} \leq Q$.

7/Fourier Methods and Transmission Lines

In Chapter 4 we discussed the response of simple ac circuits to a sudden transient change in the driving emf and in Chapters 5 and 6, the steady state response to a sinusoidal driving emf. These are especially interesting and important cases but they do not enable us to treat the response of a circuit to an emf that is an arbitrary function of the time. In this chapter we shall outline how one approaches the problem of calculating the current in a circuit driven by an arbitrary emf.

7.1 The Fourier Series Representation of a Periodic Function

In electronic circuits one often encounters emf's and currents that are periodic but that are not sinusoidal functions of the time. The analysis of such a problem may be accomplished by the use of a Fourier series expansion.

A periodic function $f(t)$ is a function that repeats itself indefinitely with some period T. In other words, $f(t) = f(t + nT)$ where n is any integer. For example, $\sin \omega t$ and $\cos \omega t$ are periodic functions with the period $T = 2\pi/\omega$. We shall discuss how to express any periodic function $f(t)$ as a linear superposition of sine and cosine functions of time. If the driving emf for a circuit can be expressed as a linear superposition of sine and cosine functions, then it is a straightforward application of the superposition theorem to express the response of an ac circuit to such an emf.

Fourier's theorem, which we shall not prove, states that a periodic function $f(t)$ with period T can be expanded in a series of the form

$$f(t) = \frac{a_0}{2} + \sum_{n=1}^{\infty} \left[a_n \cos \left(\frac{2\pi nt}{T} \right) + b_n \sin \left(\frac{2\pi nt}{T} \right) \right] \qquad (7.1)$$

If we can find the constants a_0, a_n, and b_n for all n, we have completely described $f(t)$. We shall discuss in a moment how this is done but first we note that the significance of a_0 is easy to see. Let us integrate both sides of Eq. 7.1 over a complete period. We find

$$\int_0^T f(t)\, dt = \frac{a_0 T}{2} \qquad (7.2)$$

Only the term involving a_0 remains because all the sine and cosine terms average to zero over a complete period. Thus $a_0/2$ is just the average value of $f(t)$ or, in electrical terms, the dc component of $f(t)$.

Now just as complex quantities provide great simplifications of ac circuit theory, they are also useful in Fourier series. We shall now express the series of Eq. 7.1 in complex notation. Consider the series

$$f(t) = \sum_{n=-\infty}^{\infty} c_n e^{2\pi jnt/T} \qquad (7.3)$$

where $c_n = \frac{1}{2}(a_n - jb_n)$ is a complex number, $c_{-n} = c_n{}^* = \frac{1}{2}(a_n + jb_n)$, and $c_0 = a_0/2$. Summing the two terms for n and $-n$ we have

$$\frac{1}{2}(a_n - jb_n)e^{2\pi jnt/T} + \frac{1}{2}(a_n + jb_n)e^{-2\pi jnt/T} = a_n \cos \left(\frac{2\pi nt}{T} \right) + b_n \sin \left(\frac{2\pi nt}{T} \right)$$

$$(7.4)$$

Thus Eq. 7.1 and 7.3 really give identical series expressions for $f(t)$. The complex form is frequently more convenient to use. Note, however, that this is not a case of working with a complex quantity of which we eventually use only the real part. Because we put $c_{-n} = c_n{}^*$, the sum of the entire series of Eq. 7.3 is in fact real.

Now let us return to the problem of evaluating c_n (or a_n and b_n). The strategy is somewhat similar to that used in evaluating a_0. To evaluate c_m, for instance, we multiply both sides of Eq. 7.3 by $e^{-2\pi jmt/T}$ and integrate over a complete period. We obtain

$$\int_{-T/2}^{T/2} f(t)e^{-2\pi jmt/T}\, dt = \sum_{n=-\infty}^{\infty} \int_{-T/2}^{T/2} c_n e^{2\pi j(n-m)t/T}\, dt = c_m T \qquad (7.5)$$

since all the terms on the right integrate to zero over a complete period except that one for which $n = m$. This is easily seen if we remember that the exponential can be expressed in terms of sines and cosines which average to zero. We finally obtain

$$c_m = \frac{1}{T} \int_{-T/2}^{T/2} f(t) e^{-2\pi jmt/T} \, dt$$

or in the notation of Eq. 7.1

$$a_m = \frac{2}{T} \int_{-T/2}^{T/2} f(t) \cos \frac{2\pi mt}{T} \, dt \qquad (7.6)$$

$$b_m = \frac{2}{T} \int_{-T/2}^{T/2} f(t) \sin \frac{2\pi mt}{T} \, dt$$

As long as the integration is over a complete period, the starting point is a matter of convenience. Sometimes the range 0 to T will be handier.

Many (but by no means all) of the functions we treat can be classified as even or odd, and for these there are important simplifications of the Fourier series. A function $g(t)$ is even if $g(t) = g(-t)$. A function $u(t)$ is odd if $u(t) = -u(-t)$. Clearly $\cos \omega t$ is even and $\sin \omega t$ is odd. The product of two even functions is even, the product of two odd functions is even, and the product of an even and an odd function is odd. Clearly also the integral of an odd function is zero if the midpoint of the range of integration is at the origin. With these results in mind, let us examine the formulas for the Fourier coefficients, Eq. 7.6. We see that if $f(t)$ is even, the integrand of each of the b_n is odd and all $b_n = 0$. If $f(t)$ is odd, all a_n must equal zero. The average value of an odd function is zero and therefore no odd function can have a dc component. The reader should realize that even and odd functions are rather special and that many functions are neither even nor odd; however, any function can be expressed as the sum of an even function and an odd function.

7.2 The Fourier Analysis of Two Simple Waveforms

As an example of the use of Fourier methods we shall analyze two waveforms that are useful in treating the behavior of a half-wave and a full-wave rectifier. The construction and use of rectifiers will be discussed in Chapter 11. The smoothing and filtering of the output of a rectifier is an important problem in the design of dc power supplies.

Figure 7.1 shows the voltage output of a half-wave rectifier across a resistive load. We can describe the waveform $V(t)$ as follows. For those t for which $\cos \omega t$ is positive we define

$$V(t) = V_0 \cos\left(\frac{2\pi t}{T}\right) = V_0 \cos \omega t$$

and for those t for which $\cos \omega t$ is negative (7.7)

$$V(t) = 0$$

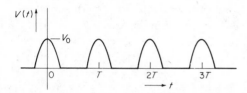

Figure 7.1 The voltage output from a half-wave rectifier connected to a resistive load.

Because $V(t)$ is even, the Fourier coefficients b_n equal zero for all n. We shall calculate the c_n for this series from Eq. 7.6. The final formula will verify that the b_n equal zero. Equation 7.6 gives

$$2c_n = \frac{2}{T} \int_{-T/4}^{T/4} V_0 \cos\left(\frac{2\pi t}{T}\right) e^{-2\pi jnt/T} \, dt \qquad (7.8)$$

We use the exponential form $\cos 2\pi t/T = \frac{1}{2}[e^{2\pi jt/T} + e^{-2\pi jt/T}]$ to simplify the integration. Note that the integration extends only from $-T/4$ to $T/4$ since the integrand is zero over the rest of the period. The integration is straightforward. We outline it briefly below.

$$2c_n = \frac{V_0}{T} \int_{-T/4}^{T/4} (e^{2\pi jt/T} + e^{-2\pi jt/T}) e^{-2\pi jnt/T} \, dt$$

$$2c_n = \frac{V_0}{2\pi} \left(\frac{e^{-2\pi j(n-1)t/T}}{-j(n-1)} + \frac{e^{-2\pi j(n+1)t/T}}{-j(n+1)}\right)\Bigg|_{-T/4}^{T/4} \qquad (7.9)$$

$$2c_n = \frac{V_0}{\pi} (j^n + j^{-n})\left(\frac{1}{1-n^2}\right)$$

In writing the final form we have made use of the fact that $e^{(\pi/2)jn} = j^n$. The factor $(j^n + j^{-n})$ is zero for odd n. It is $+2$ if n is divisible by 4 and -2 if n is divisible by 2 but not by 4. Because the expression for c_n has no imaginary part, we verify that $b_n = 0$ and therefore $2c_n = a_n$. The formula is indeterminate for $n = 1$ but we calculate directly

$$a_1 = \frac{2}{T} \int_{-T/4}^{T/4} V_0 \cos^2\left(\frac{2\pi t}{T}\right) dt = \frac{V_0}{2} \qquad (7.10)$$

We can now write the Fourier expansion of $V(t)$:

$$V(t) = \frac{V_0}{\pi}\left(1 + \frac{\pi}{2}\cos\omega t + \frac{2}{3}\cos 2\omega t - \frac{2}{15}\cos 4\omega t \cdots\right) \qquad (7.11)$$

where $\omega = 2\pi/T$ is the angular frequency of the fundamental (or first harmonic) of the waveform. The output of a half-wave rectifier operated from the 60-cycle lines has in addition to the 60-Hz component ($n = 1$) a dc component and also all the even multiples of 60-Hz, that is, 120-Hz, 240-Hz, and so on.

Now from the superposition principle we know that if the voltage $V(t)$ is applied to a circuit, each of the Fourier components of $V(t)$ produces its own separate current and the total current is the sum of these separate currents. The coefficients in the series expression for the total current are a_n/Z_n where Z_n is the complex impedance for the frequency $n\omega$. Because Z is usually a function of frequency, the current waveform may be quite different from the voltage waveform. Suppose the voltage $V(t)$ is applied to the LR circuit of Fig. 7.2. We ask, "What is the output voltage, $V_{out} = IR$, appearing across the resistor"? The Fourier component $a_n \cos n\omega t$ of $V(t)$ contributes $(V_{out})_n$ to the output voltage where

$$(V_{out})_n = \frac{R}{\sqrt{R^2 + (n\omega L)^2}}\, a_n \cos(n\omega t - \phi_n) \qquad (7.12)$$

and $\phi_n = \arctan(n\omega L/R)$.

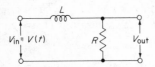

Figure 7.2   An LR circuit to which is applied the output voltage waveform $V(t)$ shown in Fig. 7.1.

Thus we may write for $V_{out} = \sum_n (V_{out})_n$

$$V_{out} = \frac{V_0}{\pi}\left[1 + \frac{\pi}{2}\frac{R}{\sqrt{R^2 + (\omega L)^2}}\cos(\omega t - \phi_1)\right.$$

$$\left. + \frac{2}{3}\frac{R}{\sqrt{R^2 + (2\omega L)^2}}\cos(2\omega t - \phi_2)\cdots\right] \qquad (7.13)$$

For given values of R, L, and ω this series could be summed and the shape of the output waveform determined. We note that the phase factors ϕ_n will introduce sine terms into the expansion and the output voltage will be neither even nor odd.

If $\omega L \ll R$, only the very high-frequency Fourier components of $V(t)$ are attenuated by the LR network. These components are already small since $a_n \propto 1/n^2$ (Eq. 7.9). Thus $V(t)$ and V_{out} have almost the same waveform. If

$\omega L \gg R$, then only the dc term remains and $V_{out} \simeq V_0/\pi$. The LR network has filtered out the ac terms.

Equation 7.13 gives the steady state output voltage. When the input voltage is connected to the filter, there is an initial transient that is damped out with a time constant equal to L/R. It can be shown that the current in the circuit is unidirectional (but not dc) both during the transient and in the steady state. The same holds for the output voltage $V_{out} = IR$. If $L > 0$, the current and output voltage are never zero.

We warn the reader that our results are not a good approximation to the behavior of an actual half-wave rectifier connected directly to an LR filter. In our analysis we have assumed that the voltage source $V(t)$ has zero internal resistance at all times. A half-wave rectifier, on the other hand, is a sinusoidal voltage source in series with a rectifying element. The rectifying element is essentially a highly nonlinear resistance that offers a low resistance to currents in one direction but a high resistance to currents in the opposite direction. The load resistance must have a value intermediate between the forward and reverse resistance of the rectifying element in order to produce the voltage of Fig. 7.1 across the load. Because the voltage source changes sign during the off half cycle, instead of staying at zero as does our $V(t)$, the output current must go to zero at some time during each off half cycle. A problem such as an LR filter connected directly to a half-wave rectifier cannot be treated using linear methods because of the nonlinearity of the rectifying element. However, the voltage across a load resistor R_L, which is driven by a half-wave rectifier can act as a voltage source $V(t)$ for the RL filter if R_L is not too large. The analysis leading to Eq. 7.13 is then appropriate.

The Fourier method works reasonably well for a full-wave rectifier connected directly to an LR filter because the total emf in the circuit is always in the forward direction and the nonlinear element always has a small resistance. A full-wave rectifier across a resistive load provides the waveform shown in Fig. 7.3. It can be written $V(t) = V_0 |\cos(2\pi t/T)|$. A glance at Fig. 7.1 and 7.3 establishes the following interesting relation between the full-wave rectifier voltage $V_{full}(t)$ and the half-wave rectifier voltage $V_{half}(t)$.

$$V_{full}(t) = 2V_{half}(t) - V_0 \cos \frac{2\pi t}{T} \tag{7.14}$$

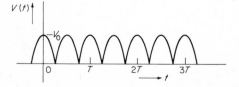

Figure 7.3 The voltage waveform from a full-wave rectifier connected to a resistive load

Thus the Fourier expansion for the full-wave rectified voltage is

$$V(t) = \frac{2V_0}{\pi} \left(1 + \frac{2}{3} \cos 2\omega t - \frac{2}{15} \cos 4\omega t \cdots \right) \qquad (7.15)$$

The frequencies that are present in a full-wave rectified voltage waveform are 0 Hz, $2/T$ Hz, $4/T$ Hz, ..., and so on. This differs from the frequencies present in a half-wave voltage waveform in that the frequency $1/T$ Hz is missing. This result is obvious if one notes that the period of the full-wave voltage is $T/2$, whereas the period of the half-wave voltage is T.

7.3   The Analysis of Nonperiodic Currents and Voltages

Thus far in our study of ac circuits we have treated the transient response of the circuit to a sudden change in the applied emf (step functions) and also the steady state response to periodic emf's. Now we discuss non-repetitive driving emf's that extend over only a finite period of time or at least approach zero sufficiently rapidly for very large times. The emf might for instance be a rectangular pulse as shown in Fig. 7.4 or a finite burst of oscillations or the exponentially decaying pulse from the RC discharge of a condenser. The Fourier series methods can be extended to nonrepetitive phenomena as follows. First we produce a periodic emf by repeating the actual emf periodically with period T where T is such that in each period the emf, whatever its shape, is over in a time τ that is much less than T. Thus our periodic function $f(t)$ is zero or very small for nearly all the period T. The reader's intuition is good if it tells him that under these circumstances it doesn't make much difference what T we choose. We can therefore evaluate the Fourier series in the limit $T \to \infty$ so that the function is actually not periodic at all.

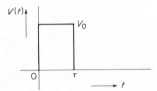

Figure 7.4 A rectangular voltage pulse of length τ.

Consider the formula, Eq. 7.6, for the Fourier amplitudes

$$c_n = \frac{1}{T} \int_{-T/2}^{T/2} f(t) e^{-2\pi jnt/T} \, dt \qquad (7.16)$$

We note that the integrand is not a function of n and T separately but only of the ratio n/T. Thus we can write the exponential as $e^{-j\omega_n t}$ where $\omega_n =$

$2\pi n/T$. The limits of integration are indeed proportional to T but if $f(t)$ is finite only over a time $\tau \ll T$ (or small enough outside the interval τ), then the limits of integration can be extended to ∞ without changing value of the integral. We may write

$$\bar{f}(\omega_n) = c_n T = \int_{-\infty}^{\infty} f(t) e^{-j\omega_n t}\, dt \qquad (7.17)$$

where $\omega_n = 2\pi n/T$ is the angular frequency of the nth Fourier component. From the integral form of $\bar{f}(\omega_n)$ in Eq. 7.17 it is obvious that $\bar{f}(\omega_n)$ is independent of the period T. The separation in frequency of successive Fourier components is given by $\Delta\omega = \omega_{n+1} - \omega_n = 2\pi/T$. As $T \to \infty$, $\Delta\omega \to 0$ and Fourier components of every frequency exist. It is then appropriate to drop the subscript from ω_n. The quantity $\bar{f}(\omega)$, considered as a function of the continuous variable ω, is called the *Fourier transform* of $f(t)$.

In the limit $T \to \infty$, it no longer makes any sense to talk about the amplitude c_n of one Fourier component. This amplitude must go to zero since there are an infinite number of Fourier components even in a very narrow frequency range. We see this from $c_n = \bar{f}(\omega_n)/T$ (Eq. 7.17) where because $\bar{f}(\omega_n)$ is finite, $c_n \to 0$ as $T \to \infty$.

Now let us see what form is taken by the Fourier expansion of $f(t)$ in the limit $T \to \infty$. The Fourier expansion is

$$f(t) = \sum_{-\infty}^{\infty} c_n e^{2\pi j n t/T} \qquad (7.18)$$

We multiply and divide each term by T and substitute $1/T = \Delta\omega/2\pi$. This gives

$$f(t) = \frac{1}{2\pi} \sum_{-\infty}^{\infty} (T c_n) e^{2\pi j n t/T}\, \Delta\omega = \frac{1}{2\pi} \sum_{n=-\infty}^{\infty} \bar{f}(\omega_n) e^{j\omega_n t}\, \Delta\omega \qquad (7.19)$$

where $\Delta\omega = 2\pi/T$ is the frequency separation of successive Fourier components in the sum. As $T \to \infty$ and $\Delta\omega \to 0$, the summation goes into the integral

$$f(t) = \frac{1}{2\pi} \int_{-\infty}^{\infty} \bar{f}(\omega) e^{j\omega t}\, d\omega = \frac{1}{2\pi} \int_{-\infty}^{-\infty} \left[\int_{-\infty}^{\infty} f(t) e^{-j\omega t}\, dt \right] e^{j\omega t}\, d\omega \qquad (7.20)$$

where again we have omitted the subscript from ω_n to show that ω is now a continuous variable. Equation 7.20 gives the Fourier integral expression of $f(t)$, a nonperiodic function, in terms of a continuum of Fourier components. The amplitude function is $\bar{f}(\omega) = \int_{-\infty}^{\infty} \bar{f}(t) e^{-j\omega t}\, dt$. Comparing the Fourier integral expression for $f(t)$ in Eq. 7.20 with the Fourier series of Eq. 7.3 we see that $(1/2\pi)\bar{f}(\omega)\, d\omega$ is the total Fourier amplitude associated with the angular frequency interval $d\omega$.

The variables ω and t have quite different physical meanings but there is a great similarity in the integral expressions for $\bar{f}(\omega)$ (Eq. 7.17) and $f(t)$ (Eq. 7.20). For this reason $\bar{f}(\omega)$ and $f(t)$ are called a *Fourier transform pair*. The function $\bar{f}(\omega)$ is the Fourier transform, or the direct transform, of $f(t)$ and $f(t)$ is the inverse transform of $\bar{f}(\omega)$.

The functions $f(t)$ and $\bar{f}(\omega)$ really give the same information, $f(t)$ in the time domain and $\bar{f}(\omega)$ in the frequency domain. As we have seen in the discussion of ac circuits and of Fourier series, however, it is often much easier to work in the frequency domain.

Our derivation of Eq. 7.20 is not rigorous but the result is generally true if $f(t)$ is sufficiently smooth to be integrated from $-\infty$ to $+\infty$ and if $f(t)$ vanishes fast enough at $\pm\infty$ to assure convergence of the integral. The functions involved in circuit analysis usually satisfy these conditions.

At this point it is useful to compute the Fourier transforms of a few simple functions before appyling the Fourier transforms to circuit analysis. As a first example we compute the Fourier transform of the rectangular voltage pulse $V(t)$ shown in Fig. 7.4. Mathematically $V(t)$ may be expressed as

$$V(t) = V_0 \quad \text{if } 0 \le t \le \tau$$
$$V(t) = 0 \quad \text{for all other } t \tag{7.21}$$

The Fourier transform of $V(t)$ is

$$\bar{V}(\omega) = \int_{-\infty}^{+\infty} V(t)e^{-j\omega t}\, dt = \int_0^\tau V_0 e^{-j\omega t}\, dt$$
$$= -V_0 \left(\frac{e^{-j\omega\tau} - 1}{j\omega} \right) \tag{7.22}$$
$$= 2V_0 e^{-j\omega\tau/2} \left[\frac{\sin(\omega\tau/2)}{\omega} \right]$$

The factor $e^{-j\omega\tau/2}$ is a phase factor and $|\bar{V}(\omega)| = 2V_0 |(\sin \omega\tau/2)/\omega|$ gives the frequency spectrum or the amplitude per unit frequency interval in the Fourier integral representation (Eq. 7.20) of the rectangular pulse. Figure 7.5 shows a plot of $|\bar{V}(\omega)|$. Most of the amplitude (or spectral density) is contained in a range $-2\pi/\tau < \omega < 2\pi/\tau$. If only these frequencies are used, one will obtain a reasonable but by no means perfect representation of the pulse. For instance, if one needs to amplify without serious distortion a pulse 1 μsec long, the amplifier should have a constant amplification (and constant phase shift) for all frequencies from dc to at least 10^6 Hz.

Let us next consider the double pulse shown in Fig. 7.6. Mathematically this double pulse is given by

$$V(t) = +V_0 \quad \text{if } -\tau < t < 0$$
$$V(t) = -V_0 \quad \text{if } 0 < t < \tau \tag{7.23}$$
$$V(t) = 0 \quad \text{for all other } t$$

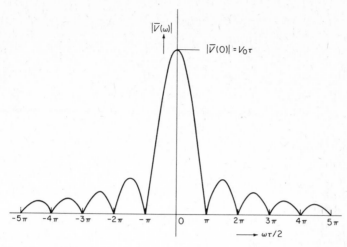

Figure 7.5   The magnitude of the Fourier transform $|\overline{V}(\omega)|$ of the rectangular pulse shown in Fig. 7.4.

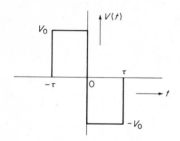

Figure 7.6   A double pulse with equal positive and negative parts.

The Fourier transform of this pulse is

$$\overline{V}(\omega) = \int_{-\tau}^{0} V_0 e^{-j\omega t}\, dt - \int_{0}^{\tau} V_0 e^{-j\omega t}\, dt$$

$$= \frac{2V_0}{j\omega}(\cos \omega\tau - 1)$$

(7.24)

A plot of $|\overline{V}(\omega)|$ is given in Fig. 7.7.

The double pulse with equal positive and negative parts has no dc component and has maxima that fall off as $1/\omega$ for large ω. The maximum intensity of the frequency spectrum of the double pulse occurs near $\omega \simeq 3\pi/4\tau$ and is approximately $|\overline{V}(\omega)|_{\max} \simeq 1.46\tau V_0$. An amplifier that is to amplify the double pulse does not need to be able to amplify dc signals, that is, frequencies near $\omega = 0$. This is occasionally an advantage as ac amplifiers are sometimes easier to build than dc amplifiers. Consequently some amplifiers employ pulse shaping devices that change single pulses into double pulses of opposite polarity.

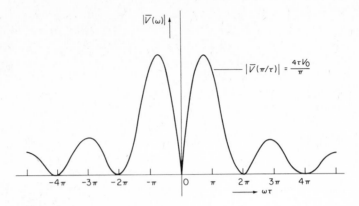

Figure 7.7 The magnitude of the Fourier transform $|\bar{V}(\omega)|$ of the double pulse shown in Fig. 7.6.

As another example, let us calculate the Fourier transform of a signal that is a sinusoidal function from $t = 0$ to $t = \tau$ and zero for times before $t = 0$ and for times greater than τ. Mathematically a function such as this can be expressed

$$V(t) = V_0 e^{j\omega_0 t} \qquad 0 \leq t \leq \tau$$
$$V(t) = 0 \qquad \text{for all other times}$$

The Fourier transform is

$$\bar{V}(\omega) = V_0 \int_0^\tau e^{j\omega_0 t} e^{-j\omega t}\, dt = \frac{V_0}{j(\omega_0 - \omega)} [e^{j(\omega_0 - \omega)\tau} - 1]$$

$$= 2V_0\, e^{j(\omega_0 - \omega)\tau/2} \left[\sin \frac{(\omega_0 - \omega)\tau}{2} \middle/ \omega_0 - \omega \right] \qquad (7.25)$$

It is interesting to note that this is the same frequency spectrum as the one obtained for the rectangular pulse, Eq. 7.22, except that it is centered about $\omega = \omega_0$ rather than $\omega = 0$.

7.4 Some Experimental Results Using a Spectrum Analyzer

Before we calculate other Fourier transforms or use Fourier transforms in circuit problems, we present some experimental results on Fourier transforms. This material will help clarify the nature of Fourier transforms for the reader. We have used a spectrum analyzer, a device which plots out the absolute value of the amplitude of each component of the Fourier series of a periodic voltage which is presented to it. Typically an analyzer might contain a high Q resonant circuit whose natural frequency can be swept over a broad range of frequencies. During the sweep the analyser will respond sequentially to the various Fourier components of the input signal. We illustrate some experiments that provide very convincing demonstrations of the theory we have developed.

In the first of these experiments a burst of eight cycles of a sine wave at 50 kHz was turned on and gated into the spectrum analyzer. This required a total time $\tau = 1.6 \times 10^{-4}$ sec. After the burst was completed, the voltage input to the spectrum analyzer was zero for a time 7τ. The sine wave burst followed by the off time was then repeated time after time. The input to the spectrum analzyer is shown in Fig. 7.8. The output from the spectrum analyzer is shown in Fig. 7.9. The input to the spectrum analyzer is periodic with period 8τ so that the Fourier series contains only frequencies $f = \omega/2\pi$ that are multiples of $1/8\tau$. The spikes show the individual Fourier components.

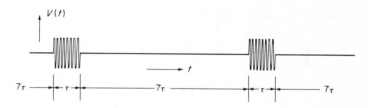

Figure 7.8 The sequence of bursts of sine waves that are used together with a spectrum analyzer to produce the Fourier spectrum shown in Fig. 7.9. The period of $V(t)$ is 8τ.

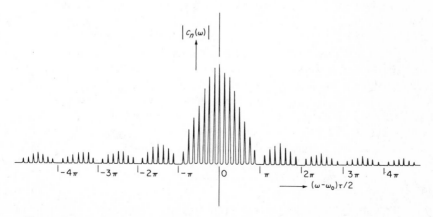

Figure 7.9 The amplitudes $|c_n(\omega)|$ in the Fourier spectrum of the $V(t)$ of Fig. 7.8.

Because the period of the input signal is long, compared to the on time for an individual burst of the 50-kHz sine wave, the spectrum is approaching the Fourier transform. The envelope of the individual Fourier components forms the Fourier transform of an individual burst. The pattern is centered about 50 kHz. The transform is that calculated in Eq. 7.25. This figure clearly shows the relationship between a Fourier series and a Fourier transform. Because the frequency separation of two adjacent individual components of the Fourier series is $\frac{1}{8}$ the frequency $1/\tau$, it is clear that there should be 8 individual Fourier components between successive minima of the Fourier transform. This is correct except that only 7 of the individual Fourier components appear; the eighth Fourier component occurs where the Fourier transform of a single burst is zero; thus that component is missing. The central maximum of the Fourier transform is of course twice as wide as the other maxima, containing 16 individual components of the Fourier series.

Numerically the fundamental frequency of the Fourier series, and therefore the frequency separation of successive components, is $\frac{50}{64}$ kHz. All Fourier coefficients c_n for $\omega_n = 2\pi f_n = 2\pi(8n)\frac{50}{64}$ are zero except for $n = 8$ since at these frequencies the Fourier transform is zero. The dc term $c_0 = a_0/2$ is zero because the average value of $f(t)$ is zero.

Let us summarize our results. We have a burst of sinusoidal oscillations of angular frequency ω_0 that lasts for a time τ. It is followed by a much longer time in which the signal is zero. The pattern of a sinusoidal burst and an off time is repeated over and over again. The burst plus the off time take up a total time T.

The frequencies of the individual components of the Fourier series depend only on T. They are $\omega_n = 2\pi n/T$. The frequency separation of successive Fourier components is $\Delta\omega = 2\pi/T$.

The envelope of the Fourier amplitudes $|c_n|$ is the Fourier transform of an individual burst. The shape of the Fourier transform depends only on τ. Successive zeros of the Fourier transform occur at the frequency intervals $\delta\omega = 2\pi/\tau$. The transform is centered at the oscillation frequency $\omega = \omega_0$.

If τ is varied but not T, the envelope expands or contracts but the frequencies of the Fourier components remain fixed. The amplitudes c_n vary of course since they are determined by the enveloping Fourier transform. If T is changed but not τ, the Fourier components move closer together or farther apart but the envelope remains fixed. The reader should note that there need not be a Fourier component whose frequency is precisely ω_0. This occurred in our example only because ω_0 was an integral multiple of $2\pi/T$. There is also no requirement that the zeros of the Fourier transform coincide with, and eliminate, some of the Fourier components. This occurred in our example only because T/τ was an integer.

Figure 7.10 shows the effect of making the period of repetition very long compared to the time that the burst is on. In this case the individual components of the Fourier series are so close together that the recorder cannot respond to the individual Fourier components and a plot of the output of the spectrum analyzer is simply the envelope of the individual Fourier components. This is of course the Fourier transform of the individual burst.

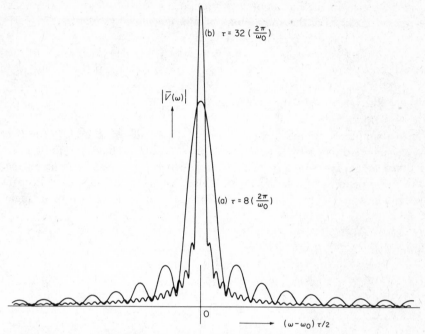

Figure 7.10   (a) The Fourier transform of a sequence of bursts of sine wave such as shown in Fig. 7.8 except that the spacing of the pulses is 255τ instead of 7τ. (b) The Fourier transform of a sequence of pulses like curve (a) except that the number of oscillations in a given burst of sine waves is 32 instead of 8.

The two Fourier transforms plotted on this sheet show the effect of leaving the burst on for a longer time. The wide Fourier transform, labeled (a), was the result of 8 cycles of a sine wave in a burst, whereas the narrower Fourier transform, labeled (b), was the result of 32 cycles of a sine wave per burst. It is obvious that the longer a burst is left on, the more the Fourier components are peaked at ω_0. If the sine wave were on all the time, then the Fourier transform would contain only the frequency ω_0. The discussion of the previous paragraph is useful in clearly understanding these results. The minima of the Fourier transforms should be zero but the response time of the recorder is too slow to show this.

Figure 7.11 shows again the Fourier transform [the curve labeled (a)] for a burst of 8 cycles of a sine wave. The period of repetition is very much greater than the time the burst is on. The curve labeled (b) is the Fourier transform of the two bursts, each 8 cycles of a sine wave lasting a total time τ, separated by a time 3τ. The period of repetition is again very long compared to τ. The pair of bursts is shown in Fig. 7.12. The Fourier transform of the pair of bursts shows interference between the bursts. A similar interference pattern occurs in the Fraunhofer diffraction due to two slits in optics. The Fourier transform of the double burst can be calculated as follows:

$$\overline{V}(\omega) = \int_{-\infty}^{\infty} V(t)e^{-j\omega t}\, dt = V_0 \int_{0}^{\tau} e^{j\omega_0 t}e^{-j\omega t}\, dt + V_0 \int_{4\tau}^{5\tau} e^{j\omega_0 t}e^{-j\omega t}\, dt$$

$$|\overline{V}(\omega)| = 4V_0 \left| \{\cos\,[(\omega - \omega_0)2\tau]\} \left\{ \frac{\sin\,(\omega - \omega_0)\tau/2}{(\omega - \omega_0)} \right\} \right|$$

(7.26)

This is simply the Fourier transform of a single burst of 8 cycles of a sine wave but modulated by a term $2|\cos\,[(\omega - \omega_0)2\tau]|$. Thus the interference pattern should have 4 interference maxima between successive minima of the Fourier transform of a single burst. This is correct; however, every fourth interference maximum falls at a zero for the Fourier transform of a single

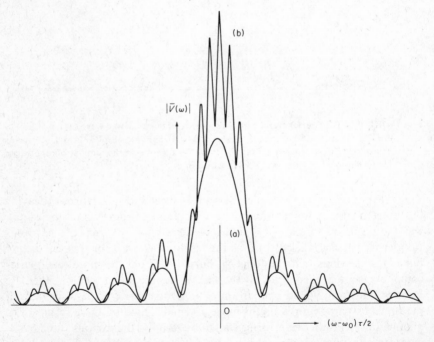

Figure 7.11   (a) The magnitude of the Fourier transform of a burst of eight oscillations of a sine wave. (b) The magnitude of the Fourier transform of the two sine wave bursts shown in Fig. 7.12.

burst and so this maximum is not seen. The central maximum for the Fourier transform of a single burst is twice as wide as the other maxima so that there are 8 interference maxima in it. The results are clearly shown in Fig. 7.11.

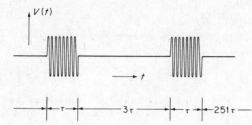

Figure 7.12 Two bursts of sine wave oscillations. The Fourier transform of this waveform is shown as curve (b) in Fig. 7.11.

Note that the interference pattern in Fig. 7.11 [curve labeled (b)] is shifted slightly in frequency from the pattern of the single burst in Fig. 7.11 [curve labeled (a)] because of a drift in the frequency ω_0 from the time when one Fourier transform was recorded to the next. Again the minima of the interference pattern do not quite go to zero because of the slow response time of the recorder.

Let us now return to the calculation of Fourier transforms for simple voltage waveforms. As a final example we briefly consider a damped harmonic oscillation such as occurs in the transient excitation of a series RLC circuit. We present only the formal analysis since we do not have spectrum analyzer results on this waveform. Let the waveform be given by

$$V(t) = 0 \qquad \text{if } t < 0$$
$$V(t) = V_0 e^{-t/\tau} e^{j\omega_0 t} \qquad \text{if } t > 0 \tag{7.27}$$

The Fourier transform is

$$\overline{V}(\omega) = \int_{-\infty}^{\infty} V(t)e^{-j\omega t}\, dt = V_0 \int_0^{\infty} e^{j(\omega_0 - \omega)t} e^{-t/\tau}\, dt$$
$$= \frac{V_0}{j(\omega - \omega_0) + 1/\tau} \tag{7.28}$$

Thus

$$|\overline{V}(\omega)| = \frac{V_0}{[(\omega - \omega_0)^2 + (1/\tau)^2]^{1/2}}$$

$|\overline{V}(\omega)|$ is plotted in Fig. 7.13. At $\omega - \omega_0 = \pm 1/\tau$ the Fourier transform is reduced to $1/\sqrt{2}$ of its maximum value $V_0\tau$, which occurs at $\omega = \omega_0$. In Chapter 5 we discussed the transient and steady state behaviour of the series RLC circuit using slightly different notation. Our present ω_0 is identical to the ω^t of Eq. 5.26. This is not precisely the resonant frequency of the driven RLC circuit although $\omega^\dagger$ is very close to the resonant frequency for a high Q circuit and it is for this reason that we have used the notation ω_0 rather than $\omega^\dagger$. The damping factor γ was used in Eq. 5.26 instead of the decay time τ. We have

$$\gamma = \frac{R}{2L} = \frac{\omega_0}{2Q} = \frac{1}{\tau} \qquad (7.29)$$

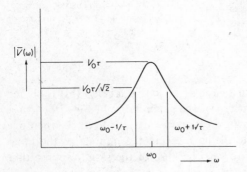

Figure 7.13 The magnitude of the Fourier transform $|\bar{V}(\omega)| = V_0[(\omega - \omega_0)^2 + 1/\tau^2]^{-1/2}$ of the damped sine wave $V(t) = V_0 e^{j\omega t} e^{-t/\tau}$.

Thus our present result $\omega - \omega_0 = \pm 1/\tau$ can be written $(\omega - \omega_0)/\omega_0 = \pm 1/2Q$. It was just this equation (see Eq. 5.31) which determined the half-power frequencies in the steady state response of the driven series RLC circuit. The Fourier transform of a damped oscillation in an RLC circuit contains just those frequencies to which a driven RLC circuit responds. In other words a driven RLC circuit responds with sympathetic oscillations to the same frequencies with which a free damped RLC circuit oscillates. It is now clear why we found in Chapter 5 that the range of frequencies to which a driven RLC circuit responds is equal to one over the lifetime of the response of an RLC circuit to a transient excitation.

Fourier transforms are extremely important. We have seen that one can talk about the frequencies that are present in an arbitrary nonperiodic function of the time. The most general result of our several examples is that the Fourier transform of a pulse of length τ contains frequencies concentrated in the frequency range from dc to $\omega = 2\pi/\tau$ and that the Fourier transform of a burst of oscillations at frequency ω_0 and lasting for a time τ contains frequencies concentrated from $\omega = \omega_0 - 2\pi/\tau$ to $\omega = \omega_0 + 2\pi/\tau$.

Fourier methods have very great generality and power. Any periodic function $f(t) = f(t + nT)$ of physical interest can be expanded in a Fourier series and any nonperiodic function $f(t)$ can be expressed as a Fourier integral provided that $f(t)$ approaches zero at $t = \pm\infty$ rapidly enough so that its Fourier transform converges. The Fourier transform

$$\bar{f}(\omega) = \int_{-\infty}^{\infty} f(t)e^{-j\omega t}\, dt$$

places the entire responsibility for convergence on $f(t)$. The sinusoidal expansion functions themselves, the $e^{j\omega t}$, have amplitudes independent of time. It is also possible to expand a function $f(t)$ in terms of damped sinusoidal waves $e^{j\omega t}e^{-\delta t}$ in which case $f(t)$ may remain finite as $t \to +\infty$ although it must be zero for times $t < 0$. The Laplace transform provides such an expansion. With the Laplace transform it is possible to treat a circuit driven by an emf such as $\varepsilon = \varepsilon_0 \cos \omega t$ for $t > 0$ and $\varepsilon = 0$ for $t < 0$. This emf does not vanish at $t = +\infty$ so that one cannot use a Fourier transform. The Laplace transform has important applications in advanced circuit analysis but we shall not discuss them in this text.

7.5 The Use of Fourier Methods in Circuit Analysis

In Fig. 7.14 we redraw the series LR circuit. It has been one of our favorites and at one time or another we have driven it with a wide variety of emf's, $\varepsilon(t)$. Now finally all we require of $\varepsilon(t)$ is that it possess a Fourier transform. It will come as no surprise to the reader that we can calculate the resulting current. Kirchhoff's voltage law gives us

$$L \frac{dI(t)}{dt} + RI(t) = \varepsilon(t) \tag{7.30}$$

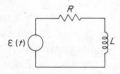

Figure 7.14 A series RL circuit driven by a nonperiodic voltage $\varepsilon(t)$.

We take the Fourier transform of both sides of the equation

$$\int_{-\infty}^{\infty} L \frac{dI(t)}{dt} e^{-j\omega t}\, dt + \int_{-\infty}^{\infty} RI(t)e^{-j\omega t}\, dt = \int_{-\infty}^{\infty} \varepsilon(t)e^{-j\omega t}\, dt$$

or $\tag{7.31}$

$$(j\omega L + R)\bar{I}(\omega) = \bar{\varepsilon}(\omega)$$

where $\bar{I}(\omega)$ and $\bar{\varepsilon}(\omega)$ are the Fourier transforms of $I(t)$ and $\varepsilon(t)$, respectively. In performing the first integral on the left side we have used the fact that the Fourier transform of the derivative of a function is $j\omega$ times the Fourier transform of the function. This can be seen at once upon integrating by parts. We have

$$\int_{-\infty}^{\infty} \frac{dI(t)}{dt} e^{-j\omega t}\, dt = I(t)e^{-j\omega t}\Big|_{-\infty}^{\infty} + j\omega \int_{-\infty}^{\infty} I(t)e^{-j\omega t}\, dt = j\omega\bar{I}(\omega) \tag{7.32}$$

The first term in the partial integration is equal to zero because $I(t)$ is zero at $\pm\infty$ and the second term is just $j\omega\bar{I}(\omega)$. Now let us rewrite Eq. 7.31. We have

$$\bar{I}(\omega) = \frac{\bar{\varepsilon}(\omega)}{j\omega L + R} = \frac{\bar{\varepsilon}(\omega)}{Z(\omega)} \tag{7.33}$$

where $Z(\omega)$ is the impedance of the series LR circuit as a function of ω. This result is almost obvious if we review the Fourier series arguments that led to Eq. 7.12. $\bar{\varepsilon}(\omega)$ is the continuous distribution of Fourier amplitudes that make up the voltage pulse $\varepsilon(t)$. The voltage amplitudes in a narrow range of frequencies $d\omega$ produce a current response whose amplitude is $\bar{I}(\omega)\,d\omega = \bar{\varepsilon}(\omega)\,d\omega/Z(\omega)$. By superposition all of these current responses add up to produce the total current $I(t)$. This superposition is expressed in the inverse Fourier transform

$$I(t) = \frac{1}{2\pi}\int_{-\infty}^{\infty} \bar{I}(\omega)e^{j\omega t}\,d\omega = \frac{1}{2\pi}\int_{-\infty}^{\infty} \frac{\bar{\varepsilon}(\omega)e^{j\omega t}}{Z(\omega)}\,d\omega \tag{7.34}$$

The evaluation of the inverse Fourier transform is usually quite difficult even for a simple circuit and a relatively simple driving function $\varepsilon(t)$. For this reason we have not given an actual example. The situation is often saved by the availability of extensive tables of Fourier transforms that can be used for many circuit problems. It should be clear that the use of Fourier transforms enables one to solve any ac circuit problem however complicated the time dependence of the driving emf may be. Fourier transforms always reduce a circuit problem to the calculation of the Fourier transform of the driving function $\bar{\varepsilon}(\omega)$, the algebraic problem of calculating the Fourier transform of the response function $\bar{I}(\omega) = \bar{\varepsilon}(\omega)/Z(\omega)$, and finally the evaluation of the Fourier inversion integral to obtain the response function $I(t)$ from $\bar{I}(\omega)$.

7.6 Transmission Lines

Although the subject of transmission lines (also called *delay* lines) is not related directly to Fourier transforms, it is a very important subject and some knowledge of Fourier transforms is useful in describing some of the effects that occur with transmission lines. Consequently we shall use the final section of this chapter to discuss transmission lines.

A transmission line is used to carry an electrical signal from one circuit to another. As the frequency of an electrical signal increases, stray capacitive and inductive effects become increasingly important compared to ordinary electrical resistance. Thus if one tries to use simply a copper wire for a lead to carry a signal from one circuit to another, the signal may be altered or even lost due to stray capacity, and so on. Energy will also be lost by radiation into space. Even when this is tolerable, the accompanying signal may be picked up either capacitatively or inductively in parts of the circuit where it is not wanted. This effect is called *intercircuit cross talk*. For these reasons high-frequency signals are often transmitted through coaxial transmission lines (Fig. 7.15). Electrons flow along the inside surface of the outer cylinder and the outside surface of the inner cylinder. All electric and magnetic fields are confined to the space between the cylinders.

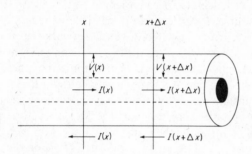

Figure 7.15 A coaxial transmission line. The current I flows in one direction in the inner conductor and returns in the opposite direction in the outer conductor. The voltage V is the potential difference between the inner and the outer conductors. Both I and V are functions of the position x along the transmission line and of the time.

Another important factor in high-frequency circuits is the finite velocity of propagation of electromagnetic signals. This must be explicitly included in the analysis of transmission line behavior. For example, suppose we wish to take a signal from one circuit to another using a transmission line. Suppose further that the signal moves with a speed v on the transmission line. The time for a signal to go from the output of one circuit to the input of the next circuit is $t = l/v$ where l is the length of the transmission line between the two circuits. Now obviously if we are interested in very low frequencies (that is, $\omega \ll v/l$), then the voltage at the output of one circuit is the same as the voltage at the input to the next circuit. As ω becomes comparable to or larger than v/l, however, the voltage at the two ends of the transmission line must be different. We must study the properties of electrical waves moving on a transmission line in order to understand correctly the delays that occur on transmitting a signal on a transmission line. In addition we must find under what conditions the signal is transmitted without being altered severely in its shape. We shall also find out how transmission line delays can be used in shaping short pulses and in introducing a pulse to a circuit at a time when the circuit is ready to process it. Although currents propagate as functions of time along a transmission line, there are no net sources or sinks of charge so that Kirchhoff's current law still holds for the line. In the same way

Kirchhoff's voltage law holds for a transmission line; that is, the sum of the increases in voltage must be equal to the sum of the voltage drops.

For some purposes coaxial transmission lines may be desirable at frequencies of a few kilohertz. They are a necessity in the megahertz region and remain useful into the hundreds or even thousands of megahertz.

Now let us analyze the behavior of the coaxial transmission line of Fig. 7.15. All the situations we have previously considered were assumed to be at such low frequencies that we could isolate our capacitors, inductors, and resistors from one another. This lumped circuit approximation is no longer possible at high frequencies and we must consider the distributed capacitance and inductance of the transmission line. We assume that the line has an inductance per unit length L_0 and a capacitance per unit length C_0. The line is driven at one end by a voltage source. Later we shall discuss the impedance of the source. From Kirchhoff's current law if one has a network that is connected at two terminals to another network, then the instantaneous current flowing out of one of the terminals must be equal to the current into the other terminal. One can break a transmission line at any point and consider it as a two-terminal network. Thus the instantaneous current at any point along the line is the same in magnitude but opposite in direction on the two conductors. The current is not the same at all points along the transmission line, however, since there will be local accumulations of equal and opposite charge on the two conductors. These produce the potential difference between the conductors. The currents and charges on the line are functions of the time. Further let us suppose that the resistance of the conductors is negligibly small and that the resistance between the two conductors is infinite. We assume also that the wavelengths of the Fourier components of the signal are very long compared to the diameter of the cable. This assumption enables us to use Kirchhoff's laws to treat a transmission line.

Now the voltage V between the conductors and the current I carried by either conductor are in general functions both of time and of position along the wire. At a particular time t, consider the voltage change in a very short length of cable Δx. We have $V(x + \Delta x) = V(x) + \Delta V$. Because ohmic drops are negligible ($R = 0$), the voltage drop along the cable $V(x) - V(x + \Delta x) = -\Delta V$ must be due to the distributed self-inductance and a time-changing current. The self-inductance of the element Δx of cable is $L_0 \Delta x$ and we may write $-\Delta V = L_0 \Delta x \, \partial I/\partial t$, or as $\Delta x \to 0$

$$-\frac{\partial V}{\partial x} = L_0 \frac{\partial I}{\partial t} \tag{7.35}$$

Now assume that the current is changing with position so that $I(x + \Delta x) = I(x) + \Delta I$. This means that the charge ΔQ on each conductor in the segment Δx is changing at the rate $\partial \Delta Q/\partial t = -\Delta I$. The capacity of the length Δx

is $C_0 \, \Delta x$. Therefore the time variation of the voltage between the conductors given by $\partial V/\partial t = \partial \, \Delta Q/\partial t (1/C_0 \, \Delta x) = -\Delta I/C_0 \, \Delta x$, or as $x \to 0$

$$\frac{\partial I}{\partial x} = -C_0 \frac{\partial V}{\partial t} \tag{7.36}$$

Equations 7.35 and 7.36 are simply an expression of Kirchhoff's voltage law applied to a segment of the transmission line Δx long. If we differentiate Eq. 7.35 with respect to x and Eq. 7.36 with respect to t and use $\partial^2 I/\partial x \, \partial t = \partial^2 I/\partial t \, \partial x$, we find

$$\frac{\partial^2 V}{\partial x^2} - L_0 C_0 \frac{\partial^2 V}{\partial t^2} = 0$$

and similarly $\tag{7.37}$

$$\frac{\partial^2 I}{\partial x^2} - L_0 C_0 \frac{\partial^2 I}{\partial t^2} = 0$$

The solutions of these equations are very general in form. Direct substitution shows that a solution is $V(x, t) = V_1(x + vt) + V_2(x - vt)$ where V_1 and V_2 are any functions of the variables $x + vt$ and $x - vt$, respectively, and where $v = 1/(L_0 C_0)^{1/2}$. These solutions represent electromagnetic waves moving on the transmission line with a velocity of v. Consider the function $V_2(x - vt)$. This function represents a wave moving toward positive x with the velocity v and is illustrated in Fig. 7.16. The function $V_2(x - vt)$ has the same basic shape at all times but is displaced in position by an amount $x = v\tau$ after a time τ. In a similar fashion $V_1(x + vt)$ represents a disturbance moving toward negative x with a velocity v. The solutions $I(x, t)$ are similar in form.

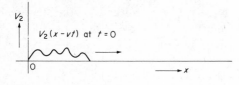

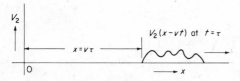

Figure 7.16 The position of a voltage wave V_2 on a transmission line at time $t = 0$ and at a later time $t = \tau$.

It can be shown that the velocity of propagation v is that of an electromagnetic wave in the medium between the two conductors. In Prob. 7.18 the reader is asked to show that the capacitance and inductance per unit length for a coaxial cable are given by $C_0 = 2\pi\varepsilon/\ln (b/a)$ and $L_0 = (\mu/2\pi) \ln (b/a)$ where

a and b are, respectively, the radii of the inner and outer conductors and where ε and μ are the susceptibility and the permeability, respectively, of the material filling the space between the conductors. From these results it follows that $1/C_0 L_0 = 1/\varepsilon\mu = v^2$. Using electromagnetic theory the speed of light in a material can be shown to be $1/\sqrt{\varepsilon\mu}$ so that we have verified the stated result that the velocity of the wave on a transmission line is the speed of light in the material filling the space between the conductors. If the line has air between the conductors, then the velocity of the wave is nearly $c = 1/\sqrt{\varepsilon_0 \mu_0} = 3 \times 10^8$ m/sec, the speed of light in vacuum.

The electromagnetic wave moving on the transmission line has energy stored in both the electric field and in the magnetic field. The energy stored in the electric field in the length Δx of the transmission line is $\frac{1}{2}C_0 V^2 \Delta x$. Similarly the energy stored in the magnetic field in the length Δx is $\frac{1}{2}L_0 I^2 \Delta x$. We shall show (look forward to Eq. 7.38) that $I = (C_0/L_0)^{1/2}V$ for a wave running in the $+x$ direction. This equation together with the energy expressions indicates that the total energy stored in the electromagnetic field of the wave moving on a transmission line is stored half in the electric field and half in the magnetic field.

The transmission lines need not be the coaxial type we have used. For example, two long parallel wires (twin lead) also have a distributed inductance and capacitance and the same analysis applies. Twin lead is extensively used as a transmission line. The fields from such a transmission line extend indefinitely into space, however, and radiation and intercircuit cross talk may be problems unless adequate shielding is used.

Now let us consider the impedance that a line offers to the driving emf. Let us suppose that the source provides a sinusoidal emf of frequency ω and that the voltages and currents on the line are running waves in the $+x$ direction and are given by $V(x, t) = V_0 e^{j(\omega t - kx)}$ and $I(x, t) = I_0 e^{j(\omega t - kx)}$ where $v = \omega/k = 1/(L_0 C_0)^{1/2}$ is the wave velocity. We assume the line is very long and that no reflected wave returns in the opposite direction. If we insert the expressions for $V(x, t)$ and $I(x, t)$ in Eq. 7.36, we obtain

$$-jkI_0 = j\omega C_0 V_0$$

or   (7.38)

$$I_0 = \frac{\omega}{k} C_0 V_0 = \left(\frac{C_0}{L_0}\right)^{1/2} V_0$$

The impedance of the line is

$$Z_0 = \frac{V(0, t)}{I(0, t)} = \frac{V_0}{I_0} = \left(\frac{L_0}{C_0}\right)^{1/2}$$ (7.39)

The impedance Z_0 is real and independent of ω. The line behaves as a resistive load but the energy supplied by the source actually goes into a running electromagnetic wave traveling in the $+x$ direction on the line and is not dissipated as heat. As we have shown, half the energy in the running wave is stored in the electric field and half in the magnetic field. This can go on indefinitely if the line is infinitely long. Because both the line impedance and the wave velocity are independent of ω, the waveform $V(x, t)$ at any point x in the line is a faithful replica of $V(0, t)$, the waveform supplied by the source. Of course it is delayed by the time $t = x/v$. Note that because the line impedance is real, the voltage and current waves are in phase.

Next we consider a number of important properties of lines of finite length. Depending on how the line is terminated an incident wave may be reflected, absorbed, or partially reflected and partially absorbed. Consider first the line of Fig. 7.17 that is terminated by a short between the inner and outer conductor. At the short the voltage $V(x, t)$ between the conductors must always be zero. We shall show how the superposition of an incident wave and an inverted reflected wave keeps the voltage zero for an incident wave of any shape. Figure 7.17 shows two waves. The one on the left is the real wave. The wave on the right is a fictitious wave moving with the same velocity as the real wave but in the opposite direction. The fictitious wave has the same shape as the real wave but the opposite voltage sign. As these two waves pass the shorted end, the sum of the two waves is zero. As the real wave moves past the short, it ceases to exist, and the fictitious wave becomes the real inverted and reflected wave. Obviously the wave solution shown satisfies the wave equation and the boundary condition that $V = 0$ at the short circuit. It is therefore the correct solution to the problem.

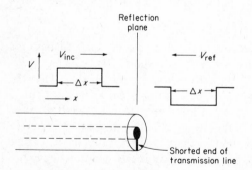

Figure 7.17 A rectangular incident voltage wave V_{inc} approaches the shorted end of a transmission line from the left. V_{ref}, the inverted mirror image of V_{inc}, approaches from the right. As they pass each other, their sum at the short is $V_{\text{inc}} + V_{\text{ref}} = 0$. The sum is zero at any time t and for any shape of the incident wave.

The mutual cancellation of an incident and a reflected wave is used in the process of delay line clipping. Consider the circuit in Fig. 7.18(a). Let us ask how the voltage between terminals A and B will vary as a function of the time after the switch S is closed. The solution is shown in Fig. 7.18(b). The delay line looks like an impedance $Z_0 = (L_0/C_0)^{1/2}$ to an incident voltage wave (Eq. 7.39). Consequently at $t = 0$ the voltage V is divided by the resistor Z_0 and the delay line so that at $t = 0$ the voltage at A is $V/2$. After the voltage wave has propagated down the delay line and been reflected by the short, it propagates back to A and cancels the applied voltage so that the voltage at A is zero for $t > 2l/v$. The function of the source resistor labeled Z_0 in Fig. 7.18 is to prevent reflections at the point A of the wave returning from the short.

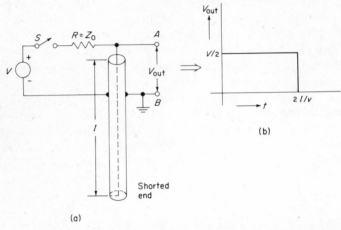

(a)

Figure 7.18   (a) A shorted transmission line of length l and characteristic impedance Z_0 is connected to a voltage source V through a resistor Z_0. At $t = 0$ the switch is closed and the voltage $V_{out} = V/2$ appears across the input to the transmission line. The other half of V appears across the resistor Z_0. After a time $2l/v$ a voltage $V_{ref} = -V/2$ that was reflected from the short appears back at the input. It cancels the input and for all times $t > 2l/v$ the output $V_{out} = 0$. (b) The time dependence of the voltage V_{out} observed across AB.

This will be explained later in this section. A shorted delay line of length l is commonly used to alter a very long pulse into a short pulse of length $2l/v$. For a shorted line the current is reflected without change in sign. That the current is reflected without a change in sign can be seen using the conservation of energy as follows. Consider the superimposed incident and reflected pulses of Fig. 7.17 at the instant in time when exactly half the incident pulse has been reflected from the short; that is, there is $\Delta x/2$ of the incident pulse left and the reflected pulse is $\Delta x/2$ long where Δx is the length of the entire incident pulse before any reflection occurs. The voltages of the incident and the reflected pulses exactly cancel at this instant in time and the total energy stored in the electric field must be zero since V is zero everywhere. At this instant in time the total energy is now entirely stored in the magnetic field. The total energy in the magnetic field is given by $\frac{1}{2}L_0(I_{inc} + I_{ref})^2 \Delta x/2$. By the conservation of energy this must be equal to the energy in the incident pulse, which is $\frac{1}{2}C_0 V_{\text{inc}}^2 \Delta x + \frac{1}{2}L_0 I_{\text{inc}}^2 \Delta x = L_0 I_{\text{inc}}^2 \Delta x$. These expressions for the energy can be the same only if $I_{\text{ref}} = I_{\text{inc}}$.

For a line terminated by an infinite impedance (open circuited), the current is zero at the open circuited end and so the current is reflected with a change in sign. For an open circuited line the voltage is reflected without change in sign.

Let us now try to calculate the reflection of waves from a transmission line terminated by an arbitrary impedance Z. This problem is complex enough that the simple pictorial solutions used for a short circuited or open circuited line (that is, $Z = 0$ or $Z = \infty$, respectively) are not adequate.

We consider a line of finite length l terminated by the impedance Z as shown in Fig. 7.19. Since we must consider both incident and reflected waves, we write the sinusoidal trial solutions

$$I(x, t) = I_{\text{inc}} e^{j(\omega t - kx)} + I_{\text{ref}} e^{j(\omega t + kx)}$$
$$V(x, t) = V_{\text{inc}} e^{j(\omega t - kx)} + V_{\text{ref}} e^{j(\omega t + kx)} \qquad (7.40)$$

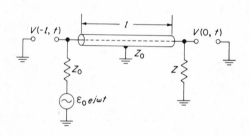

Figure 7.19 A transmission line of length l and characteristic impedance Z_0 is driven by a sinusoidal source whose internal impedance is purely resistive and equal to Z_0. The terminating impedance Z is not necessarily resistive. To simplify certain expressions the origin $x = 0$ has been placed at the termination of the line. Thus $V(-l, t)$ is the voltage at the input and $V(0, t)$ is the voltage at the termination.

Substituting in Eq. 7.36 and using $\omega/k = 1/(L_0 C_0)^{1/2}$ and $Z_0 = (L_0/C_0)^{1/2}$, we obtain

$$-kjI_{\text{inc}} e^{j(\omega t - kx)} + kjI_{\text{ref}} e^{j(\omega t + kx)} = -C_0 j\omega V(x, t)$$

or

$$V(x, t) = Z_0[I_{\text{inc}} e^{j(\omega t - kx)} - I_{\text{ref}} e^{j(\omega t + kx)}] \qquad (7.41)$$

If we explicitly write $V(x, t)$ in terms of V_{inc} and V_{ref} and equate the coefficients of the exponentials describing the incident and reflected waves or if we substitute the trial solutions in Eq. 7.35 instead of Eq. 7.36, we obtain

$$I(x, t) = \frac{1}{Z_0}[V_{\text{inc}} e^{j(\omega t - kx)} - V_{\text{ref}} e^{j(\omega t + kx)}] \qquad (7.42)$$

Now by definition the terminating impedance is $V(x, t)/I(x, t) = Z$ where the voltage and current are to be evaluated at the termination. We have not yet chosen an origin for the x in our trial solutions. It is convenient to place it at the termination, which is where the reflection takes place, rather than at the voltage source. We then have

$$Z = \frac{V(0, t)}{I(0, t)} = Z_0\left(\frac{I_{inc} - I_{ref}}{I_{inc} + I_{ref}}\right)$$

$$= Z_0\left(\frac{V_{inc} + V_{ref}}{V_{inc} - V_{ref}}\right)$$

(7.43)

The two expressions for Z are equivalent. The first comes from using Eq. 7.41; the second, from Eq. 7.42.

An important result of Eq. 7.43 is that if we want the reflected wave to be zero, we must have $Z = Z_0$. Thus reflections from the end of a finite transmission line can be eliminated if it is terminated with a resistor whose value is equal to the characteristic impedance Z_0 of the infinite line. Next we note that if the line is shorted ($Z = 0$), we have $I_{inc} = I_{ref}$ and $V_{inc} = -V_{ref}$. Thus the current is reflected without change of phase, and the voltage is reflected with a 180° change of phase as we stated in the qualitative discussion of the shorted line. Finally if $Z = \infty$, the open termination, we have $I_{inc} = -I_{ref}$ and $V_{inc} = V_{ref}$, which we also stated earlier.

The ability to eliminate reflected waves is very important especially for pulse work where a pulse after several reflections can be easily confused with an independent incident pulse. It is common practice to terminate a transmission line with a resistive impedance equal to its characteristic impedance if one wishes to transport a signal from one circuit to another without the alteration in the waveform that reflections can produce. In the previous discussion we went rather quickly and easily from a discussion of the reflection of a sinusoidal wave to a discussion of pulses. The reason that a pulse is not reflected when a transmission line is terminated by a resistance equal to the characteristic impedance of the line is simply that none of the Fourier components that make up the pulse is reflected.

At this point we discuss why in Fig. 7.19 the internal impedance of the ac driving source was selected to be equal to Z_0, the characteristic impedance of the transmission line. In order to understand this, consider a wave that starts from the generator. This wave runs down the transmission line to the end where it is reflected from the impedance Z. The reflected wave now runs back to the source. At the source this returning wave will be reflected a second time unless the internal impedance of the driving source is Z_0. Thus we see it was because the internal impedance of the generator was chosen to be Z_0 that we can talk simply of an incident wave and a reflected wave. If the internal impedance of the generator had not been Z_0, then the wave traveling from small x to large x would have been made up of the incident wave plus a reflection from the reflected wave, and so on.

Next let us calculate the impedance seen by the driving voltage source when the line of length l is terminated by the impedance Z. Since we have placed the x origin at the termination, the source is at $x = -l$. We must evaluate $Z_{-l} = V(-l, t)/I(-l, t)$. If we use Eq. 7.41, we obtain

$$Z_{-l} = Z_0 \frac{I_{inc}\, e^{jkl} - I_{ref}\, e^{-jkl}}{I_{inc}\, e^{jkl} + I_{ref}\, e^{-jkl}} \tag{7.44}$$

We would like to eliminate the current amplitudes and introduce Z the terminal impedance. This can be done by solving for I_{ref}/I_{inc} from Eq. 7.43 and substituting in Eq. 7.44. This gives

$$Z_{-l} = Z_0 \left[\frac{(Z_0 + Z)e^{jkl} - (Z_0 - Z)e^{-jkl}}{(Z_0 + Z)e^{jkl} + (Z_0 - Z)e^{-jkl}} \right] \tag{7.45}$$

First note that if $Z = Z_0$, then $Z_{-l} = Z_0$; that is, the finite line has the same impedance as the infinite line. Now let us consider the special case of the shorted line $(Z = 0)$. In this case the impedance seen by the source is

$$Z_{-l} = Z_0 \left(\frac{e^{jkl} - e^{-jkl}}{e^{jkl} + e^{-jkl}} \right) = Z_0 j \tan kl \tag{7.46}$$

Suppose l is a quarter wavelength long, then $kl = \pi/2$ and $\tan kl = \infty$. This fact is often used to enable one to connect loosely a transmission line onto another line without altering the system significantly. Next let us consider the special case of an open circuited line. In this case $Z_{-l} = -Z_0 j \cot kl$. Therefore the input impedance of an open circuited line times the input impedance of the same line short circuited is $Z_0{}^2$.

We are now in a position to discuss more quantitatively the recombination of the incident and reflected waves. To calculate the sum of the incident and reflected voltages, we need the ratio V_{ref}/V_{inc} from Eq. 7.43. The ratio is

$$\frac{V_{ref}}{V_{inc}} = \frac{Z - Z_0}{Z + Z_0}$$

so that $\tag{7.47}$

$$V(x, t) = V_{inc} \left[e^{j(\omega t - kx)} + \left(\frac{Z - Z_0}{Z + Z_0} \right) e^{j(\omega t + kx)} \right]$$

If we evaluate $V(x, t)$ at the source $(z = -l)$, we see that there is a phase factor e^{-2jkl} between the incident and the reflected wave. This phase factor is produced by the extra path length $2l$ that the reflected wave must travel. The behavior of the transmission line has been calculated for pure sinusoidal waves but our results are independent of frequency providing the terminating impedance Z is not a function of frequency. Thus a waveform of any shape is reflected without change of shape. In Fig. 7.20(a)–(f) we show the waveforms observed at the source when a step function voltage is applied to the line at $t = 0$. The termination Z is assumed to be purely resistive.

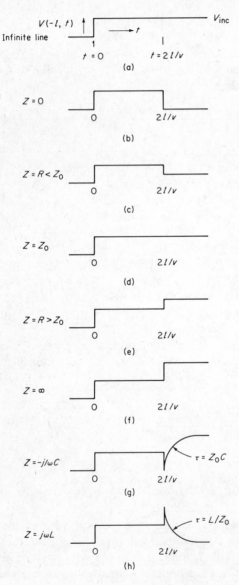

Figure 7.20 In Fig. 7.19 the sinusoidal voltage source is replaced by a dc source $\varepsilon_0 = 2V_{inc}$ that is switched on at $t = 0$. The voltage $\varepsilon_0/2 = V_{inc}$ then appears across the input to the line. It would persist indefinitely as in (a) if no wave is reflected back to the input. Parts (b)–(h) show $V(-l, t)$, the voltage at the input when the finite line of length l is terminated by various impedances Z.

Figure 7.20(g) and (h) shows what happens if Z is a capacitance or an inductance. A capacitance has an impedance $-j/\omega C$ which is zero for very high frequencies and infinite for low frequencies. The sharp leading edge of the step function contains very high-frequency components; thus the leading edge of the step voltage is reflected out of phase with the incident step. The flat top of the step function is essentially dc, however. Thus after a short time the voltage is reflected back in phase. This is illustrated in Fig. 7.20(g). An inductive termination produces just the opposite behavior as shown in Fig. 7.20(h).

Thus far in our discussion an impedance Z at the end of a line has been thought of as a terminating impedance. By this we mean that the energy carried by the incident wave is completely accounted for by reflection and by absorption in Z. Nothing gets beyond the end of the line. In the limiting cases $Z = 0$ and $Z = \infty$ all the energy is reflected; for a resistive termination $R = Z_0$ all the energy is absorbed and for other cases (if the termination is at least partially resistive) there will be partial reflection and partial absorption.

Now a change in the impedance of a line need not be a terminating impedance. Two lines of different characteristic impedances Z_0 and Z_1 may be joined as shown in Fig. 7.21. The junction is shown at $x = 0$. The lines may differ in size or in the material between the conductors or both. If neither line has any resistive losses, then a wave incident on the junction may be partially reflected and partially transmitted but the total energy carried by the incident wave must appear in the reflected plus the transmitted waves.

Now let us consider the appropriate boundary conditions on the currents and voltages at the junction. These are analogous, for instance, to $V_{\text{inc}} = -V_{\text{ref}}$ and $I_{\text{inc}} = I_{\text{ref}}$, which must hold at the end of a shorted line. Consider an infinitesimal length of line dx that includes the junction. We assume a wave is incident from the left. No net charge can build up on dx on either the inner or the outer conductor (because dx is infinitesimal). Therefore Kirchhoff's current law applies and the total current entering the junction is zero. This gives $I_{\text{inc}} + I_{\text{ref}} = I_{\text{tr}}$ where I_{tr} is the current transmitted into the line on the right. Now apply Kirchhoff's voltage law to the path shown in Fig. 7.21 that is of length dx and goes from the outer conductor to the inner conductor of one transmission line and back again in the other line. There can be no voltage drops along the line because the inductance associated with dx approaches zero as $dx \to 0$. Therefore the potential difference between the conductors must be the same in each line or $V_{\text{inc}} + V_{\text{ref}} = V_{\text{tr}}$. These arguments do not contradict our earlier use of Kirchhoff's laws with infinitesimal lengths Δx of transmission line to derive the differential equations relating I and V (Eq. 7.37). What we have shown here is that there can be no finite discontinuities in current or voltage as we pass from one line to the other.

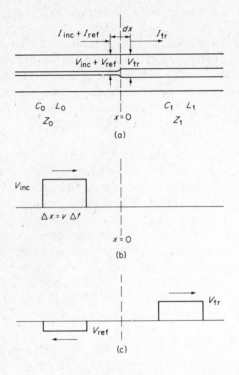

(a)

(b)

(c)

Figure 7.21 (a) The junction of two different coaxial cables and the boundary conditions on the current and ·voltage at the junction. The cable on the left has the smaller inner conductor, therefore $Z_0 > Z_1$ (see Prob. 7.17). (b) A rectangular voltage pulse of length $\Delta x = v\,\Delta t$ approaches the junction from the left. (c) The incident pulse divides into a reflected and a transmitted pulse. Since $Z_0 > Z_1$, the reflected pulse is inverted with respect to the incident pulse. If the dielectric between the conductors is the same in the two cables, the transmitted pulse will have the same velocity as the incident and reflected pulses.

If we divide $V_{inc} + V_{ref} = V_{tr}$ by V_{inc}, we can write

$$1 + R = T$$

or   (7.48)

$$R = T - 1$$

where $R = V_{ref}/V_{inc}$ and $T = V_{tr}/V_{inc}$ are, respectively, the reflection coefficient and the transmission coefficient for the voltage. In Eq. 7.47 we have already calculated $V_{ref}/V_{inc} = R = (Z_1 - Z_0)/(Z_1 + Z_0)$. Note that this calculation is independent of whether the change in impedance of the line from Z_0 to Z_1 is due to a terminating impedance or to joining a new line of characteristic impedance Z_1. All the currents and voltages know is that the ratio $V/I = Z$ must change at a certain point in the line. This is accomplished by superimposing a reflected wave on the incident wave. Now we substitute this value of R in Eq. 7.48 and calculate immediately

$$T = \frac{V_{tr}}{V_{inc}} = \frac{2Z_1}{Z_0 + Z_1}$$ (7.49)

We now have the amplitude of the reflected voltage and of the transmitted voltage in terms of the incident voltage, and we turn our attention to energy considerations. Suppose a rectangular voltage pulse of amplitude V_{inc} and duration Δt is incident from the left in Fig. 7.21. Its length is $\Delta x = v \, \Delta t$ where $v = 1/(L_0 C_0)^{1/2}$ is the velocity of the pulse. The energy in the electric field per unit length of line is $V_{inc}^2 C_0/2$ and for the entire pulse of length $v \, \Delta t$ the energy is $\Delta t V_{inc}^2 C_0/2(L_0 C_0)^{1/2}$. Remembering that $Z_0 = (L_0/C_0)^{1/2}$ this becomes

$$U_E = \frac{\Delta t \, V^2_{inc}}{2Z_0} \tag{7.50}$$

where U_E is the total energy stored in the electric field of the pulse. An equal amount is stored in the magnetic field but to simplify things we shall consider only the electrical half. After the incident pulse has gone through the junction, only the reflected and transmitted pulses remain. Conservation of energy demands

$$\frac{\Delta t V_{inc}^2}{2Z_0} = \frac{\Delta t V_{ref}^2}{2Z_0} + \frac{\Delta t V_{tr}^2}{2Z_1}$$

or $\hspace{8cm}$ (7.51)

$$\frac{V_{inc}^2}{Z_0} = \frac{V_{ref}^2}{Z_0} + \frac{V_{tr}^2}{Z_1}$$

It is important to note that the duration Δt of the pulses in the two transmission lines is the same but their length $\Delta x = v \, \Delta t$ may not be since v may change.

Now if we substitute $V_{ref} = V_{inc}(Z_1 - Z_0)/(Z_1 + Z_0)$ in Eq. 7.51, we may calculate $V_{tr} = V_{inc} 2Z_1/(Z_0 + Z_1)$, the same result we obtained from the boundary condition $V_{inc} + V_{ref} = V_{tr}$. It is reassuring to know that energy is conserved but there are some puzzling things. If $Z_1 \gg Z_0$, then $V_{ref} \simeq V_{inc}$ and $V_{tr} \simeq 2V_{inc}$. Because V_{ref} and V_{inc} are in the same transmission line they have the same length and therefore almost the same energy. That nearly all the energy of the incident pulse appears in the reflected pulse despite the fact that the transmitted pulse has the greater amplitude may seem mysterious. The puzzle disappears when we remember the factors of pulse length and energy unit per length. Their product is proportional to V^2/Z. The transmitted pulse in this case is short (because its velocity is small) or has a low energy per unit length (because $C_1 \ll C_0$) or both.

Now let us briefly consider the common junction of three transmission lines whose characteristic impedances are Z_0, Z_1, and Z_2. A pulse approaches the junction along the line whose impedance is Z_0. A review of our arguments on boundary conditions gives

$$I_{inc} + I_{ref} = I_1 + I_2$$
$$V_{inc} + V_{ref} = V_1 = V_2$$

(7.52)

where I_{inc} and I_{ref} refer to pulses on line Z_0 and I_1, I_2, V_1, and V_2 refer to the transmitted pulses on lines Z_1 and Z_2. If we solve Eq. 7.52 for V_1 and V_2, we obtain

$$V_1 = V_2 = 2V_{inc} \frac{Z_p}{Z_0 + Z_p}$$

(7.53)

where $Z_p = Z_1 Z_2/(Z_1 + Z_2)$ is the impedance equivalent to the parallel combination of Z_1 and Z_2. Note that if Z_1 or Z_2 is infinite, this reduces to Eq. 7.49 for the transmitted voltage at the junction of two lines. Now Eq. 7.53 says that $V_1 = V_2$ is the voltage across Z_1 and Z_2 when their parallel combination is in series with an impedance Z_0 and the series parallel combination is driven by an emf $2V_{inc}$. The temptation to draw an equivalent circuit is, of course, irresistible. We succumb to the temptation in Fig. 7.22. The currents in Z_1 and Z_2 are given by $I_1 = V_1/Z_1$ and $I_2 = V_2/Z_2$. The circuit shows immediately that the current entering the junction is $I_{inc} + I_{ref} = 2V_{inc}/(Z_0 + Z_p) = I_1 + I_2$. It does not give I_{inc} and I_{ref} separately but this can be managed if we remember that $I_{ref} = 0$ when $Z_p = Z_0$. We then have, as expected, $I_{inc} = 2V_{inc}/2Z_0 = V_{inc}/Z_0$.

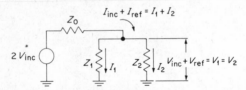

Figure 7.22 An equivalent circuit that provides the currents and voltages at the junction of three coaxial cables. A voltage pulse of amplitude V_{inc} approaches the junction along the cable Z_0. Transmitted pulses V_1 and V_2 enter Z_1 and Z_2 and a reflected pulse V_{ref} returns on Z_0. The voltage transmission and reflection coefficients are easily calculated from the circuit. $T = V_1/V_{inc} = V_2/V_{inc} = 2Z_P/(Z_0 + Z_P)$ and $R = V_{ref}/V_{inc} = (Z_P - Z_0)/(Z_P + Z_0)$ where Z_P is the impedance equivalent to the parallel combination of Z_1 and Z_2.

We conclude this section with a word of caution. In many actual applications of transmission lines either the resistivity of the conductors or the conductivity of the material between the conductors may be appreciable. If this is the case, then signals propagating on the line will be altered by attenuation due to resistive losses, and our simple analysis that ignores these effects must be enlarged to include the resistive losses.

Problems

7.1   Three periodic current waveforms, each of period T, are shown in Fig. 7.23. Calculate the rms value of the current for each waveform and the average power dissipated in a resistor R through which each current flows.

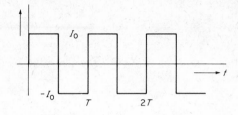

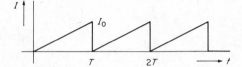

Figure 7.23

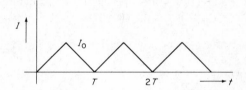

7.2   Calculate the Fourier series expansion for the voltage waveform $V(t)$ shown in Fig. 7.24.

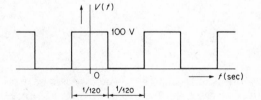

Figure 7.24

7.3   Calculate the Fourier series expansion for the voltage waveform $V(t)$ shown in Fig. 7.25.

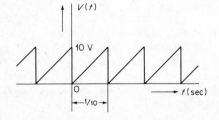

Figure 7.25

7.4   Suppose that the voltage waveform of Prob. 7.2 is applied as the input voltage to the RC network shown in Fig. 7.26. What is the output voltage?

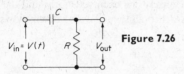

Figure 7.26

7.5   If the voltage waveform $V(t)$ given in Prob. 7.3 is applied to the RL network shown in Fig. 7.27, what current flows?

Figure 7.27

7.6   Suppose the RC network of Prob. 7.4 is such that $RC \ll \frac{1}{60}$ sec. The voltage waveform $V(t)$ of Prob. 7.2 is applied to the input of the RC circuit. Sketch the output voltage of the circuit.

7.7   Calculate the Fourier transform of the voltage pulse $V = V_0 e^{-(t/\tau)^2}$.

7.8   Calculate the Fourier transform of the voltage waveform $V(t)$ of the double pulse shown in Fig. 7.28.

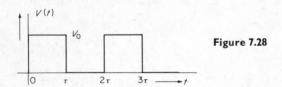

Figure 7.28

7.9   A voltage signal consists of a burst of 100 kilohertz sine wave oscillation which is on for 10 cycles (10^{-4} sec) followed by zero voltage for 99×10^{-4} sec. This signal is repeated indefinitely. Describe in detail the Fourier series of this periodic voltage waveform and describe the relationship of the Fourier transform of the burst of oscillation to the Fourier series of the repeated bursts of oscillation.

7.10   If the voltage pulse given in Prob. 7.7 is to be amplified without serious distortion and if $\tau = 10^{-7}$ sec, what band of frequencies must be amplified by an amplifier?

7.11   Show that any function can be expressed as the sum of an even function and an odd function.

7.12 A coaxial cable has a capacitance of 29.5×10^{-12} farad/ft and an inductance of 7.5×10^{-8} henry/ft. What is the velocity of a wave on this cable?

7.13 If the coaxial cable described in Prob. 7.12 is to be terminated so as to eliminate reflected waves, what impedance should be used?

7.14 A coaxial transmission line has a characteristic impedance of 75 ohms and a capacitance per unit length of 10^{-10} farad/meter. What is the velocity of an electromagnetic wave on the line?

7.15 A rectangular pulse 10 volts high and 10 microsec long is propagating along a 75-ohm coaxial transmission line with a velocity 3×10^8 m/sec. What is the magnitude of the current in the pulse? What is the energy in the pulse?

7.16 The pulse of Prob. 7.15 moves along the transmission line in the positive x direction. A similar but inverted pulse ($V = -10$ volt) moves toward it in the negative x direction. At the moment when they completely superpose, the voltages are zero everywhere. What happened to the energy in the two pulses? Use the conservation of energy to show that $V = Z_0 I$ for a pulse moving in the positive x direction and $V = -Z_0 I$ for a pulse moving in the negative x direction.

7.17 Four 50-ohm coaxial cable transmission lines meet at a junction point. A rectangular voltage pulse 10 volts in magnitude and 10 microsec in length propagates on one of the transmission lines toward the junction. After the pulse reaches the junction, describe the reflected voltage pulse and the voltage pulses transmitted on each of the other three transmission lines. Show that energy is conserved in the process.

7.18 Show that the capacitance and inductance per unit length of a coaxial cable are given by $C_0 = 2\pi\varepsilon/\ln(b/a)$ and $L_0 = (\mu/2\pi)\ln(b/a)$, respectively, as stated in the text. The radius of the inner conductor is a. The radius of the outer conductor is b. (*Hint.* This problem is most easily solved by calculating the electric and magnetic fields in the coaxial cable and then calculating the energy stored in these fields per unit length of cable. L_0 and C_0 are related to the magnetic and electric energy per unit length.)

7.19 A rectangular voltage pulse propagates on a transmission line of characteristic impedance Z_0 that is terminated by a resistor R. Show that the energy of the incident pulse minus the energy of the reflected pulse is the energy dissipated in the resistor.

7.20 A coaxial transmission line is joined to a loop of identical line as shown in Fig. 7.29. A rectangular pulse of magnitude 10 volts and length 10^{-8} sec moves toward the junction. Show that the pulse sequence at B is two pulses each 5 volts high and 10^{-8} sec long separated by a time $\tau = l/v$ where l is the length of the loop of cable. What pulse sequence propagates back toward A as a result of the sequence of reflections and transmissions at the junction?

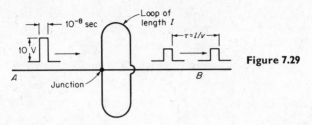

Figure 7.29

7.21 A continuous sinusoidal wave $A \sin \omega_0 t$ has only one Fourier component, namely $A \sin \omega_0 t$. However, it may be considered to be a burst of n waves, that is repeated over and over again. Discuss this waveform in terms of the Fourier transform of a group of n waves and the Fourier series of a function whose period is $T = n2\pi/\omega_0$. Show that the only Fourier component obtained is $A \sin \omega_0 t$.

7.22 Let the input V to the delay line of Fig. 7.18 be a rectangular pulse of magnitude V_0 and duration $\tau = 2l/v$. Show that the output is a double pulse with equal positive and negative parts each of magnitude $V_0/2$ and that the total duration of the double pulse is $2\tau = 4l/v$. The output pulse has the same shape as the pulse of Fig. 7.6.

Sometimes the rectangular input pulse is produced by delay line clipping of a step voltage. The entire process is then called *double delay line clipping*.

7.23 A continuous transmission line or delay line can be approximated by a number of lumped circuit elements as shown in Fig. 7.30. The network is assumed to extend to infinity in the positive x direction and each section consists of a finite inductance $L_0 \Delta x$ and a finite capacitance $C_0 \Delta x$.

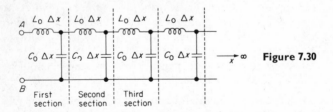

Figure 7.30

Calculate the impedance Z_0 between terminals A and B. This is most easily done by recognizing that when one adds an additional section to the left of A and B, the impedance seen at the new terminals is still Z_0. Show that in the limit $\Delta x \to 0$ one obtains $Z_0 = (L_0/C_0)^{1/2}$. Calculate the ratio of the voltage at the end of the first section to the input voltage. Show that if Δx is very small, the amplitude of this ratio is unity but that the voltage at the end of the first section is retarded in phase by an amount $\phi = \omega (L_0 C_0)^{1/2} \Delta x = \omega \Delta x/v = \omega \Delta t$ where $v = 1/(L_0 C_0)^{1/2}$ is identified as the velocity of the signal and $\Delta t = \Delta x/v$ is the time delay in reaching the output of the first stage. In the limit $\Delta x \rightarrow 0$, the lumped delay line and the continuous delay line obey similar equations. In a lumped delay line, however, the quantities $L_0 \Delta x$ and $C_0 \Delta x$ represent actual inductors and capacitors whose values can be separately chosen. The values are not linked by the properties of the medium between the wires as are the L_0 and C_0 of a continuous line. Thus very small effective wave velocities can be achieved and therefore large delay times in a relatively short line.

7.24 Design a delay line using lumped circuit components (see Prob. 7.23). The delay line is to have a total of 100 sections, a total delay of 10 microsec, and characteristic impedance of 500 ohms. How long would an ordinary delay line have to be to have a delay of 10 microsec?

8/Vacuum Tubes and Nonlinear Circuits

Nearly all the passive circuit components we have discussed in the first seven chapters can be characterized by a constant impedance; in other words the ratio of applied voltage to the current through the component is independent of both voltage and current. Kirchhoff's laws applied to circuits containing emf's and constant impedances lead either to linear algebraic equations or to linear differential equations with constant coefficients. We have treated these equations in some detail and developed a great many generalizations and applications. To a large extent the essential features of linear circuit theory are expressed in the superposition theorem and its consequences.

It is nonlinear components, however, that have made possible the modern electronic circuits used in communication and instrumentation. Kirchhoff's laws remain valid but the second law frequently does not lead to useful equations. Voltages and currents are not proportional. Mathematically intractable current-voltage characteristics are the rule and superposition is lost.

The vacuum tube was the first important nonlinear circuit component and it remains important in a few specialized applications. Because the properties of vacuum tubes are relatively easy to understand, they are useful for a first introduction to nonlinear circuits. In low power, low voltage applications vacuum tubes are now almost completely replaced by solid state devices. These we shall discuss in succeeding chapters.

8.1 Thermionic Emission

Let us digress briefly on the particle nature of matter and electricity and on the distribution of the internal energy of a system among the particles making up the system. Such a discussion must be statistical. We can never measure, even approximately, the energy or momentum of each of the large number of particles that make up a macroscopic system. The number of particles involved is of the order of Avogrado's number, 6.02×10^{23} particles per gram mole. For example, there are 8.5×10^{22} free electrons per cubic centimeter in copper metal.

Let us consider a gas of monatomic particles, for instance helium or neon atoms. The internal energy of the system U is proportional to the absolute temperature T and is stored almost entirely in the kinetic energy of the atoms. The result of a statistical analysis is that

$$U = \tfrac{1}{2}Nm\langle v^2 \rangle = \tfrac{3}{2}NkT \tag{8.1}$$

where N is the number of particles in the system, m is the mass of each particle, $\langle v^2 \rangle$ is the average squared velocity of a particle, and the constant k (called *Boltzmann's constant*) has a magnitude $k = 1.38 \times 10^{-23}$ joule/°K. For diatomic or polyatomic gases, liquids, and solids the internal energy of the system is usually of the order of magnitude of NkT but not necessarily $\tfrac{3}{2}NkT$. For these cases the internal energy is usually partly kinetic and partly potential energy.

We see from Eq. 8.1 that the average energy of a gas molecule in a monoatomic gas is $\tfrac{3}{2}kT$. The quantity kT is a most important parameter. For many systems it is a fairly accurate measure of the internal energy per particle although sometimes this is not true as we shall see when we discuss the conduction electrons in a metal. However, kT is always a good measure of the average energy exchanged when the different particles of a system at temperature T collide or interact with one another.

If a gas is a mixture of particles of different masses, the average kinetic energy of each particle is the same, independent of the mass of that particle. For example, the center of mass of a speck of dust suspended in a gas undergoes a Brownian motion from the random bombardment of gas atoms. The average translational kinetic energy of the dust particle is $\tfrac{3}{2}kT$. Even so macroscopic a quantity as the angular position of a galvanometer coil has associated with it a "kT jitter" that results from the exchange of thermal energy between the coil and its surroundings. Galvanometers can be built whose ultimate sensitivity is limited by just this effect.

We now ask, "What is the distribution of particle energies about the average energy?" To answer this question, we discuss briefly some of the basic ideas of statistical mechanics. This discussion is simpler if we adopt the language and viewpoint of quantum mechanics. We shall assume that interactions among the particles can be neglected compared to much stronger forces that act in the same way on each of the particles. Given the forces that act on a particle or, what is more directly useful, the potential energy of the particle as a function of its position, one can calculate from quantum mechanics the energies of the allowed quantum states. When the particle is confined in a finite volume, there will be only a finite number of allowed quantum states in any finite energy interval. The simplest problem of this type is that of a particle in a box. The box is a region of space in which the potential energy of the particle is constant. Thus the particle moves inside the box without change in momentum or energy. At the walls of the box the potential energy of the particle suddenly increases to a very large value. The wall, therefore, acts as an almost impenetrable barrier that reflects or scatters the particle back into the box. We shall see in the next chapter that the particle in a box provides a very useful model for the behavior of the conduction electrons of a metal.

Now we do not yet have a system of particles. This we achieve by putting a large number N of particles in the box. At any instant each particle must be found in one or another of the allowed quantum states. We are now able to give a statistical description of the system. This is essentially a statement of the average number of particles that will be found in each quantum state.

In statistical mechanics one asks for the most probable distribution of particles among the various quantum states. In finding the most probable distribution one usually permits the system to be in thermal contact with a large heat reservoir at an absolute temperature T. One requires that the total energy of the system and the heat reservoir together be constant and that the number of particles in the system be constant. If the number density of particles is low and the temperature is high, the most probable distribution is very nearly that for which n_r, the average number of particles in the rth quantum state, is given by $n_r = Ce^{-E_r/kT}$ where E_r is the energy of the rth quantum state, k is Boltzmann's constant, and C is a constant depending on the parameters of the particular system being considered. The quantity $e^{-E_r/kT}$ is called the *Boltzmann factor*, and the statistical system described is said to obey Maxwell-Boltzmann statistics. The constant C can be determined by the fact that

$$N = \sum_r n_r = C \sum_r e^{-E_r/kT} \quad \text{or} \quad C = \frac{N}{\sum_r e^{-E_r/kT}} \tag{8.2}$$

If the total number of particles in the system is large, then the fluctuations in the system will never cause the system to deviate substantially from the most probable distribution. In fact if there are N particles in the system, the fluctuations of some quantity about the average value of the quantity will be of the order of $\sqrt{N}$ Thus if $\langle E \rangle$ is the average value of the energy of a single particle in the system, the average total internal energy of the entire system will be $U = N\langle E \rangle$ and the thermal fluctuations in this quantity will be $\Delta U = \sqrt{N}\langle E \rangle$. It is seen that $\Delta U/U = 1/\sqrt{N}$ is very small if N is large.

Since the most probable population of the rth energy level is $n_r = Ce^{-E_r/kT}$, it is obvious that the population of an energy level is very small if $E_r \gg kT$. The average energy of a single particle is obviously not greatly different from $E \simeq kT$, and the average internal energy of a system will be of the order of magnitude $U \simeq NkT$. For a particular system $\langle E \rangle$ may be somewhat larger or smaller than kT. For example, a solid at a high temperature has $\langle E \rangle = 3kT$, and as we have already mentioned, a monoatomic gas has $\langle E \rangle = \frac{3}{2}kT$.

Although the most probable population of the rth level $n_r = Ce^{-E_r/kT}$ is constant in time, the situation is not a static one. A given particle will be in one energy level at one time and another energy level at a later time. The motion of a given particle from one energy level to another is due to interactions between particles. In the cases we shall treat, these interactions can be assumed to be small but they are necessary to ensure the thermal equilibrium. On the average a given particle will spend a fraction n_r/N of the time in the rth energy level of the system. Since energy levels with $E_r \gg kT$ are essentially unpopulated, a given particle will have fluctuations in its energy that are comparable to kT.

The Maxwell-Boltzmann statistics are often useful but they give only an approximation to the actual distribution functions. In particular they do not apply to a dense electron gas such as the electrons in a metal. In treating the electrons in a metal one must treat the electrons as indistinguishable particles, and because of the Pauli exclusion principle one must permit no more than two electrons, one for each value of the spin, in a given energy level. Combining these requirements together with considerations such as discussed for the Maxwell-Boltzmann statistics, one can calculate the most probable distribution of electrons among the quantum energy levels of a system. The most probable distribution of electrons among the quantum energy levels is given by the Fermi-Dirac distribution function.

$$f = \frac{1}{e^{(E-E_F)/kT} + 1} \tag{8.3}$$

where E_F is a quantity called the *Fermi energy level* of the system. The Fermi level is a function of T, of N, and of the distribution in energy of the quantum states of the system under consideration. We have assumed that each spin level is a different state so that the maximum value of $f = 1$. If we had allowed two electrons per state, then the maximum value of f would have been 2. At the absolute zero of temperature, $f = 1$ if $E < E_F$ and $f = 0$ if $E > E_F$. At a finite temperature $f \simeq 1$ for $E \ll E_F$, $f \simeq 0$ for $E \gg E_F$, and the distribution function f changes from $f \simeq 1$ to $f \simeq 0$ in a band of energies approximately equal to kT in width and centered at E_F. Figure 8.1 shows both the Maxwell-Boltzmann distribution function and the Fermi-Dirac distribution function as functions of the energy at a given absolute temperature T.

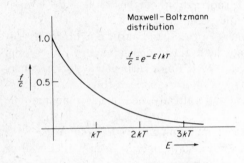

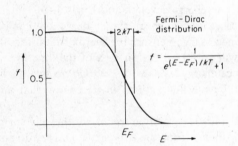

Figure 8.1 The Maxwell-Boltzmann and Fermi-Dirac distribution functions.

At finite temperatures if $(E - E_F)/kT$ is greater than 4 or 5 then $e^{(E-E_F)/kT} \gg 1$ and the 1 in the denominator of Eq. 8.3 can be ignored. Under these conditions a good approximation to the Fermi-Dirac distribution is

$$f \simeq e^{-(E-E_F)/kT} = e^{E_F/kT} e^{-E/kT} \simeq C e^{-E/kT} \qquad (8.4)$$

where C is the constant in Eq. 8.2. It can be shown that $e^{E_F/kT} \simeq C$ if $e^{(E-E_F)/kT} \gg 1$. Thus under the conditions stated the Fermi-Dirac distribution goes over into the Maxwell-Boltzmann distribution. This occurs because $f \simeq 0$ for $e^{(E-E_F)/kT} \gg 1$ and it is not necessary to worry about limiting the population of a given level by the Pauli exclusion principle. For an electron gas E_F depends on the density of the electrons and on the temperature. If the density is sufficiently low and the temperature sufficiently high $E_F \ll kT$.

Thus if E is a few times kT the condition $e^{(E-E_F)/kT} \gg 1$ is satisfied and for such energies E a low density electron gas obeys Maxwell-Boltzmann statistics. We stress the fact that the really fundamental difference between the Maxwell-Boltzmann distribution and the Fermi-Dirac distribution is imposed by the Pauli exclusion principle, which, no matter how low the temperature or how high the number density of particles, permits no more than one particle per energy state.

In the preceding discussion we have presented some of the results of statistical mechanics that are needed for an understanding of the behavior of electrons in metals and semiconductors. The next two chapters amplify the discussion but here we limit ourselves to the thermionic emission of electrons from heated metals. The conduction electrons in a metal have a Fermi-Dirac distribution of energies. The most energetic electrons still present in appreciable numbers have approximately the Fermi energy E_F. Beyond this the distribution drops exponentially. To remove an electron of energy E_F through the surface of the metal requires an energy ϕ, called the *work function* of the metal. The magnitude of ϕ varies from about 1.5 eV (electron volts) for certain barium and strontium coatings on nickel to 4.5 eV for pure tungsten. The electron volt is a convenient unit of energy when dealing with individual electrons. It is the kinetic energy gained by an electron falling through a potential difference of 1 volt. From $E = QV$ and the electronic charge $e = 1.6 \times 10^{-19}$ coulomb we have 1 eV $= 1.6 \times 10^{-19}$ joule. At room temperature (300°K) kT is 0.026 eV or about $\frac{1}{40}$ eV. For metals E_F is a few electron volts. Electrons that can escape from the metal must have an energy E such that $E - E_F > \phi$. Their number can be shown to be proportional to $e^{-\phi/kT}$.

The complete expression for the current density leaving a heated metal surface can be shown to be given by Richardson's equation

$$j = AT^2 e^{-\phi/kT} \tag{8.5}$$

where A is a constant. For commonly used thermionic emitters ϕ/kT may be about 10 at the operating temperature. Thus the exponential dominates the temperature dependence and the current rises very rapidly with increasing temperature. It is possible to achieve emission currents of a few amperes per square centimeter of filament surface. The thermionic emission of electrons from a heated metal is analogous to the evaporation of molecules from a liquid. The work function is the latent heat of evaporation per electron.

8.2 Vacuum Tubes

The simplest tube is the vacuum diode. It consists of a hot cathode which emits electrons and a nearby plate or anode which, when at a positive potential with respect to the cathode, collects some or all of the electrons emitted by the cathode. If the cathode and plate are at the same potential, the emitted electrons form a cloud of space charge close to the cathode. Within the space charge there is an electric field tending to return electrons to the cathode. This is the fate of most of them. Only a small fraction reach the plate. The vacuum diode is enclosed in an evacuated container so that collisions of the electrons with the background gas do not occur often enough to affect the behavior of the diode.

As the plate to cathode potential difference is made positive, how-ever, more electrons reach the plate. This current which is essentially the rate at which the field of the plate can snatch electrons from the space charge is described by Child's law.

$$I = CV^{3/2} \qquad (8.6)$$

Here C is a constant and V is the potential difference between the plate and the cathode. This current must not be confused with that described by Richardson's equation. The current from Child's law is the difference be-tween the total thermionic emission, given by Richardson's equation, and the current returning to the cathode from the space charge. As V becomes larger, the space charge is depleted and the current to the plate falls below that predicted by Child's law. Eventually the current saturates at the Richard-son's equation value. Further increases in potential do not increase the current. Figure 8.2 shows the current-voltage curve for a diode. Currents in the $V^{3/2}$ region are said to be *space charge limited*, those in the saturation region are *temperature limited*. The constant C in Child's law depends on the geometry of the diode but whatever the geometry, C increases as the distance between the plate and cathode decreases. Thus, for a given V, larger currents reach the plate if it is placed close to the cathode. Most vacuum tubes are operated in the space charge limited region although if high energy electrons are needed, as in an X-ray tube, the current from the filament may be tem-perature limited.

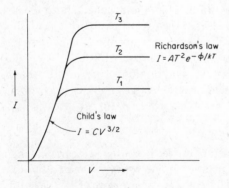

Figure 8.2 The current-voltage charac-teristics of a diode at three different cathode temperatures, $T_3 > T_2 > T_1$. At sufficiently high plate voltage the current saturates at the value given by Richardson's equation.

The $I = CV^{3/2}$ relationship of Child's law is most easily understood if we treat a one-dimensional model of the diode. Let the cathode and plate be parallel planes each perpendicular to the x axis. We assume that the electric field and all electron velocities are perpendicular to the two planes and therefore along the x axis. Very close to the cathode, as we have mentioned, the electric field pushes electrons back toward the cathode. Closer to the plate the field reverses direction and electrons are accelerated toward the plate. Thus in between there must be a position of zero field and a maximum in the electron potential energy (a minimum of the electrostatic potential). In most cases this minimum is close to the cathode. The potential of the cathode is only a few times kT/e greater than the minimum. We therefore confine our attention to the region between the potential minimum and the plate. Nearly all the potential drop across the tube occurs in this region and all the electrons in this region move toward, and eventually reach, the plate.

Let us place the potential minimum at $x = 0$ and the plate at $x = d$. We shall be interested in $V(x)$, the potential; $\rho(x)$, the charge density; and $v(x)$ the electron velocity, all as functions of x. We put $V(0) = 0$. Since the electric field E is zero at $x = 0$, we have also $-(dV/dx)_{x=0} = E(0) = 0$. $V(d)$ is the potential drop across the tube which appears in Child's law. Now the current that passes through any plane perpendicular to x is a constant. We may write $I = \rho(x)v(x)$. Thus I is the current at any point in the tube and also the current that reaches the plate. We omit multiplicative constants, for instance an area factor multiplying $\rho(x)(v(x))$, since we are not interested in the actual value of C in Child's law.

The equation $I = \rho(x)v(x)$ is the central equation in the discussion. It is important to introduce $V(x)$ and this can be done by noting that since the work done by the field appears as the kinetic energy of the electrons, we have $v(x) \propto V^{1/2}(x)$. In addition $\rho(x) = -\varepsilon_0 \, d^2V/dx^2$. This follows if we apply Gauss's law to a surface consisting of two parallel planes each of unit area, one at x and one at $x + dx$. The surface integral of the field is $E + dE - E = dE$ and the volume integral of the charge is $\rho(x) \, dx$. Thus $dE = \rho(x) \, dx/\varepsilon_0$ or $dE/dx = -d^2V/dx^2 = \rho(x)/\varepsilon_0$. Our fundamental equation now reads $I = V^{1/2}(x)d^2V(x)/dx^2$ where we ignore as before the constants of proportionality such as $-\varepsilon_0$. This is a nonlinear second-order differential equation for $V(x)$. The solution is $V(x) = V_0 x^{4/3}$ where V_0 is a constant. Direct substitution in the differential equation gives $I = \frac{4}{9}V_0^{3/2}$. The boundary conditions at $x = 0$ are also satisfied by $V(x) = V_0 x^{4/3}$. Since $V(d) = V_0 d^{4/3}$, we have finally $I = (2/3d)^2 V^{3/2} (d)$. This is Child's law. No numerical significance should be attached to the constant $(2/3d)^2$ since we have neglected constants of proportionality in our derivation.

The solution also makes clear why moving the plate closer to the cathode permits one to draw the same plate current at a smaller plate voltage.

If the plate is put at some $x < d$ and the plate voltage is reduced to $V(x)$, then no changes in $v(x)$ or $\rho(x)$ occur and the same current still flows.

From our solution we see that $v(x) \propto V^{1/2}(x) = V_0^{1/2} x^{2/3}$ and $\rho(x) = -\varepsilon_0 \, d^2 V(x)/dx^2 = -\varepsilon_0 \frac{4}{9} V_0 x^{-2/3}$. Thus it is obvious why $I = \rho(x)v(x)$ is independent of x. At $x = 0$ the electron velocity goes to zero and therefore the charge density must become infinite. Physically this is distressing. It results from our assumption that at any x all electrons have the same velocity. In a real diode electrons leave the cathode with a distribution of velocities. The more energetic electrons surmount the potential energy maximum at $x = 0$ and go on to provide the current to the plate. Our derivation presents an essentially accurate description of the source of the $V^{3/2}$ dependence, however.

The vacuum diode is useful because a current flows when the plate to cathode voltage is positive but no current flows when the plate to cathode voltage is negative. The plate itself emits no electrons because it is cold. Electrons emitted by the cathode must return to the cathode when the plate to cathode voltage is negative. A device that conducts a current when the voltage across the device has one sign but does not conduct a current for the opposite sign voltage is called a *rectifier*. The uses of rectifiers are discussed in Chapter 11.

Further capabilities of vacuum tubes are realized when additional electrodes are introduced. In Fig. 8.3 we illustrate a diode and also the introduction of a grid to form a triode. The cathode of a low power triode is usually a thin nickel tube, coated with oxides to reduce the work function and heated internally by a tungsten filament. The cathode is surrounded by a grid that is an open mesh of wire and, finally, by the plate. Because it is close to the cathode a given potential on the grid affects the current drawn from the cathode more than does the same potential on the plate. But because the grid is an open mesh, most of the current leaving the cathode goes through the grid and reaches the plate. Thus the triode can be used as an amplifier. Since the current to the grid is small, the power in the grid circuit is small. The grid circuit can control the much larger power expended in the plate circuit, however.

Triodes may be given a variety of current-voltage characteristics depending on the size and relative spacing of the electrodes. The characteristics are usually presented as curves of plate current versus plate to cathode potential difference for each of a range of grid to cathode potential differences. Figure 8.4 shows such curves for the 6SN7, a small general-purpose twin triode (two triodes in the same vacuum envelope) once widely used in TV sets.

When the grid to cathode voltage is zero, the dependence of the plate current on the plate to cathode voltage is quite close to that predicted by Child's law. As the grid to cathode voltage is made negative, higher positive plate to cathode voltages are necessary before the plate current begins to increase. If both the grid to cathode and plate to cathode voltages are given, however, the field near the cathode and consequently the plate current is determined. The grid current is also determined but it is usually negligible if the grid is negative. Current to a grid that is negative with respect to the cathode can arise from positive ions provided by residual gas in the tube, from photoelectrons ejected by photons that strike the grid, or of course from leaky insulation. Where ac signals are involved, the capacitive coupling of the grid to other electrodes will also produce a grid current.

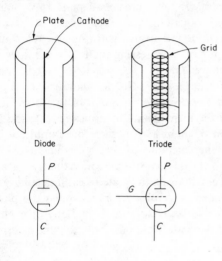

Figure 8.3 Cutaway views and the standard circuit symbols for a diode and a triode.

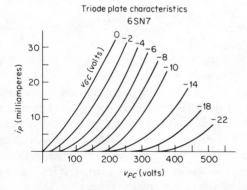

Figure 8.4 Average plate current-plate voltage characteristics of a 6SN7 triode. The plate to cathode and grid to cathode voltages are v_{PC} and v_{GC} respectively. (*Characteristic curves courtesy RCA Corporation.*)

Of the two voltages and two currents that can be associated with the three terminals of a triode any two may be chosen as independent variables. The customary and convenient choices are the plate to cathode voltage and the grid to cathode voltage. The dependence of the currents on the voltages is highly nonlinear and the general theorems of linear circuit analysis do not apply.

Before proceeding further we must discuss the notation we shall use in the remainder of the text. In electronic circuits it is often important to treat the voltages and currents as made up of two parts, a large dc or quiescent current or voltage on which is superimposed a small incremental current or voltage. We shall use a small letter with a capital subscript as a total quantity. Thus v_{PC} is the total plate to cathode voltage. We shall use a capital letter with a capital subscript as the large dc voltage so that V_{PC} is the dc plate to cathode voltage. An incremental quantity will be represented by a small letter and a small subscript. Thus v_{pc} is the incremental plate to cathode voltage. From the definitions it is obvious that $v_{PC} = V_{PC} + v_{pc}$.

Now let us assume that a triode is operated with total currents and voltages that vary by only small amounts. Expressing the total quantities in the notation just described we have $i_P = I_P + i_p$, $v_{PC} = V_{PC} + v_{pc}$, and $v_{GC} = V_{GC} + v_{gc}$ where i_p, v_{pc}, and v_{gc} are small. The dc quantities, I_P, V_{PC}, and so on, are called the *operating point* or *quiescent point* of the tube. In a great many electronic circuits only the incremental quantities i_p, v_{pc}, and so on, carry the signal or information of interest. It is therefore important to develop equations which relate the incremental variables but which avoid as much as possible the involvement of the dc quantities. From a purely mathematical viewpoint this is easily done. Choosing the plate to cathode and grid to cathode voltages as the independent variables we may express the plate current and grid current as a Taylor series expansion around any operating point (fixed values of V_{PC} and V_{GC}) we wish. Retaining only the linear terms we have

$$i_p = \left(\frac{\partial i_P}{\partial v_{GC}}\right)v_{gc} + \left(\frac{\partial i_P}{\partial v_{PC}}\right)v_{pc}$$

$$i_g = \left(\frac{\partial i_G}{\partial v_{GC}}\right)v_{gc} + \left(\frac{\partial i_G}{\partial v_{PC}}\right)v_{pc}$$

(8.7)

where the partial derivatives are evaluated at the operating point selected. The partial derivatives depend on the structure of the tube but it must be remembered that for a given tube the partial derivatives depend also on what operating point is chosen. If this were not so, higher derivatives would be zero and the tube would be a linear component. Finally we note that because Eq. 8.7 neglects the higher-order terms in the Taylor expansion, it is accurate only for sufficiently small values of the incremental variables.

Despite these limitations Eq. 8.7 are the basic equations for the treatment of the incremental variables. The equation for i_g is seldom needed because both the total grid current i_G and the incremental grid current i_g are very small. Special names and symbols are given to the partial derivatives in the equation for i_p. Using these symbols the first of Eq. 8.7 may be rewritten

$$i_p = g_m v_{gc} + \frac{1}{r_p} v_{pc} \tag{8.8}$$

The quantity $g_m = \partial i_P / \partial v_{GC}$ is the grid-plate transconductance, a transconductance because it relates to a current in one branch to a voltage in a different branch. The quantity $r_p = (\partial i_P / \partial v_{PC})^{-1}$ is the plate resistance. These are called the *small signal parameters* of the tube. Another useful parameter is the amplification factor μ which is given by the ratio of the change in plate voltage to the change in grid voltage which leaves the plate current unchanged. If we put $i_p = 0$ in Eq. 8.8, we have

$$\mu = -\frac{v_{pc}}{v_{gc}} = g_m r_p \tag{8.9}$$

The amplification factor is a measure of the relative efficiencies of the grid and the plate in controlling the current from the cathode. Because this is determined mainly by the positioning of the electrodes, the value of μ is relatively insensitive to the values of the dc current and voltage or the operating point.

The manufacturer will give values of the small signal parameters for a typical operating point and sometimes also curves showing their dependence on i_P. These values can also be estimated from the i_P versus the v_{PC} characteristic curves. Referring to Fig. 8.4 we note that along a horizontal line ($i_p = 0$) a change in v_{GC} of 2 volts corresponds to a change in v_{PC} of about 40 volts. Thus for the 6SN7, $\mu \simeq 20$. The reciprocal of the slope of a characteristic curve ($v_{gc} = 0$) is r_p. We see that r_p increases as i_p decreases and as v_{PC} increases. A typical value is $r_p \simeq 7000$ ohms. The corresponding value of the transconductance is $g_m = \mu / r_p \simeq 2900$ micromhos. A mho is an inverse ohm.

Many of the properties of the triode can be improved by the use of additional grids. In a pentode two additional grids are inserted leading to the five-electrode structure shown in Fig. 8.5. The control grid exerts the primary control on the tube current. It has the same function as the single grid in a triode. The screen grid is kept at a fixed positive potential with respect to the cathode. One of its functions is to shield the cathode from the electric field of the plate. Field lines from the plate terminate on the screen and on the suppressor grid and thus the plate current is almost independent of the plate voltage. Put another way, the pentode can have a very high amplification factor and a very high plate resistance. To understand the role of the suppressor grid we must say a word about secondary electron emission. When an electron hits a metal surface, for instance, the plate of a vacuum tube, it may eject one or more secondary electrons, and these may have an appreciable fraction of the energy of the incident electron. In a triode the secondaries must return to the plate since it is the only positive electrode available. In a pentode they could reach the positive screen grid. This would introduce undesirable anomalies in the plate current, particularly at low plate voltages. The suppressor grid, however, if it is kept at or near the cathode potential, forces the secondaries back to the plate and eliminates this problem.

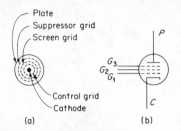

Figure 8.5 (a) Top view of a pentode. (b) The standard circuit symbol for a pentode. The control, screen, and suppressor grids are frequently labeled G_1, G_2, and G_3 in the order of their separation from the cathode.

One might ask why a single additional grid at cathode potential could not serve as both a screen and a suppressor. If this were attempted and if this grid were held at the cathode potential, then the only field at the cathode would be that provided by the control grid and no appreciable plate current would flow except with the control grid positive. This mode of operation is undesirable. Grid currents flow and power must be expended in the grid circuit. Thus we see that an important function of the positive screen is to provide an accelerating field at the cathode and therefore a large plate current even with a negative control grid.

In Fig. 8.6 we show the characteristic curves of the 6AU6, a once popular small pentode. The curves shown are with the screen at 100 volts and the suppressor at 0 volt (connected to the cathode). It is obvious that the plate current is almost independent of the plate to cathode voltage. This implies that r_p and μ are large. Typical values for the 6AU6 are $r_p = 10^6$ ohms, $g_m = 5200$ μmhos, and $\mu = 5200$. We point out that on characteristic curves such as appear in Fig. 8.4 and 8.6 the tube manufacturers often label the plate to cathode voltage v_{PC} and grid to cathode voltage v_{GC} simply as the plate voltage v_P and grid voltage v_G, respectively. It is assumed that voltages are measured with respect to the cathode.

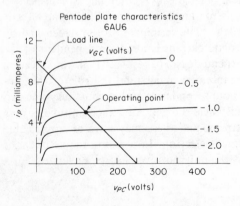

Figure 8.6 Average plate characteristics of the 6AU6 pentode. The screen grid is held at +100 volts; the suppressor grid, at 0 volt. All voltages are measured with respect to the cathode. (*Characteristic curves courtesy RCA Corporation.*)

8.3 Vacuum-Tube Circuits

Let us consider a simple one-tube amplifying circuit. The principles illustrated are those we shall use later in transistor circuits. Figure 8.7 shows a 6AU6 pentode connected in a circuit with external plate, cathode, and grid resistors R_P, R_C, and R_G and a constant plate supply voltage V. Typically V might be 200 or 300 volts. Connections for the screen and suppressor grids are shown but because these grids are held at fixed voltages they do not enter in an analysis of incremental quantities. For an incremental analysis, Fig. 8.7 could also represent a triode amplifier.

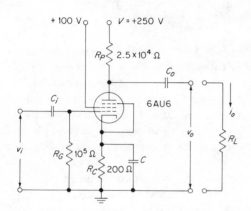

Figure 8.7 A single tube voltage amplifier using the 6AU6 pentode.

A small incremental signal v_i is applied between grid and ground across the resistor R_G. The incremental output voltage v_o is measured between the plate and ground. If the input signal v_i contains no very low-frequency components, it is usually advantageous to use the blocking capacitors C_i and C_o shown in the input and output circuits. These must be large enough so that very little impedance is offered to the signal voltages. These capacitors block dc voltages at the input from the grid and prevent the dc plate to ground potential from appearing across the output load R_L. If one wishes to amplify very low-frequency components in v_i, however, the capacitors must be omitted. In the former case the amplifier is said to be ac coupled. In the latter case the amplifier is said to be dc coupled. The function of the capacitor C that shunts R_C will be described later.

We shall first discuss the circuit in the absence of incremental signals; therefore, $v_i = v_o = 0$. We shall also limit our present discussion to the situation where $R_L = \infty$ so that $i_o = 0$ even if v_i and v_o are not zero. A dc current I_P flows from the voltage supply V to ground through the tube and through the resistors R_P and R_C. Because the grid current is negligible, the electron current from the cathode is all collected by the plate.

We see that the plate to cathode potential difference V_{PC} must satisfy the equation

$$V_{PC} = V - I_P(R_P + R_C) \tag{8.10}$$

where V is the constant supply voltage. This relation is imposed by the external circuit. It is in addition to and quite independent of the dependence of I_P on V_{PC} and V_{GC} that is given by the characteristic curves of Fig. 8.6. In Fig. 8.6 we have plotted Eq. 8.10 on the same coordinates as the characteristic curves themselves. The straight line given by Eq. 8.10 is called the *load line*. The intercept with the voltage axis is at $V_{PC} = V$, and the intercept with the current axis is at $I_P = V/(R_P + R_C)$. For the particular line drawn, $V = 250$ volts and $R_P + R_C = 2.5 \times 10^4$ ohms. Since $R_C \ll R_P$, the value of R_C has little effect on the load line.

Where on the load line the tube finds itself (for $v_i = 0$), however, is a sensitive function of the grid to cathode potential and therefore of the cathode resistor R_C. Since the grid current is zero, the quiescent potential of the grid is zero and consequently the grid to cathode potential is given by $V_{GC} = -I_P R_C$. This equation together with Eq. 8.10 and the characteristic curves completely determine I_P, V_{PC}, and V_{GC}. The plate current and plate to cathode voltage so determined are called the *dc operating point* or *quiescent point* of the tube.

Given the circuit parameters, the load line is easily drawn but finding the operating point is frequently a matter of guesswork and successive aproximations. For example, let us take $R_C = 200$ ohms, and let us assume that $I_P = 2$ milliamp. This gives $V_{GC} = -0.4$ volt, but a glance at Fig. 8.6 shows that these values of I_P and V_{GC} are not on the load line. A better guess is $I_P = 5$ milliamp which gives $V_{GC} = -1.0$ volt and $V_{PC} = 125$ volts, a point on the load line and therefore the operating point. With triodes the process can be systematized by plotting the points where the current I_P crosses the characteristic curve whose grid to cathode voltage is $V_{GC} = -I_P R_C$. These points will fall on a reasonably straight line whose intersection with the load line gives the operating point. In circuit design one is apt to reverse the process. An operating point and load line will be chosen that make the best use of the tube characteristics. These choices then determine the resistance values and the supply voltage. The dc potentials at the various electrodes are usually called the *bias voltages*.

Now let us consider what happens when a slowly varying incremental input signal v_i is applied. In order to apply a slowly varying signal we assume for the moment that the blocking capacitors C_i and C_o are shorted out, but we leave the capacitor C in the circuit. The grid to cathode voltage is no longer $V_{GC} = -I_P R_C$. It becomes instead $(V_{GC} + v_{gc}) = -(I_P + i_p)R_C + v_i$. Both i_p and v_{PC} move away from their dc values but the load line equation (Eq. 8.10), which is determined solely by the supply voltage and the circuit resistors, remains valid for the total variables i_P and v_{PC}. Thus we may write

$$V_{PC} + v_{pc} = V - (I_P + i_p)(R_P + R_C) \qquad (8.11)$$

An input signal simply produces excursions of v_{PC} and i_P along the load line. If we subtract Eq. 8.10 from Eq. 8.11, we obtain

$$v_{pc} = - i_p(R_P + R_C) \qquad (8.12)$$

relating the incremental plate current and the incremental plate to cathode potential difference. It is well to remember that Eq. 8.11 and 8.12 are not small signal approximations. They hold for any values of the variables.

Thus far we have assumed that v_i was a slowly varying signal. Let us now assume that v_i varies so rapidly that the capacitor C, which shunts R_C, is a short circuit for all the important frequency components of v_i. For this high-frequency v_i, the blocking capacitors do not have to be shorted but can be in the circuit as shown in Fig. 8.7. The incremental quantities v_{pc} and i_p will have the same frequency components as v_i. We see that Eq. 8.11 is no longer correct because the cathode potential is simply $I_P R_C$ and not $(I_P + i_p)R_C$. The rapidly varying current i_p is shorted to ground through the capacitor and produces no potential drop between the cathode and ground. Instead of Eq. 8.11 we have

$$V_{PC} + v_{pc} = V - I_P(R_P + R_C) - i_p R_P$$

and subtracting Eq. 8.10   (8.13)

$$v_{pc} = -i_p R_P$$

Comparing Eq. 8.12 and the second of Eq. 8.13 we note that the incremental quantities v_{pc} and i_p lie on a straight line that goes through the operating point and whose slope is $i_p/v_{pc} = -(R_P + R_C)^{-1}$ if v_i varies slowly but whose slope is $i_p/v_{pc} = -R_P^{-1}$ if v_i varies rapidly. The slope $-(R_P + R_C)^{-1}$ is that of the load line drawn in Fig. 8.6. It is called the *dc load line*. The steeper line of slope $-R_P^{-1}$ is called the *ac load line*. Because $R_P \gg R_C$, the two load lines are almost indistinguishable for the circuit Fig. 8.7. In many circuits, however, the two load lines differ greatly and the distinction must be kept in mind. The capacitor C must be large enough so that $|-j/\omega C| \ll R_C$ for all frequencies of interest. Otherwise v_{pc} and i_p may fall between the two load lines and in addition there will be a phase difference between v_{pc} and i_p. We remind the reader that the blocking capacitors C_i and C_o can be used when v_i is a high-frequency signal but not when v_i is a low-frequency signal.

8.4 Voltage Amplification

By the voltage amplification of a circuit we mean $A = v_o/v_i$ where v_o and v_i are the incremental output and input voltages, respectively (Fig. 8.7).

In calculating A it is usually assumed that there is no output load (that is, $R_L = \infty$) and therefore no incremental output current flows (that is, $i_o = 0$). Thus, as assumed in the last section, all the plate current through the tube also flows through the plate resistor R_P. Let us also assume that the capacitor C is a short for all signal frequencies of interest. We can now write the following relations among the incremental quantities: $v_i = v_{gc}$; $v_o = v_{pc}$ and $v_o = -i_p R_P$. These are used to eliminate i_p, v_{gc}, and v_{pc} from the basic equation $i_p = g_m v_{gc} + v_{pc}/r_p$. After some algebraic manipulation we obtain

$$A = \frac{v_o}{v_i} = \frac{-r_p g_m}{1 + (r_p/R_P)} = \frac{-\mu}{1 + (r_p/R_P)} \tag{8.14}$$

The negative sign indicates that input and output are 180° out of phase. With a triode it is possible to use plate resistors R_P that are somewhat larger than r_p. Thus amplifications approaching μ are possible. This cannot be done with a pentode since values of R_P of the order of a megohm would be required. If plate currents I_p of a few milliamperes are to flow through such large plate resistors, the plate voltage supply V would have to be in the kilovolt range. Thus with $R_P \ll r_p$ a good approximation to the amplification of a pentode is $A = -R_P g_m$. For the pentode circuit of Fig. 8.7,

$$A = -R_P g_m = -2.5 \times 10^4 \times 1.25 \times 10^{-3} \simeq -30$$

To see the real reason for the use of the shunt capacitor C, let us calculate the amplification in its absence. The cathode is no longer an ac ground. By an ac ground we mean any point whose potential with respect to ground is constant. For instance, the supply voltage V is always an ac ground and this gives immediately the relation $v_o = -i_p R_P$, which we have used.

With the shunt capacitor removed the equations relating the incremental quantities are $v_i = v_{gc} + i_p R_C$; $v_o = -i_p R_P$ and $v_o = v_{pc} + i_p R_C$. In the last of these equations we may usually neglect $i_p R_C$ in comparison with v_{pc}. If we now solve for the amplification, we obtain

$$A = \frac{v_o}{v_i} = \frac{-r_p g_m}{1 + (r_p/R_P)(1 + g_m R_C)} \tag{8.15}$$

With $R_C = 200$ ohms and $g_m = 5000$ micromhos we have $g_m R_C = 1$. Thus our amplification is only -15 instead of -30. Physically the reason is that without the capacitor the actual grid signal is $v_{gc} = v_i - i_p R_C$. With the capacitor it is the larger value $v_{gc} = v_i$. This reduction in the effective value of the input signal by the subtraction of a term proportional to an output signal ($i_p R_C$) is called *negative feedback*. It is eliminated in this case by making the cathode an ac ground. Negative feedback is frequently very beneficial and is often deliberately used rather than eliminated. We shall discuss it in detail in later chapters.

The amplifier is of necessity a three- or four-terminal device with separate input and output loops. However multielectrode vacuum tubes may also be used as two-terminal nonlinear components where they provide a valuable separation of ac and dc parameters. This is accomplished by setting the control grid at some fixed potential. For instance, a glance at the 6AU6 characteristic curves shows that the tube can be operated at $V_{PC} = 200$ volts, $V_{GC} = -2$ volts, and I_P a little less than 2 mamp. At this operating point the dc resistance is $V_{PC}/I_C \simeq 10^5$ ohms. However the small signal ac resistance $r_p = v_{pc}/i_p$ is several megohms. Such a nonlinear resistance is just what is needed for the plate resistor of a pentode amplifier. The low dc resistance allows one to use an advantageous operating point without the necessity of an inconveniently high plate supply voltage. The high ac resistance means that amplifications close to μ are possible. Actually such uses of tubes are limited because of the complications of providing the correct bias voltages to a number of electrodes. We shall see later that transistors are simpler in this respect and that field effect transistors have pentode-like characteristics.

8.5 Linear Circuit Analysis in Tube Circuits

Let us review our discussion of vacuum tubes as nonlinear circuit components. Our purpose is to see what parts of the general circuit theory developed in earlier chapters can be salvaged for application to tube circuits.

We have found that the total currents and voltages in a vacuum tube are related in a very complex nonlinear way. The functional relationships are best presented graphically as plate current versus plate to cathode voltage characteristics for each of a family of fixed grid to cathode voltages. The plate current i_P is determined once v_{PC} and v_{GC} are given. In principle the grid current i_G is also determined but it is usually too small to be of interest. In setting the dc operating point for a tube in an actual circuit, additional independent relations will be imposed (the load line and a relationship between V_{GC} and I_P) and a solution (the operating point) satisfying simultaneously the characteristic curves, the load line, and the relationship between V_{GC} and I_P must be found graphically.

In the use of a tube circuit the total currents and voltages may have small signal deviations i_p, v_p, and so on, from the quiescent values at the operating point. Often the entire purpose of a circuit is to achieve some desirable relation among the small signals, for instance, a voltage amplification. Thus it is advantageous to be able to treat the small signals independently of the total currents. This is possible since we have linear relations of the type $i_p = i_P - I_P$ where both i_P, the total current, and I_P, the dc component at the operating point, obey Kirchhoff's laws. Thus their differences, the incremental components, must independently obey the same laws.

Now the fundamental equation (Eq. 8.8) governing the small signals is in fact an expression of linearity and superposition. The equation says that the incremental plate current is a linear superposition of two partial currents; one of them, $g_m v_{gc}$, is directly proportional to a voltage v_{gc} between grid and cathode and the other, v_{pc}/r_p is directly proportional to a voltage v_{pc} between the plate and the cathode. As long as the additional constraints imposed by external circuit components, for instance, R_P, R_C, C and V of Fig. 8.7, are also linear, then the incremental currents in the entire circuit also satisfy the superposition theorem.

In Fig. 8.8 we redraw the amplifier of Fig. 8.7 eliminating everything that does not enter into the incremental analysis and including the load resistor R_L that we no longer assume is infinite. The battery terminal $+V$ is an ac ground. Thus the ac potential differences between the battery and ground must sum to zero. Now since Thevenin's theorem follows directly from superposition, we can ask, "What are the Thevenin parameters of the input and output terminal pairs"? From the input one sees a passive circuit of internal resistance $R_G = 10^5$ ohms for the circuit of Fig. 8.7. Because almost no current flows to the grid, the effective grid resistance of the tube itself is very great.

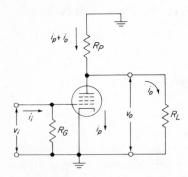

Figure 8.8 The circuit of Fig. 8.7 as seen by the incremental voltages and currents.

We have already calculated v_o, the output voltage for zero output current ($R_L = \infty$). This is the Thevenin emf of the amplifier viewed from the output. We have

$$\varepsilon_{\text{Th}} = v_o(i_o = 0) = Av_i = \frac{-\mu v_i}{1 + (r_p/R_P)} = \frac{-g_m v_i}{[(1/R_P) + (1/r_p)]} \qquad (8.16)$$

Now if the output is shorted ($R_L = 0$), we have $v_p = 0$. No ac current flows through the plate resistor R_P and the plate current depends only on v_i. The short circuit output current is given by

$$i_s = -i_p = -g_m v_i = -\frac{\mu}{r_p} v_i = i_o \ (R_L = 0) \qquad (8.17)$$

Dividing Eq. 8.16 by Eq. 8.17, we find

$$\frac{\varepsilon_{\text{Th}}}{i_s} = r_{\text{Th}} = \frac{r_p}{1 + (r_p/R_P)} = \frac{R_P r_p}{R_P + r_p} \qquad (8.18)$$

r_{Th}, the output resistance of the amplifier, is just the parallel combination of R_P and r_p. This is obvious from Fig. 8.8 where between the plate terminal and ground, R_P and the tube itself are in parallel. For the circuit of Fig. 8.7 using the 6AU6 pentode $r_p \gg R_P$ so that $r_{Th} \simeq R_P = 2.5 \times 10^4$ ohms. We point out that the amplification A is customarily defined for zero output load ($i_o = 0$) and in this case $v_o = \varepsilon_{Th}$. However, the quantity v_o/v_i approaches zero for sufficiently large loads ($R_L \simeq 0$), since $v_o = \varepsilon_{Th} - i_o r_{Th}$. The incremental output current drawn from the amplifier by a load resistor R_L is $i_o = \varepsilon_{Th}/(r_{Th} + R_L)$.

We have shown that for small ac signals the amplifier behaves as a linear three-terminal network and that Thevenin parameters can be assigned to the input and output terminal pairs. In one important respect, however, the linear circuit theory of earlier chapters fails. We do not keep the reciprocity theorem, that is, the interchange of an emf and an ammeter without alteration of the ammeter reading. In the amplifier circuit of Fig. 8.7 an emf at the input will produce a current in the output but if the same emf is moved to the output, no current flows in the input.

Reciprocity was used in Chapter 2 to show that only three, not four, parameters are needed to characterize a passive three-terminal resistive network. In the absence of reciprocity we expect that four parameters will be required, and this is indeed the case. Considering only the nonlinear component itself (tube or transistor) the four small signal parameters are related in some simple way to the four partial derivatives of Eq. 8.7, or to their equivalents if another choice of independent variables is made. If the nonlinear component is only a part of a three-terminal network, the small signal parameters must, of course, also depend on the other components in the network. For a tube, only two of the four parameters are important since the grid current is negligible, but for transistors sometimes all four parameters are required.

Let us make one or two additional comparisons to the results of earlier chapters. An amplifier is a passive network in the sense that the input and output currents depend on only the input and output voltages. There are no partial currents from internal sources. It is an active network, however, in the sense that more ac power leaves the network than enters it. This is because the network contains a dc emf that is really the source of the ac power.

We emphasize again that the partial derivatives are functions of the operating point and that for a given operating point the linear incremental analysis we have presented demands that the incremental signals be small. Equations 8.7 are just the linear terms in a power series expansion about the operating point. If the excursions from the operating point are too great, the entire linear analysis, superposition, Thevenin's theorem, and so on, breaks down. The concepts of the Thevenin input resistance and the Thevnien output emf and resistance of an amplifier are small signal concepts and are not valid for large incremental signals.

We shall see in later chapters that in many important applications of tubes and transistors, for instance, in oscillators and pulse and digital circuits, the current and voltage swings are so great that linear analysis cannot be used.

It is frequently helpful, particularly in circuits more complex than the one-tube amplifier we have treated, to replace the tubes by formally equivalent linear circuit elements such as incremental or ac sources and resistors. It then easier to apply Kirchhoff's laws and solve for the currents by the familiar procedures. The starting point is to rewrite the basic small signal equation in the form $v_{pc} = -\mu v_{gc} + i_p r_p$. The plate to cathode voltage appears as the series combination of an incremental or ac emf $-\mu v_{gc}$ and an iR drop of magnitude $i_p r_p$. We can then redraw the circuit of Fig. 8.8 in the form shown in Fig. 8.9(a). The latter is the incremental or ac circuit equivalent to the amplifier of Fig. 8.8. The application of Kirchhoff's laws leads to the same algebraic manipulations and the same results for the amplification and the Thevenin parameters that we have already obtained. For a pentode circuit, since r_p is usually very large, it is often useful to look on the pentode as a current source and a resistor in parallel. The Norton equivalent output circuit for the pentode tube only (that is, not including the external circuit components such as the plate resistor R_p) is simply a current source $i_N = -g_m v_{gc}$ and a resistance r_p in parallel. This is also shown in Fig. 8.9(b). This equivalent circuit also gives the same Thevenin output parameters for the circuit of Fig. 8.8 that we have already obtained.

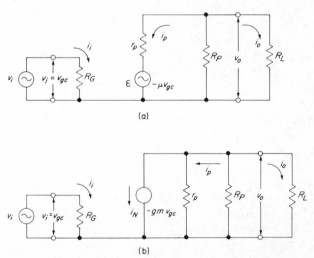

(a)

(b)

Figure 8.9 Linear circuits equivalent to the circuit of Fig. 8.8. In (a) the output of the tube itself is represented by a constant voltage source $\varepsilon = -\mu v_{gc}$ in series with an internal resistance r_p. In (b) the tube output is represented by a constant current source $i_N = -g_m v_{gc}$ in parallel with an internal resistance r_p.

Problems

8.1   In the equation for thermionic emission $j = AT^2 e^{-\phi/kT}$ the constants are $A = 60.2 \times 10^4$ amp/meter^2 °K^2 and $\phi = 4.5$ eV if the filament is made of tungsten metal. Calculate the total thermionic emission from a filament that is 1 cm long and 10^{-1} mm in diameter as a function of the temperature. What temperature is necessary to obtain 10^{-6} amp? What temperature is necessary to obtain 10^{-3} amp?

8.2   A vacuum diode is operating in the space charge limited region. The plate current is 10^{-4} amp and the plate to cathode voltage is 50 volts. What voltage must exist between the plate and the cathode to raise the current to 10^{-3} amp? You may assume that the temperature limited current is 10^{-1} amp. What will the plate current be if the plate to cathode voltage is raised to 100 volts? If the area of both the cathode and anode of the diode are doubled without altering the spacing or shape of these structures, what current will flow with a plate to cathode voltage of 50 volts?

8.3   For the 12AX7 triode circuit shown, see Fig. 8.10 and 8.11, plot the load line on the characteristic curves for the 12AX7, and find the quiescent operating point. (*Characteristic curves courtesy General Electric Company.*)

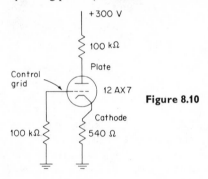

Figure 8.10

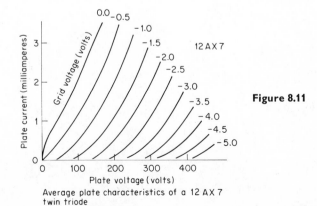

Figure 8.11

Average plate characteristics of a 12 AX 7 twin triode

8.4   For the circuit of Prob. 8.3, find the new position on the load line if an incremental voltage of $+0.2$ volt is applied to the grid. What is the ratio $\Delta v_p/\Delta v_G$ for $\Delta v_G = +0.2$ volt?

8.5   Suppose a 1-microfarad capacitor is used to shunt the 540-ohm resistor in the circuit of Prob. 8.3. Is the quiescent operating point altered? If so, by how much? If a 1-microfarad capacitor is used to shunt the 10^5-ohm resistor in the plate circuit, is the quiescent operating point altered?

8.6   An amplifier circuit using a 12AX7 is shown in Fig. 8.12. The basic dc biasing is identical with the circuit of Prob. 8.3. From the characteristic curves for the 12AX7, find g_m, μ, and r_p at the quiescent operating point of this circuit. Calculate the voltage gain of the circuit assuming that the signal frequencies involved are such that $1/\omega C$ is negligible.

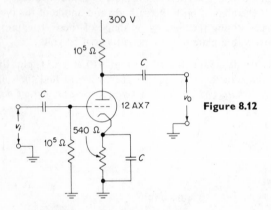

Figure 8.12

8.7   For the circuit shown using a 12AU7 triode in Fig. 8.13, draw the load line and find the operating point. From the characteristic curves in Fig. 8.14 find r_p, g_m, and μ for the 12AU7 at the operating point. This circuit is called a cathode follower circuit. (*Characteristic curves courtesy General Electric Company.*)

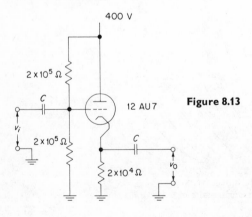

Figure 8.13

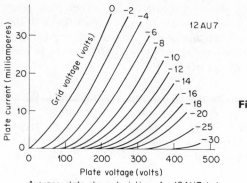

Figure 8.14

Average plate characteristics of a 12AU7 twin triode. Maximum plate dissipation = 2.75 watts.

8.8   For the cathode follower circuit of Prob. 8.7, calculate the voltage gain v_o/v_i for incremental signals. You may assume that for the frequencies in the incremental signals $1/\omega C$ is negligibly small.

8.9   You have probably noted from your calculations that the voltage gain of the amplifier circuit of Prob. 8.6 is much greater than one, whereas the voltage gain of the cathode follower circuit of Prob. 8.7 and 8.8 is slightly less than one. You might be wondering what is the purpose of a cathode follower circuit. A cathode follower is normally used to drive a low impedance. It can do this satisfactorily because the cathode follower has a low output impedance. An ordinary amplifier circuit has a high output impedance and therefore is not suitable for transferring power to a small load impedance. Calculate the output impedance of both the amplifier shown in Prob. 8.6 and the cathode follower of Prob. 8.7. Calculate also the input impedance of both circuits.

8.10   The characteristic curves for a 6AU6 pentode are shown in Fig. 8.15 for a screen voltage of $+150$ volts and a suppressor voltage of 0 volt. For a plate current of 8 milliamp and a plate to cathode voltage of 250 volts,

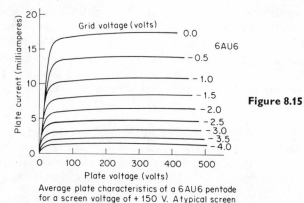

Figure 8.15

Average plate characteristics of a 6AU6 pentode for a screen voltage of + 150 V. A typical screen current is 3 milliamps.

find r_p, g_m, and μ from the characteristic curves. How much have the values of r_p, g_m, and μ changed from the values found in the text for a 6AU6 with a screen voltage of 100 volts, a plate current of 5 milliamp, and a plate to cathode voltage of 125 volts? (*Characteristic curves courtesy General Electric Company.*)

8.11 An amplifier circuit using a 6AU6 is shown in Fig. 8.16. What value must the resistor R_P have if the dc load line is to go through the point where the plate current is 8 milliamp and the plate to cathode voltage is 250 volts? Use the characteristic curves of Prob. 8.10. Note that the screen current is 3 milliamp and is not negligible.

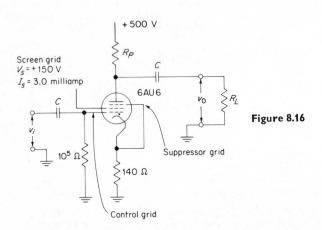

Figure 8.16

8.12 For the circuit given in Prob. 8.11, calculate the small signal gain v_o/v_i. You may assume that for the frequencies in v_i the impedance $1/\omega C$ is negligibly small. If the 140-ohm cathode resistor were shorted for incremental ac signals by a large capacitor, how would the gain be altered?

8.13 The input voltage v_i in the circuit of Prob. 8.11 is divided by the capacitor C and the $R = 10^5$ ohm resistor so that the actual voltage at the grid is given by $v_g = Rv_i/(R - j/\omega C)$. For what value of C is the magnitude of v_g/v_i equal to $1/\sqrt{2}$ if the frequency of v_i is 10^3 Hz? You may assume that the internal resistance of the voltage source v_i is negligible. This problem illustrates why the gain of an ac coupled amplifier falls off at low frequencies.

8.14 For the circuit of Prob. 8.11, calculate the input and the output resistance of the pentode amplifier. You may assume for the frequencies in v_i that $1/\omega C$ is negligible. What load resistance R_L at the output will reduce the output voltage by a factor $\frac{1}{2}$ from the Thevenin output voltage, that is, from the open circuit output?

8.15   Design a cathode follower using a 6AU6. The circuit should operate with a plate current of 8 milliamp and a plate to cathode voltage of 250 volts. What is the voltage gain, input resistance, and output resistance of your circuit?

8.16   Derive Eq. 8.14.

8.17   Derive Eq. 8.15.

8.18   Show that the amplification of a pentode $A = -g_m R_P$ can be derived using only the dependence of i_p on v_{gc} and ignoring the dependence of i_p on v_{pc}.

9/The Electronic Structure of Crystals

In this chapter we discuss the solid state physics that one needs for an understanding of semiconductor devices. These devices have almost completely replaced vacuum tubes in modern electronic circuitry. Considered only as circuit components, vacuum tubes and their semiconductor counterparts have many similarities. Both are nonlinear components that can be described by a set of input, output, and transfer current-voltage characteristics. The physical processes, however, that lead to these characteristics are quite different in the two cases. In a semiconductor, electric charges move and interact in a single crystal, not in a vacuum. We shall see that there are essential quantum mechanical aspects to the behavior of electrons in semiconductors and in crystals generally. Fortunately the understanding of these aspects requires no more than a few wave mechanical concepts now treated in most introductory physics and chemistry courses.

9.1 Electrons in a Potential Well

Wave mechanics associates with any material particle of momentum p, a wavelength $\lambda = h/p$ where h is Planck's constant ($h = 6.62 \times 10^{-34}$ joule-sec). If the particle moves in a region of constant potential energy, its momentum remains constant and the associated wave is a simple sine or cosine function that in a one-dimensional problem might have the form

$$\psi(x) = A \sin \frac{2\pi x}{\lambda} = A \sin \frac{2\pi p x}{h} \qquad (9.1)$$

If we twice differentiate Eq. 9.1, we find that $\psi(x)$ satisfies the equation

$$\frac{d^2\psi(x)}{dx^2} = -\frac{4\pi^2 p^2}{h^2} \psi(x) \qquad (9.2)$$

Let us relax the requirement that p be constant and write more generally $p^2(x)/2m = E - V(x)$. Here $p^2(x)/2m$ is the kinetic energy of the particle, E is the total energy, and $V(x)$ is the potential energy, now a function of x. Note that $V(x)$ is the potential energy of the particle, not an electrostatic potential. Substituting for $p(x)$ in Eq. 9.2, we obtain

$$\frac{d^2\psi(x)}{dx^2} + \frac{8\pi^2 m}{h^2} [E - V(x)]\psi(x) = 0 \qquad (9.3)$$

This is the Schrödinger wave equation. The solutions $\psi(x)$ are the wave functions of the particle. If $\psi(x)$ is properly normalized, the quantity $|\psi(x)|^2 \, dx$ is the probability of finding the particle between x and $x + dx$. Of course our discussion has been conspicuously incomplete. Its purpose is simply to show how the association of a wavelength and a momentum connects with the Schrödinger equation.

In an important class of problems the particle is trapped in a potential energy well. By this we mean that for the total energies E of importance in the problem, the kinetic energy $E - V(x)$ is positive only in some finite volume of space. The classical particle cannot leave this volume. When a particle in such a potential is treated wave mechanically, that is, by the solution of the Schrödinger equation, one demands that the wave function be large only in the volume of the classically allowed motion. The wave function must decrease exponentially upon entering regions of negative kinetic energy. Thus boundary conditions are imposed on the solutions of the Schrödinger equation and we find that only for certain values of the energy E can these conditions be met. The mathematics is the same as that which predicts the existence of discrete resonance frequencies for the vibrations of a finite stretched string, membrane, or elastic solid.

Let us treat the particular case of an electron in a one-dimensional potential well or "box." This example is a reasonable first approximation of the wave mechanical behavior of the free electrons in a metal. In Fig. 9.1 we show a rectangular potential energy well extending from $x = 0$ to $x = L$. Within this region the potential energy of an electron is zero. Outside the box the potential energy is V_i. The length L is a macroscopic dimension, for instance, the distance between opposite surfaces of a small block of metal.

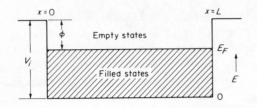

Figure 9.1 The energy levels of the conduction electrons of a metal as described by the electron in a box model. E_F is the Fermi energy, V_i is the inner potential, and ϕ is the work function.

The quantity V_i is called the *inner potential* of the metal but it is actually an energy. It is characteristically of the order of 10 eV. It arises primarily from the attractive forces of the positive atomic nuclei of the metal. When an electron is outside the metal, these forces are compensated by the electrons of the metal but because atomic electron clouds extend beyond the nuclei, the compensation is incomplete as the electron enters the metal.

Our electron in a box model permits a very simple solution of the Schrödinger equation. Since we are interested in energies $E < V_i$, we assume that the wave function is finite only within the box and zero outside. Thus the allowed wave functions inside the box must be zero at $x = 0$ and $x = L$. Otherwise they cannot be joined continuously to the outside wave function. We are still in some trouble since there is now a discontinuity in $d\psi(x)/dx$ at $x = 0$ and $x = L$. Continuity of both $\psi(x)$ and $d\psi(x)/dx$ can be maintained only if the wave function is finite at the boundary and there joins smoothly onto an exponentially decaying tail extending into the region of negative kinetic energy. But our approximation is really very good. We shall ignore the exponential tails.

Since $V_i = 0$ within the box, Schrödinger's equation (Eq. 9.3) reduces to Eq. 9.2 whose solutions are given in Eq. 9.1. To satisfy the boundary conditions $\psi(x) = 0$ at $x = 0$ and $x = L$, we must have

$$\frac{1}{\lambda} = \frac{p}{h} = \frac{n}{2L} \tag{9.4}$$

where n is a positive integer. The allowed energies are then

$$E_n = \frac{p^2}{2m} = \frac{n^2 h^2}{8mL^2} \tag{9.5}$$

and the corresponding wave functions

$$\psi_n(x) = A \sin \frac{n\pi x}{L} \tag{9.6}$$

The integer n is a quantum number labeling the allowed states of the system. In Fig. 9.2 we show the three wave functions of lowest energy.

Now our model is meant to represent the behavior in the metal of the most loosely bound atomic electrons, those outside the last closed shell of the atom. If it is to represent a metal crystal of macroscopic size, the box must contain a very large number of electrons, perhaps as many as three or four electrons per atom. We ask how these electrons are distributed among the various allowed energy states. At the absolute zero of temperature they will fill the states of lowest energy but with the limitation imposed by the Pauli exclusion principle that no more than two electrons, one with spin up and one with spin down, can be placed in the same quantum state. Therefore, if the box contains N_0 electrons, the quantum number of the highest filled state is $n = N_0/2$ and the energy of this state, called the *Fermi energy* E_F, is $E_F = N_0^2 h^2/32mL^2$.

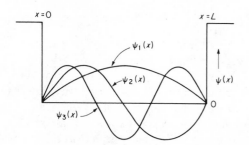

Figure 9.2 The three wave functions of lowest energy for the electron in a box model.

The quantity $V_i - E_F$ is the smallest energy that will remove an electron from the metal. It is related to the work function ϕ of the metal, introduced in Chapter 8, by the equation $V_i - E_F = \phi$. The work function can be determined experimentally from either the thermionic or the photoelectric emission of electrons. For metals the Fermi energy is in the range 2 to 7 or 8 eV. The work function is usually somewhat smaller than the Fermi energy.

Even our one-dimensional model will give the right order of magnitude for E_F. For example, the separation of nearest neighbor copper atoms in the crystal is 2.5Å. A close packed line 1 cm long in a copper crystal contains 4×10^7 atoms. Let us assume each atom contributes one free electron to the box. Thus we have $N_0 = 4 \times 10^7$ and we may calculate $E_F = N_0^2 h^2/32mL^2 = 1.5$ eV. Because N_0 and L enter the formula in the ratio $(N_0/L)^2$, the Fermi energy depends not on the size of the crystal but only on the density of conduction electrons. It is a characteristic property of the metal itself. On the other hand the average separation of successive energy levels does depend on L but this separation is always very small indeed, approximately 10^{-7} eV in our example where over 10^7 states are compressed into an energy interval of 1.5 eV. For all practical purposes we may consider that the energy levels are a continuum. In fact in the three-dimensional problem the number of independent states having an energy less than some given energy is approximately the cube of that obtained in one dimension. Thus the average separation of states is really about 10^{-21} eV. But in three dimensions the number of electrons to be placed in the box is also cubed. This is why we get about the right value for E_F in our one-dimensional calculation.

Our treatment of electrons in a box has also been incomplete in ignoring the time dependence of the wave function. A proper treatment would show that the $\psi_n(x)$ must be multiplied by a factor $\exp(-2\pi jE_n t/h)$ where E_n/h is a frequency coming from the familiar relation $E_n = hf$. Because our complete wave function is the product of independent space and time parts, we see that the $\psi_n(x)$ must represent standing waves. Frequently it is advantageous to use as the basic wave functions the two oppositely directed running waves that combine to form a standing wave. Such functions are usually written

$$\psi_k(x) = Ae^{j(kx - 2\pi E_k t/h)} \tag{9.7}$$

Here k is a quantum number describing the state, but both positive and negative values of k are permitted, corresponding to oppositely directed running waves. Comparison with Eq. 9.6 shows that $k = \pm 2\pi n/2L$. In two or three dimensions k becomes a vector in the direction of the running wave and its components k_x, k_y, k_z are the quantum numbers describing the state. In the two- or three-dimensional problem, k is called the *wave vector* and its magnitude (as it is also in one dimension) is 2π times the reciprocal of the wavelength.

9.2   The Band Structure of Crystals

The electron in a box model of the conduction electrons provides a good understanding of the inner potential, the Fermi energy, and the work function of a metal but gives no clue as to why some materials should be excellent conductors of electricity and others insulators. In our discussion we simply postulated the existence of free electrons. In insulators, however, they seem not to exist.

We must treat a more realistic model that recognizes that the potential seen by a moving electron inside a metal is not constant but has the periodicity of the lattice. An electron which penetrates the ion core of an atom has a lower potential energy than one which is halfway between two atoms. In Fig. 9.3 we illustrate a potential well with an internal periodic potential. The nuclei of the atoms are located at the potential minima and a_0, the distance between neighboring atoms, is the period of the potential. Mathematically this means that the potential energy at any point x satisfies the equation $V(x) = V(x + ma_0)$ where m is any integer. The inner potential V_i is now measured from the average value of the periodic potential.

Let us see how the existence of a periodic potential might influence our earlier results. The wave functions we obtained were standing waves with an internodal distance equal to L/n where n is the quantum number of the state.

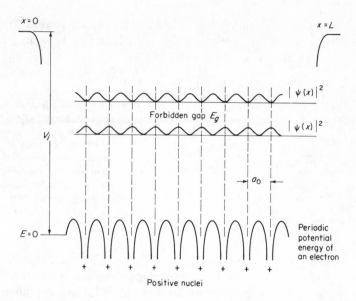

Figure 9.3 Schematic of the periodic potential energy produced by a row of atoms and the position of $|\psi(x)|^2$ in the high energy and low energy standing waves. In a plot of $\psi(x)$, alternate maxima of the wave function would be of opposite sign as in Fig. 9.2.

As long as there is no special relationship between L/n and a_0, the electron is as likely to be found in the positive region of the periodic potential as in the negative and the energy of the state is determined by the average value of the potential. Thus our treatment that ignores the periodic potential is reliable. If $L/n = a_0$, however, things are very different. As shown in Fig. 9.3, we may, by slightly translating the standing wave, either place all the nodes at ion cores or place them all between the ion cores. These states must have different energies, contrary to the predictions of the simple electron in a box model. The difference in energy E_g is greater, the greater the amplitude of the periodic potential. Within the energy gap E_g there are no allowed energy states.

If instead of standing waves we use running waves, we gain another useful physical picture of the origin of forbidden energy gaps. An electron wave moving through a crystal is partially scattered by the ion cores. Usually the various scattered waves are out of phase and no strong scattered wave is built up. However, if the wavelength λ is twice the atomic separation, then all backscattered waves are in phase. Physically an electron of such a wavelength attempting to enter the crystal is reflected out again. This is entirely equivalent to the Bragg reflection of X rays by a crystal. We see that the condition $\lambda = 2a_0$ is the same as $L/n = a_0$ since the wavelength of the running wave is twice the internodal distance. In Fig. 9.4 we plot the energy E_k

against the quantum number $k = 2\pi/\lambda$ used for running waves. The first gap occurs at $k = \pm\pi/a_0$. Higher-order reflections give gaps at $\pm 2\pi/a_0$, $\pm 3\pi/a_0$, and so on.

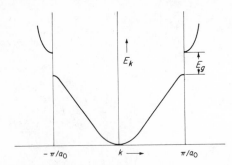

Figure 9.4 E_k as a function of k in a periodic potential. Except near the band gap $E_k = k^2h^2/8\pi^2m$ as in the electron in a box model. The states at the bottom and the top of the gap cannot be represented by running waves. They are the standing waves of Fig. 9.3.

Finally we mention another approach, the atomic approximation, which also can be very useful in understanding the electronic structure of solids. We start with isolated atoms, for example, of lithium. A lithium atom in its ground state has three electrons: two electrons filling the 1s shell and one electron that half fills the 2s shell. The higher levels, 2p, 3s, and so on, are not filled. Suppose we have a large number N of lithium atoms placed on a regular lattice, as they would be in the metal, but with a lattice spacing a very much greater than a_0, the actual lattice spacing. The widely separated atoms are independent. The energy levels of each are just those of the isolated atom. Now let a gradually decrease toward a_0. As the electron clouds of adjacent atoms begin to overlap, there is an energy of interaction among the atoms that splits the allowed energy states into a band about the original atomic level. The splitting is greater, the greater the overlap of the wave functions. Thus 1s states, whose wave functions are small, are split much less than the 2s states. In Fig. 9.5 we show the gradual splitting of the atomic levels into bands as the separation of the atoms decreases. Our final result is quite like that deduced from the periodic potential. There exist allowed bands of very closely spaced energy states, and these bands may be separated from one another by large forbidden energy gaps. Frequently bands are referred to in terms of the atomic levels from which they arise. For instance, it is the half-filled 2s band of lithium that is responsible for the electrical conductivity of the metal.

Some things come out particularly clearly in the atomic approximation. Consider the question of the number of electrons needed to fill a band. Since bringing the atoms together does not change the total number of energy states but merely redistributes them in energy, the total number of electrons that can be accommodated in a band must be the same as the number in the corresponding states of the N widely separated atoms. By the Pauli principle this number is just $2N$.

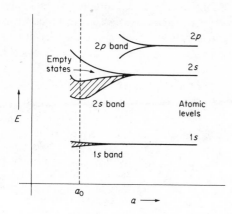

Figure 9.5 The splitting of the $1s$ and $2s$ bands of lithium as the atoms are brought together toward a_0, the atomic separation in the metal. Note that the filled states of the $2s$ band are on the average lower in energy than the $2s$ atomic level. This is the source of most of the binding energy of the metal. The shape and positioning of the curves is only qualitatively correct.

We can deduce the same result from the periodic potential model. The first forbidden gap occurs for the quantum number $n = L/a_0$. But $L/a_0 = N$ is the number of atoms in the linear crystal. Thus there are N allowed states in the first band and $N_0 = 2N$ electrons are needed to fill them.

The atomic approximation also makes clear that different bands may overlap in energy. This will happen when the spacing between atomic levels becomes less than the widths of the bands into which these levels are spread in the solid. In many crystals the bottom of a band arising from empty atomic levels may be at a lower energy than the top of a band arising from occupied atomic levels. In this case $2N$ electrons filling the lowest available energy states will occupy some states of the upper band and leave vacant an equal number of states in the lower band.

From the viewpoint of the periodic potential model the overlapping of bands is more complicated. Overlapping of bands is more easily understood in two or three dimensions than in a one-dimensional problem. Suppose we have a two-dimensional square array of atoms with a lattice spacing a_0. The running electron waves are described by the quantum numbers k_x and k_y. In Fig. 9.6 we show a plot in k space of the limits of the first band determined by the lines $k_x = \pm \pi/a_0$ and $k_y = \pm \pi/a_0$. The region of k space that includes a given band is called a *Brillouin zone*, and the lines we have drawn are the boundaries of the first zone. As we might guess from the results for one dimension, the energy of a state is proportional to $(k_x^2 + k_y^2)$ provided we are not too close to zone boundaries. Thus lines of constant energy in k space, for small energies, are circles about the origin. Let us place $2N$ electrons in the crystal, however, and ask how the energy of the highest filled state, the Fermi energy, depends on k_x and k_y. We sometimes find the complex curve of Fig. 9.6 that shows some filled states in the second zone and some empty states in the first zone. Stated in more physical terms, an electron traveling in a given direction in a crystal has available a quasi-continuum of states of increasing energy until its k vector contacts the zone boundary. There is then an energy gap that must be surmounted before the electron can again travel in that direction. But this energy gap begins at different energies for different directions of travel and in particular begins at the lowest energy for electrons traveling along the x or y axes.

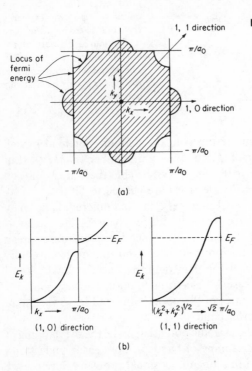

(a)

(b)

Figure 9.6 (a) The locus of the Fermi energy in a two-dimensional crystal containing $2N$ electrons and having some overlap between the first and second Brillouin zones. (b) Plots of the energy E_k versus k for two different directions in k space. In both plots all the states for $E_k < E_F$ are filled and all the states for $E_k > E_F$ are empty. In the $k_y = 0$ (1, 0) direction the boundary of the first Brillouin zone occurs at $k_x = \pi/a_0$. Some of the lowest energy states of the second Brillouin zone are below E_F and these states are filled. In the $k_x = k_y$ (1, 1) direction the boundary of the first Brillouin zone is at $(k_x^2 + k_y^2)^{1/2} = \sqrt{2}\pi/a_0$. Some of the states still in the first Brillouin zone are above E_F and these states are empty.

We emphasize that while bands may overlap in the periodic potential model, they do not have to. The band gap will always vary in position and width for different direction of electron travel but it may be so wide that the highest energy state of the first zone has a lower energy than the lowest energy state of the second zone. In this case $2N$ electrons will just fill the first zone. We shall see in the next section that such a material is an electrical insulator.

9.3 Conductors, Semiconductors, and Insulators

We have found that the electrons of a solid go into the potential well produced by the ion cores located at the lattice sites. Because the potential well in which the electrons move is periodic, there are regions of both allowed and forbidden energy as shown in Fig. 9.7. The more tightly bound atomic electrons are very little perturbed by neighboring atoms but the outer electrons overlap in the solid and lead to the bands that determine the electrical properties and, in fact, many other properties of the crystal.

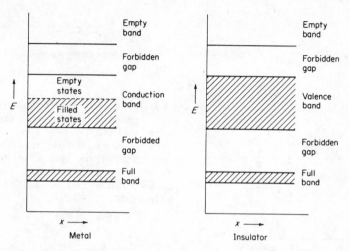

Figure 9.7 The band structure of a metal and of an insulator. If the topmost band that contains any electrons is only partly filled, it is called the *conduction band* and the material is a metal. If it is full, it is called the *valence band* and the material is an insulator Lower bands are always filled and narrow and do not influence properties of interest to us.

Let us consider how electrical conductivity arises in a partially filled band. We must use running waves, described by the quantum numbers $k = 2\pi/\lambda$, since only a running wave can carry a current. Now the absence of a net current implies that for every occupied state with a given k vector there is an occupied state with the opposite k vector. This is the expected equilibrium situation. One obtains a net current by filling more states corresponding to motion in one direction than one fills for motion in the opposite direction. Thus some energy must be supplied to the electrons in the band. However, ·because the energy states are so very densely packed within the band this energy can be furnished even by a weak electric field. In Fig. 9.8 we show for the two-dimensional case the occupation of k states corresponding to the presence and absence of a current. The displacement of the Fermi surface by the electric field has been greatly exaggerated.

We next consider a solid (the insulator of Fig. 9.7) whose bands are either completely full or entirely empty. The full bands, although they contain mobile electrons, cannot carry a current. One cannot increase the flow of electrons in a given direction because all the states corresponding to motion in that direction are already occupied. One cannot easily decrease the flow in the opposite direction since to remove an electron from such a state requires that it be excited to at least the bottom of the first empty band. This may take an energy of several electron volts, which neither an applied electric field nor thermal fluctuations normally provide.

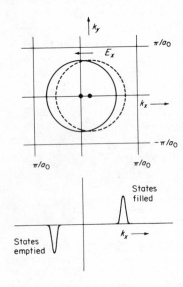

Figure 9.8 The locus of the Fermi level in a partially filled zone. Solid circle, center at origin, is in the absence of a field; dotted circle, center to the right, is in the presence of an electric field E_x to the left. The lower figure shows, in section along k_x, the levels that are emptied and filled by the electric field.

    Thus within the framework of a single structure, the electron band model of a crystal, we have a natural explanation of both conductors and insulators. But we mention again the importance of energy overlap of different bands. The isolated atoms of beryllium and magnesium have two $2s$ and two $3s$ electrons, respectively, in their outermost occupied shells. The solids are metals, not insulators, because the $2s$ and $3s$ bands overlap with higher unfilled bands. Thus there is available the required high density of empty states immediately above the filled states.

    In some pure materials the gap between the top of a filled band and the bottom of the next empty band may be so small that appreciable thermal excitation of electrons across the band gap occurs even at room temperature. Such materials are called *intrinsic semiconductors*. They show a weak conductivity that increases rapidly with increasing temperature. Silicon and germanium are the two most widely used semiconductors. Their band gaps are 1.1 and 0.70 eV, respectively.

    At thermal equilibrium the number of electrons (or occupied states) in the almost empty band must equal the number of unoccupied states in the almost full band. Both bands contribute to the conductivity although not necessarily equally since the mobility of the charge carriers is usually not the same in the two bands. The current in both bands is in fact due to the motion of electrons but it is convenient to discuss the current in the almost full or valence band in the following quite different way. The removal of an electron from a band creates a region of excess positive charge. This is called a *positive hole* or more frequently just a *hole*. Experiment shows (for instance, the Hall effect) that the hole moves in response to electric and magnetic fields just as if it had an electric charge equal but opposite in sign to that of an electron. The correct explanation of this behavior is complex. It involves details of band theory that we cannot cover. In semiconductors the current in an almost full band is called a *hole current*, while that in an almost empty band is called an *electron current*.

We shall not continue the discussion of intrinsic semiconductors. Their practical applications are few. Much more useful properties are found if the semiconductor is doped; that is, small amounts of certain foreign atoms are deliberately added. With respect to their electrical properties such crystals are said to be impurity or extrinsic semiconductors. Let us consider the doping of silicon and germaniun. Both are chemically similar to carbon. There are four valence electrons outside a closed shell and these electrons occupy the uppermost filled or valence bands. These are usually spoken of as a single band. In the silicon or germanium crystal, as in diamond, each atom is surrounded tetrahedrally by four other atoms. This is indicated in Fig. 9.9. The impurity atoms that are used replace or substitute for a silicon or germanium atom at one of the regular lattice sites. Thus there is little interference with either the lattice or the band structure of the crystal but, depending on the impurity used, there may be an increase or a decrease in the number of electrons available to fill the band. Pentavalent impurities, for instance, phosphorus, arsenic, or antimony, provide an extra electron that cannot be accommodated in the filled valence band. However, such impurities also provide an ion core whose effective positive charge is one unit greater than that of silicon or germanium. The additional electron is weakly attracted to this positive charge. On the other hand trivalent impurities, for instance, boron or indium, produce a positive hole in the valence band. This hole is weakly attracted to the excess negative charge of the impurity ion core.

The excess electron, or the hole, and the corresponding impurity ion core are systems similar to the electron and proton of the hydrogen atom but the binding energies are much less. This is due in part to the high dielectric constants of germanium and silicon, about 16 and 12, respectively. In Fig. 9.10 we show the energy level diagrams for a doped semiconductor. The localized impurity levels are within the forbidden energy region either just below the bottom of the conduction band (pentavalent impurities) or just above the top of the valence band (trivalent impurities). The energy separations from the bands are only a few hundredths of an electron volt or less.

At temperatures near $0°K$ a lightly doped extrinsic semiconductor is an insulator. The excess electrons, or the holes, are bound to the impurity centers. However, if the impurity doping is heavy then the impurity levels begin to form a band due to the overlapping of one impurity atom's wave function with another. Therefore even at temperatures near 0°K a heavily doped semiconductor behaves like a conductor. Thermal ionization of the impurity levels is very easy, however, and at room temperature it may be almost complete. Because the number of current carriers is almost constant, extrinsic semiconductors do not exhibit, at room temperature, the rapid increase of conductivity with temperature that characterizes intrinsic semiconductors.

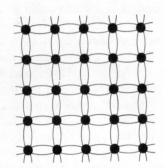

Figure 9.9 A schematic diagram of the bonding in a silicon or germanium crystal. The black circles represent the silicon or germanium atoms. The lines represent the electron bonds. Each silicon or germanium atom shares its four valence electrons with the four near neighbor atoms. Each pair of near neighbor atoms shares two electrons, one from each atom. There are no single electron bonds. Going from the atomic picture to the band picture this means that the valence band is completely full (at $0°K$) because the energy states of the double bonded electrons fill the valence band. The actual three-dimensional structure has the four near neighbor atoms in a tetrahedron about the silicon or germanium atom and not in a plane as shown.

The impurity levels near the bottom of the empty band are called *donor levels* and those near the top of the filled band, *acceptor levels*. The corresponding extrinsic semiconductor materials are termed *n type* and *p type* since the carriers of current are negative electrons or positive holes, respectively. Note that the terms *donor* and *acceptor* each refer to what happens to an electron, not a hole. At absolute zero the donor states are occupied by electrons and the acceptor states are empty. At higher temperatures a donor state can donate its electron to the conduction band and the acceptor state can accept an electron from the valence band, thus releasing the hole to become a current carrier in the valence band.

Silicon and germanium single crystals of extremely high purity can be grown from the melt. The atomic fraction of electrically active impurities can easily be made less than 10^{-10}. Such highly purified crystals are, of course, intrinsic semiconductors. One of the ways of producing the commercial extrinsic semiconductor material is to add enough impurity to the melt to give a crystal whose resistivity at room temperature is about 1 or 2 ohm-cm. This requires roughly 10^{-7} atomic fraction of impurities. The room temperature resistivity of the intrinsic material is approximately 50 ohm-cm (Ge) and 50,000 ohm-cm (Si). It is interesting that even the doped crystals are very pure indeed by ordinary chemical standards.

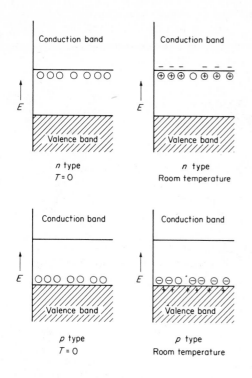

Figure 9.10 The localized impurity levels of n- and p-type extrinsic semiconductors. At very low temperatures the impurity is neutral. At higher temperatures an electron or hole is released.

As we shall discuss later, most semiconductor devices are doped differently in different parts of the same crystal. This can be accomplished in several different ways. For instance, additional impurities may be diffused part way into a crystal after it is grown. In another method, that of epitaxial growth, new single crystal material is deposited from the vapor onto the surface of a crystal. The new material is an extension of the original single crystal but may contain different impurities. Very sharp boundaries between one impurity content and another can be produced.

9.4 The Fermi-Dirac Distribution

So far in this chapter we have discussed really just one question, namely, "Where are the allowed energy levels for electrons in a crystal"? The problem of how these levels are filled by the available electrons must now be given more attention. As we have seen, the answer is simple at the absolute zero of temperature. The lowest available energy states are filled, two electrons to each state, until the supply of electrons is exhausted. At higher temperatures we expect some excitation of electrons into levels which are empty at absolute zero but thermal excitation will be unlikely to fill levels which are more than a few times kT above the uppermost filled levels at $T = 0°K$.

The quantitative answer to the question posed in the previous paragraph is given by the Fermi-Dirac distribution

$$f = \frac{1}{e^{(E - E_F)/kT} + 1} \tag{9.8}$$

A general discussion of the Fermi-Dirac distribution was given in Chapter 8. There is some deliberate repetition in what follows. In Eq. 9.8, f is the probability that a state of energy E will be occupied when the system is at the temperature T (Kelvin). E_F is the Fermi energy of the system. We shall discuss E_F in more detail in a moment but we remember that for the electron in a box problem, E_F was just the energy of the topmost filled level. In Fig. 9.11 we plot f versus E. We see that for $E \ll E_F$, $f \simeq 1$ and for $E \gg E_F$, $f \simeq 0$. The transition from one to zero takes place in an energy range $\Delta E \simeq kT$ centered about E_F. That the maximum value of f is 1, not 2, means that we are treating as separate states the two Pauli spin states associated with each energy level.

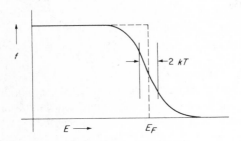

Figure 9.11 The Fermi-Dirac distribution $f = [e^{(E - E_F)/kT} + 1]^{-1}$ The curve has inversion symmetry through the point where it crosses the line $E = E_F$ (assuming f to be defined for negative values of E). The dashed line shows the behavior for $T = 0$.

At very low temperatures the transition from filled states to empty states takes place in a very narrow range of energies centered at the Fermi energy. Thus at absolute zero E_F must be somewhere between the topmost filled level and the lowest empty level. In a metal this position is well defined and may be taken to be the energy of the topmost filled level. There is, after all, an empty level within the next 10^{-21} eV. For an insulator or intrinsic semiconductor, E_F is somewhere in the forbidden energy gap between the valence band and the conduction band although just where depends on details of the band structure. Extrinsic semiconductors at very low temperatures have the Fermi energy pinned in the narrow gap between the donor levels and the bottom of the conduction band (n type) or between the acceptor levels and the top of the valence band (p type).

Besides depending on the distribution of energy levels and the number of electrons in the system, the Fermi energy is also a function of temperature. The change in the Fermi energy between absolute zero and room temperature will be no more than a few times kT but even small shifts can have interesting implications. Consider an n-type extrinsic semiconductor. In an n-type semiconductor, E_F must decrease as T increases. This can be seen as follows. If E_F stayed fixed at its $T = 0$ position between the donor levels and the bottom of the conduction band, the probability of occupancy of a donor level could never be less than one-half. This comes about because the quantity $E_D - E_F$ is negative and therefore $\exp[(E_D - E_F)/kT]$ is less than 1. E_D, the energy of the donor level, is substituted for E in Eq. 9.8. But in fact the donor levels of a typical n-type semiconductor may be almost completely ionized (empty) at room temperature. The Fermi-Dirac distribution predicts this only if $E_D - E_F$ is positive and at least a few times greater than kT. Therefore E_F must decrease as the temperature increases.

The probability of occupancy is always smaller for a state in the conduction band than for a donor level because the energy of the former is greater. That in spite of this most of the electrons end up in the conduction band at room temperature is due to the very large number of states available. The ratio of the number of band states to the number of donor states is the same as the ratio of semiconductor atoms to impurity atoms, about 10^7 for a typical doping. In a p-type semiconductor, E_F increases as the temperature increases. This is necessary because most of the acceptor levels contain electrons at room temperature. For a normal metal, E_F decreases but usually by an amount less than kT.

It would take us too far afield to give a more quantitative treatment but perhaps enough has been said to show that the Fermi energy is an important thermodynamic parameter of the crystal at equilibrium. It is of central importance in understanding semiconductor devices because the relative positions of the Fermi levels determine which way electrons will flow when two different substances are brought into electrical contact. Electrons move from the material with the higher E_F to that with the lower E_F until the resulting electric field prevents further transfer of charge. At equilibrium the two Fermi levels must match. Those who have studied some physical chemistry will recognize that the Fermi energy is just the chemical potential of the electrons of a crystal.

In Fig. 9.12 we illustrate a simple example. Two dissimilar metals A and B are a very short distance apart. Both are neutral and there is no electrical field between them. But this is not an equilibrium situation. An electron can reach a lower energy by jumping from A to B. Electrons will transfer until the Fermi levels are brought into coincidence. There is then an electric field between the two and a potential difference, called a *contact potential*, whose value is $(\phi_B - \phi_A)/|e|$ where e is the charge on the electron and ϕ_B and ϕ_A are the work functions of the two metals. At room temperature, equilibrium is reached rapidly only if the metals are brought into contact. At the areas of contact the electric field is concentrated in a very thin layer 1 to 2Å thick. Only a very small fraction of the available electrons need transfer to establish the required electrostatic potential difference.

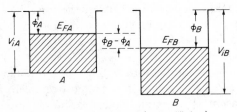

(a) Before tranfer of charge (nonequilibrium)

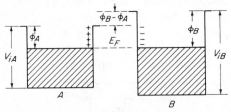

(b) After transfer of charge, $E_{FA} = E_{FB} = E_F$
(equilibrium)

Figure 9.12   The upper picture (a) shows metals A and B when each is electrically neutral (nonequilibrium). The lower picture (b) is after transfer of enough electrons from A to B to match the Fermi levels (equilibrium). There is now an electric field and a potential difference between the metals.

9.5   Summary

The single crystal is a medium in which charges can move and interact. Compared with the vacuum of a vacuum tube, the crystal can contain an enormous density of current carriers. Appreciable currents demand only a small average drift velocity of the carriers. In a vacuum tube similar currents result from a very low density of carriers moving together at high speeds.

The high density of carriers in a crystal is possible only because the positive ion cores neutralize the Coulomb repulsions among the electrons. It is this high density of electrons that forces us to treat the problem using wave mechanics. We found, for the electron in a box model, that the average energy of an electron could be a few electron volts and almost independent of temperature. This is in sharp contrast to the classical prediction that assigns to each electron an average energy of $3kT/2$.

The periodic potential of the ion cores breaks the quasi-continuum of energy levels into bands or zones separated from one another by gaps in which there exist no allowed energy levels. The partially filled band of a metal can carry a current; the completely filled or entirely empty bands of an insulator cannot.

If the forbidden energy gaps are not too wide, a material that is an insulator at absolute zero may become an intrinsic semiconductor at room temperature from the thermal excitation of a few electrons across the gap. The number of electrons in an almost empty band, or holes in an almost full band, may be greatly increased and controlled by doping, that is, the deliberate addition of a few impurity atoms. Such materials are called *extrinsic semiconductors*. The impurities provide localized energy levels from which electrons or holes are easily released. The silicon and germanium used in commercial semiconductor devices has been doped to varying degrees. Although the density of current carriers is very much less in semiconductors than in metals (commonly a factor of 10^{-7}), it is still much greater than any electron density ever achieved in a vacuum tube. This is one of the factors that makes possible the miniaturization of solid state circuitry.

The probability of an electron filling the different allowed energy states of a crystal is given by the Fermi-Dirac distribution. Below the Fermi energy E_F, the states are almost completely filled; above the Fermi energy, the states are almost completely empty. The transition takes place in an energy interval about kT in width. The Fermi energies of different materials must match if they are in thermal and electrical equilibrium. The matching occurs in a barrier layer between the two materials across which a contact potential difference appears. We shall see in the next chapter that barrier layers in a semiconductor may be several thousand angstroms thick. The voltages across the layers and the currents through them can be externally controlled. From this come the important nonlinear circuit characteristics of semiconductor devices.

In succeeding chapters we shall frequently use a more familiar classical language in our discussion of the motion of electrons. This is quite accurate and much more convenient. Only in a perfect crystal at absolute zero is the electron wave spread throughout the crystal. In real crystals, electron waves are scattered by thermal vibrations, imperfections, and impurities. The waves have a mean free time between scattering events and a mean free path. The scattering process gives energy to the lattice and leads to Joule heating just as in the classical picture. What the classical picture could not provide was the all important distribution of energy levels, their grouping into allowed bands and forbidden gaps, and the Fermi-Dirac distribution of electrons in the allowed levels.

Problems

9.1   Calculate the wavelength of an electron whose energy is equal to the Fermi energy in a one-dimensional model of a solid where the solid is 1 cm long and adjacent atoms are separated by 2.5Å. Assume each atom contributes one free electron to the solid. What is the wavelength if the solid is 2 cm long?

9.2   What is the separation in electron volts of adjacent energy levels near the Fermi energy for the hypothetical solid of Prob. 9.1?

9.3   For the one-dimensional solid of Prob. 9.1, find the velocity of an electron with the Fermi energy.

9.4   For a three-dimensional solid it was stated that the average energy separation of states is about 10^{-21} eV. Over what distance does an electric field of 1 volt/m have to act on an electron to give the electron an energy of 10^{-21} eV?

9.5   For the one-dimensional solid of Prob. 9.1, write the Bragg's law condition for the diffraction of a wave in the backward direction, that is, the wave is incident in a direction that is parallel to the one-dimensional chain of atoms and is diffracted toward the opposite direction. Show that those electrons in a solid having a wavelength such that they satisfy the Bragg's law condition are the electrons at the band gaps. The Bragg's law reflection of running electron waves sets up standing electron waves. This happens only when the electron has a wavelength that satisfies Bragg's law.

9.6   For the one-dimensional solid lattice of Prob. 9.1, find the wavelength and the k vector at which the lowest energy band gap will occur.

9.7   Write a brief discussion of the interaction that is responsible for the existence of forbidden energy gaps in the electron energy levels in a solid.

9.8   Write a brief discussion of the differences among a conductor, a semiconductor, and an insulator.

9.9   Why is it often advantageous to dope silicon or germanium with valence 3 or valence 5 impurities?

9.10   For an insulator at room temperature we know that $f \simeq 1$ for the valence band and $f \simeq 0$ for the conduction band since the conductivity is very small. Show that E_F must be in the forbidden gap and well removed (in terms of kT) from either band. Show further that if the average separation of states is greater at the bottom of the conduction band than at the top of the valence band, then E_F must be closer to the conduction band than to the valence band.

9.11   When two different metals are brought into contact, some electrons spill over from one metal to the other so that the Fermi levels in the two metals are the same. The electrons that are transferred form a double layer that results in a contact potential. Assuming that the surface density of charge in the double layer is about 10^{-2} electrons/Å^2, estimate the separation of the two oppositely charged sheets in the double layer if the contact potential is 2 eV.

9.12 In order to appreciate how high the density of charge carriers is in a solid conductor, estimate the electrostatic potential at the edge of a sphere of charge 1 mm in diameter and containing the same number of free charges as copper metal but without the ion cores to neutralize the total charge. Copper metal contains 8.5×10^{22} atoms/cm^3 and each atom contributes one free electron.

9.13 Write a brief discussion of why sodium is a conductor using the ideas of the band structure of solids. You should understand carefully the atomic structure of sodium before attempting to answer this question.

9.14 Now that you have an understanding of the band structure of solids, review the material from Chapter 2 on the temperature dependence of the resistivity of metals and semiconductors, and write a brief discussion of the role the band structure plays in determining the temperature dependence of the resistivity of a typical metal, a typical intrinsic semiconductor, and a typical extrinsic semiconductor.

9.15 For very pure intrinsic silicon the band gap is $E_g = 1.1$ eV at room temperature. Assume the Fermi level is midway between the top of the valence band and the bottom of the conduction band. What is the probability that a state at the bottom of the conduction band is occupied? It is because this probability is so small that the doping of silicon with acceptor or donor impurities is useful.

9.16 Previously we spoke of an intrinsic semiconductor as one with no impurities. A somewhat more useful definition might be that an intrinsic semiconductor is a material that has equal numbers of holes and electrons. Thus one way to produce an intrinsic semiconductor is simply to have very pure semiconductor material. Another way to produce an intrinsic semiconductor is to have equal numbers of p- and n-type impurities. Discuss how you think the temperature dependence of the resistivity of a very pure intrinsic semiconductor might differ from the temperature dependence of a heavily doped but still intrinsic semiconductor for temperatures near room temperature.

10/Semiconductor Devices

In this chapter we discuss the structure and functions of typical semiconductor diodes and transistors. We shall defer until later chapters, however, the details of their behavior as circuit components. Here we are more interested in what goes on inside the semiconductor. Nearly all the action is at junctions between different semiconductor materials. We begin with a simple example.

10.1 Metal-Semiconductor Contacts

In Fig. 10.1 we show the contact between a metal and an n-type semiconductor. The matching of the Fermi levels is brought about by a transfer of electrons from the conduction band of the semiconductor to the surface of the metal. A volume distribution of ionized donor atoms is left behind in a thin barrier layer near the surface. Almost no free electrons exist in this volume. This contrasts with the bulk material where the donor atoms are also largely ionized but the free electrons they have contributed to the conduction band remain nearby and neutralize the material. Let us calculate the width of an ionized or barrier layer across which a potential difference V_0 appears. The origin $x = 0$ is at the beginning of the barrier layer in the semiconductor. The electric field is zero at $x = 0$ and within the barrier is perpendicular to the metal-semiconductor boundary. Let N be the number density of donor atoms. At any point x the electric field is the same as in a parallel plate capacitor with a surface charge density of magnitude Nex.

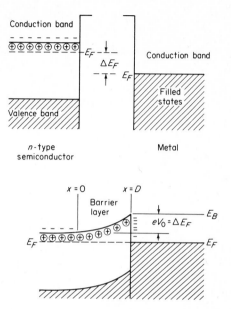

Figure 10.1 The contact between a metal and an n-type semiconductor. The top picture is before contact and the establishment of equilibrium. The bottom picture is after equilibrium. Between $x = 0$ and $x = D$ is a region of positive space charge. The electrons that normally shield the ionized donor impurities have been removed and form a surface charge at the contact between the metal and the semiconductor.

The magnitude of the field is therefore

$$E = \frac{Nex}{\varepsilon_r \varepsilon_0} \tag{10.1}$$

where ε_r is the dielectric constant of the semiconductor. The potential difference across a layer of thickness x is the integral of Eq. 10.1. The total potential drop across the entire barrier layer of thickness D becomes $V_0 = NeD^2/2\varepsilon_r\varepsilon_0$. Solving for D, we find

$$D = \left(\frac{2\varepsilon_r \varepsilon_0 V_0}{Ne}\right)^{1/2} \tag{10.2}$$

As an example, let us take $V_0 = 1$ volt, $\varepsilon_r = 16$, and $N = 4.4 \times 10^{21}/\text{meter}^3$. The latter values correspond to germanium containing about 10^{-7} donor atoms per germanium atom. We find $D = 6000\text{Å}$ and the maximum electric field within the barrier, $E = NeD/\varepsilon_r\varepsilon_0$, is about 3×10^4 volts/cm.

We next consider the movement of charge through the barrier layer. In the absence of an externally applied voltage the net current must be zero. At equilibrium, however, equal and opposite currents of electrons flow from the metal into the semiconductor and from the semiconductor into the metal.

These currents are carried by thermally excited electrons whose energy is greater than E_B (Fig. 10.1), the energy necessary to pass over the top of the barrier. The number of such electrons is proportional to the probability that states of energy greater than E_B are filled. This is given by the Fermi-Dirac distribution, Eq. 9.8. Usually for $E \geq E_B$ we have $(E - E_F)/kT \gg 1$ in which case Eq. 9.8 reduces to $f = \exp\left[-(E - E_F)/kT\right]$. Thus we write for the current density in either direction

$$j = j_0 \, e^{-(E_B - E_F)/kT} \tag{10.3}$$

A complete justification of this formula requires much more detailed arguments. For instance, one must calculate the fraction of the electrons that are moving toward the barrier. The number of states whose energy is greater than E_B is also involved in addition to the probability of their occupancy; however, Eq. 10.3 gives correctly the important exponential dependence of the currents on the height of the barrier and on the temperature. It can also be shown that the parameter j_0 is the same for both currents and varies only slowly with temperature. Note that if the Fermi energy were not the same in the metal and the semiconductor, the currents would not balance. Mutually compensating currents of holes also pass through the barrier. They will be treated in the next section but are here omitted in order to simplify the discussion.

Now if external emf's are available, a net current may flow through the junction. If we place a metal-semiconductor junction in a simple circuit [Fig. 10.2(a)] with a variable source of emf and a milliammeter, we observe the highly nonlinear current-voltage characteristic shown in Fig. 10.2(b). The junction is a rectifier. Let us see how this comes about.

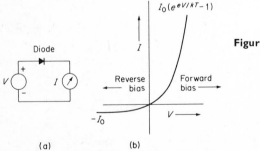

Figure 10.2 The current-voltage characteristic of a metal-semiconductor contact. The arrow in the diode symbol is always in the direction of easy flow of conventional positive current.

(a)   (b)

Because the conduction electrons have been drained away from the barrier layer, it is often called the *depletion layer*. Lacking current carriers, the depletion layer has a high resistance. Therefore if the externally applied emf V is not too great, most of it will appear across the barrier and will raise or lower one Fermi level with respect to the other as shown in Fig. 10.3.

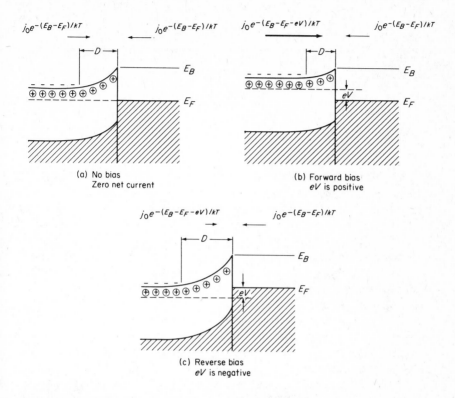

(a) No bias
Zero net current

(b) Forward bias
eV is positive

(c) Reverse bias
eV is negative

Figure 10.3   Metal n-type semiconductor contacts under different bias voltages V. The E_F in the exponential is always the E_F of the metal. The barrier height $E_B - E_F$ seen from the metal remains constant. The arrows are the directions of electron flows. Under forward bias the conventional positive current flows from the metal to the n-type semiconductor.

Note that the barrier width D also depends on V. In Eq. 10.2, V_0 must be replaced by $V + V_0$. For reverse bias, V and V_0 have the same signs and for forward bias V and V_0 have opposite signs. Thus the barrier becomes thicker under reverse bias and thinner under forward bias. The barrier layer across which the potential V appears is entirely in the semiconductor. The top of the barrier that electrons must surmount is right at the metal-semiconductor interface. Rectification occurs because electrons moving from the metal to the semiconductor see the constant barrier height $E_B - E_F$, independent of V, while electrons moving from the semiconductor to the metal see the variable barrier height $E_B - (E_F + eV)$. In all that follows, the quantity eV is positive for forward bias and negative for reverse bias.

We find for the net current density

$$j = j_0 \, e^{-(E_B - E_F - eV)/kT} - j_0 \, e^{-(E_B - E_F)/kT}$$
$$= j_0 \, e^{(-E_B - E_F)/kT} (e^{eV/kT} - 1)$$

(10.4)

Usually the interest is in the actual current rather than the current density and we write

$$I = I_0 (e^{eV/kT} - 1)$$

where

(10.5)

$$I_0 = j_0 \, A e^{-(E_B - E_F)/kT}$$

Here A is the area of the metal-semiconductor junction. Note that I_0 is not just j_0 times an area. We have included an exponential term in the definition of I_0. Thus I_0 is a strong function of temperature even though j_0 is not. The expression for the current is plotted in Fig. 10.2(b). When the junction is conducting (forward bias), the net flow of electrons is from the semiconductor to the metal. However, the arrow of the standard diode symbol, Fig. 10.2(a), is always in the direction of easy flow of the conventional positive current. It is important to recognize that the resistance of the junction under reverse bias must be large compared to the resistance of the rest of the circuit. Almost any contact between unlike materials has some rectifying properties. But if large currents can be passed in either direction at small voltage drops, then an externally applied voltage appears across other (ohmic) parts of the circuit and no net rectification is observed.

Had we treated instead the properties of a metal-p-type semiconductor contact we would have found similar rectifying properties but the currents through the barrier would be carried primarily by holes.

In the next section we shall study the rectifying properties of the junction between p- and n-type semiconductors. These are more widely used than metal-semiconductor junctions but the basic theory is similar. Actually an important practical problem is to avoid the rectifier action of metal-semiconductor junctions. Wherever a semiconductor component is attached to the metal wire of a circuit such a junction exists; however, we want it to be a low resistance ohmic contact. One way of achieving this is to use a solder that heavily dopes the semiconductor. The barrier layer is then very thin and electrons (or holes) can easily tunnel through in either direction. Tunneling is a wave mechanical effect for which there is no classical analogy. A particle without enough energy to go over the top of a potential barrier can sometimes tunnel right through it. Consider two regions of positive kinetic energy separated by a narrow forbidden region (the barrier) in which the kinetic energy of the particle would be negative. The exponential tail of the particle's wave function penetrates the barrier and may connect the two regions if the barrier is thin. The probability of tunneling increases very rapidly as the energy of the particle increases and as the width and height of the barrier decrease. Because of the tunneling the voltage drop across a heavily doped metal semiconductor junction is nearly zero.

10.2 The *pn* Junction

In Fig. 10.4 we show a junction between *n*-type and *p*-type semiconductors. Let us assume for simplicity that the boundary between the two different dopings is sharp. On the *n* side of the barrier layer there are ionized positively charged donor centers but no free electrons and on the *p* side there are negatively charged occupied acceptor centers but no free holes. There are almost no free charge carriers in the layer to neutralize the ionized donors or acceptors. Matching of the Fermi levels is achieved by the transfer of electrons from the *n*-type to the *p*-type material.

On either side of the boundary the electric field and potential will vary with position in just the way we calculated for the metal-semiconductor junction. The number of densities of ionized impurity atoms N_n and N_p may be quite different, however. We must write separately

$$D_n = \left(\frac{2\varepsilon_r \varepsilon_0 V_n}{N_n e}\right)^{1/2} \quad \text{and} \quad D_p = \left(\frac{2\varepsilon_r \varepsilon_0 V_p}{N_p e}\right)^{1/2} \tag{10.6}$$

The total width and potential drop are given by

$$D_0 = D_n + D_p \quad \text{and} \quad V_0 = V_n + V_p \tag{10.7}$$

Because the total number of nonneutralized positive donors in the barrier on one side of the boundary must equal the number of nonneutralized negative acceptors on the other side, we have the additional relation

$$N_p D_p = N_n D_n \tag{10.8}$$

From Eq. 10.6 and 10.8 we see that most of the barrier, whether measured in width or in potential change, is in the more lightly doped material. This is illustrated in Fig. 10.4(b).

We now consider the various compensating currents that flow at equilibrium, that is, in the absence of an externally applied voltage. In Fig. 10.4 the quantity ΔE_n is the separation of the Fermi level from the bottom of the conduction band in the *n*-type material and ΔE_p is the separation of the Fermi level from the top of the valence band in the *p*-type material. E_g is the width of the forbidden gap between the valence band and the conduction band. It is the same in *n*-type and *p*-type material. In the *n*-type material only those electrons whose energy E satisfies the relation $(E - E_F) > (E_g - \Delta E_p)$

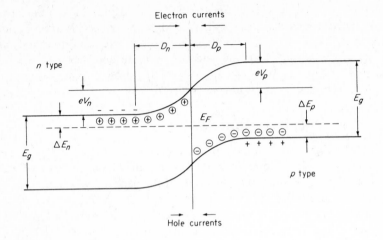

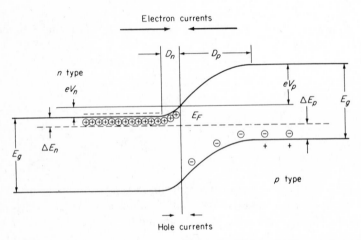

Figure 10.4 Energy level diagrams of *pn* junctions at equilibrium. (a) The *n* and *p* materials have the same doping levels, $V_n \simeq V_p$; $D_n \simeq D_p$ and $\Delta E_n \simeq \Delta E_p$. The two compensating electron currents have about the same magnitude as the two compensating hole currents. (b) The *n* material is more heavily doped than the *p* material, $V_n < V_p$; $D_n < D_p$ and $\Delta E_n < \Delta E_p$. The electron currents are larger than the hole currents.

can pass over the barrier and penetrate the *p*-type material. The number of such electrons is proportional to $\exp[-(E_g - \Delta E_p)/kT]$ by the argument given in the previous section. A similar electron population exists in the conduction band of the *p* material and these can move (without barrier hindrance) into the *n* material. Equal and opposite electron currents exist at equilibrium.

Next we investigate the hole currents but some introduction is necessary. We must be able to calculate the probability of a hole, that is, the probability that a given state is empty. This is easily done using the Fermi-Dirac distribution. We calculate in the usual way the probability f that the state is occupied. The quantity $1 - f$ is then the probability that the state is empty. But it is important to remember that the energy E to be used in the Fermi-Dirac expression is the energy of an electron in the given state. It follows that holes are more probable at the top of the valence band than at the bottom. Intuitively we expect this because it takes less energy to remove an electron from the top of the valence band.

In Fig. 9.11 we mentioned the inversion symmetry of the Fermi-Dirac distribution. As a consequence of this, the probability of an electron in a state a certain distance above the Fermi energy is the same as the probability of a hole in a state an equal distance below the Fermi energy. We leave the proof to a problem.

Let us return to the hole currents and consider first Fig. 10.4(a) where the n and p doping is the same. In this case $\Delta E_n \simeq \Delta E_p$. Therefore $(E_g - \Delta E_n) \simeq (E_g - \Delta E_p)$ and the population of holes in the p material which can cross the barrier into the n material is about the same as the population of electrons in the n material which can pass into the p material. There exist mutually compensating hole currents just as there are compensating electron currents. These compensating hole currents are about the same magnitude as the electron currents. In Fig. 10.4(b) we show the junction of a heavily doped n material and a lightly doped p material and indicate that in this case most of the current through the barrier is carried by electrons. Perhaps this is obvious but some important points become clearer if we ask how doping affects the position of the Fermi energy. As we add donor impurities to a semiconductor, the population of electrons in the conduction band goes up. This is possible only if the Fermi energy increases with an increase in n-type doping. But such a shift of the Fermi level moves it farther away from the top of the valence band; therefore the population of holes must go down.

A more physical argument leads to the same result. We make the reasonable assumption that the rate at which thermal fluctuations excite electrons across the band gap E_g is independent of the amount of doping. At equilibrium electrons must drop back into the valence band (that is, recombine with holes) at the same rate that they are produced. Recombination depends on an electron finding a hole, however. For each hole the probability of recombination is therefore proportional to n_n, the number density of electrons. The total rate of recombination is also proportional to n_p. the number density of holes. Finally then

$$n_n n_p = K(T) = n_i{}^2 \tag{10.9}$$

where $K(T)$ is proportional to the rate of excitation of electrons across the band gap and is a function of temperature only, and n_i is the number density of electrons or holes in the intrinsic material.

A similar result obtains for p-type material. With increasing p doping the Fermi level decreases, that is, moves closer to the top of the valence band. An increase in the density of holes therefore depresses the density of electrons in the conduction band. The product of the two remains constant. In an extrinsic semiconductor the particle responsible for most of the current is called the *majority carrier*; the other, the *minority carrier*. Thus in an n-type semiconductor, electrons are the majority carriers; in p type, the majority carriers are holes. Normal dopings may provide a majority carrier density 10^2 to 10^4 times greater than exists in the intrinsic material. Equation 10.8 then tells us that the ratio of majority to minority carriers in the doped material is 10^4 to 10^8.

In summary we have shown in this section that the pn junction is an ionized insulating barrier layer partly in the n doped crystal and partly in the p doped crystal. On either side of the boundary the fields and potentials are like those found in the semiconductor part of a metal-semiconductor contact. Most of the barrier is in the more lightly doped material. At equilibrium there are equal and opposite electron currents through the barrier and equal and opposite hole currents through the barrier. One type of current may be much larger than the other. The larger current is from the majority carriers of the more heavily doped material.

In the next section we discuss rectification at the pn junction and some common diode circuit components.

10.3 Diode Devices

When an external voltage is applied to a pn junction, a net current flows. The current-voltage characteristic is similar to that of a metal-semiconductor contact. The applied voltage shifts one Fermi level with respect to the other and changes the height of the barrier seen by the forward current. The forward current depends exponentially on V. Figure 10.5 shows the barrier under forward and reverse bias.

Comparison of Fig. 10.3 and 10.5 shows that the reverse current in a pn junction is not quite comparable to that in a metal-semiconductor contact. In a pn junction the carriers of the reverse current, electrons from p into n and holes from n into p, see no barrier. The reverse current is very small and independent of V because it is proportional to the population of minority carriers.

From Fig. 10.4 and 10.5 and the discussion preceding Eq. 10.4 we may write for the electron and hole current densities through a pn junction

$$j_n = j_{0n} e^{-(E_g - \Delta E_p)/kT}(e^{eV/kT} - 1)$$
$$j_p = j_{0p} e^{-(E_g - \Delta E_n)/kT}(e^{eV/kT} - 1)$$

$$(10.10)$$

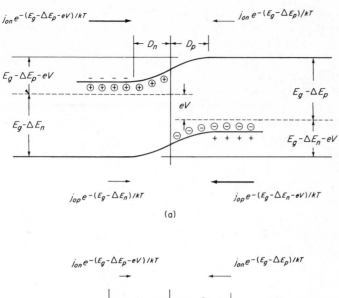

(a)

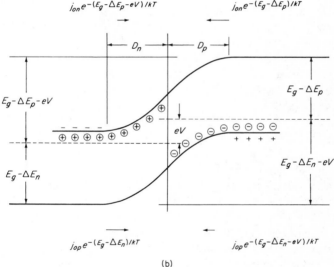

(b)

Figure 10.5 Energy level diagrams and current flows for a *pn* junction under (a) forward and (b) reverse bias. The quantity eV is positive for forward bias and negative for reverse bias. About equal doping levels on the *p* and *n* side have been assumed. Note that electrons moving toward the right and holes moving toward the left both represent current flow to the left.

The parameters j_{0n} and j_{0p} are of the same order of magnitude. That the majority carriers of the more heavily doped material carry most of the current is expressed in the equations through the quantities ΔE_p and ΔE_n. If the *n* side is more heavily doped, then $\Delta E_p > \Delta E_n$ and $j_n > j_p$. Whatever the relative doping, for forward bias the p side of the junction is positive with respect to the n side and the direction of easy flow of the conventional positive current is from the p side to the n side.

On the other hand the greater contribution to the reverse current is from the minority carriers of the less heavily doped material. To a good approximation the ratio of the maximum usable forward current to the reverse current is the ratio of the majority carrier density on the heavily doped side to the minority carrier density on the lightly doped side. If this ratio is to be large, the doping should be reasonably heavy and of the same magnitude on the two sides of the barrier. One must remember, of course, that the majority carriers on one side of the junction are the same kind of particle, electrons or holes, as the minority carriers on the other side of the junction.

The actual currents as opposed to the current density are proportional to the contact area of the junction. This area can vary greatly from the small diodes used in instrumentation to the large diodes used as rectifiers in power supplies. A small silicon diode at room temperature can have a reverse current of less than 10^{-10} amp and under forward bias show an accurate exponential dependence of I on V from 10^{-10} to 10^{-4} amp or more. The remarkable range and accuracy of this relationship is used in analog computers for the generation of exponential and logarithmic functions. In the equation $I = I_0[\exp(eV/kT) - 1]$, which relates the current and voltage, the quantity kT/e has the value 2.58×10^{-2} volt at room temperature (300°K). Therefore the exponentially increasing forward current is multiplied by the factor 2.718 (the base of the natural logarithms) by an increase in the voltage across the junction of only 2.58×10^{-2} volt. An increase of 5.96×10^{-2} volt increases the current by a factor of 10. Most small germanium diodes are manufactured so that they will give a forward current of about 1 milliamp at a forward voltage of about 0.25 volt. Small silicon diodes give the same current at about 0.6 volt.

The exponential dependence fails at low currents when $\exp(eV/kT)$ is no longer large compared to 1. It fails at high currents when the barrier is so reduced in height that most of the majority carriers can pass over it. The resistance of the barrier layer is then comparable to that of the rest of the diode structure and ohmic resistance controls the current. The forward current is then said to be saturated. The ability of the diode to dissipate heat is the eventual limiting factor.

Both the forward and reverse currents are highly temperature sensitive since kT appears in the exponential. This is a difficulty with all semiconductor devices and necessitates precautions in circuit design. Because currents rise with increasing temperature, Joule heating can lead to disastrous thermal runaway in an overloaded component. Temperature problems are more acute with devices made from germanium, which has a smaller band gap E_g than does silicon.

If very low reverse currents are desired, some attention must be given to the surface cleanliness of the diode. Surface leakage shunts the barrier and frequently shows up as an ohmic reverse bias current much greater than that predicted by Eq. 10.10.

Another important parameter of junction diodes is the reverse voltage that can be applied before breakdown occurs. In the example calculated from Eq. 10.2 we saw that very high electric fields can exist in the barrier. The maximum field increases as $(V + V_0)^{1/2}$ where V_0 is the potential difference across the barrier at zero net current (Fig. 10.4). Eventually the carriers of the reverse current acquire energy from the field faster than they transfer it to the lattice. Breakdown then results from ionization and current avalanching. Another breakdown mechanism was pointed out by Zener. At high reverse voltage the valence band of the p material is lifted well above the conduction band of the n material. Electrons may tunnel through the barrier. The tunnel current as well as the avalanche current increases very rapidly with increasing reverse voltage and a quite well-defined breakdown voltage results. Zener breakdown tends to occur before avalanche breakdown across narrow barriers and thus at lower voltages. Wide barriers sustain higher voltages and breakdown occurs by avalanching.

It is of great practical importance that the width of the barrier and therefore the field strength can be controlled independently of the voltage across the barrier. This comes about because N, the density of donors or acceptors, occurs in the denominator of Eq. 10.2. A lightly doped pn junction has, for a given back voltage, a wide barrier and a low field. A heavily doped junction for the same voltage has a narrow barrier and a high field strength. The reverse breakdown voltage can be adjusted from a few to several hundred volts. Sometimes when high reverse voltages are required, as in power rectifiers, a layer of intrinsic material is inserted between the p and n sides of the barrier. Such a device is called a *pin diode*.

The rapid increase in reverse current at the breakdown voltage is widely used to provide a constant reference voltage. In such an application the diode is called a *Zener diode*. Figure 10.6(a) shows the current voltage characteristic of a pn junction including the breakdown current and Fig. 10.6(b) shows a simple circuit providing a constant reference voltage. The Zener diode when used at a voltage that is sufficient to cause breakdown will operate with some fairly large dc voltage and current I. The dc resistance V/I can be large. However, if the current is varied by a small incremental amount i about the large dc value I, then v, the incremental change in the voltage produced by the incremental change in the current, is very small indeed. Thus the resistance for incremental signals v/i is very small. Consequently if the source voltage in Fig. 10.6(b) changes, the change in the voltage mostly appears across the resistance R and the voltage across the Zener diode remains nearly constant. The Zener diode has an ac resistance that is much smaller than its dc resistance.

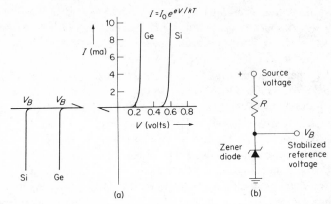

Figure 10.6   (a) Current-voltage characteristics of typical small germanium (Ge) and silicon (Si) *pn* diodes. A considerable range of reverse breakdown voltages can be achieved for both germanium and silicon diodes. On a linear current scale of milliamperes the current is indistinguishable from zero over most of the voltage range. (b) Circuit using a *pn* diode at reverse breakdown to provide a stable reference voltage V_B.

In addition to its applications as a rectifier and as a constant voltage source, the *pn* junction is also used as a variable condenser. When reverse biased, the barrier supports a potential difference and stores a charge. These are the functions of a condenser. The width of the barrier D varies as $(V + V_0)^{1/2}$. Thus the capacitance C is proportional to $(V + V_0)^{-1/2}$. The capacitance can be changed electrically by changing V. Commercially such devices are called *varicaps* or *varactors*.

In the next sections we describe two widely used three-terminal semiconductor devices, the bipolar transistor and the field effect transistor (FET).

10.4   The Bipolar Transistor

We have neglected some important aspects of the behavior of current carriers in semiconductors simply to be able to focus the discussion on one thing at a time. In particular we have not concerned ourselves with the rate at which a nonequilibrium concentration of carriers returns to equilibrium. Such nonequilibrium concentrations are always present in and near a junction between nonidentical materials when a net current flows. In a *pn* junction under forward bias, large currents of electrons move from the *n* material into the *p* material where they are the minority carrier. The excess concentration of electrons gradually disappears by recombination with holes. Hole currents move in the opposite direction through the barrier and recombine with electrons in the *n* material. How far a nonequilibrium concentration of carriers moves before it is dissipated by recombination influences the behavior of most semiconductor devices. The effect is especially important in the bipolar transistor.

The mechanisms of recombination are not completely understood. The energy is usually released as heat (phonons) but in some cases photon emission can be important. It seems likely that recombination is usually a two-step process. An electron or hole is trapped for awhile, but not at an ordinary impurity center; then a carrier of the opposite sign comes along and completes the recombination. The total process is, of course, the transition of an electron from the conduction band to the valence band with the release of an amount of energy equal to the band gap.

What is experimentally clear is that the lifetime before recombination is large compared to the time between successive scatterings. Thus an excess population of minority carriers undergoes a random diffusive motion before it is lost by recombination. If an electric field is present, there is an additional drift velocity in the direction of the force on the carriers.

Although actual values vary considerably, it is useful to have a rough idea of the magnitudes of the various parameters we have introduced. The recombination lifetime τ of an excess population of minority carriers in lightly doped material ranges from 10^{-3} to 10^{-6} sec. Heavy doping, because it increases the population of majority carriers, shortens the recombination lifetime. The mean free time a carrier moves between successive scattering events is of the order of 10^{-12} sec. Mean free paths for scattering are of the order of 0.1 micron. The diffusion length L, that is, the distance a carrier can diffuse during a recombination lifetime τ, is typically 100 microns. Because L is proportional to $\tau^{1/2}$, diffusion lengths are much less variable than recombination lifetimes. We shall see the importance of large diffusion lengths in our discussion of the bipolar transistor.

The bipolar transistor is essentially a sequence of two oppositely directed pn junctions built into one single crystal. It can be a sandwich of p material between layers of n or of n material between layers of p. These are shown in Fig. 10.7. In normal operation the emitter-base junction is forward biased and the collector-base junction is reverse biased. The current-voltage characteristics of the emitter-base junction are the same as for a pn diode. Over several orders of magnitude the forward current depends exponentially on the forward bias. Currents in the milliampere range begin at a forward bias of about 0.25 volt for germanium and 0.6 volt for silicon. The collector-base reverse bias is limited by the reverse breakdown voltage. Reverse biases of 10 or 20 volts are common and they may be higher.

The majority carriers of the emitter flow into the base where they are minority carriers. Frequently the process is referred to as minority carrier injection. Those that reach the collector-base junction are immediately drawn into the collector since a pn junction offers no barrier to minority carriers. Let us write $I_C = \alpha I_E$. Alpha, called the total forward current gain of the transistor, is the fraction of the emitter current that reaches the collector. Actually the collector current I_C is made up of two parts. One part is the reverse current that flows across the back-biased base to collector junction.

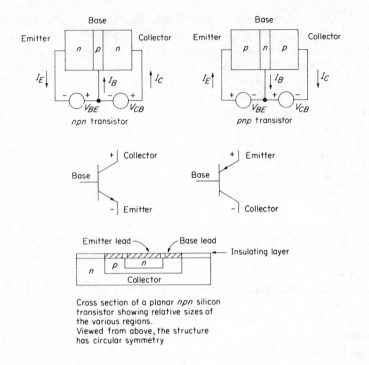

Cross section of a planar *npn* silicon
transistor showing relative sizes of
the various regions.
Viewed from above, the structure
has circular symmetry

Figure 10.7 Circuit configurations and standard symbols for bipolar transistors. The arrow
on the emitter is in the direction of the conventional positive current when the
emitter-base junction is forward biased. Although this figure indicates that the emitter
and the collector of a transistor are equivalent, many transistors are constructed with
a larger area collector than emitter in order to ensure the efficient collection of
minority carriers injected into the base from the emitter. One usually cannot interchange the collector and the emitter leads and obtain useful transistor action. A
planar transistor is shown in cross section and roughly to scale.

We call this current I_{CO}. The other part is the fraction α of the emitter current
that is collected. Thus $I_C = \alpha I_E + I_{CO}$; however, often I_{CO} is very small and
may be ignored. This is especially true for silicon transistors that have very
small reverse currents. Transistors are most useful if α is close to unity and
values as high as 0.95 to 0.995 or even higher are achieved in commercial
transistors. Values of α near unity require a thin base and a lightly doped
base. The base must be appreciably thinner than the diffusion length of the
injected minority carriers so that few carriers are lost by recombination in the
base. Light doping of the base helps to reduce recombination losses and
therefore increases the diffusion length, but, more important, it ensures that
most of the forward current across the emitter-base junction is carried by the
majority carriers of the emitter. Assume, for example, that the emitter and

the base were doped about equally; then half the emitter current might be due to the majority carriers of the base. This latter current cannot penetrate the reverse bias into the collector and α could not exceed one-half.

Base thicknesses of less than a micron can be achieved although very thin bases are a necessity only if high-frequency operation of the transistor is required. The reciprocal of the diffusion time of the injected minority carriers across the base is an upper limit to the frequency that the transistor will amplify (see Chapter 12 for a discussion of this effect). The lateral dimensions of the junctions are usually large compared to the base thickness. Thus the diffusion of carriers is essentially one-dimensional. In Fig. 10.8 we show the band structure and energy levels of an *npn* transistor under normal biases.

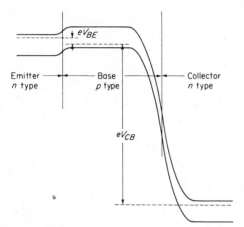

Figure 10.8 Energy level diagram of an *npn* bipolar transistor under normal biasing. V_{BE} is the forward emitter base bias and V_{CB} is the reverse collector base bias. For simplicity the barriers are drawn for equal doping of emitter, base, and collector.

From Kirchhoff's law the currents (Fig. 10.7) must satisfy the relation $I_E = I_B + I_C$. Using $I_C = \alpha I_E$, we obtain $I_B = (1 - \alpha)I_E$. If we had used the complete expression $I_C = \alpha I_E + I_{CO}$, then we would have obtained $I_B = (1 - \alpha)I_E - I_{CO}$. Thus I_B is very small when α is close to unity. We shall not consider circuit configurations in any detail until the next chapter but we show in Fig. 10.9 two frequently used transistor circuits, the common base configuration and the common emitter configuration. The term *common base* comes from the fact that the base is part of both the input and output circuits. As a typical case, let us assume that $\alpha = 0.99$ and that a change in the emitter-base voltage of 1 millivolt will produce a change in the collector current of 0.1 milliamp. In the common base circuit the input impedance is then 10 ohms. Almost the entire current change appears in the output. The corresponding output voltage change is 0.5 volt if R is taken to be 5000 ohms. Thus the current gain is approximately unity, and the voltage and power gains are both 5×10^2. We remind the reader that the current-voltage characteristic of a forward-biased *pn* junction is $I_E = I_0 \exp{(eV/kT)}$ where V

is the voltage across the base to emitter junction. Thus if a change in the emitter-base voltage of 1 millivolt produces a change in the collector current of 0.1 milliamp, the total collector current must be about 2.6 milliamp. The calculation is left to a problem.

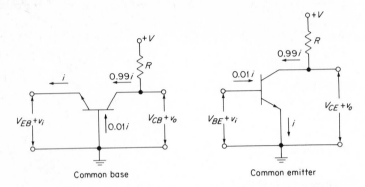

Figure 10.9 An *npn* bipolar transistor used as an amplifier in two standard circuit configurations. In the example illustrated $\alpha = 0.99$ and $\beta = \alpha/(1-\alpha) = 99$. The quantities v_i and v_o are the incremental input and output voltages and i is the incremental emitter current.

Let us compare this with the common emitter circuit. The same input voltage signal appears between the base and the emitter but the input current is now only 10^{-2} times the collector current. The output characteristics are essentially unchanged. We thus find for the common emitter arrangement an input impedance of 1000 ohms. The current gain is 10^2 and the voltage gain remains 5×10^2. The power gain is increased to 5×10^4, however.

The common emitter circuit is the more widely used of the two illustrated because of its greater current and power gain. For this circuit the forward current gain is called β and is defined by the equation $I_C = \beta I_B$. From $I_E = I_C + I_B$ and $I_C = \alpha I_E$ we calculate $\beta = \alpha/(1-\alpha)$ and $I_E = (\beta+1)I_B$. Actually if we had used the exact expressions $I_C = \alpha I_E + I_{CO}$ and $I_E = I_C + I_B$, we would have obtained $I_C = \beta I_B + (\beta+1)I_{CO}$ and $I_E = (\beta+1)I_B + (\beta+1)I_{CO}$ where $\beta = \alpha/(1-\alpha)$. The β of a transistor is one of the parameters most useful in circuit design. In the discussion of the two circuits we have assumed the incremental relations $i_c = \beta i_b$ and $i_c = \alpha i_e$ although α and β were actually defined in terms of the large dc currents. Over much of the operating range of a bipolar transistor α and β are independent of the current and the assumption is justified.

Before leaving this subject we emphasize again that the carriers from the emitter must diffuse through the base with a minimum of loss and enter the collector. Let us assume the opposite, that is, that the base is thick and emitter carriers recombine without reaching the collector. In this case the relation $I_E = I_C + I_B$ still holds but I_C is the usual reverse current of a back-biased *pn* junction. It is several orders of magnitude smaller than I_E or I_B and almost independent of I_E. All transistor action is lost; the amplification is zero. We simply have two independent *pn* junctions that happen to share the base as a common connector.

10.5 The Field Effect Transistor (FET)

An example of the richness and variety of solid state electronics is provided by another semiconductor device, the field effect transistor (FET). The FET utilizes an entirely different principle of operation than the bipolar transistor although structurally it is also an *npn* or *pnp* sandwich.

In Fig. 10.10 we show an FET of the *pn* junction type and the appropriate biases. Ohmic contacts are made at the two ends of the *n*-doped channel. The negative electrode is called the *source*; the positive end is the *drain*. Within the channel electrons flow from source to drain. The channel is surrounded by *p*-type material called the *gate*. The two halves of the gate are at the same potential and reverse biased with respect to the *n* channel. Thus only very small currents flow in the gate circuit.

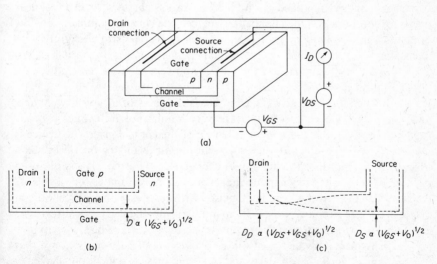

Figure 10.10 (a) Schematic of an *n* channel junction FET. The two halves of the gate are shorted together (not shown). (b) Section of the FET showing the channel and barrier layer when V_{DS} is small. For a fixed V_{GS} and V_{DS} not too large the channel behaves as an ohmic resistance. I_D is proportional to V_{DS}. The width of that part of the barrier that appears in the channel is $D \propto (V_{GS} + V_0)^{1/2}$. (c) The channel and barrier layer in the pinch off region. V_{DS} and I_D are large and I_D is almost independent of V_{DS}.

The essential point in the operation of the FET is that the thickness of the n channel, between the two layers of p material, be of the same order of magnitude as the thickness of the pn barrier layer. In Fig. 10.10(b) the dashed lines show the edge of the barrier layer in the n material. As indicated in Fig. 10.10, V_{GS} and V_{DS} are, respectively, the gate to source and the drain to source voltages. Because the barrier layer is depleted of carriers, only that part of the channel between the dashed lines is available as a conduction path from the source to the drain. The width of the barrier layer varies as $(V_{GS} + V_0)^{1/2}$ (Eq. 10.6), however. Thus the width of the channel and therefore its resistance can be controlled by the gate voltage. For a fixed V_{GS} and V_{DS} not too large, we have an Ohm's law relation

$$V_{DS} = R(V_{GS})I_D \qquad (10.11)$$

where R is a function of V_{GS}.

Let us now investigate what happens as V_{DS} increases. In Fig. 10.10(c) are shown the channel and barrier layer if $V_{DS} > V_{GS}$. The voltage V_{GS} is applied between the gate and the source. Therefore near the source the potential drop across the barrier is $V_{GS} + V_0$. If V_{GS} is not too large, the barrier is thin and the channel is wide. Near the drain the potential drop across the barrier is $V_{GS} + V_{DS} + V_0$. The barrier is thicker and the channel narrower. Thus even at fixed V_{GS} an increase in V_{DS} narrows the channel near the drain and increases the channel resistance. The current I_D must fall below the Ohm's law value (Eq. 10.11). Eventually I_D increases only slightly with increasing V_{DS}. This phenomenon is called *pinch off*. If $V_{DS} + V_{GS}$ is too large, breakdown occurs across the barrier and very large currents flow from the gate into the drain.

In Fig. 10.11 we give the characteristic curves of a typical n channel junction FET. The characteristics are very similar to those of a pentode vacuum tube with the gate playing the role of the control grid. The device is most useful as an amplifier when V_{DS} is in the pinch off region. Then small changes in V_{GS} produce the maximum change in I_D. Positive V_{GS} is not used because of the large gate currents that would be drawn under forward bias; that is, the gate is always kept back biased with respect to the source. Silicon rather than germanium is used in manufacturing the FET because of its higher barrier breakdown voltages and much lower reverse bias currents.

The major advantage of the FET over the bipolar transistor is that the input impedance of the FET is that of a back-biased pn junction. Typically this may be 10^9 ohms for dc. For the bipolar transistor the input impedance is much lower.

FET's are also made with a p channel and an n gate. The polarities of the biases must be reversed. The source is always the electrode at which the majority carrier enters the channel. Thus conventional current flows from source to drain in a *p* channel and from drain to source in an *n* channel. The customary symbols are shown in Fig. 10.12.

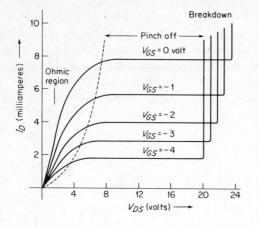

Figure 10.11 Drain characteristics of an *n* channel junction FET.

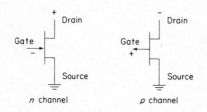

Figure 10.12 Standard symbols and polarities for *n* channel and *p* channel FET's.

The dc input impedance of the FET is further improved if one part of the gate is insulated from the channel by a thin layer of silicon oxide. The actual gate electrode is a metallic film above the oxide. The remainder of the gate is called the *base* and a separate electrical lead is brought out. These devices are called *metal-oxide-semiconductor field effect transistors* or *MOSFET*'s. They are really four-terminal components.

Figure 10.13 shows two common types in cross section. The operation of the depletion type is similar to the junction FET. With no gate voltage a low resistance path connects the source and the drain. A negative gate voltage produces a barrier layer in the channel next to the insulating layer and reduces the channel width. Voltages applied to the base also affect the channel width. Sometimes a dc bias applied to the base is used to pick the correct operating point of the MOSFET and the incremental signal to be amplified is applied to the gate. Because of the insulating oxide layer the input impedance of the gate is very high, typically 10^{14} ohms. Also the gate may be forward biased without producing appreciable gate currents.

The enhancement-type MOSFET is normally cut off. There is no channel through the *p*-type base. Depending on the sign of the potential difference between source and drain, either the source-base barrier or the drain-base barrier is reverse biased. But now suppose a positive voltage is placed on the gate. The electric field penetrates into the base, and all the energy levels of the base near the insulating layer are lowered. As soon as the conduction band of the p-type base is lowered to the level of the conduction bands of the n-type source and drain, electrons can flow. An n-type channel has been induced by the field. There is very little field penetration into the source and drain since they are deliberately heavily doped. The resulting characteristics can be most useful. No drain current flows until the gate potential reaches some positive value. The current then increases rapidly with increasing gate voltage.

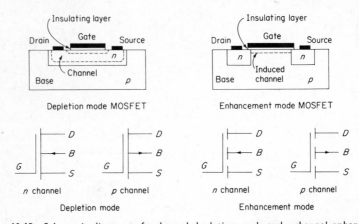

Depletion mode MOSFET   Enhancement mode MOSFET

n channel p channel n channel p channel

Depletion mode   Enhancement mode

Figure 10.13   Schematic diagrams of n channel depletion mode and n channel enhancement mode MOSFET's. Similar devices with p channels are also made. The standard symbols for all four types are shown.

Both types of MOSFET's are also made with the p and n regions interchanged and used with the biases reversed. Again because of the insulating oxide layer the gate current of a MOSFET is not only very small, but it is relatively free of temperature effects. The MOSFET is very widely used, particularly in large-scale integrated circuitry (LSI).

Before leaving this chapter we should like to comment on one point. We have written currents and voltages using capital letters and capital subscripts. As discussed in Chapter 8, this is our notation for large dc quantities. We here use this notation to emphasize that we are treating the properties of semiconductor devices and not discussing linear amplifier circuits in this chapter. Some things such as the drain current versus drain to source voltage characteristic curves might, in the notation of Chapter 8, be better written as i_D versus v_{DS} to emphasize that this is valid for total currents and voltages. We shall try to use the notation of Chapter 8 in the remaining chapters of this text that deal with the circuit application of semiconductor devices.

Problems

10.1 At thermal equilibrium the population of quantum states by electrons everywhere in an *n*-type semiconductor-metal junction is given by the Fermi-Dirac distribution function. Using this, discuss why there are many fewer free charge carriers in the conduction band in the depletion layer than in the bulk material. Estimate the ratio of electrons in the conduction band in bulk material to the number in the depletion layer. You may take the energy from the Fermi level to the conduction band to be 0.15 eV, and you may take $eV_0 = 1$ eV.

10.2 Calculate the width of the depletion layer at a metal-semiconductor junction if $V_0 = 1.5$ volts, $\varepsilon_r = 12$, and $N = 5 \times 10^{23}/\text{meter}^3$.

10.3 Using the Fermi-Dirac distribution function, show that the probability of an electron filling a state of energy ΔE above the Fermi energy is the same as the probability of a hole filling a state an energy ΔE below the Fermi energy.

10.4 A silicon *pn* junction has a back bias saturation current of 10^{-13} amp. Graph on semilog paper the forward bias current from $V = 0.25$ volt to $V = 0.65$ volts.

10.5 For the diode described in Prob. 10.4, plot the current-voltage characteristic from $V = -5$ volts to $V = +0.05$ volt.

10.6 For intrinsic silicon, $n_i = 1.5 \times 10^{10}/\text{cm}^3$. If silicon is doped with boron so that at room temperature there are 5×10^{14} holes/cm^3 in the valence band, what is the density of electrons in the conduction band? If the silicon is doped with phosphorus so that there are 10^{13} electrons/cm^3 in the conduction band, what is the density of valence band holes?

10.7 For intrinsic germanium, $n_i = 2.4 \times 10^{13}/\text{cm}^3$. What doping of arsenic (a pentavalent impurity) will make the ratio of the density of electrons in the conduction band to the density of holes in the valence band equal to 10^6?

10.8 A varicap is biased with a dc voltage of 100 volts. You may assume that 100 volts $\gg V_0$. The varicap has a capacitance of 10^{-11} farad at this bias voltage. The varicap is in parallel with an inductor of 10^{-6} henry. If one wishes to vary slowly the resonant frequency of the parallel resonant circuit by 1%, how large an amplitude ac voltage must be applied to the varicap in addition to the dc bias?

10.9 If a bipolar transistor has $\alpha = 0.995$, what is β? If $\beta = 50$, what is α?

10.10 The current-voltage characteristic of a diode is shown in Fig. 10.14. What is the Zener breakdown voltage? Is the diode made of silicon or germanium?

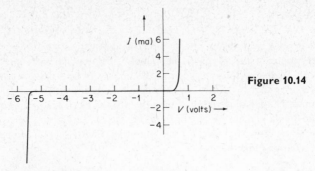

Figure 10.14

10.11   A bipolar transistor is connected in a common emitter configuration as shown in Fig. 10.9. The base current is measured to be 1.5×10^{-5} amp and the collector current is measured to be 3×10^{-3} amp. What is the forward current gain? What is α?

10.12   A bipolar transistor is connected in a common base configuration as shown in Fig. 10.9. The base current is measured to be 5×10^{-5} amp and the collector current is measured to be 10^{-3} amp. What is the forward current gain? What is α? What is β?

10.13   A bipolar transistor has an α of 0.99 and a base-emitter junction back bias saturation current of 10^{-13} amp. Write an expression for the collector current as a function of the base to emitter voltage.

10.14   For the FET whose characteristic curves are shown in Fig. 10.11, plot the resistance $R(V_{GS}) = V_{DS}/I_D$ as a function of V_{GS}. $R(V_{GS})$ is to be obtained from the ohmic region only of course.

10.15   For the FET whose characteristic curves are shown in Fig. 10.11, plot the transconductance $g = (\partial I_D / \partial V_{GS})_{V_D}$ in the pinch off region as a function of V_{GS}.

10.16   A change in base to emitter voltage of 1 millivolt produces a change in emitter current of 0.1 milliamp in a given bipolar transistor operating at room temperature. Show that the total emitter current must be 2.6 milliamp.

11/Semiconductor Circuit Components

In the last chapter we tried to explain how various semiconductor devices are constructed and why they behave the way they do. Here we begin the discussion of their circuit applications. Diodes are important only because of their nonlinear behavior. Their most common uses include the conversion of an ac signal to a dc signal and various applications as switches that are controlled by the dc voltage across the diode. Transistors on the other hand have important uses both as linear and as nonlinear components. The linear behavior is easier to analyze and in this chapter we confine our discussion of transistors to circuits involving a single transistor used as a linear component. Amplification and impedance matching are the properties of greatest interest.

11.1 Diodes as Rectifiers

In Fig. 11.1 we show a diode in a simple half-wave rectifier circuit. Let us assume the ac input voltage is 110 volts rms, 60 Hz. The turns ratio of the transformer is chosen to provide whatever particular output voltage is needed. For instance, many transistor circuits operate from dc supplies in the 5- to 30-volt range. The resistor R_L is the load of the power supply. Efficient rectification requires that the forward bias resistance of the diode be much less than R_L and that the back bias resistance be much greater than R_L. Thus for the half of the cycle when the diode is forward biased, the output voltage of the secondary appears across R_L and A and B are at almost the same potential. During the other half of the cycle the diode is back biased, the output appears across the diode, and B remains at almost ground potential. The potentials at A and B as a function of time are shown in Fig. 11.2. The potential at B has a dc component as well as ac components at 60 Hz and even integral multiples of 60 Hz.

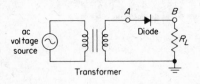

Figure 11.1   A half-wave rectifying circuit.

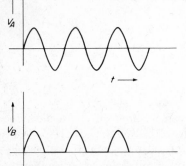

Figure 11.2   The potentials at A and B (Fig. 11.1) as a function of time.

The half-wave rectifier is seldom used because the output of the rectifier is difficult to filter, that is, it is difficult to remove all the ac components leaving only the dc component. Figure 11.3 illustrates a full-wave rectifier and the resulting potential difference across the load. The large 60-Hz ac component is now missing. Only 120 Hz and its harmonics remain. Note that the peak output voltage is only half that available from the same transformer used in a half-wave rectifying circuit. At any given moment only one-half of the secondary is furnishing voltage and current to the load. The other half sees the load in series with a reverse biased diode.

The circuit of Fig. 11.3 is widely used in dc supplies of low or medium power output. If high voltage or power is required, the bridge circuit of Fig. 11.4 may be preferable. Such a circuit provides full-wave rectification, utilizes the full voltage output of the secondary winding, and permits the use of a smaller and cheaper transformer than does the circuit of Fig. 11.3.

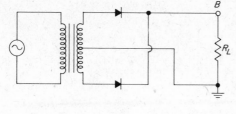

Figure 11.3 A full-wave rectifying circuit and the output waveform.

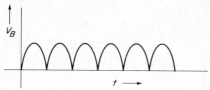

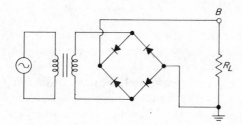

Figure 11.4 The circuit of a full-wave bridge rectifier.

For most applications the ac ripple of even a full-wave rectifier is much too great and additional smoothing must be provided. Arrangements of capacitors and inductances are used that give a high ac impedance in series with the load R_L and low ac impedance shunting the load. Typical is the π filter of Fig. 11.5. With such filters it is difficult to reduce the ratio of the ripple amplitude to the average dc voltage below 0.01 if appreciable current must be supplied by the rectifier. The power supplies of high quality instrumentation circuits are almost always electronically regulated after some filtering. The regulation makes use of the principles of negative feedback, which we discuss in the next chapter. An electronic regulator for a dc power supply will be discussed in Chapter 14. Ripple amplitudes can be made less than 10^{-5} of the output voltage. In addition, the dc output voltage can be highly stabilized against changes of the output current and fluctuations of the 60-Hz input line voltage.

Although the dc power supply circuits we have discussed are very important, we are also interested in information processing in this text. Let us consider a half-wave rectifier from such a viewpoint. In Fig. 11.6 we show a high-frequency ac signal of slowly varying amplitude entering the rectifier.

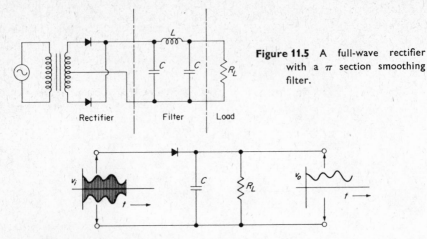

Figure 11.5 A full-wave rectifier with a π section smoothing filter.

Figure 11.6 Rectification (or detection) of an amplitude modulated ac signal.

The output is a slowly varying dc voltage whose magnitude changes with time in the same way as the amplitude of the high-frequency ac signal. If the amplitude modulation of the high-frequency ac signal is to be translated accurately to a low-frequency variation of the dc level, the time constant $R_L C$ must be large compared to the period of the high-frequency ac wave but small compared to any periods of the slow modulation. In communications electronics the modulation of the ac signal might be an audio or video signal imposed on the broadcast carrier frequency. Rectification is then usually referred to as detection or demodulation.

Instrumentation circuits sometimes use the methods just described. A slowly varying dc signal is chopped, that is, converted to an amplitude modulated ac wave, amplified as ac, and then rectified to recover an amplified output of the original dc input. This roundabout procedure is useful because stable high gain amplifiers are easier to design for ac than for dc. The rectification of the ac to dc is often done with a circuit called a *phase-sensitive rectifier* which preserves the information on the polarity of the dc input. These problems are discussed in more detail in Chapter 14 where dc amplifiers are treated. There are, of course, other uses of diodes in instrumentation too numerous to catalog. For instance, the standard laboratory volt-ohm-milliammeter utilizes diodes in a rectifier in series with the microammeter when the instrument is used to measure ac voltages or currents.

Another common use of diodes is illustrated in Fig. 11.7. The diodes in this circuit form a clipper. It is assumed that forward resistance of the diodes is small compared to R and the reverse resistance is large compared to R. A time-varying input voltage v_i is shown in Fig. 11.7(b). As long as $-V \leq v_i \leq +V$, the output voltage v_o is the same as v_i. If $v_i > V$, however, the diode D_1 conducts and the output is clipped at $+V$. If $v_i < -V$, diode D_2 conducts and the output is clipped at $-V$. Actually if the diodes are made of silicon, the clipping occurs at approximately $+(V + 0.6)$ volts and $-(V + 0.6)$ volts.

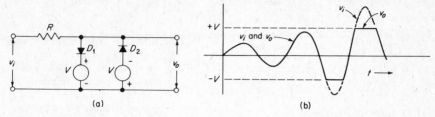

Figure 11.7 Two diodes used as voltage clippers. The output voltage v_o must remain between $+V$ and $-V$. As long as $|v_i| \leq V$, we have $v_o = v_i$. If $|v_i| > V$, then $v_o = +V$ or $v_o = -V$.

Clipping is used for a number of different purposes. One of them is to protect a circuit against destructive voltage surges. We shall see many other uses of diodes as we discuss individual circuits.

11.2 Bipolar Transistor Parameters

In the previous chapter we described very briefly two circuits in which a bipolar transistor can be used as an amplifier and in Chapter 8 we discussed in general terms the properties of active nonlinear circuits making use of vacuum tubes. Let us begin our more detailed treatment of transistor circuits with some of the standard terminology for transistor parameters. Figure 11.8 shows an *npn* bipolar transistor in a common emitter configuration. We have used the same notation for the currents and voltages that was described in Chapter 8. Total currents and voltages are indicated by lowercase letters with capital subscripts; the large dc component of the total quantity, by capital letters with capital subscripts; and small incremental quantities, by lowercase letters with lowercase subscripts. Thus in Fig. 11.8 the total base current i_B is provided by two current sources I_B and i_b in parallel. The total collector to emitter voltage v_{CE} is shown as two voltage sources V_{CE} and v_{ce} in series. Equations of the form $v_{CE} = V_{CE} + v_{ce}$, $i_B = I_B + i_b$, and so on, can be written for all the current and voltage variables of the circuit. The dc quantities I_B, I_C, and V_{CE} give the operating point or quiescent point of the transistor. The incremental quantities i_b, i_c, v_{ce}, and so on, are the signal variables, and it is usually only in terms of these that we calculate circuit characteristics such as amplification and input impedance.

The notation can be remembered more easily by keeping in mind that capital letters indicate dc quantities and lowercase letters indicate time-varying quantities. The key to the subscripts is that capital subscripts indicate large quantities while lowercase subscripts indicate small incremental quantities. Our convention that i_E flows out of the emitter as shown in Fig. 11.8 is nonstandard. Most books will have all three currents i_C, i_B, and i_E flowing into the transistor. We believe that our notation although nonstandard is easier for students to grasp since it implies that $i_C \simeq i_E$ rather than $i_C \simeq -i_E$.

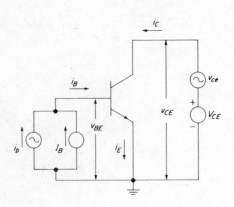

Figure 11.8 An *npn* bipolar transistor driven by ideal voltage and current sources in the collector and base circuits, respectively. Because the sources are ideal, they introduce no constraints among the currents and voltages. The incremental quantities i_b and v_{ce} can be chosen independently. Once they are chosen, the other incremental quantities are determined by Eq. 11.2. These equations depend only on the properties of the transistor.

Now let us return to the analysis of the circuit of Fig. 11.8. There are four variables to the problem, i_B, i_C, v_{BE}, and v_{CE}. We assume that i_E has been eliminated using $i_E = i_B + i_C$ and v_{CB} has been eliminated using $v_{CB} = v_{CE} - v_{BE}$. There are two independent loops that from Kirchhoff's voltage law give us two relationships among the four variables. Thus only two of the variables are independent. It is customary to take i_B and v_{CE} as the independent variables. It was for this reason that we made explicit the dc and incremental parts of i_B and v_{CE} in Fig. 11.8. We then write for small incremental changes in the dependent variables

$$\Delta v_{BE} = \left(\frac{\partial v_{BE}}{\partial i_B}\right)_{V_{CE}} \Delta i_B + \left(\frac{\partial v_{BE}}{\partial v_{CE}}\right)_{I_B} \Delta v_{CE}$$

$$\Delta i_C = \left(\frac{\partial i_C}{\partial i_B}\right)_{V_{CE}} \Delta i_B + \left(\frac{\partial i_C}{\partial v_{CE}}\right)_{I_B} \Delta v_{CE}$$

(11.1)

where $(\partial v_{BE}/\partial i_B)_{V_{CE}}$ means that the partial derivative is evaluated while holding $v_{CE} = V_{CE}$. These equations are simply the first terms in a Taylor series expansion of v_{BE} and i_C about the values V_{BE} and I_C. The small incremental variations are in our notation $\Delta v_{BE} = v_{be}$, $\Delta i_C = i_c$, $\Delta v_{CE} = v_{ce}$ and $\Delta i_B = i_b$. The derivatives of Eq. 11.1 are evaluated at and depend strongly on the operating point of the transistor.

If we translate Eq. 11.1 into the conventional transistor notation, we have

$$v_{be} = h_{ie} i_b + h_{re} v_{ce}$$
$$i_c = h_{fe} i_b + h_{oe} v_{ce}$$

(11.2)

where the symbols h_{ie}, h_{re}, h_{fe}, and h_{oe} are used for the partial derivatives. The partial derivatives h_{ie}, and so on, are characteristic of the particular transistor. Their values must either be measured or obtained from the manufacturer. If the incremental signals are sufficiently small, higher terms in the Taylor series expansion of v_{be} and i_c may be ignored. The incremental currents and voltages are then linearly related and it is in this sense that we speak of the transistor as a linear circuit component.

The first subscript identifies the role of the h parameter in the circuit, while the second subscript simply identifies the circuit as the common emitter configuration. The emitter is common to both input and output. The parameter h_{ie} relates an input voltage and current, and the parameter h_{oe} relates an output voltage and current. The parameters h_{re} and h_{fe} are transfer characteristics relating the input and the output. The former, h_{re}, is a reverse voltage characteristic since it gives the rate of change of an input parameter v_{be} with an output parameter v_{ce}. The latter, h_{fe}, is a forward current transfer characteristic relating an input current i_b to an output current i_c. Because one of the independent variables is a voltage and the other is a current, the h parameters do not all have the same physical dimensions. For this reason they are called *hybrid parameters*.

From the definitions one sees that h_{fe} is just β, the base-collector forward current gain. Note, however, that $h_{fe} = i_c/i_b$, while in the previous chapter we defined $\beta = I_C/I_B$. Actually β is relatively independent of I_C and the incremental and total values do not differ greatly. Another important characteristic is the input impedance $h_{ie} = v_{be}/i_b$. We shall make frequent use of a closely related parameter $r_{tr} = h_{ie}/\beta$. In a later section we show how r_{tr} and h_{ie} can be estimated. The parameter h_{ie} depends strongly on the operating point at which it is evaluated. In particular, h_{ie} increases rapidly with decreasing I_C.

The coefficients of v_{ce} (h_{re} and h_{oe}) are usually small and a great deal of useful circuit analysis can be done on the assumption that they are zero. That they are not zero is due to the modulation of the base width by the collector-emitter voltage. An increase in v_{CE} increases the width of base-collector barrier layer and decreases the base width. As a result a larger fraction of the emitter current can reach the collector. In addition, the base current sees a larger ohmic resistance in the narrower base. Thus a change in v_{CE} demands changes in both i_C and v_{BE} if i_B is to remain constant ($i_b = 0$).

Because some of the h parameters may vary appreciably with I_C and because the nonlinear properties cannot be predicted using h parameters, a graphical presentation of transistor characteristics is necessary, This is usually in the form of plots of i_C versus v_{CE} for a series of different constant base currents. Figure 11.9 shows such plots for the 2N5183, a widely used general-purpose silicon npn transistor. The typical values of the h parameters given by the manufacturer are

$$h_{fe} = 175$$
$$h_{ie} = 600 \text{ ohms}$$
$$h_{oe} = 75 \text{ micromhos} \qquad (11.3)$$
$$h_{re} = 125 \times 10^{-6}$$

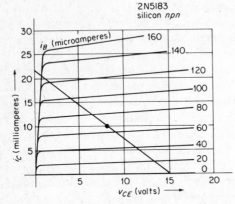

Figure 11.9 The collector characteristics of the silicon npn transistor No. 2N5183. This transistor can also be used at much higher currents (characteristics not shown). The load line and operating point shown are for the circuit of Fig. 11.14. (*Characteristic curves courtesy RCA Corporation.*)

These have been measured at $V_{CE} = 12$ volts and $I_C = 10$ milliamp. The parameters determining i_c can be read directly from the characteristic curves. To find h_{fe}, simply read the change in i_C (at constant v_{CE}; that is, $v_{CE} = V_{CE}$ so that $v_{ce} = 0$) from one i_B curve to the next. Then $\Delta i_C / \Delta i_B = (i_c/i_b)_{V_{CE}} = h_{fe}$. The slope of any one of the curves, since i_B is constant, gives $\Delta i_C / \Delta v_{CE} = (i_c/v_{ce})_{I_B} = h_{oe}$.

11.3 Small Signal Amplification

We shall discuss briefly a simple common emitter amplifier using the 2N5183 of Fig. 11.9 and the h parameters given in Eq. 11.3. For the operating point we choose $V_{CE} = 8$ volts and $I_C = 10$ milliamp. The lower value of V_{CE} (8 volts instead of 12 volts) is somewhat more convenient in terms of supply voltages and biasing resistors. Since we keep $I_C = 10$ milliamp, there are no important changes in the h parameters. Referring to Fig. 11.8 the dc power for the transistor would be provided by an 8-volt battery (V_{CE}) in the collector circuit and a current source I_B in the base circuit. The magnitude of I_B is chosen so that a collector current I_C of 10 milliamp is obtained.

Under these conditions the base must be about 0.6 volt positive with respect to the emitter since the transistor is made of silicon. Small signal variations around the operating point are then related by Eq. 11.2. It is important to realize, however, that Fig. 11.8 does not represent an amplifier circuit. The voltage and current sources, both incremental and dc, are ideal. The voltage sources have zero internal resistance and the current sources have infinite internal resistance. Thus the relationships between the voltages and currents involve only the properties of the transistor and do not involve IR drops across external impedances. There are two independent variables. The value of one of the independent variables alone does not uniquely determine the value of either of the dependent variables. In order to define an amplification one must constrain one of the independent variables so that the value of the other independent variable uniquely determines the values of the dependent variables. Under these conditions, for example, the value of the input voltage v_{be} might uniquely determine the output voltage v_{ce} and hence the voltage amplification v_{ce}/v_{be} is determined.

In a real amplifier external impedances furnish additional relations between currents and voltages and the number of independent variables is reduced to one. The amplification, input impedance, and output impedance of the amplifier will then depend on the external impedances as well as the h parameters of the transistor.

Figure 11.10 shows an actual amplifier circuit based on the configuration of Fig. 11.8. A resistor R_C is connected to the collector and a dc power supply provides a constant voltage V between the top of R_C and ground. For the present we assume that the output is open, that is $R_L = \infty$ and $i_o = 0$.

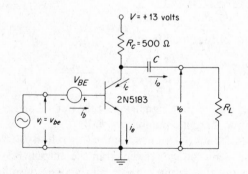

Figure 11.10 A voltage amplifier circuit based on the common emitter configuration of Fig. 11.8. In the numerical work it is assumed that the transistor is operating at $I_C = 10$ milliamps, $V_{CE} = 8$ volts.

The voltage V satisfies the equation $V = i_C R_C + v_{CE} = (I_C + i_c)R_C + (V_{CE} + v_{ce})$. Now since V is independent of time, the time-dependent terms on the right-hand side of the equation must sum to zero. Thus we obtain the two equations $0 = i_c R_C + v_{ce}$ and $V = I_C R_C + V_{CE}$. The second of these equations relates I_C and V_{CE}, the operating point parameters of the transistor, to the external circuit parameters R_C and V. If $I_C = 10$ milliamp, $V_{CE} = 8$ volts, and we choose $R_C = 500$ ohms, then V must be $+13$ volts. The first of the equations $v_{ce} = -i_c R_C$ is an important additional relation among the incremental variables that reduces the number of independent variables from two to one.

Before we calculate the properties of the amplifier, let us consider the input and output circuits. Because we wish to discuss the voltage amplification of the circuit, the base is shown driven by voltage sources rather than the current sources used in Fig. 11.8. V_{BE} is determined once I_C and V_{CE} are chosen. For a silicon transistor, however, V_{BE} will be close to 0.6 volt. The incremental input signal is $v_i = v_{be}$. We shall assume that v_i is from a source with zero internal resistance. The incremental output voltage is $v_o = v_{ce}$. The capacitor C prevents the large dc voltage V_{CE} from appearing in the output. The capacitor may be used only if v_i is an ac signal. In addition, C must be large enough so that for the important frequency components of v_i the impedance $1/\omega C$ is very small. We shall also assume initially that the output is an open circuit. Only in this case does the entire incremental current i_c flow through R_C. The relation $v_o = v_{ce}$ is always true if $1/\omega C$ is small compared to any output load impedance but $v_{ce} = -i_c R_C$ holds only if the output draws no current. Now let us calculate the voltage amplification $A = v_o/v_i$ of the amplifier of Fig. 11.10. Substituting $v_{ce} = -i_c R_C$ in Eq. 11.2, we obtain

$$v_{be} = h_{ie} i_b - h_{re} i_c R_C$$
$$i_c = h_{fe} i_b - h_{oe} i_c R_C \tag{11.4}$$

We eliminate i_b, solve for i_c, and find

$$i_c = \frac{h_{fe} v_{be}}{h_{ie}(1 + h_{oe} R_C) - h_{fe} h_{re} R_C} \tag{11.5}$$

The voltage amplification with no output load is then

$$A = \frac{v_o}{v_i} = \frac{v_o}{v_{be}} = \frac{-i_c R_C}{v_{be}} = \frac{-1}{[h_{ie}(1 + h_{oe} R_C)/R_C h_{fe}] - h_{re}} \tag{11.6}$$

The -1 indicates that the input and output signals are $180°$ out of phase. Note that the input and the output share a common ground.

Because i_c was calculated for an open output circuit, it follows that $v_o = -i_c R_C$ is the Thevenin equivalent voltage viewed from the output. It is important to remember that i_c is not the output current. In the open output calculations we have done so far, the output current is zero. The output impedance R_o of the transistor amplifier is the Thevenin equivalent impedance seen looking into the two output terminals of the amplifier. To determine the output impedance R_o we must divide v_o by the short circuit output current i_s. This current flows, for example, when a large condenser is put across the output, shorting all incremental ac signals to ground. All the incremental collector current flows through the condenser, not R_C, and no incremental ac voltage appears in the collector circuit. Thus both v_{ce} and v_o are zero. In this case i_s satisfies the equations

$$v_{be} = h_{ie} i_b$$
$$i_s = h_{fe} i_b$$

or (11.7)

$$i_s = \frac{h_{fe} v_{be}}{h_{ie}}$$

We find for the output impedance

$$R_o = \frac{v_o}{i_s} = \frac{i_c R_C}{i_s} = \frac{R_C}{(1 + h_{oe} R_C) - (h_{fe} h_{re} R_C / h_{ie})}$$ (11.8)

The input impedance $R_i = v_{be}/i_b$ can be calculated by eliminating i_c from Eq. 11.4. When the output is shorted, $v_{ce} = 0$, we find $R_i = h_{ie}$. This remains a good approximation for any output load. In fact from our earlier statement that the terms involving v_{ce} were usually small we expect that the expressions for A and R_o also can be simplified. Let us check this with the h parameters given for the 2N5183 and for $R_C = 500$ ohms. We find $h_{oe} R_C = 0.0375$; thus $1 + h_{oe} R_C$ can be replaced by unity. In the denominator of i_c (Eq. 11.5), h_{ie} is 600 ohms and $h_{fe} h_{re} R_C$ is 0.022 ohms. Therefore to an excellent approximation $i_c = h_{fe} v_{be}/h_{ie}$, which is just the result if the terms in v_{ce} are neglected. We then obtain $A = -h_{fe} R_C/h_{ie} = -146$ and $R_o = R_C = 500$ ohms. In the next section we develop simplified methods of circuit analysis based on the parameters h_{ie} and h_{fe} (or β).

11.4 The Transresistance

Along with their many and obvious virtues, certain shortcomings of transistor amplifiers must be mentioned. One is nonlinearity. The exponential relation $i_E = I_0 \exp(eV/kT)$, where V is the voltage across the base-emitter pn junction, introduces a major nonlinearity between v_{BE} and i_E and also between v_{be} and i_e. Note that v_{BE} and V are not identical because there are ohmic voltage drops in both the emitter and in the base. The relation between v_{be} and i_c is also nonlinear since $i_c = \alpha i_e$ and α remains relatively constant independent of v_{be}. Suppose, for example, that a transistor operating with a quiescent collector current $I_C = 10$ milliamp is driven by a sine wave voltage input, which produces across the base-emitter junction a maximum voltage variation of ± 0.018 volt. Knowing that $kT/e = 0.026$ volt at room temperature, it is easily calculated that such an input during its positive cycle will double the collector current and on its negative cycle will halve it. Thus the forward swing of i_c is 10 milliamp and the reverse swing is only 5 milliamp. The output voltage $v_o = -i_c R_C$ is greatly distorted. It should be remembered that with an input signal voltage as large as the one we have just used Eq. 11.2 do not accurately describe the relations among the incremental variables.

It is possible to use a current drive in the input and avoid much of this difficulty since $i_c = h_{fe} i_b \simeq \beta i_b$ and β is fairly constant independent of the dc current I_C. This constancy is only with respect to variations of the operating point of a given transistor, however. It is not easy to control β from one transistor to the next in the manufacturing process. The 2N5183 typically has $\beta = 175$ but the manufacturer guarantees a minimum β no greater than 70. Thus the replacement of a transistor may greatly alter circuit behavior and several transistors, presumably identical, may differ appreciably.

In actual circuit design these and similar problems are handled by the use of negative feedback. A detailed discussion is given in the next chapter with one or two examples in the next sections. What negative feedback does is to make the amplification of the circuit almost independent of the transistor parameters. The price one pays is a reduction in total amplification, but high amplification is easily obtained and is willingly sacrificed for increased stability and linearity.

In summary, nearly all transistor amplifiers use some form of negative feedback so that circuit behavior is almost independent of the transistor parameters. It is therefore pointless to use difficult and complicated methods of circuit analysis such as a full exploitation of the h parameters. A simple semiquantitative approach might better be used since inadequacies in the description of the transistor are not going to affect the circuit behavior very much anyway.

Such an approach uses the transresistance r_{tr}, which is defined as

$$r_{tr} = \left(\frac{\partial v_{BE}}{\partial i_C} \right)_{V_{CE}} = \left(\frac{v_{be}}{i_c} \right)_{V_{CE}} \tag{11.9}$$

The quantity $r_{tr} = v_{be}/i_c$ is called a *transresistance* since it relates an input voltage v_{be} to an output current i_c. Since $\alpha \simeq 1$, we have $i_c \simeq i_e$ and therefore $r_{tr} = v_{be}/i_c \simeq v_{be}/i_e$ or $v_{be} = r_{tr} i_c \simeq r_{tr} i_e$. Let us consider separately the resistances associated with the emitter-base junction, the emitter, and the base. The first we find by differentiating $i_E = I_0 \exp (eV/kT)$. We obtain

$$di_E = \frac{e}{kT} I_0 e^{eV/kT} dV = \frac{e}{kT} i_E dV$$

or $\tag{11.10}$

$$r_d = \frac{dV}{di_E} = \frac{kT}{eI_E} = \frac{0.026}{I_E}$$

where r_d has been evaluated at the quiescent point so that $i_E = I_E$ and where I_E is in amperes or $r_d = 26/I_E$ where I_E is in milliamperes. This is the dynamic resistance of the junction; dV is that part of v_{be} that appears across the junction. There are, in addition, ohmic resistances r_e' and r_b' in the emitter and the base. These two terms introduce an incremental voltage drop $v_{ohmic} = [r_e' + r_b'/(1 + \beta)]i_e$. The term $r_b'/(1 + \beta)$ enters because only the current $i_b = i_e/(1 + \beta)$ flows in the base. We then write

$$v_{be} = \left(\frac{0.026}{I_E} + r_{ohmic}\right)i_e \simeq r_{tr} i_e$$

where   (11.11)

$$r_{ohmic} = r_e' + \frac{r_b'}{1 + \beta}$$

Now the usefulness of this approach depends on the fact that r_{ohmic} is remarkably constant for a wide variety of transistors. Power transistors may have an r_{ohmic} of a few tenths of an ohm. For low performance transistors, r_{ohmic} may exceed 20 ohms. But for the great majority of the high performance low and medium power transistors used in instrumentation circuits r_{ohmic} will be found between 1 ohm ($\beta \simeq 300$) and 5 ohms ($\beta \simeq 40$). Note that since $r_{ohmic} = r_e' + r_b'/(\beta + 1)$, the value of r_{ohmic} and hence r_{tr} decreases as β increases, other things being equal. High current transistors usually have a low value of r_{ohmic}.

We shall also use r_{tr} for computations where $v_{ce} \neq 0$. In this case there is a contribution to v_{be} from the base width modulation effect, but it is small and will be ignored. To the approximation that terms in v_{ce} can be neglected and for $\beta \gg 1$ so that $i_e \simeq i_c$, we have

$$v_{be} = h_{ie} i_b = \frac{h_{ie}}{\beta} i_c \simeq \frac{h_{ie}}{\beta} i_e$$

and   (11.12)

$$r_{tr} = \frac{v_{be}}{i_c} = \frac{h_{ie}}{\beta} \simeq \frac{v_{be}}{i_e}$$

Let us reconsider the circuit of Fig. 11.10. The input voltage is given by $v_i = r_{tr} i_c$ and the output voltage is still given by $v_o = -i_c R_C$. The amplification is given by $A = v_o/v_i = -R_C/r_{tr}$. The output impedance is still given approximately by R_C and the input impedance is given by $R_i = v_{be}/i_b = \beta r_{tr}$. These results are of course equivalent to those obtained using the h parameters. The nonlinearity of a transistor amplifier with voltage drive is attributable to the rapid variation of h_{ie} or r_{tr} with I_E.

To summarize, we have expressed r_{tr} as the sum of two terms, an ohmic resistance $r_{\text{ohmic}} = r_e' + r_b'/(1 + \beta)$ and a dynamic resistance $r_d = 26/I_E$ where I_E is in milliamperes. For $I_E > 26$ milliamp the transresistance is approximately constant and only slightly greater than the ohmic resistance. For $I_E \ll 26$ milliamp the dynamic resistance makes the main contribution to r_{tr} and the transresistance can become much larger than the ohmic resistance.

11.5   Equivalent Circuits for Bipolar Transistors

Thus far in this chapter we have treated the bipolar transistor as a three-terminal black box. Considering only the small signal behavior, the transistor is a linear active network—an active betwork because more ac power can come out of the box than is put in. Our results for the common emitter configuration of an *npn* transistor are summarized in Fig. 11.11 and Eq. 11.13.

$$v_{be} = h_{ie} i_b + h_{re} v_{ce}$$
$$i_c = h_{fe} i_b + h_{oe} v_{ce}$$

$$(11.13)$$

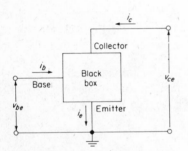

Figure 11.11   The common emitter configuration of a transistor represented as a black box. Knowing the four h parameters for the transistor, one can completely describe the behavior of the circuit with respect to the incremental variables.

As we have seen, Eq. 11.13 may be used to calculate all the relevant properties of any circuit in which the transistor occurs. However, there are sometimes advantages to having an actual circuit, that is, an arrangement of interconnected components, resistors, and voltage sources or current sources that reproduce the behavior of the black box. We saw an example of this in Chapter 2 where a three-terminal passive network was reduced to three actual resistances in either the Δ or Y connection. In our present case four parameters are necessary because the network is active and the reciprocity theorem does not hold.

An actual circuit representing the black box is called an *equivalent circuit*. What we demand is that the application of Kirchhoff's laws to the equivalent circuit must lead to a set of equations equivalent to Eq. 11.13.

We might guess (correctly) from the Δ, Y example that several different circuits may be equivalent to the same black box. One of these, the hybrid equivalent, is suggested by the form of Eq. 11.13.

The first of these equations says that v_{be} is made up of two voltage drops in series. The first is an iR drop of magnitude $h_{ie}i_b$, and the second is a voltage source of magnitude $h_{re}v_{ce}$. The second of these equations says that i_c divides into two currents in parallel; the first is represented by a current source $h_{fe}i_b$ and the second by the current in the resistance $(h_{oe})^{-1}$ across which the voltage v_{ce} appears. The resulting hybrid equivalent circuit is shown in Fig. 11.12.

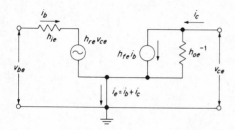

Figure 11.12 The hybrid h equivalent circuit for a transistor in the common emitter configuration.

It is immediately obvious that the application of Kirchhoff's laws to the input loop (v_{be}) and the output loop (v_{ce}) gives Eq. 11.13. At first glance the circuit appears to have isolated the output from the input. The necessary interaction is actually provided by the voltage source in the input which depends on an output variable and the current source in the output which depends on an input variable.

Another useful and physically more appealing network is the T equivalent circuit shown in Fig. 11.13. In this circuit the interaction between input and output is provided by the voltage $r_e(i_b + i_c)$ that is common to both. Now we must show that Kirchhoff's laws lead to equations equivalent to Eq. 11.13 and then find the relationships between r_e, r_b, r_c, β, and the four h parameters.

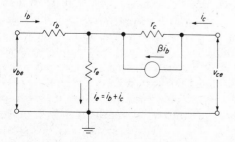

Figure 11.13 The T network equivalent circuit for a transistor in the common emitter configuration.

Applying Kirchhoff's voltage law to the input and output loops, we obtain

$$v_{be} = i_b r_b + (i_b + i_c)r_e$$
$$v_{ce} = (i_c - \beta i_b)r_c + (i_b + i_c)r_e \qquad (11.14)$$

The second of the equations above gives i_c as a function of i_b and v_{ce} and is equivalent to the second of Eq. 11.13. We may use the second equation to eliminate i_c from the first equation. This gives v_{be} as a function of i_b and v_{ce} and is equivalent to the first of Eq. 11.13. The algebraic skirmish is straightforward, brief, and decisive. We find

$$v_{be} = \left[r_b + \frac{r_c r_e(\beta + 1)}{r_c + r_e} \right] i_b + \frac{r_e}{r_c + r_e} v_{ce}$$

$$i_c = \left(\frac{\beta r_c - r_e}{r_c + r_e} \right) i_b + \left(\frac{1}{r_c + r_e} \right) v_{ce} \qquad (11.15)$$

We equate similar coefficients in Eq. 11.13 and 11.15 and obtain the following table relating the h parameters and the T equivalent parameters.

Table 11.1

$$h_{ie} = r_b + \frac{r_c r_e(\beta + 1)}{r_c + r_e} \simeq r_b + r_e(\beta + 1) \simeq \beta r_{tr}$$

$$h_{re} = \frac{r_e}{r_c + r_e} \simeq \frac{r_e}{r_c} \simeq \frac{r_{tr}}{r_c}$$

$$h_{fe} = \frac{\beta r_c - r_e}{r_c + r_e} \simeq \beta$$

$$h_{oe} = \frac{1}{r_c + r_e} \simeq \frac{1}{r_c}$$

The final results use the approximation $r_c \gg r_e$ or r_b, which is always very good. That $r_c \gg r_e$ can be seen from the equation $1/h_{oe} = r_c + r_e$. For the 2N5183 transistor $1/h_{oe} = 1.3 \times 10^4$ ohms. Now $r_e = r_d + r_e' < r_{tr} \simeq 5$ ohms so that $r_c \gg r_e$ and $h_{oe} \simeq 1/r_c$. In these equations $r_d = 26/I_E$ where I_E is in milliamperes. We have earlier defined $r_{tr} = r_d + r_e' + r_b'/(1 + \beta)$ where r_e' and r_b' are ohmic resistances associated with the emitter and the base. The r_e of the T equivalent circuit is $r_d + r_e'$. The T equivalent r_b is the same as r_b'. We have also assumed in Table 11.1 that $r_e \simeq r_{tr}$. This is, of course, equivalent to assuming that $r_b'/(1 + \beta)$ is small compared to $r_e = r_d + r_e'$. This is a good approximation if I_E is small but may not be too good if I_E is greater than about 5 milliamp. The reader is warned that many authors use r_c for the r_c'' of our Problem 11.22. Thus their r_c is approximately β times our r_c.

11.6 Properties of Single-Stage Amplifiers

We first consider, using the circuit of Fig. 11.14, how one supplies the proper bias voltages to a transistor. By the bias voltages we mean the dc quiescent voltages V_{CE} and V_{BE}. Figure 11.14 is essentially the circuit of Fig. 11.10 redrawn to use a single 15 volt power supply to supply the bias voltages. We shall assume the power supply has zero internal resistance.

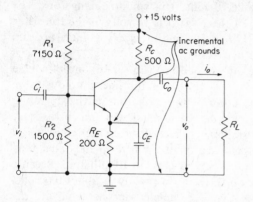

Figure 11.14 The amplifier circuit of Fig. 11.10 redrawn to use a single 15-volt power supply. The load line and operating point are shown in Fig. 11.9.

Given $I_E \simeq I_C = 10$ milliamp, we have a 5 volt drop across R_C and a 2 volt drop across R_E. This gives the desired operating point $V_{CE} = 8$ volts. The resistances R_1 and R_2 are so chosen that the base is at 2.6 volts. This gives $V_{BE} = 0.6$ volt, which is required for a silicon transistor. R_1 and R_2 must not be too small since they shunt the input to ground and decrease the input resistance. They must not be too large lest the base current flowing through them unduly affect V_{BE}. The current through the divider, in this case, $\frac{15}{8650} = 1.7$ milliamp, should be large compared to I_B, which in this case is about 60 microamp.

As we shall see, the emitter resistor R_E serves as a source of feedback voltage if the ac bypass capacitor C_E is removed. R_E is also important in stabilizing the dc operating point. If any change in transistor characteristics causes I_E to increase, for instance, an increase in temperature, then the emitter potential rises and V_{BE} decreases, which of course tends to decrease I_E. The voltage divider biasing for the base of the transistor as shown in Fig. 11.14 has considerable stability and is widely used.

Once the supply voltage and R_C and R_E are chosen, Ohm's law imposes a linear relationship between I_C (or I_E) and V_{CE}. Assuming $I_C \simeq I_E$, the relation is

$$V_{CE} = 15 - I_C(R_C + R_E) \tag{11.16}$$

This is called the dc load line. It is drawn through the operating point in

Fig. 11.9. The slope of the load line is $-(R_C + R_E)^{-1}$ and the intersection with the V_{CE} axis is the supply voltage. The load line must be considered in nearly all design problems. It relates the extreme values of I_C, V_{CE}, and I_B and these must be known if one is to remain within a range of useful transistor characteristics. It may enforce trade-offs between R_C and the supply voltage. Had we wished to double the amplification by choosing $R_C = 10^3$ ohms and still keep the same operating point, $I_E = 10$ milliamp, $V_{CE} = 8$ volts, the supply voltage would have to be 20 volts. This could be dangerous. The minimum v_{CE} breakdown voltage of the 2N5183 is 18 volts. If the transistor is accidentally driven to cutoff ($i_C = 0$), the entire supply voltage appears across the transistor and damage may result.

Now I_C and V_{CE} must lie on the load line (Eq. 11.16) and of course the point $I_C = I_E = 10$ milliamp, $V_{CE} = 8$ volts is on this line. But how do we know that this rather than some other point on the line is the operating point (or quiescent point) when all the incremental voltages and currents are zero? Here we see the important stabilizing effect of the emitter resistor R_E. If the base is put roughly 2.6 volts above ground and if $R_E = 200$ ohms, then I_E, in the absence of a signal v_i, must be very close to 10 milliamp. Because a forward-biased silicon pn junction turns on so rapidly at about 0.6 volts, any other value of I_E gives a V_{BE} that is completely inconsistent with the assumed I_E.

The capacitors C_i and C_o are used to block the dc voltages of the amplifier from the input and output circuits. They should not, however, attenuate the signal voltages. Thus for the important frequencies in v_i the impedance $1/\omega C_i$ must be small compared to the input resistance of the amplifier and $1/\omega C_o$ must be small compared to the output resistance.

The reasons for using the capacitor C_E are somewhat more subtle. When we calculate the amplification of the circuit, we shall find that the amplification is greatly reduced if the incremental voltage $i_e R_E$ appears between the emitter and ground. This can be avoided if C_E is large enough so that $1/\omega C_E \ll R_E$ for the important frequencies in v_i and therefore in i_e. Thus for the incremental variables the emitter is shorted to ground. We remind the reader that R_E contributes greatly to the simplicity and stability of the dc biasing so the solution to the amplification problem is not to put $R_E = 0$.

However, the capacitor C_E forces us to reconsider Eq. 11.16. Let us rewrite Eq. 11.16 including not just the dc but also the incremental parts of the total voltages and currents. The result is $V_{CE} + v_{ce} = 15 - (I_C + i_c)R_C + I_C R_E$. Because $1/\omega C_E \ll R_E$, the incremental voltage from the emitter to ground is zero. Using the fact that the dc quantities satisfy $V_{CE} = 15 - I_C(R_C + R_E)$, we obtain from Eq. 11.16 the relation $v_{ce} = -i_c R_C$ between the incremental variables. We have assumed $I_C \simeq I_E$ so that $I_E R_E \simeq I_C R_E$ in this analysis.

Thus the incremental signals produce excursions of i_C and v_{CE} from the quiescent point along a line whose slope is $i_c/v_{ce} = -1/R_C$. This is called the *ac load line*. In the absence of C_E the excursions would follow the dc load line whose slope is $-1/(R_C + R_E)$. The reader is referred to Chapter 8 where these points were discussed in somewhat greater detail. The discussion of the ac load line has assumed that the output is an open circuit, that is $R_L = \infty$ and $i_o = 0$.

Let us now, with all our biases adjusted, return to the use of r_{tr} in the calculation of amplifier performance. In the following analysis we assume that the capacitors C_i, C_o, and C_E provide a negligibly small ac impedance at the frequencies of the incremental signals. In Fig. 11.14 we note that C_E is connected so that the emitter is an ac ground and the circuit is precisely that of Fig. 11.10. We have $v_i = v_{be} = r_{tr} i_c$ and $v_o = -i_c R_C$. The voltage amplification is $A = v_o/v_i = -R_C/r_{tr}$. The dynamic contribution to r_{tr} is $r_d = \frac{26}{10} = 2.6$ ohms. Since $\beta = 175$ is rather high, let us take $r_{\text{ohmic}} = 3$ ohms. We obtain $r_{tr} = 5.6$ ohms and $A = -500/5.6 = -89$. The value calculated in the previous section using the h parameters is $A = -146$. The gain we calculate with r_{tr} is somewhat low. From $r_{tr} = h_{ie}/\beta = \frac{600}{175} = 3.4$ ohms and $r_d = 2.6$ ohms we see that r_{ohmic} of this transistor is really a little less than 1 ohm. This is not unexpected. The 2N5183 may be used at currents 10 times higher than those of Fig. 11.9. High current transistors usually have low values for r_{ohmic}, as previously discussed. Estimates of r_{ohmic} will usually be much better than this but we shall see that even in this "worst case" our estimated r_{tr} gives excellent results in a circuit using negative feedback.

The input impedance for the transistor only is $R_i = v_i/i_b = v_{be}\beta/i_e = \beta r_{tr} = 175 \times 5.6 = 980$ ohms. For the more accurate calculation we previously obtained $R_i = h_{ie} = 600$ ohms. Both the resistors R_1 and R_2 go to an incremental ac ground, that is, to a fixed dc potential. Including the base bias resistors in the input an additional resistance $R_{eq} = R_1 R_2/(R_1 + R_2) = 1240$ ohms shunts the input resistance of the transistor itself. Thus the total input resistance seen by v_i is $600 \cdot \frac{1240}{1840} = 400$ ohms. The calculation of the output resistance is the same as in the previous section and $R_o = R_C = 500$ ohms.

Now let us repeat our calculations assuming that the shunt capacitor C_E has been removed. We now have $v_i = v_{be} + i_e R_E = r_{tr} i_c + R_E i_e$ and as before $v_o = -R_C i_c$. In this circuit the resistor R_E is said to produce a negative feedback voltage, $i_e R_E$. Since $i_e \simeq i_c$, the voltage gain is

$$A = \frac{v_o}{v_i} = -\frac{R_C}{r_{tr} + R_E} = -2.5$$

This is a rather low amplification so we shall separate the dc and incremental ac functions of R_E by placing the shunt capacitance C_E across a portion of R_E. Assume we shunt 150 ohms with C_E and leave 50 ohms as a feedback resistance. We then have a higher amplification

$$A = -\frac{500}{5.6 + 50} = -9$$

The transistor parameters enter only through the transresistance $r_{tr} = 5.6$ ohms. A variation of 50% in r_{tr} changes the amplification by only 5 or 6%. Had we used the more accurate value $r_{tr} = 3.4$ ohms, the gain would be $A = -9.4$. Uncertainties in r_{tr} can arise from deliberate approximations in the calculation of r_{tr}, from the parameter fluctuation introduced in manufacture, from change in position on the load line (nonlinearity), or from other causes. The effect of these uncertainties in r_{tr} on the amplification is very small in a properly designed circuit using negative feedback.

We now calculate the input resistance R_i with feedback. We have for the input resistance of the transistor only

$$R_i = \frac{v_i}{i_b} = \frac{(r_{tr} + 50)i_e}{i_b} = \beta 55.6 = 9700 \text{ ohms}$$

The input resistance has been greatly increased. The actual input resistance is now determined primarily by the bias voltage divider $R_{eq} = 1240$ ohms, which is in parallel with the input resistance of the transistor itself. If the higher input resistance is needed, a different bias supply must be used.

In the next chapter we give a more complete and formal treatment of feedback but the subject is so important, and sometimes elusive, that one must have a clear and simple picture of what is going on. Let us look at the circuit of Fig. 11.14 from a different point of view.

Without feedback we have $v_{be} = v_i$ and $A_0 = -146$. With feedback we have $v_{be} = v_i - i_e R_E'$ and $A_F = -9$. The resistance R_E' is that part of the emitter resistor R_E that is used for feedback (50 ohms). From the viewpoint of the transistor the input is always v_{be} and the gain is always $A_0 = -146$. As seen by the transistor the feedback simply cuts down the input signal by a factor of

$$\frac{v_i - i_e R_E'}{v_i} = \frac{A_F}{A_0} \simeq \frac{1}{16}$$

in our example. Now where does the stabilization against transistor variation come from? Suppose we have to replace the transistor and we get one with $r_{ohmic} \simeq 4$ ohms rather than $r_{ohmic} \simeq 0.8$ ohm. A_0 would be -73 in this case. If this is to be compensated for in the amplifier with feedback, v_{be} must be doubled. In order to double v_{be}, however, the quantity $i_e R_E'$ need only drop from $\frac{15}{16} v_i$ to $\frac{14}{16} v_i$, a change of only 6%. In a feedback amplifier, large changes in transistor characteristics, which appear as large changes in A_0, are bucked out by compensating but necessarily also large fractional changes in v_{be}. This is accomplished with only small changes in the output of the amplifier with feedback. The reason can be put in very familiar terms. Given two large and almost equal quantities (v_i and $i_e R_E'$), a small fractional change in one of them can produce a very large fractional change in their small difference (v_{be}). This simple fact encompasses a good bit of feedback theory. We must also remember that $v_i \simeq i_e R_E' \gg v_{be}$ requires $A_0 \gg A_F$.

We have already seen that negative feedback makes the gain insensitive to the properties of the transistor. Let us now consider the improvement of linearity by the use of feedback. Students sometimes view the claim of improved linearity with suspicion, assuming that the comparison of the amplifier with and without feedback is made for the same v_i. In this case linearity is also improved simply because the lower amplification with feedback limits operation to a smaller region of the ac load line. The comparison should be made for the same output signal and therefore the same v_{be}. Suppose the circuit without feedback is driven by a pure sine wave voltage input $v_i = v_{be}$. If the part of v_{be} appearing across the junction is as great as 0.01 volt, there will be considerable nonlinearity in the output voltage and current. In Fig. 11.15 we show the input $v_i = v_{be}$ and the output i_c. Without feedback, i_c is badly distorted; the positive swing is much greater than the negative. We now introduce feedback. The input voltage v_i remains a pure sine wave but its amplitude must be increased by the factor A_0/A_F to provide the same output. The output is almost but not quite sinusoidal. The positive swing is somewhat larger than the negative. Now $v_{be} = v_i - i_e R_E'$ is badly distorted but in the opposite sense. Its negative swing is much larger than the positive. This is just the input signal needed if the transistor is to provide an almost sinusoidal output. Thus we see that feedback does, in fact, greatly improve the linearity. The distortion of v_{be} in a feedback amplifier is easily observed if one displays the base-emitter voltage on an oscilloscope.

Let us now examine Fig. 11.16, which illustrates an emitter follower circuit. The output is taken between the emitter and ground. It is obvious that the dc quiescent point is the same as for the previous amplifier circuit. However, the collector resistor R_C is shorted to ground for incremental ac signals by the capacitor C_C. The emitter resistor R_E is not bypassed by a capacitor.

For the small signal analysis of this circuit we can write the input voltage as $v_i = v_{be} + R_E i_e \simeq (r_{tr} + R_E)i_e$. The output voltage is $v_0 = i_e R_E$. The amplificator of the circuit is given by

$$A = \frac{v_o}{v_i} = \frac{R_E}{r_{tr} + R_E} = \frac{200}{5.6 + 200} = 0.97$$

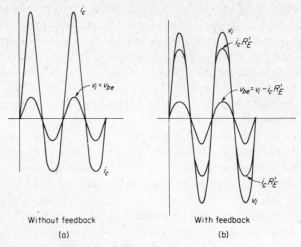

Without feedback
(a)

With feedback
(b)

Figure 11.15 Input and output waveforms showing how the use of negative feedback reduces the distortion in the output. In (a) $v_i = v_{be}$ is sinusoidal; the output i_c is badly distorted. In (b) the input v_i is sinusoidal; the output i_c is slightly distorted but the actual input to the transistor $v_{be} = v_i - i_c R'_E$ is badly distorted.

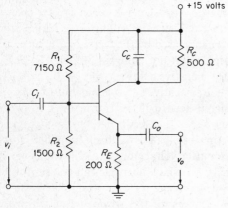

Figure 11.16 An emitter follower circuit. Each of the capacitors is assumed to have negligible impedance for the incremental signals.

The gain of an emitter follower is close to $+1$ provided $r_{tr} \ll R_E$. There is no phase inversion with an emitter follower circuit. The input resistance of the transistor is given by

$$R_i = \frac{v_i}{i_b} = \frac{(r_{tr} + R_E)i_e}{i_e/\beta} = \beta(r_{tr} + R_E) \simeq \beta R_E = 3.5 \times 10^4 \text{ ohms}$$

The total input impedance of the circuit is dominated entirely by the 7150- and 1500-ohm resistors used to bias the base. These appear in parallel with R_i for the ac input signal, and the actual total input resistance of the circuit is about 1240 ohms.

The output impedance of the emitter follower can be calculated as follows. The Thevenin open circuit voltage at the output is $\varepsilon_{Th} = i_e R_E$. If the output is shorted to ground for ac signals by, for example, putting a large capacitor across R_E, we find that the short circuit current that flows is

$$i_s = \frac{v_i}{r_{tr}} = \frac{(r_{tr} + R_E)i_e}{r_{tr}}$$

Therefore the output resistance is given by

$$R_o = \frac{\varepsilon_{Th}}{i_s} = \frac{i_e R_E}{i_e[(R_E + r_{tr})/r_{tr}]} = \frac{r_{tr} R_E}{R_E + r_{tr}} \simeq r_{tr} = 5.6 \text{ ohms}$$

The output impedance of the emitter follower acts as if it were made up of two resistors r_{tr} and R_E in parallel. This is of course an obvious result if one thinks of the circuit in terms of the equivalent T circuit. If the voltage source v_i had not been an ideal one but had been a voltage source and a resistor R_g in series, the output impedance of the emitter follower would have been given by R_E in parallel with $r_{tr} + R_g/(\beta + 1)$ as seen by considering the equivalent T circuit. The term $R_g/(\beta + 1)$ occurs because the current in the base circuit is less than i_e by a factor $1/(\beta + 1)$.

An emitter follower is normally used as a buffer stage to match a high impedance source so that it can drive a low impedance load. The gain of the emitter follower is $+1$ so that it does nothing but change the impedance level.

As a final comment we point out that an emitter follower is usually built with $R_C = 0$. We shorted R_C with the capacitor C_C for ac signals but left R_C in for dc signals only because this paralleled our previous discussion and enabled us to avoid calculating a new operating point.

11.7 FET Circuits

The structure and properties of junction FET's and MOSFET's were described in the preceding chapter. They are characterized by very high input impedances and pentode-like characteristics. Like the bipolar transistor, they are three-terminal nonlinear devices whose small signal behavior can be described by four parameters. The four small signal variables are i_g, i_d, v_{ds}, and v_{gs} where the subscripts refer to the gate, the drain, and the source. Any two of the variables may be taken as independent but the customary choices are v_{ds} and v_{gs}. Because the input impedance is high, i_g is very small and its variation with v_{ds} and v_{gs} can be ignored. It must be understood that these approximations may not hold at high frequencies where gate to channel capacitances become important. We therefore have at low frequencies only one important equation.

$$i_d = \left(\frac{\partial i_D}{\partial v_{GS}}\right)_{V_{DS}} v_{gs} + \left(\frac{\partial i_D}{\partial v_{DS}}\right)_{V_{GS}} v_{ds}$$

or   (11.17)

$$i_d = g_{fs} v_{gs} + \frac{1}{r_{os}} v_{ds}$$

The parameter g_{fs} is a forward transconductance and r_{os} is an output resistance.[1] The second subscript s refers to the common source configuration of Fig. 11.17. Since i_g is neligible, $i_s = i_d$. Note that i_s is the incremental source current and not a short circuit current here. The common source circuit is generally used if voltage gain is desired. The source is common to both the input and the output.

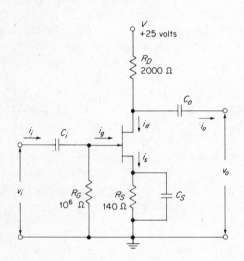

Figure 11.17 An amplifier circuit using an n channel FET. The capacitors present a negligible impedance to the incremental signals.

Characteristic curves, Fig. 11.18, are usually presented as i_D versus v_{DS} for a set of constant values of v_{GS}. Both g_{fs} and r_{os} can be read from such curves. In the drain current pinch off region, r_{os} is large and the term v_{ds}/r_{os} frequently can be ignored. Let us calculate the voltage gain of the circuit of Fig. 11.17 on the assumption that C_S is large and therefore the source is an ac ground. We also assume that C_i and C_o are large enough to be very small impedances for the ac frequencies of the incremental signal v_i.

[1] Manufacturers of semiconductor products often use the notation $y_{fs} = g_{fs}$ and $y_{os} = 1/r_{os}$.

We have $v_i = v_{gs}$ and $v_o = -i_d R_D$ so that the amplification is given by $A = v_o/v_i = -i_d R_D/v_{gs}$. In the approximation that $i_d = g_{fs} v_{gs}$, this reduces to $A = -g_{fs} R_D$. Typical values are $g_{fs} = 4000$ micromhos and $R_D = 2000$ ohms, giving $A = -8$. As in the corresponding common emitter bipolar transistor circuit, the input and output are $180°$ out of phase.

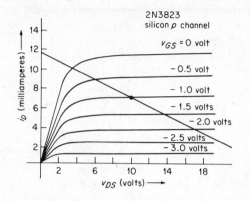

Figure 11.18 The drain characteristics of the 2N3823, a silicon n channel junction FET. The load line and operating point are for the circuit of Fig. 11.17. (*Characteristic curves courtesy Texas Instruments Incorporated.*)

If we do not neglect v_{ds}/r_{os}, we must use Eq. 11.17 and the additional relation $v_{ds} = -i_d R_D$. This relation is valid because the ac voltage between the power supply V and ground and the ac voltage across the capacitor C_S are both zero. We eliminate v_{ds} from Eq. 11.17 and with some rearrangement we find $i_d/v_{gs} = g_{fs}/(1 + R_D/r_{os})$. Usually r_{os} is in the range of 10^4 and 10^5 ohms.

Our calculations have assumed that the output circuit is open. Thus $v_o = -g_{fs} R_D v_{gs}$ is the Thevenin equivalent voltage of the amplifier viewed from the output. If we place a large condenser between v_o and ground, the short circuit output current $i_{sc} = g_{fs} v_{gs}$ will flow. This result is not approximate since on short circuit $v_o = v_{ds} = 0$. We now have for the output impedance $R_o = v_o/i_{sc} = R_D$ or more accurately $R_o = R_D/(1 + R_D/r_{os}) = r_{os} R_D/(r_{os} + R_D)$. The output resistance looks like R_D in parallel with r_{os}. This is similar to a bipolar transistor where the output resistance looks like R_C in parallel with $1/h_{oe}$.

Because the characteristic input impedance of the FET is very high, perhaps 10^9 ohms, the actual input impedance of the circuit is simply equal to the resistance R_G.

If R_S is not shunted by C_S, a feedback voltage $i_d R_S$ exists and the discussion of the previous section applies. Negative feedback may not be as essential in FET circuits as in bipolar transistor circuits because the FET is somewhat more linear than the bipolar transistor. Variation of parameters from one transistor to another remains a problem, however. The forward transconductance g_{fs} decreases as v_{GS} becomes more negative, as seen in

Fig. 11.18, but the variation is not exponential as it is for a bipolar transistor. Ideal linearity would correspond to equal spacings of the pinch off currents for equal increments of v_{GS}.

The calculation of load lines and operating points presents no special difficulties. If an operating point has been chosen, that is, I_D, V_{GS}, and V_{DS} are fixed, then in the circuit of Fig. 11.17 R_S is determined by the relation $V_{GS} = -I_D R_S$. Thus if $I_D = 7$ milliamp and $V_{GS} = -1$ volt are chosen, the value of R_S must be 140 ohms. The supply voltage V and R_D may vary but not independently. They must satisfy the dc load line equation $V_{DS} = V - I_D(R_D + R_S)$. For instance, if in addition to $I_D = 7$ milliamp we require $V_{DS} = 10$ volts, then the choice $R_D = 2000$ ohms demands a power supply voltage $V = 25$ volts. The corresponding load line is shown in Fig. 11.18. If linearity is important, the ac excursions along the ac load line must stay within the region of pinch off currents. The voltage gain is greatest at less negative V_{GS} and high I_D. It must also be remembered that when the shunt capacitor C_S is used, the dynamic and static load lines have different slopes. With the capacitor C_S in the circuit of Fig. 11.17 the ac load line is the equation $v_{ds} = -i_d R_D$.

Because of the very high input impedance of the FET, it is frequently used as an impedance matching stage between a source of high internal impedance and a load of low internal impedance. A clever circuit for impedance matching is the source follower shown in Fig. 11.19. Fundamentally it does the same job as the emitter follower described in the preceding section. The symmetry of the present circuit provides certain advantages, however.

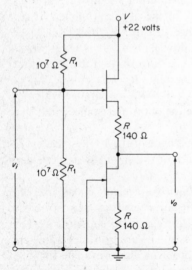

Figure 11.19 A direct coupled source follower circuit. The symmetry greatly reduces output drift due to temperature changes.

First let us consider the dc properties of this circuit. We shall assume that the two FET's are both 2N3823's, and we shall assume that the two FET's have been selected so that the actual characteristics of the FET's are as nearly identical as possible.

For convenience we choose for each FET the same operating point that was used for the circuit of Fig. 11.17. Thus $I_D = 7$ milliamp, $V_{DS} = 10$ volts, and $V_{GS} = -1$ volt. This requires that each $R = 140$ ohms and that V, the supply voltage, be $+22$ volts. The voltage drop across each FET plus its source resistor R is 11 volts. That $V_{GS} = -1$ volt for both the upper and lower FET's is guaranteed if the two resistors R_1 are identical. This places the gate of the upper FET 11 volts above ground. Its source is 12 volts above ground so $V_{GS} = -1$ volt as required.

Now let us see what is accomplished by using a symmetrical circuit with matched transistors. In any transistor circuit there are slow variations of the operating point that result from slow changes in transistor characteristics. Temperature changes are the principle cause of these variations. For instance, the dc potential drop across R_E in the emitter follower circuit of Fig. 11.16 might easily change by a few hundredths of a volt as the transistor warms up. Such variations do little harm if the incremental signal frequencies are high enough so that blocking capacitors can be used in the input and output. The slow variations do not get through the blocking capacitors. Therefore the incremental behavior of the circuit is affected only if the operating point moves by so large an amount that incremental parameters such as r_{tr} and β change appreciably.

Unfortunately one frequently needs a circuit that will amplify very slowly varying input signals. Blocking capacitors must then be omitted as they have been in the source follower of Fig. 11.19. In this case slow changes in the dc potentials cannot be distinguished from the incremental signals, and they may easily be as large as the incremental signal voltages. Drift problems can be greatly reduced, however, if the two FET's of Fig. 11.19 are carefully matched and mounted symmetrically and close together so that they are always at the same temperature. A change in temperature may change I_D but the same I_D, whatever its value, flows through each FET. Thus the output voltage remains at $+11$ volts as long as the input is at $+11$ volts. A change in I_D changes $I_D R$ but an equal and opposite change occurs in V_{DS}. Thus the voltage across each FET plus its source resistor R remains at 11 volts as symmetry demands.

Amplifiers for slowly varying signals that do not separate the dc and incremental voltages in the input and output are called *direct coupled amplifiers*. They almost always use some form of symmetry to reduce drifting. A more general discussion will be given in Chapter 14.

We now consider the behavior of the source follower when an incremental input voltage v_i is applied. By v_i and v_o we shall mean the incremental parts of the input and output voltages, in other words, the deviations from +11 volts. We first treat the lower FET. It is essentially a two-terminal passive device serving as a source resistor for the upper FET. The incremental voltage between the drain of the lower FET and ground is $v_{ds} + i_d R$. Dividing by the incremental current i_d we find that the lower FET behaves as a resistor of magnitude $r_{inc} = (v_{ds}/i_d) + R$. The quantity v_{ds}/i_d can be calculated by eliminating v_{gs} between $i_d = g_{fs} v_{gs} + v_{ds}/r_{os}$ (Eq. 11.17) and $v_{gs} = -i_d R$. We obtain

$$r_{inc} = \frac{v_{ds}}{i_d} + R = (1 + g_{fs} R)r_{os} + R$$

or   (11.18)

$$r_{inc} \simeq \frac{v_{ds}}{i_d} = (1 + g_{fs} R)r_{os}$$

With $g_{fs} = 4000$ micromhos, $r_{os} = 5 \times 10^4$ ohms, and $R = 140$ ohms, we have $v_{ds}/i_d \simeq 8 \times 10^4$ ohms. Thus the approximation $r_{inc} = v_{ds}/i_d$ is very good. Note that r_{inc}, the incremental resistance of the FET, is much greater than its dc resistance, which is given by $V_{DS}/I_D \simeq 1.4 \times 10^3$ ohms.

Now let us calculate $A = v_o/v_i$, the amplification of the source follower. We replace the lower FET by its incremental resistance r_{inc}. We ignore the source resistance R of the upper FET since $R \ll r_{inc}$, and we assume that no output current is drawn. We have the relations $i_d = g_{fs} v_{gs} + v_{ds}/r_{os}$; $v_o = i_d r_{inc} = -v_{ds}$ and $v_i = v_{gs} + i_d r_{inc} = v_{gs} + v_o$. These give

$$A = \frac{v_o}{v_i} = \frac{r_{inc} g_{fs}}{1 + (r_{inc}/r_{os}) + r_{inc} g_{fs}}$$ (11.19)

The quantity $r_{inc} g_{fs} = 320$ and r_{inc}/r_{os} is less than 2. Thus the amplification is very close to $+1$; it is in fact greater than 0.99.

To calculate the output impedance we follow the usual procedure of dividing the open circuit output, or Thevenin, emf by the short circuit output current. The open circuit output voltage is $v_o \simeq v_i$. The short circuit current i_{sc} is calculated assuming that the output is held at $+11$ volts, for instance, by a constant voltage source. All the incremental current through the upper FET then flows through the voltage source. In this calculation we may not neglect the source resistor R of the upper FET compared to r_{inc} of the lower FET because r_{inc} is shorted out by the voltage source. In addition to $i_d = i_{sc} = g_{fs} v_{gs} + v_{ds}/r_{os}$ we have the relations $v_i = v_{gs} + i_d R$ and $v_{ds} = -i_d R$. From these we obtain the output resistance

$$R_o = \frac{v_o}{i_{sc}} = \frac{v}{i_d} = \frac{1}{g_{fs}} + R + \frac{R}{g_{fs} r_{os}}$$ (11.20)

Of these terms, $1/g_{fs} \simeq 250$ ohms, $R = 140$ ohms, and $R/g_{fs}r_{os} \simeq 1$ ohm. Thus the total output impedance is about 400 ohms. It is essentially $1/g_{fs}$ in series with the source resistor R.

The input impedance of the source follower is just that of the network furnishing the gate bias since this resistance, 5×10^6 ohms, is much less than the 10^9-ohm input resistance of the FET's themselves.

In introducing the source follower circuit that we have just analyzed we mentioned that the operating point of each FET was $I_D = 7$ milliamp, $V_{DS} = 10$ volts. In closing we point out that linear load lines through the operating points do not exist. Let V_{DS1} and V_{DS2} be the drain to source voltages of the upper and lower FET's, respectively, of Fig. 11.19. The only linear load line relation we can write is $(V_{DS1} + V_{DS2}) + 2I_D R = V$. Of course, both V_{DS1} and V_{DS2} are determined if I_D is given but the dependence of each on I_D involves the incremental parameters of the FET's and is linear only for small excursions from the operating points. For the upper FET the slope of the incremental load line near the operating point is $i_d/v_{ds1} = -1/(r_{inc} + R)$. Because r_{inc} is very large, such a line drawn on the characteristic curves of Fig. 11.18 would be almost horizontal.

Problems

11.1 Sketch a half-wave rectifier circuit. The input to the transformer is 110 volts rms, 60 Hz. The rectified half-wave output is to have a 24-volt dc component. What is the turns ratio for the transformer? The rectified output feeds a 100-ohm load. Sketch the current through the load as a function of time. Assume the ac source and the transformer have negligible internal resistance.

11.2 Discuss the output waveform of the bridge rectifier shown in Fig. 11.4. Sketch the voltage across the output of the transformer and the voltage across the load resistor.

11.3 The current-voltage characteristics of the diode in the half-wave rectifier circuit of Fig. 11.1 are given in Fig. 11.20. Devise a graphical method for determining the current when the load resistor and the instantaneous transformer voltage are given. This can be done by plotting on the accompanying figure the relationship between the voltage drop across the diode, the transformer voltage, and the IR_L drop across the load. This relation is called the *load line*. What current flows when the transformer voltage is 1.0 volt, 1.5 volts, and 2.0 volts if the load resistor R_L is 100 ohms?

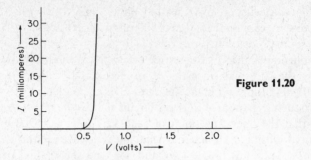

Figure 11.20

11.4   In the circuit of Fig. 11.21 the transistor has a $\beta = 300$ and is made of silicon. Calculate I_B, I_C, and I_E. Calculate the following voltages with respect to ground: V_B, V_C, and V_E. For small signals, calculate r_{tr}, $A = v_o/v_i$ (assuming no output current is drawn), R_i, and R_o. The capacitors offer a negligible impedance at the signal frequencies. What must be the sign of the supply voltage?

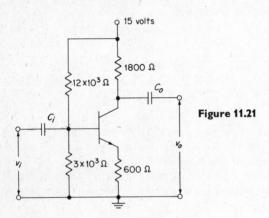

Figure 11.21

11.5   For the circuit of Prob. 11.4, draw the equivalent resistive network seen by the incremental input signal v_i.

11.6   Repeat Prob. 11.4 assuming that the transistor is a germanium transistor and that $\beta = 150$.

11.7   The collector characteristic curves for a 2N4961 silicon transistor are shown in Fig. 11.22. Estimate h_{fe} and h_{oe} if the transistor operating point is $I_C = 20$ milliamp, $V_{CE} = 27$ volts. (*Characteristic curves courtesy Fairchild Semiconductor.*)

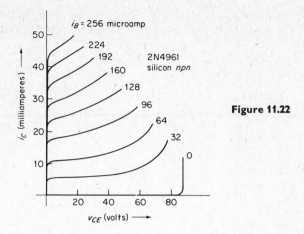

Figure 11.22

11.8   Design a dc bias circuit for the 2N4961 silicon transistor. The operating point is that given in Prob. 11.7. Use an emitter resistor $R_E = 200$ ohms, a collector resistor $R_C = 0$ ohms, and a single power supply. The input resistance of the base bias circuit should be greater than 1000 ohms.

11.9   An emitter follower circuit using a 2N4961 is shown in Fig. 11.23. The collector characteristics are given in Prob. 11.7. Find the dc load line and the quiescent point for the circuit shown. Calculate I_C, I_E, and I_B. Calculate the voltages V_C, V_E, and V_B. For small signals, calculate r_{tr}, R_i, and R_o. Calculate the amplification $A = v_o/v_i$ assuming no output current is drawn. The capacitors offer a negligible impedance at the signal frequencies.

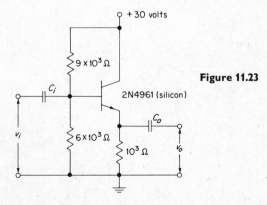

Figure 11.23

11.10   For the amplifier circuit shown in Fig. 11.24, calculate the same quantities requested in Prob. 11.9. The transistor is the 2N4961 used in Prob. 11.7 and 11.9.

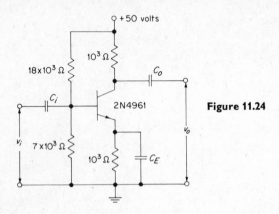

Figure 11.24

11.11   Repeat Prob. 11.10 with the circuit altered so that the bypass capacitor C_E shunts only 950 ohms of the 1000-ohm emitter resistor. This leaves 50 ohms as a feedback resistor.

11.12   An amplifier circuit and the collector characteristic curves for the 2N3565 transistor are shown in Fig. 11.25 and 11.26. Analyze the circuit, and calculate the same quantities requested in Prob. 11.9. (*Characteristic curves courtesy Fairchild Semiconductor.*)

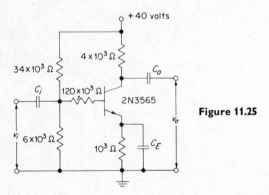

Figure 11.25

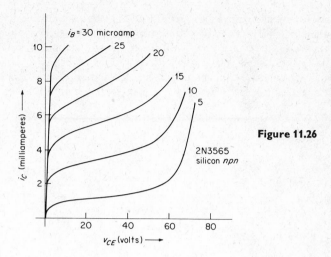

$i_B = 30$ microamp

25

20

15

10

5

2N3565
silicon *npn*

i_C (milliamperes)

v_{CE} (volts) ⟶

Figure 11.26

11.13 In the circuit of Prob. 11.12 a load resistor of 2×10^3 ohms is connected across the output terminals. By what factor is the output voltage reduced from the open circuit output voltage?

11.14 The drain characteristic curves for a 2N4360 *p* channel junction FET are shown in Fig. 11.27. If the FET is operating with a drain to source voltage of 6 volts and a gate to source voltage of 1.5 volts, what is the drain current? What is the forward transconductance g_{fs}, and what is the output resistance r_{os}? (*Characteristic curves courtesy Fairchild Semiconductor.*)

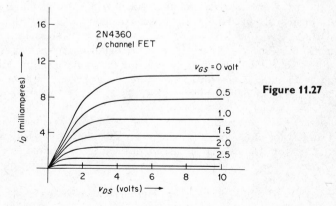

2N4360
p channel FET

$v_{GS} = 0$ volt

0.5

1.0

1.5

2.0

2.5

i_D (milliamperes)

v_{DS} (volts) ⟶

Figure 11.27

11.15   An FET amplifier circuit is shown in Fig. 11.28 using the p channel 2N4360 FET of Prob. 11.14. Draw the dc load line and the quiescent point. What is the sign of the 10-volt power supply? Compute the currents I_D and I_S. Find the voltages with respect to ground V_D, V_G, and V_S. Estimate g_{fs} and r_{os}. Assuming that the capacitors are short circuits for the signal frequencies, compute $A = v_o/v_i$ (no output current), R_i, and R_o.

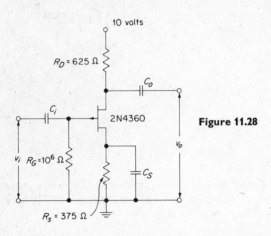

Figure 11.28

11.16   For the circuit of Prob. 11.15, what is the small signal open circuit gain of the amplifier if the capacitor C_S is removed?

11.17   For the circuit of Prob. 11.15, let us assume that $C_i = 10^{-9}$ farad. Both C_o and C_S can be assumed to be very large compared to C_i, that is assume that $1/\omega C_o \ll R_D$ and $1/\omega C_S \ll R_S$. Calculate the small signal gain of the amplifier as a function of the frequency. For this problem you may assume that the voltage signal between the gate and the source is amplified as in Prob. 11.15 but of course the gate to source voltage is not the same as v_i. What is the phase angle between v_i and v_o?

11.18   A load resistor of 625 ohms is placed across the output terminals of the circuit in Prob. 11.15. What is the gain of the loaded circuit?

11.19   The circuit of Prob. 11.15 is modified into a source follower as shown in Fig. 11.29. The three capacitors are short circuits for the signal frequencies of interest. Calculate I_D, $A = v_o/v_i$ (no output current is drawn), R_i, and R_o.

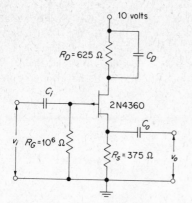

10 volts

$R_D = 625\ \Omega$ C_D

C_i 2N4360

Figure 11.29

C_o

v_i $R_G = 10^6\ \Omega$ $R_S = 375\ \Omega$ v_o

11.20   Compare the advantages and disadvantages of the source follower shown in Fig. 11.19 and discussed in the text and the source follower of Prob. 11.19.

11.21   Transistors are sometimes used in a common base circuit as discussed in Chapter 10. A transistor in a common base configuration is shown in Fig. 11.30. For this configuration where the base is common to the input and output circuits, i_e and v_{cb} are taken as the independent parameters and the dependent parameters are taken as v_{eb} and i_c. The relationships between these incremental parameters are

$$v_{eb} = h_{ib}\, i_e + h_{rb}\, v_{cb}$$

and

$$i_c = h_{fb}\, i_e + h_{ob}\, v_{cb}$$

Using the relations $i_e = i_c + i_b$ and $v_{ce} = v_{cb} + v_{be} = v_{cb} - v_{eb}$ and Eq. 11.2, show that

$$h_{ie} = \frac{-h_{ib}}{(1 - h_{fb})(1 - h_{rb}) - h_{ob}h_{ib}} \simeq \frac{h_{ib}}{1 - h_{fb}}$$

$$h_{re} = \frac{h_{rb}(1 - h_{fb}) - h_{ob}h_{ib}}{(1 - h_{fb})(1 - h_{rb}) - h_{ob}h_{ib}} \simeq \frac{-h_{ob}h_{ib}}{1 - h_{fb}} - h_{rb}$$

$$h_{fe} = \frac{h_{fb}(1 - h_{rb}) + h_{ob}h_{ib}}{(1 - h_{fb})(1 - h_{rb}) - h_{ob}h_{ib}} \simeq \frac{h_{fb}}{1 - h_{fb}}$$

$$h_{oe} = \frac{h_{ob}}{(1 - h_{fb})(1 - h_{rb}) - h_{ob}h_{ib}} \simeq \frac{h_{ob}}{1 - h_{fb}}$$

It should be noted that h_{fb} is simply the quantity α except that $h_{fb} = \Delta i_C/\Delta i_E = i_c/i_e$, whereas $\alpha = I_C/I_E$. Since α is nearly independent of I_C, we have the result $h_{fb} \simeq \alpha$. We should point out that if we had adopted the more conventional notation of i_E flowing into rather than out of the transistor, then the parameters h_{ib} and h_{fb} would have a sign opposite to the one given here. In this case $h_{fb} = -\alpha$.

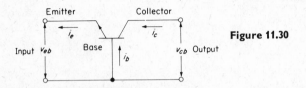

Figure 11.30

11.22   An equivalent T circuit for a transistor used in a common base configuration is shown in Fig. 11.31. The parameters of the T equivalent circuit for the common base configuration are r_e'', r_b'', r_c'', and α. By comparison with the equivalent T circuit for the common emitter configuration (Fig. 11.13), and using Table 11.1, show that

$$r_b \simeq r_b''$$

$$r_e \simeq r_e'' \simeq r_{tr}$$

$$r_c \simeq r_c''(1 - \alpha)$$

$$\beta = \frac{\alpha}{(1 - \alpha)}$$

where r_b, r_e, r_c, and β are the parameters of the T equivalent circuit for the common emitter configuration. This shows clearly the important results that in the common emitter configuration the forward current gain β is much larger than the forward current gain α in the common base configuration and that the output resistance r_c in the common emitter configuration is much lower than the output resistance r_c'' in the common base configuration.

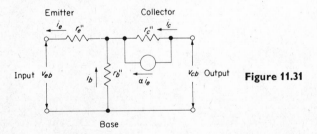

Figure 11.31

11.23 In the full-wave rectifier circuit of Fig. 11.3 a capacitor C is inserted in parallel with the load resistor R_L. Show that if R_L is sufficiently large, the output voltage is equal to the peak voltage across half the transformer. Show that the two diodes can be simultaneously back biased during parts of the cycle. Sketch the output voltage when $R_L C = \frac{1}{120}$ sec.

Show that if the capacitor is removed and an inductor L is placed in series with R_L, then at any time in the cycle at least one of the diodes is forward biased.

Note that linear analysis, for instance, Fourier series methods, is appropriate for the $R_L L$ filter but not for the $R_L C$ filter. Since both diodes may be back biased in the $R_L C$ case, the circuit is highly nonlinear.

11.24 For the circuit shown in Fig. 11.32 demonstrate that the output voltage v_o is a dc voltage whose magnitude is twice the maximum value of the ac voltage v_i.

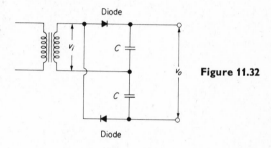

Figure 11.32

12/Amplifiers— Some General Considerations

Many properties and problems are common to nearly all amplifier circuits. It is useful to give a general discussion before we treat particular amplifiers in detail. In this chapter we shall discuss negative feedback at some length and more briefly the subjects of noise and the high-frequency response of a transistor.

12.1 Negative Feedback

We have previously pointed out that transistors as used in electronic circuits (1) are very nonlinear if voltage drive is used because of the exponential current-voltage characteristic $i_E = I_0\, e^{eV/kT}$; (2) are not identical, the β of one transistor of a given type will differ from the β of others of the same type; and (3) have properties that depend on the temperature and other operating conditions. It is therefore necessary to design circuits whose gain, input impedance, and output impedance are independent of the properties and operating conditions of the individual transistors in the circuit and to design circuits that reduce to a minimum the distortion of the signal due to nonlinear properties of the transistors in the circuits. Negative feedback enables us to accomplish these goals. Of course negative feedback was used long before the introduction of transistors but it was a less essential element in vacuum-tube circuits.

There are various types of negative feedback. We shall classify feedback by two parameters. The first parameter is determined by whether we sample the output voltage or the output current in order to obtain our signal for feedback. Which method is used depends on whether we want to control the output voltage across a load or the output current through a load. The second parameter states whether the signal fed back to the input is a voltage signal or a current signal. The names for the various types of feedback are shown in Table 12.1.

Table 12.1

Output Signal Sample	Input Feedback Signal	Name of Feedback
v	v	Voltage feedback
i	v	Current feedback
v	i	Operational Voltage feedback
i	i	Operational Current feedback

We shall consider the properties of each of these types of feedback but first let us be clear about the amplifier without feedback.

For the amplifier without feedback we assume the properties illustrated in Fig. 12.1. Viewed from the input (v_i) the amplifier is a two-terminal passive network with the internal resistance R_i. Viewed from the output (v_o) the amplifier is a two-terminal active network containing a voltage source $A_0 v_i$ and the internal resistance R_o. The voltage amplification A_0 is always the amplification at zero output current. The output parameters are related thus:

$$v_o = A_0 v_i - i_o R_o \qquad (12.1)$$

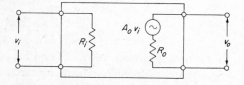

Figure 12.1 Schematic showing the Thevenin equivalent of the amplifier without feedback.

The output depends on the input through the term $A_0 v_i$ but the input is assumed independent of the output. Feedback, of course, deliberately introduces a strong dependence of the input on the output parameters.

The input and output voltages and currents are small signals superimposed on dc biases that we do not illustrate. The dc level of the input and output may be quite different although in much of what follows we shall assume, for simplicity, that they share a common ground.

12.2 Voltage Feedback

We consider an amplifier whose voltage gain without feedback is A_0. We return to the input a fraction β of the output voltage. Figure 12.2 is a schematic of such a circuit that illustrates the important elements in our calculation. Actual circuits achieve the sampling of the output and the feedback to the input in a wide variety of ways, some of which we shall discuss.

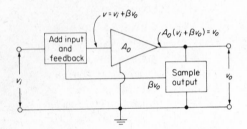

Figure 12.2 Schematic of a voltage feedback circuit. A_0 represents the amplifier without feedback shown in Fig. 12.1. The input and output share a common ground.

We assume that the feedback network does not load the input nor draw current from the output. In other words βv_o is an ideal voltage source that adds to v_i. The amplifier A_0 sees the actual input

$$v = v_i + \beta v_o \tag{12.2}$$

and delivers the output

$$v_o = A_0 v = A_0(v_i + \beta v_o) \tag{12.3}$$

The feedback network can represent either ac or dc coupling but we assume that v_i and v_o are small signals. From the equation $v_o = A_0(v_i + \beta v_o)$ we solve for $v_o/v_i = A_F$, the actual amplification of the circuit with feedback. We obtain

$$\frac{v_o}{v_i} = A_F = \frac{A_0}{I - \beta A_0} \tag{12.4}$$

First we note that if βA_0 is positive and near unity, A_F can be as large as one wishes. This case is called *positive feedback*. It leads to circuit instability and self-sustained oscillations. We shall discuss later how positive feedback is deliberately used in the design of oscillators. In the present discussion βA_0 must be negative. A_0 may be negative because of inversion of the output signal. If A_0 is positive, then the feedback network must invert the signal, which means β is negative. Given that βA_0 is negative, then $|A_F| < |A_0|$ and the actual input $|v_i + \beta v_o|$ that A_0 sees is less than $|v_i|$. Because the denominator of Eq. 12.4 is positive, A_F necessarily has the same sign as A_0.

Now the essential point of the discussion is that β and A_0 are so chosen that $|\beta A_0| \gg 1$. Then we have

$$A_F \simeq -\frac{A_0}{\beta A_0} = -\frac{1}{\beta} \tag{12.5}$$

Our amplifier with feedback is almost independent of the properties of the amplifier without feedback. The β of the feedback network is usually determined only by passive, and therefore stable, components.

There is a very simple way of looking at Eq. 12.5. A_0 is greater than A_F simply because v_i, the input without feedback, is greater than v, the actual input with feedback. We have, without approximation, $A_F/A_0 = v/v_i$. If $|A_0| \gg |A_F|$, then $|v_i| \gg |v|$ or $v_i \simeq -\beta v_o$. This last equation gives immediately $v_o/v_i = A_F \simeq -1/\beta$.

Although the dependence of A_F on A_0 is small, it is important to have a quantitative estimate of how strongly A_F depends on A_0. Let us calculate to the first order how A_F is affected by changes in A_0. Taking the logarithm of both sides of Eq. 12.4 and differentiating, we find $dA_F/A_F = dA_0/A_0 + \beta \, dA_0/(1 - \beta A_0)$, which reduces to

$$\frac{dA_F}{A_F} = \left(\frac{A_F}{A_0}\right)\frac{dA_0}{A_0} \tag{12.6}$$

This is the quantitative measure of the elimination of variations in A_0. If, for instance, $A_0 = 10^3$ and $\beta = -\frac{1}{10}$, we have $A_F = 10$ and $A_F/A_0 = 10^{-2}$. In this case a 10% variation in A_0 gives only a 0.1% variation in A_F. The variation in A_0 can be from any cause. For example, A_0 can vary depending on the magnitude of the output voltage. The effect is called *distortion*. The value of A_0 can also vary due to temperature changes, signal frequency changes, or transistor nonuniformity.

At the risk of pedagogical overkill we emphasize that the improvement in stability of amplification is not because the amplification is reduced but instead because negative feedback is used. We could, without feedback, reduce A_0 by a factor of 10^2 just by feeding v_i into an input voltage divider that sends only $10^{-2}v_i$ into the amplifier. In this case we find not Eq. 12.6 but rather $dA_R/A_R = dA_0/A_0$ where A_R is the divider reduced amplification. There is no improvement in stability.

Now there are some things that negative feedback cannot do and at least one of them is very important. Suppose that an unwanted voltage signal ε is introduced at some point in the amplifier. This is unavoidable since thermally induced random fluctuations of charge will produce voltage noise in every circuit. The amplifier cannot distinguish this from the signal input v_i and the output will contain an amplified noise voltage.

We first consider the amplifier without feedback. The output voltage is $v_o = A_0 v_i + A_0'\varepsilon$. If ε is at the input, it is fully amplified and $A_0' = A_0$. If ε is introduced within the amplifier, $A_0' < A_0$. The signal to noise ratio at the output is simply $A_0 v_i / A_0'\varepsilon$. Now let us compare this to the amplifier with feedback. With feedback the actual voltage at the input of the amplifier is $v = v_i + \beta v_o$ and the signal at the output of the amplifier is $v_o = A_0 v + A_0'\varepsilon$. Combining these equations yields $v_o = v_i A_0/(1 - \beta A_0) + \varepsilon A_0'/(1 - \beta A_0) = A_F v_i + A_F \varepsilon A_0'/A_0$. Thus in the circuit with feedback the unwanted voltage source ε makes a contribution $\varepsilon A_0' A_F/A_0$ to v_o at the output. The signal of interest v_i makes a contribution $v_i A_F$ at the output. Dividing the latter by the former we see that the signal to noise ratio is again $A_0 v_i / A_0'\varepsilon$. Thus negative feedback does not improve the signal to noise ratio. In fact, because the Fourier spectrum of most noise sources is very broad, one can actually be worse off. Negative feedback usually broadens the frequency response of an amplifier, and with feedback the signal to noise ratio may decrease.

In general, negative feedback changes both the input and output impedance of an amplifier. Frequently this is just as important to the circuit designer as stability of amplification.

Let us consider first the input resistance. Ignoring feedback the input resistance of the amplifier is $v/i_i = R_i$ where i_i is the input current. We have assumed that the feedback network draws no input current; therefore, the same current i_i flows in the presence of feedback. But with feedback the input resistance is now computed with respect to v_i, not v. Using $v_i/v = A_0/A_F$ we have $v_i/i_i = (A_0/A_F)v/i_i = (A_0/A_F)R_i$. Thus voltage feedback increases the input resistance by the factor (A_0/A_F). It is important to remember that the input resistance with feedback is not independent of the output current. If we short the output $v_o = 0$, there is no feedback voltage and the input resistance is just R_i.

Qualitatively we expect that feedback must decrease the output resistance since feedback stabilizes v_o not only against changes in A_0 but also against changes arising from variations in i_o, the output current. To make this quantitative, let us determine the output Thevenin parameters of the amplifier with feedback. Since all our calculations of amplification have been for zero output current, the Thevenin output voltage is just $v_{\text{Th}} = v_o = v_i A_F$. If we short the output, there is no output voltage and no feedback. Therefore the short circuit output current i_s satisfies the equation $i_s R_o = v_i A_0$ where R_o is the output resistance of the amplifier A_0 without feedback. Because of the absence of a feedback voltage, A_0 rather than A_F must be used. We obtain $v_{\text{Th}}/i_s = (A_F/A_0)R_o$. Thus voltage feedback reduces the output resistance by the factor A_F/A_0. In this expression the values for A_F and A_0 at zero output current must be used. The output resistances, both with and without feedback, are constants independent of i_o.

12.3 Current Feedback

Figure 12.3 shows a circuit where a voltage proportional to the output current i_o is fed back to the input. The voltage v actually seen by the amplifier is the difference between v_i and the feedback voltage, both of which are assumed large compared to v. We shall see that such a circuit tends to stabilize the output current rather than the output voltage. It is convenient to include explicitly the load resistor R_L in the circuit so that we may more easily relate the output voltage and the output current. Let R_o be the output impedance of the amplifier without feedback. At the input we have $v = v_i + \beta i_o R_F$ and at the output $A_0 v = A_0(v_i + \beta i_o R_F) = i_o(R_L + R_o + R_F)$. Solving the latter equation for i_o, we find

$$i_o = \frac{A_0 v_i}{R_L + R_o + (1 - A_0 \beta)R_F} \tag{12.7}$$

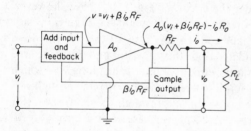

Figure 12.3 A current feedback circuit.

We require that $A_0\beta$ be negative. If in addition $-A_0 \beta R_F \gg R_L + R_o + R_F$, we have $i_o \simeq -v_i/\beta R_F$; that is, the output current is largely independent of A_0, R_o, and R_L. The amplifier behaves as a constant current source. Stated another way, the output resistance of the amplifier is greatly increased by the use of current feedback. The Thevenin output voltage is $v_{\mathrm{Th}} = A_0 v_i$ since no feedback exists when no current is drawn. The short circuit current ($R_L = 0$) is $i_s = A_0 v_i/[R_o + (1 - A_0 \beta)R_F]$. Thus the output resistance with feedback is $v_{\mathrm{Th}}/i_s = R_o + (1 - A_0 \beta)R_F$. Note that by the same argument used in the voltage feedback circuit the input resistance is also increased by current feedback.

Before we discuss operational feedback, we wish to see how some of the amplifiers we previously discussed fit into our classification scheme. Consider the circuit of Fig. 11.14. The feedback voltage is developed across R_E. We assume the capacitor C_E is removed. The feedback voltage is $v_f = -R_E i_e = [R_E i_o - R_E(i_e + i_o)]$ where i_e is the emitter current, i_o is the output current that flows through the output terminals to the load, and $i_e + i_o$ is the current through the collector resistor R_C. We have used $i_c \simeq i_e \gg i_b$. Since $v_o = -R_C(i_e + i_o)$, the feedback voltage can be written $v_f = i_o R_E + v_o R_E/R_C$. The feedback is neither pure current feedback nor pure voltage feedback, but instead the feedback signal is a linear super-position of the two. Now we know that for a circuit employing considerable negative feedback $v_i \simeq -v_f$ or $v_i \simeq -i_o R_E - v_o R_E/R_C$. Let us calculate the output resistance of this amplifier. The open circuit voltage (that is, $i_o = 0$) is given by $v_{\text{Th}} = -v_i R_C/R_E$. The short circuit current (that is, $v_o = 0$) is given by $i_s = -v_i/R_E$. Therefore the output impedance is given by $R_o = v_{\text{Th}}/i_s = R_C$, which agrees with the results of the previous chapter. The reader must remember that because the amplification is negative, i_e and i_o have opposite signs. At shorted output ($v_o = 0$), $i_e = -i_o$ and the current through R_C is zero.

It is an important point that many real amplifiers have feedback net-works that are mixtures of our four classic cases of feedback. The input and output resistances of amplifiers can be varied almost at will by utilizing the linear combinations of the four types of feedback.

In some situations it is useful to change our perspective. In the previous discussion we regarded the amplifier as consisting of the transistor and its associated resistors, including R_C. This amplifier drove an external load resistor. Let us again consider the circuit of Fig. 11.14. Now, however, let us consider R_C actually to be on the load, as it might be if the transistor were, for example, driving a speaker coil. In this case all the current flows through R_C. The amplifier is considered to be the transistor and the associated resistors in the base and in the emitter circuits, but not including R_C, which is considered to be the load. The output current i_o is the current through R_C. All this current flows through the transistor and through the emitter resistor R_E. The feedback voltage is $v_f = -i_o R_E$. This is a case of pure current feed-back. The output resistance of the transistor as seen by the collector resistor is increased above $1/h_{oe}$ due to the negative current feedback provided by R_E.

An example of pure voltage feedback with $\beta = -1$ is provided by the emitter follower and source follower circuits of the preceding chapter.

12.4 Operational Feedback

The next type of negative feedback to be discussed is what we called *operational voltage feedback*. In this type of feedback a current signal pro-portional to the output voltage is fed back in a negative sense to the input of the amplifier. This is represented schematically in Fig. 12.4. For the circuit shown it is assumed that the intrinsic input impedance of the amplifier is very large compared to R_1 or R_2. Therefore almost all the current that flows through R_1 also flows into R_2 and almost none flows into the input to the amplifier. The current that flows in R_1 and R_2 is

$$i_i = \frac{v_i - v}{R_1} = \frac{v - v_o}{R_2} \qquad (12.8)$$

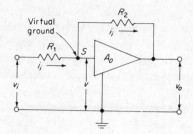

Figure 12.4 Operational voltage feedback.

We eliminate v using $v_o = A_0 v$ and solve for $v_o/v_i = A_F$. We obtain

$$A_F = \frac{v_o}{v_i} = \frac{1}{R_1/R_2(1/A_0 - 1) + 1/A_0} \tag{12.9}$$

If A_0 is very large, then $A_F = -R_2/R_1$ and once again we obtain a gain for the amplifier with feedback that is almost independent of the amplification A_0 of the unfedback circuit. Just as in the discussion of voltage feedback if $A_0 \gg A_F$, then $v_i \gg v$. Thus the point S of Fig. 12.4 must remain very close to ground. In circuits of this type, S is called a *virtual ground*. This permits an immediate calculation of A_F. From $v_i = i_i R_1$ and $v_o = -i_i R_2$ we have $v_o/v =_i A_F = -R_2/R_1$.

We point out to the reader that v_o and v_i must have opposite signs in this amplifier; that is, A_0 must be negative. If this is not the case, the voltage $i_i R_1$ adds to v_i and the feedback is positive. The amplifier will drive itself into saturation.

We also mention that the equation $A_0 v = v_o$ is not quite exact. When the output circuit is open, the current i_i must return to ground through the amplifier. A small term $i_i R_o$ has been neglected where R_o is the output resistance of the amplifier without feedback.

The final type of feedback is called *negative operational current feedback*. In this type of feedback a current signal proportional to the output current is fed back to the amplifier input. Such a circuit is shown in Fig. 12.5.

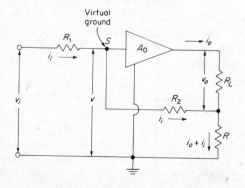

Figure 12.5 Operational current feedback.

We assume that A_0 is very large and that S is a virtual ground. The input resistance of A_0 without feedback is large so that almost none of the current i_i enters the amplifier. As with operational voltage feedback A_0, the gain without feedback must be negative. Because we want to sample the output current and not the output voltage, v_o and v_i do not share a common ground. The signal resistor R must be grounded.

With the approximations above we find $i_i R_2 = -(i_o + i_i)R$ or $i_o/i_i = -(R_2 + R)/R$. Thus the ratio of the output current to the input current is independent of the properties of the amplifier A_0. Using $v_i = i_i R_1$, we obtain $i_o = -v_i(R_2 + R)/R_1 R$, which is the appropriate form if we put a voltage source at the input rather than a current source.

In Table 12.2 we summarize the effects of the different kinds of feedback on input and output impedances. The terms *high* and *low* mean with reference to the impedances of the amplifier without feedback.

Table 12.2

Type of Feedback	Feedback Signal	Output Sample	Input Impedance	Output Impedance
Voltage	v	v	High	Low
Current	v	i	High	High
Operational Voltage	i	v	Low	Low
Operational Current	i	i	Low	High

It is easy to see how these results arise. Consider first the input impedance. If a voltage signal is fedback, the actual input to the amplifier is reduced from v_i to v. Thus less input current flows and the input impedance is increased. The feedback of a current signal to the input increases the input current and therefore reduces the input impedance. Another way of viewing the latter case is to remember that the current feedback network provides an alternate path to ground for input current. The resistance of the alternate path is usually much less than the path through the amplifier.

The key to understanding the output impedances is simply that the quantity sampled tends to be stabilized. If the output voltage is sampled, it is stabilized against variations in output current. The output behaves as a good voltage source and therefore has a low internal resistance. If the output current is sampled, it is stabilized against variations in output voltage. The output is a good current source and has a high internal resistance.

We have already calculated the actual values of the input and output impedances for voltage and current feedback. Let us briefly consider the results for the operational feedback circuits, Fig. 12.4 and 12.5. Since S is a virtual ground in both cases, we have $v_i/i_i = R_1$ and the input impedance is just R_1.

Suppose the resistance R_1 is omitted, and the input current i_i is furnished by a current source. With current drive the input resistance is v/i_i. For operational voltage feedback we have $v_o = -i_i R_2 = A_0 v$. Thus the input impedance is $v/i_i = -R_2/A_0$. For operational current feedback we have $i_i R_2 = -(i_o + i_i)R$ from the input loop and $(i_o + i_i)R + i_o(R_L + R_o) = vA_0$ from the output loop. Eliminating i_o and solving, we obtain for the input resistance $v/i_i = -(1/A_0)[(1 + R_2/R)(R_L + R_o) + R_2]$. Note that A_0 is negative so the input resistance for both operational voltage and operational current feedback is a positive quantity.

For current drive input the output impedances are approximately $-R_o/A_0$ in the operational voltage feedback circuit (Fig. 12.4) and $-A_0 R$ for operational current feedback (Fig. 12.5). These arise as follows. For the case of operational voltage feedback (Fig. 12.4) the open circuit voltage is $v_{Th} \simeq -i_i R_2$ where we have ignored the voltage drop as i_i returns to ground through R_o. Now if we short-circuit the output to ground, then the voltage v at the input is $i_i R_2$. Viewed from the output the amplifier contains the voltage source $A_0 i_i R_2$ and the internal resistance R_0. The shorted output current must satisfy the equation $i_s R_o = A_0 i_i R_2$. Therefore the output resistance is $v_{Th}/i_s = -R_o/A_0$. Note that A_0 is negative so that $-R_o/A_0$ is positive.

For the case of operational current feedback (Fig. 12.5), suppose the output is open circuited ($R_L = \infty$). In this case the voltage at the input is $i_i(R_2 + R)$ and the output voltage is $v_{Th} = A_0 i_i(R_2 + R) - i_i R \simeq A_0 i_i(R_2 + R)$. When the output is short circuited ($R_L = 0$), the output current is given by $i_s = -i_i(R_2 + R)/R$. Therefore the output resistance is $v_{Th}/i_s \simeq -A_0 R$. In this expression, note that R is the sampling resistor and not the intrinsic output resistance of the amplifier, which we have assumed is small compared to R. If input voltage drive is used, the expressions become somewhat more complicated.

12.5 Noise

In an electronic circuit there are noise voltages that tend to obscure a desired voltage signal. It is important to understand where these noise voltages come from and what limitations they place on the circuits we use.

The first question we must consider is "What signals are due to noise, and what signals are not due to noise?" We shall attempt to answer the latter part of the question first. Often signals such as the 120-Hz component of the ripple from a power supply using a full-wave rectifier are not completely eliminated from the output of the power supply. The ripple voltage may interfere with the observation of a desired signal voltage. Consequently the ripple voltage is sometimes referred to as noise. We shall *not* regard the ripple voltage as noise, however, because by the use of better filtering, shielding, or other techniques the ripple can be almost completely eliminated as a problem. In order for us to consider a signal as noise it must arise from some random fluctuation over which we have no control and which we cannot eliminate by filtering, shielding, and so on. Because of the random or statistical nature of noise the Fourier spectrum of noise must be a smooth continuous function of the frequency. In addition, the statistical nature of noise means that the noise voltage at a given Fourier frequency has a random phase so that it makes sense only to talk about the absolute magnitude of the noise voltage or the rms noise voltage and not the time dependence of a noise voltage. As a result of the random phase of a noise voltage if two noise voltages are present, the rms value of the total noise voltage will be the square root of the sum of the squares of the individual rms noise voltages; that is, the noise voltages add as if they were vectors at right angles to each other. Put another way, noise powers add algebraically but not noise voltages and currents.

There are several basic types of noise signals. We shall discuss carefully only the following types:

1. Shot noise (current noise)
2. Thermal noise (Johnson noise)
3. $1/f$ noise

We shall discuss shot noise first. Any semiconductor junction (or vacuum tube) will produce noise whenever current is flowing across the junction. This noise is called *shot noise* (or current noise). Shot noise arises because of the statistical fluctuations in the current crossing the junction. The average current crossing a junction in time τ is $I = ne/\tau$ where ne is the total charge that crosses the junction in the time τ. The current that flows across a semiconductor junction is a statistical quantity; in other words, random statistics determine how many charge carriers strike the junction with a velocity great enough to permit them to pass over the junction barrier. If the average total number of charge carriers that cross the junction in the time τ is n, then it can be shown that there will be fluctuations about this number of the magnitude $\pm\sqrt{n}$. In other words, the number of charge carriers crossing the junction in a time τ is likely to be in the range $n \pm \sqrt{n}$. Since the fluctuations are random, there are as many positive as negative fluctuations so that on the average n charge carriers cross the junction in a time τ. The fluctuations in the current crossing the junction will be

$$|\Delta I| \simeq \sqrt{n}\,\frac{e}{\tau} = \sqrt{\frac{eI}{\tau}} \qquad (12.10)$$

The fluctuations in the current are called the *current noise* or *shot noise*. The individual and randomly occurring event is the passage of one electron through the barrier. For an electron with an energy of 5 eV and for a barrier 1 micron thick the passage time is 10^{-12} sec. The total current is the sum of a large number of these very short and uncorrelated current spikes. The Fourier spectrum of such a spike has components spread uniformly from 0 to about 10^{12} Hz. The fraction of this spectrum seen by the circuit depends on the response time of the circuit. This is just the characteristic time τ of our derivation. It is the reciprocal of Δf, the frequency bandwidth of the system Thus the noise is $|\Delta I| \simeq \sqrt{eI \, \Delta f}$ so that the shot noise increases as I increases and as the frequency bandwidth of the system increases. Actually the fluctuations are not all $\pm \sqrt{n}$ in magnitude. Indeed some fluctuations can be much larger than $\sqrt{n}$; some are much smaller than $\sqrt{n}$. The $\sqrt{n}$ is simply a standard deviation. A somewhat more careful analysis of the shot noise in a *pn* junction yields the rms current noise as

$$\Delta I_{rms} = \sqrt{2eI \, \Delta f} \tag{12.11}$$

All semiconductor *pn* junctions whether in a diode or in a bipolar transistor and all vacuum tubes show this broadband shot noise. This type of noise is called *white noise* because the noise power is spread uniformly over a broadband of frequencies.

As an example, if a transistor is carrying 10^{-2} amp collector current, then the current noise is $I_{rms} = \sqrt{2 \times 1.6 \times 10^{-19} \times 10^{-2} \times \Delta f}$. Thus if Δf is determined by the remainder of the circuit to be $\Delta f = 10^5$ Hz, we find that $\Delta I_{rms} = 1.8 \times 10^{-8}$ amp. If the collector impedance is a 10^4-ohm resistor, the noise current produces a noise voltage in the collector circuit $\Delta V_{rms} = R \, \Delta I_{rms} = 1.8 \times 10^{-4}$ volt.

Often the current noise in a transistor is treated as though it arose from an equivalent noise current source in the base circuit of the transistor. The magnitude of this equivalent noise source is $\Delta I_{rms}/\beta$. It should, however, be clear that the shot noise is inherent in the current passing through a *pn* junction and is not actually generated in the base circuit of the transistor.

Thermal noise or Johnson noise arises from the thermal fluctuations of the distribution of free electrons in a resistor. In its equilibrium configuration the average potential difference from one end of an isolated resistor to the other is zero. However, because the system is at a temperature T, it is possible for thermal fluctuations about the equilibrium to occur as long as the energy of these fluctuations is of the order of kT where k is Boltzmann's constant.

Imagine a resistance R divided into a large number N of thin identical slices in series. Because of the random diffusive motion of the electrons an excess charge may momentarily flow from one slice to the next, producing a potential difference ΔV_n where n denotes one of the N slices. This transfer from thermal to electrostatic energy is probable if the electrostatic energy $q \Delta V_n$ is of the order of kT. The electrostatic energy relaxes back into the thermal bath by ordinary ohmic dissipation in a very short time Δt. We write $q \Delta V_n = \Delta t (\Delta V_n)^2 / R_n \simeq kT$ where R_n is the resistance of the nth slice.

Solving, we find $\Delta V_n = (R_n kT/\Delta t)^{1/2} = (R_n kT \Delta f_0)^{1/2}$. Now in this discussion $\Delta t = 1/\Delta f_0$ is the very short time associated with the relaxation of an elementary voltage fluctuation. Since the voltage fluctuations relax by ohmic losses, the time Δt is comparable to the time between collisions for an electron. Thus $\Delta f_0 = 1/\Delta t \simeq 10^{14}$ Hz. The time Δt is analogous to the passage time of an electron through the barrier in the shot effect. Thus Δf_0 is the entire frequency bandwidth over which the Fourier spectrum of the Johnson noise is spread. A real circuit can only sample a fraction $\Delta f / \Delta f_0$ of the noise power where Δf is the bandwidth of the circuit. The effective circuit noise is therefore $\Delta V_n = (R_n kT \Delta f)^{1/2}$.

It remains to calculate the noise for the entire resistor. This is easily done. Because the voltage fluctuations in different slices R_n are random and independent, we have $\Delta V_{\text{rms}} = \sqrt{\Sigma (\Delta V_n)^2}$ where ΔV_{rms} is the voltage noise across the entire resistor and where the sum $\Sigma (\Delta V_n)^2$ is over all the N slices. We obtain $\Delta V_{\text{rms}} = (RkT \Delta f)^{1/2}$ where $R = \Sigma R_n$ is the total resistance of the resistor. A somewhat more careful analysis would yield

$$\Delta V_{\text{rms}} = (4RkT \Delta f)^{1/2} \qquad (12.12)$$

For example, a 10^6-ohm resistor in a system with a bandwidth of 10^6 Hz has a noise voltage $\Delta V_{\text{rms}} = 1.3 \times 10^{-4}$ volt. Obviously Johnson noise like shot noise is broadband white noise.

In addition to the shot noise at the pn junctions of a transistor, Johnson noise is generated in the ohmic pathways of a transistor. For instance, the base resistance $r_b{}'$ of a bipolar transistor and the channel resistance of an FET are sources of Johnson noise.

Whether one has shot noise or Johnson noise is essentially a question of whether or not the number of current carriers is proportional to the current. In a pn junction or vacuum tube there are almost no current carriers normally present. Fluctuations must be associated with the number of current carriers that enter. In a resistor a very large and constant number of current carriers is always present. A net flow of current is a very small perturbation of the random thermal motion of the carriers. We expect the noise to be independent of current and therefore to appear as a voltage fluctuation across the resistor.

Finally, noise whose power spectrum varies as $1/f$ arises in both resistors and in transistors. The physical source of the $1/f$ noise is not known in all cases. It is found, however, that physical systems have slow changes that produce noise with a $1/f$ spectrum. For example, the structure of the surface of the base region in a transistor may change causing the recombination rate of minority carriers to vary. The changes are random and are more likely to occur at very low frequencies than at high frequencies. Such changes may give rise to $1/f$ noise. It should be noted that it is the noise power $(\Delta V_{rms})^2$ that varies as $1/f$ and not the noise voltage.

For transistors the $1/f$ increase in noise will become more important than shot noise at some critical frequency that cannot be predicted on the basis of any existing theory. Below this critical frequency the noise power will increase by 6 dB/ octave, and above this critical frequency the noise power will be roughly independent of frequency. Some low noise transistors have critical frequencies that are as low as 100 Hz.

For resistors it is found that the $1/f$ noise is roughly proportional to the maximum current density in the resistor. For this reason wire-wound or metal film resistors are less noisy than carbon composition resistors. Composition resistors have very high current densities at the small areas where the conducting particles touch, whereas a wire resistor has a constant current density throughout the wire. For a typical carbon composition resistor the $1/f$ noise power will begin to exceed the Johnson noise below about 10^5 Hz.

We have not exhausted the physical sources of noise. Any random process involving charges is a potential source of electrical noise. These include the diffusion of carriers across the base region of a bipolar transistor, the thermal generation of holes and electrons and the recombination of holes and electrons. Further discussion of these noise sources is beyond the scope of the text, however.

For an amplifier circuit it is common to define a noise figure F as

$$F = \frac{\text{input signal power/input noise power}}{\text{output signal power/output noise power}}$$

For a perfect amplifier, $F = 1$; for a real amplifier, $F > 1$. Low noise amplifiers with $F \simeq 1$ exist and are commonly used where the signal to noise ratio is important.

12.6 The High-Frequency Response of Transistors

In order to understand the high-frequency response of a bipolar transistor, let us consider the very simple common emitter transistor amplifier shown in Fig. 12.6. Only the components needed for an incremental analysis are shown. The transistor has a grounded emitter and has a resistor R_C in the collector circuit. The amplifier circuit is driven by a voltage source v_i in series with a resistance R_S. The capacitors C_{bc} and C_{be} represent unavoidable internal capacitances associated with the structure and operation of the transistor. At sufficiently high frequencies C_{bc} and C_{be} are short circuits from the base to the collector and the base to the emitter and amplification is lost.

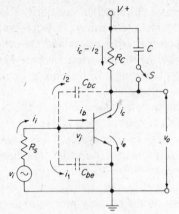

Figure 12.6 An amplifier circuit using a bipolar transistor. Only the incremental currents and voltages are shown. C_{be} and C_{bc} are internal capacitances of the transistor in parallel with the base-emitter junction and the base-collector junction, respectively.

C_{bc} is the capacitance of the back-biased base to collector pn junction. As discussed in Chapter 10, a pn junction contains unshielded ionized donor and acceptor centers. Thus there is an electric field directed from the n to the p side of the junction and a potential difference across the junction. The thickness of the junction varies approximately as the square root of the potential difference and the capacitance of the junction varies approximately as the reciprocal of the square root of the potential difference. The capacitance C_{bc} is of the order of $1 - 20 \times 10^{-12}$ farad for a high-frequency bipolar transistor. The capacitance can be appreciably less than 1 picofarad for transistors in integrated circuits.

These same considerations apply to the base to emitter pn junction. The base to emitter junction is forward biased, however, and usually operates at a much lower voltage than the base to collector junction. Thus the base to emitter junction is thinner and the junction capacitance is greater. It might be of the order of 5–100×10^{-12} farad for a high performance transistor. The junction capacitance represents only one part of the total base to emitter capacitance C_{be}.

We must now discuss briefly the flow of minority carriers across the base from the emitter to the collector since this process makes a contribution to C_{be} that is independent of, and usually larger than, the junction capacitance itself. In normal operation at not too high frequencies the base is an electrically neutral and nearly field-free region through which minority carriers diffuse from the emitter to the collector. The concentration of minority carriers is a maximum at the emitter junction and decreases, almost linearly, to zero at the collector junction. The gradient of this concentration is proportional to the diffusion current. If the concentration of minority carriers everywhere in the base is doubled so that the gradient of the concentration is doubled, then the collector current is doubled. Thus the total charge in the base due to minority carriers is proportional to the collector current. The constant of proportionality is just the time necessary for carriers to diffuse across the base. We may write $q_B = i_C \tau$ where q_B is the charge stored in the base and τ is the average time for a minority carrier to diffuse across the base. Characteristically, τ is of the order of 10^{-10} to 5×10^{-9} sec for a high-frequency transistor. It is important to note that if the base is to remain neutral, an excess of base majority carriers must also be present and these must contribute a total charge $-q_B$. Currents of base majority carriers must flow through the base lead. Very few can penetrate the base-collector junction because it is back biased and very few flow through the forward-biased base-emitter junction because the emitter is much more heavily doped than the base. The stored charge q_B implies an effective base-emitter capacitance. Suppose i_C has an incremental variation i_c so that i_C increases slowly from $i_C = I_C$ to $i_C = I_C + i_c$. The stored base charge changes by $\Delta q_B = q_b = i_c \tau$. As we discussed in Chapter 11 the incremental change in the current i_c is linearly related to the incremental change in the voltage across the base to emitter junction v_j by the relation $v_j = r_d i_c$ where $r_d = 26/I_E$ (I_E is in milliamperes) is the dynamic junction resistance. Thus there is an effective capacitance $C_d = q_b/v_j = \tau/r_d = \tau I_E/26$ (again I_E is in milliamperes). If I_E is large, then the capacitance C_d can be large compared to the $5{-}100 \times 10^{-12}$ farad assignable to the junction capacitance C_{je}.

To summarize, we have shown that C_{be} is an effective capacitance in parallel with the base to emitter junction as shown in Fig. 12.6. It is not in parallel with the ohmic resistances r_e' or r_b'. These are in series with C_{be}. We can write $C_{be} = C_{je} + C_d$ where C_{je} is the base to emitter junction capacitance (also called the transition capacitance) and C_d is the charge storage capacitance (also called a diffusion capacitance).

Now let us assume that the incremental voltage across the junction v_j has a sinusoidal time dependence. At low frequencies an incremental emitter current that is given by $i_e = v_j/r_d$ flows. The emitter current i_e is in phase with v_j. At higher frequencies an out of phase component of the emitter current becomes important. This component is the time rate of change of the charge stored in the base to emitter junction and in the base. At higher frequencies the incremental emitter current $i_e + i_1$ flows. Note in Fig. 12.6 that the part of the emitter current that flows through the junction and mostly falls over into the collector is i_e and that part of the emitter current that flows to charge C_{be} is i_1. The sum $i_e + i_1$ (not i_e alone) is the complete incremental emitter current. The current i_e is in phase with v_j but the capacitive current i_1 leads the charge by 90° and therefore leads v_j by 90° since v_j is really determined by the charge distribution in the junction and the base and not by the currents. Any charge q supplied to the junction or to the stored base charge by the out of phase emitter current i_1 must be balanced by a charge $-q$ (to assure neutrality of the junction and the base) supplied by an out of phase base current i_{b1}. Thus for the out of phase emitter current we have $i_1 = i_{b1}$, whereas for the in phase emitter current we have $i_e = (\beta + 1)i_b$ where i_b is the in phase base current. Most of the in phase emitter current i_e falls over into the collector and only that fraction of the in phase emitter current that recombines in the base produces base current. Thus if $1/\omega C_{be} \ll r_d$, then the incremental emitter current is almost equal to the incremental base current. This implies that the incremental collector current is very small and the current amplification approaches zero. It should be noted that the stored base charge q_b determines the number of minority charge carriers per unit time that fall over into the collector. This quantity is simply i_c. Thus q_b determines the in phase part of the emitter current since $i_e \simeq i_c$. q_b is supplied by the out of phase component of the emitter current i_1, however. Note also that at high frequencies the incremental collector current is $i_c - i_2$, not just i_c. Both q_b and i_c are in phase with v_j and the relationships $q_b = i_c \tau$ and $v_j = r_d i_c$ always hold. Thus our calculation of $C_d = q_b/v_j = \tau/r_d$ is not affected by the out of phase component of the emitter current i_1.

We now turn to the analysis of the circuit of Fig. 12.6. We first calculate the incremental current gain $A_i = (i_c - i_2)/i_i$ under the assumption that the collector is an ac ground. This is accomplished by closing switch S and shunting R_C with the large capacitor C. We note that i_b, i_e, and i_c in Fig. 12.6 are the in phase components of the incremental currents. The out of phase components are given by i_1 and i_2. Also note that because both the emitter and collector are ac grounds, the incremental base-emitter and incremental base-collector voltage are both given by v_j. We can now write

$$i_b = \frac{i_e}{\beta + 1} = \frac{v_j}{(\beta + 1)r_d}$$

$$i_1 = \frac{v_j}{1/j\omega C_{be}} \tag{12.13}$$

$$i_2 = \frac{v_j}{1/j\omega C_{bc}}$$

and

$$i_i = i_1 + i_2 + i_b$$

Substituting these expressions in $A_i = (i_c - i_2)/i_i$ gives

$$A_i = \frac{\beta}{1 + \beta r_d j\omega(C_{be} + C_{bc})}$$

or since usually $C_{be} \gg C_{bc}$, $\tag{12.14}$

$$A_i \simeq \frac{\beta}{1 + \beta r_d j\omega C_{be}}$$

We have used the approximation $\beta \gg 1$; thus $i_e \simeq i_c$ and $\beta \simeq \beta + 1$, and we have used the approximation $1/\omega C_{bc} \gg r_d$.

Equations 12.14 reduce to $A_i = \beta$ at zero frequency ($\omega = 0$). As ω increases, the current gain remains near β until ω becomes comparable to $1/[\beta r_d(C_{be} + C_{bc})]$. At still higher frequencies the current gain falls at the rate of 20 dB/decade. If $\omega = \omega_\beta = 1/[\beta r_d(C_{be} + C_{bc})]$, the absolute value of the current gain is $|A_i| = \beta/\sqrt{2}$. The frequency ω_β at which the current gain falls to $1/\sqrt{2}$ of its value at zero frequency is sometimes called the β cutoff frequency of the transistor or sometimes the common emitter bandwidth. As we shall illustrate in a problem, amplification extends to higher frequencies in a common base configuration than it does in a common emitter configuration. In a common base configuration the cutoff frequency (called the α cutoff frequency) can be shown to be $\omega_\alpha = \beta\omega_\beta/\alpha$ so that $\alpha\omega_\alpha = \beta\omega_\beta$. These products are sometimes called the gain-bandwidth product of the transistor and symbolized by ω_T. Another commonly used frequency parameter is $f_T = \omega_T/2\pi = 1/[2\pi r_d(C_{be} + C_{bc})]$. Thus we have introduced a number of different figures of merit for the high-frequency performance of a transistor. We see that all of them depend on the internal characteristics of the transistor and not on external circuit parameters. Probably f_T and ω_T are the most frequently used.

Let us now turn our attention to how the gain-bandwidth product of a transistor depends on the quiescent operating conditions. Figure 12.7 shows gain-bandwidth product contours for the Fairchild bipolar transistor SE6020, a general-purpose silicon npn transistor. For low values of I_C the gain band-width product is dominated by the base to emitter junction capacity, which is approximately 100×10^{-12} farad; by the base to collector junction capacity, which is approximately 15×10^{-12} farad; and by the base storage capacity, which is determined by a base transit time for minority carriers of about $\tau = 3\text{–}4 \times 10^{-10}$ sec. As the current through the transistor increases, the dynamic resistance of the transistor decreases. Therefore the gain-bandwidth product for the transistor increases as I_C increases for small values of I_C.

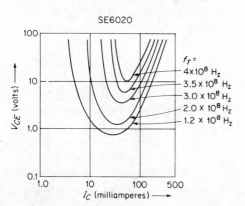

Figure 12.7 Contours of constant gain-bandwidth product (f_T) for the Fairchild SE6020 bipolar transistor. (These curves courtesy Fairchild Semiconductor Incorporated.)

This comes about because of the decrease of the terms $r_d C_{je}$ and $r_d C_{bc}$. The term $r_d C_d = \tau$ and is independent of the dynamic resistance.

We see from the curves that at high values of I_C the gain-bandwidth product begins to decrease with increasing I_C. This is due to a number of effects that we cannot discuss in detail. When I_C and I_E are large, the concentration of minority carriers in the base may exceed the base doping concentration. If the minority carrier injection is very large, the distribution of majority carriers is quite different from that found at lower injection levels, and our discussion of the base storage capacitance no longer applies. For instance, in high-frequency transistors the collector region near the base is usually only lightly doped. A high concentration of minority carriers will push the base-collector junction through this region, thus increasing the base width and the transit time τ. This increases C_d and decreases the gain-bandwidth product.

At fixed I_C the gain-bandwidth product increases as V_{CE} increases. This results from the increase in the width of the base-collector junction and the resulting decrease in the width of the base. Both C_{bc} and C_d are decreased.

We now consider the circuit of Fig. 12.6 with the switch S open and the resistor R_C in the circuit. The collector is no longer an ac ground but instead is at the output potential $v_o = -(i_c - i_2)R_C$. The voltage across the base-collector junction is now $v_j - v_o$ and the current i_2 is given by $i_2 = (v_j - v_o)j\omega C_{bc}$. We may write $v_o = A_v v_j$ where A_v is the voltage gain of the transistor. From earlier results we know that $A_v \simeq -R_C/r_{tr}$. Thus we have

$$I_2 = v_j(1 - A_v)j\omega C_{bc} = v_j j\omega C_{bc}\left(1 + \frac{R_C}{r_{tr}}\right) \qquad (12.15)$$

Now the voltage v_o enters the incremental analysis only through its effect on i_2. Comparing Eq. 12.15 to the previous expression $i_2 = v_j j\omega C_{bc}$ (Eq. 12.13), we see that the current amplification with R_C in the circuit can be obtained by simply substituting $C_{bc}(1 + R_C/r_{tr})$ for C_{bc} in Eq. 12.14. We then have

$$A_i = \frac{\beta}{1 + \beta r_d j\omega[C_{be} + C_{bc}(1 + R_C/r_{tr})]} \qquad (12.16)$$

The importance of this result is that C_{bc}, although usually considerably smaller than C_{be}, now appears multiplied by the factor $(1 + R_C/r_{tr})$, which may be large. Thus C_{bc} cannot be ignored when voltage amplification is present. The multiplication of C_{bc} by a voltage amplification factor is called the *Miller effect*.

We must point out that writing $A_v = -R_C/r_{tr}$ involves some approximations. We have ignored the fact that v_o is given by $v_o = -(i_c - i_2)R_C$

rather than $v_o = -i_c R_C$. In addition, because we evaluated v_o/v_j where v_j is just the voltage across the base-emitter junction, the appropriate voltage amplification is $-R_C/r_d$, not $-R_C/r_{tr}$. Neither approximation alters our result in any important respect.

Finally we shall calculate the voltage gain of the circuit of Fig. 12.6 with v_i as the input voltage and v_o as the output voltage. We use the approximations $v_o = -i_c R_C$ and $v = v_i - i_i R_S$. The latter equation ignores the effects of the ohmic resistances in the base and emitter. It is equivalent to the assumption that $r_{tr} \simeq r_d$. With the help of Eq. 12.13, but replacing C_{bc} by $C_{bc}(1 + R_C/r_{tr})$, we can evaluate $A_v = v_o/v_i$. The result is

$$A_v = \frac{v_o}{v_i} = \left(-\frac{R_C}{r_{tr}}\right) \frac{1}{1 + R_S\{j\omega[C_{be} + C_{bc}(1 + R_C/r_{tr})] + 1/\beta r_{tr}\}} \tag{12.17}$$

At very low frequencies the amplification is $A_v = -R_C/r_{tr}(1 + R_S/\beta r_{tr})$ or $A_v = -R_C/r_{tr}$ if $R_S \ll \beta r_{tr}$. At higher frequencies and ignoring the term $1/\beta r_{tr}$ we may write

$$A_v = \left(-\frac{R_C}{r_{tr}}\right) \frac{1}{1 + R_S j\omega[C_{be} + C_{bc}(1 + R_C/r_{tr})]}$$

or

$$A_v = \left(-\frac{R_C}{r_{tr}}\right) \frac{1}{1 + (j\omega/\omega_1)} \tag{12.18}$$

where

$$\omega_1 = \frac{1}{R_S\{C_{be} + C_{bc}[1 + (R_C/r_{tr})]\}}$$

For frequencies well above ω_1 the voltage gain falls at the rate of 20 dB/decade and the phase of the output voltage is no longer 180° with respect to the input.

The high-frequency response of an FET is limited by the gate to source capacity and by the gate to drain capacity. The Miller effect is also important for an FET.

Problems

In the following problems the term *operational amplifier* will be used. An operational amplifier is a high gain amplifier that can amplify frequencies from 0 Hz (dc) to some high frequency. Operational amplifiers often have both $+$ and $-$ inputs. In Fig. 12.8 the output voltage is given by $v_o = A_0(v_+ - v_-)$ where A_0 is the gain of the amplifier, v_+ is the voltage at the $+$ input, and v_- is the voltage at the $-$ input. We show the symbol for an operational amplifier and the connections for using an operational amplifier with the various forms of feedback. Operational amplifiers are almost always used in circuits with large negative feedback. Thus the actual amplification is much less than A_0. For this reason $v_+ - v_-$ is usually very small and the approximation $v_+ - v_- = 0$ is useful.

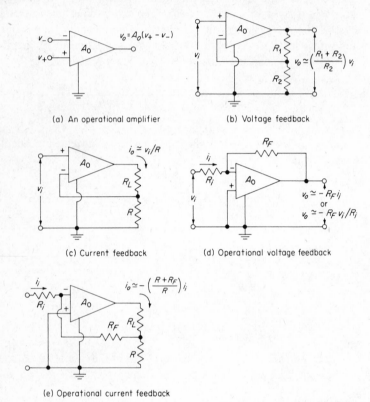

(a) An operational amplifier

(b) Voltage feedback

(c) Current feedback

(d) Operational voltage feedback

(e) Operational current feedback

Figure 12.8

12.1 Sketch a circuit using voltage feedback and providing an output voltage v_o that is 25 times as large as the input voltage v_i. In your diagram you may use an operational amplifier with both $+$ and $-$ inputs, and you may assume that the gain A_0 of the operational amplifier is very large.

12.2 For voltage feedback we have shown that $A_F = A_0/(1 - \beta A_0)$. Let us assume that the gain without feedback A_0 varies with the frequency ω as $A_0 = A/[(1 + (j\omega/\omega_0)]$ where A and ω_0 are constants. A is the dc gain and ω_0 is the bandwidth. The gain-bandwidth product of the amplifier without feedback is defined as $A\omega_0$. Show that the gain-bandwidth product of the amplifier with feedback is also given by $A\omega_0$ provided the feedback ratio β does not depend on ω.

12.3 An amplifier has a gain without feedback that is given by

$$A_0 = A/[1 + j(\omega/\omega_0)]$$

where $A = 10^4$ and where $\omega_0 = 2\pi \times 10^4$ sec^{-1}. The first stage of the amplifier consists of an FET with a 10^3-ohm resistor from the input to ground. If the noise output of the amplifier arises primarily from Johnson noise in the resistor between the input and ground of the first stage of the amplifier, what is the rms noise voltage at the output of the amplifier? If voltage feedback is employed to reduce the dc gain of the amplifier to $A_F = 100$, what is the rms noise voltage at the output of the amplifier? (*Note.* The bandwidth increases due to the feedback.)

12.4 Sketch a circuit using current feedback so that the ratio of the output current i_o to the input voltage v_i is $i_o/v_i = 10^{-2}$ mho. You should use a high gain operational amplifier with + and − inputs in your circuit.

12.5 Two different feedback circuits are shown in Fig. 12.9. The amplifiers shown are all identical without feedback having gain: $A_0 = 10^3$, input impedance $R_i = 10^5$ ohms, and an output impedance $R_o = 20$ ohms.

(a) What is the gain with feedback for each of the circuits?
(b) Which of the circuits will have the least distortion?
(c) Which of the circuits will have the highest input impedance?
(d) Which of the circuits will have the lowest output impedance?

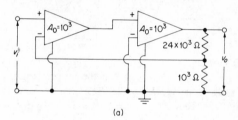

(a)

Figure 12.9

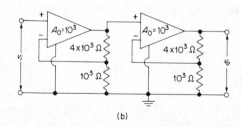

(b)

12.6 An operational amplifier without feedback has a gain $A_0 = 10^4$, an input impedance $R_i = 10^3$ ohms, and an output impedance $R_o = 10$ ohms. The operational amplifier has both + and − inputs. Sketch a follower circuit using this operational amplifier with suitable feedback. What is the gain, the input impedance with feedback, and output impedance with feedback of the circuit?

12.7   Sketch a circuit using negative operational current feedback and providing an output current i_o that is 15 times as large as the input current i_i. Your circuit should utilize a high gain operational amplifier with both $+$ and $-$ inputs.

12.8   Sketch a circuit using negative operational voltage feedback and providing a ratio $v_o/i_i = 10^3$ ohms. Your circuit should utilize a high gain operational amplifier with both $+$ and $-$ inputs.

12.9   A high gain amplifier is used with voltage feedback to produce a gain of 20 with feedback. If the gain with feedback is to vary by no more than 0.01% when the gain without feedback varies by 10%, what value of the gain without feedback A_0 is required?

12.10   An amplifier circuit employing a high gain operational amplifier and a linear combination of voltage and current feedback is shown in Fig. 12.10. Calculate the voltage gain of the circuit. Calculate v_o/i_o, the output impedance of the circuit as seen by the load resistor. The load resistor is R_L.

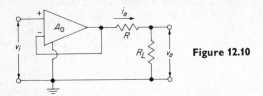

Figure 12.10

12.11   An amplifier circuit employs current feedback (Fig. 12.3). Calculate the input resistance of the circuit with feedback if the input resistance of the amplifier without feedback is given by R_i. The amplifier has gain A_0 without feedback and the feedback voltage signal is $i_o R_F$ that is $\beta = 1$.

12.12   What rms voltage due to Johnson noise is produced by a 10^6-ohm resistor in the bandwidth from 10 to 10^5 Hz?

12.13   What rms current due to shot noise is produced in a transistor with a total collector current of 10 milliamp? The bandwidth of the circuit extends from 10 to 10^6 Hz.

12.14   A 10^6-ohm resistor produces an rms noise voltage of 5 microvolts over a 10^3-Hz bandwidth centered at 10^4 Hz. Is the voltage due to Johnson noise? Can $1/f$ noise explain the voltage? If the noise is due to $1/f$ noise, at what frequency is the Johnson noise as large as the $1/f$ noise?

12.15   A transistor in a common emitter circuit has a gain-bandwidth product of $f_T = 4 \times 10^7$ Hz. The collector is an ac ground. If the dc current gain β of the transistor is 200, what is the bandwidth ω_β of the circuit?

12.16 Show that a bipolar transistor used in a common base circuit and with a collector that is shorted to ground for ac signals has a current gain that is given by

$$A_i = \frac{\alpha}{1 + j\omega r_{tr}\,\alpha(C_{be} + C_{bc})}$$

Show also that the gain-bandwidth product in the common base configuration is $\omega_T = 1/[r_{tr}(C_{be} + C_{bc})]$, the same result as in the common emitter circuit. Assume $r_{tr} \simeq r_d$.

12.17 For the circuit of Fig. 12.6, suppose that the various quantities have the following values: $I_C = 10^{-3}$ amp, $\beta = 300$, $R_C = 10^3$ ohms, $R_S = 500$ ohms, $C_{be} = 10^{-10}$ farad, and $C_{bc} = 5 \times 10^{-12}$ farad. At what frequency is the magnitude of the voltage gain of the circuit reduced by $1/\sqrt{2}$ from the dc gain?

12.18 A resistor at a temperature T can be considered to have a Thevenin equivalent circuit that consists of an emf source $\varepsilon_{rms} = \sqrt{4kTR\,\Delta f}$ and a noiseless resistor R in series. Consider a resistor R and a capacitor C in parallel as shown in Fig. 12.11. Show that the square of the rms noise voltage V across the capacitor C in the infinitesimal bandwidth $d\omega$ and centered at frequency ω is $4kTR(d\omega/2\pi)/[1 + (\omega RC)^2]$, and show that the total rms voltage squared across the capacitor integrated overall frequencies is $4kTR(1/4RC)$. This is interpreted as meaning that the bandwidth of the parallel RC network is $\Delta f = 1/4RC$. Thus the noise is just the Johnson noise from a resistor R operating into a circuit with a bandwidth $1/4RC$.

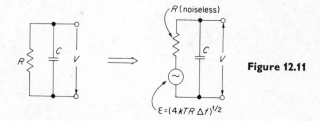

Figure 12.11

12.19 Calculate $A = v_o/v_i$ for the emitter follower amplifier shown in Fig. 12.12. Calculate $R_i = v_i/i_i$ both with and without the feedback capacitor C_F in the circuit. Show that the input resistance R_i is greatly increased by the use of the feedback capacitor. This use of positive feedback to increase the input resistance is called *bootstrapping*. All three capacitors are very low impedances for the signal frequencies of interest.

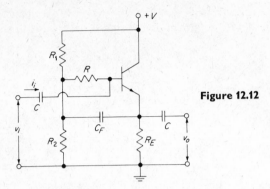

Figure 12.12

12.20   A circuit for an n channel FET is shown in Fig. 12.13. The capacities C_{gs} and C_{gd} are the stray capacities between the gate and source and between the gate and the drain, respectively. These arise due to the capacity of the back-biased pn junction between the gate and the channel. They might be of the order of 2–20 picofarads. Show that the Miller effect increases the effect of the capacity C_{gd}. Calculate the voltage gain $A_v(\omega)$ of the circuit as a function of the frequency, and show that the Miller effects limits the high-frequency response of this circuit. For this problem you should calculate $A_v(\omega) = v_o/v_i$ where v_i and v_o are the incremental signals at the input and output, respectively, that is, v_o does not include the dc contribution to the voltage at the output.

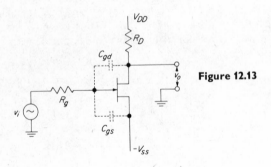

Figure 12.13

13/AC Amplifiers

In earlier chapters we have discussed the properties and circuit applications of single transistors. Only occasionally has more than one transistor been used in a single circuit. Most real circuits will employ a number of different transistors usually serving a number of different functions. For example, the amplifiers introduced in the preceding chapter in our discussion of negative feedback would employ a number of transistors in order to achieve high gain and stability. High gain is usually obtained by making the output of one stage of amplification the input for the next stage. There are important advantages if signal voltages can be transmitted through capacitors rather than by direct resistive connections. Such amplifiers cannot amplify very slowly varying signal voltages, hence the name ac amplifiers. We analyze a simple RC coupled amplifier in the next section

13.1 An RC Coupled Amplifier

As an example of an RC coupled amplifier we shall discuss the circuit shown in Fig. 13.1. This amplifier has three stages. The first two stages are amplifiers and the third stage is an emitter follower. The transistors are all 2N3643 silicon transistors that have $\beta = 100\text{--}300$. The various stages are coupled by 0.5 microfarad capacitors. These capacitors are very useful because they enable us to set the dc bias voltages independently of the ac coupling of the collector of one stage to the base of the next stage. Since $1/\omega C$ becomes very large as ω, the signal frequency, becomes small, however, the gain of the RC coupled amplifier falls off as ω decreases. Let us assume that the frequency ω of the input signal is large enough so that $1/\omega C$ is small compared to the various resistors used in the circuit. We also assume that ω is small enough so that intrinsic shunt capacitances of the transistors do not limit the gain. This is called the *midband approximation*. For this case we can just ignore $1/\omega C$ in estimating the gain of the amplifier. The 10^3 ohms that connect the emitter of the third stage to the emitter of the first stage provides negative feedback for the amplifier.

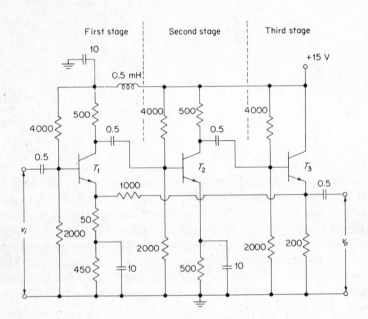

Figure 13.1   A three-stage capacitance coupled amplifier. All resistances are in ohms and all capacitances are in microfarads. The transistors are all silicon *npn* type 2N3643 transistors.

From the input voltage divider we see that the base of the first stage has a dc voltage of about 5.0 volts. Since the transistors are silicon, the emitter of the first stage is at about 4.4 volts. Because the emitter of T_3 is also at about 4.4 volts, very little dc current flows through the 1000-ohm resistor. Therefore the dc emitter current of T_1 is $I_E = 4.4/500 = 8.8 \times 10^{-3}$ amp. Hence the transresistance of the transistor is $r_{tr} = (26/8.8) + 4 = 7$ ohms where we assume the ohmic resistance is 4 ohms. The voltage gain of the first stage is therefore $A_1 = -R_C/(R_E + r_{tr}) = -\frac{500}{57} = -8.8$ where R_E is the 50-ohm resistor in the emitter of the first stage. The 450-ohm resistor in the emitter of the first stage is assumed to be short circuited for ac signals by the 10-microfarad capacitor. The output resistance of the first stage is simply the 500-ohm collector resistance. In calculating this gain we have ignored the negative feedback from the third stage. We shall incorporate the feedback later.

Let us now consider the coupling of the first stage to the second stage for this *RC* coupled amplifier. This coupling is illustrated in Fig. 13.2.

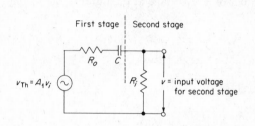

First stage | Second stage

$v_{Th} = A_1 v_i$

R_o C

R_i

v = input voltage for second stage

Figure 13.2 The coupling between the first and second stages of the amplifier of Fig. 13.1. The Thevenin output parameters of the first stage are $v_{Th} = A_1 v_i = -8.8 v_i$ and $R_{Th} = R_o = 500$ ohms. The input resistance of the second stage is $R_i = 680$ ohms and the two stages are separated by a 0.5-microfarad capacitor.

The Thevenin output parameters for the first stage are $v_{Th} = A_1 v_i = -8.8 v_i$ and $r_{Th} = R_o = 500$ ohms. The input resistance of the second stage is the resistance equivalent to a (4×10^3)-ohm resistor to ground in parallel with a (2×10^3)-ohm resistor to ground in parallel with the input resistance to the second transistor itself. The (4×10^3)-ohm resistor goes from the base of the second transistor to the 15-volt power supply, which is an ac ground, and so for the small signal analysis it is in parallel with the (2×10^3)-ohm resistor to ground. The second-stage transistor is similar to the first, and $r_{tr} = 7$ ohms for the second transistor. The input resistance to the second transistor itself is therefore $\beta r_{tr} = 200 \times 7 = 1.4 \times 10^3$ ohms. Thus for the second stage

$$\frac{1}{R_i} = \frac{1}{R_{eq}} = \left(\frac{1}{4} + \frac{1}{2} + \frac{1}{1.4}\right)10^{-3} \qquad (13.1)$$

or $R_i = 680$ ohms. The two stages are separated by a 0.5-microfarad capacitor. The incremental voltage at the input to the second stage is

$$v = \left[\frac{R_i}{R_o + R_i - (j/\omega C)}\right]v_{Th}$$

$$v = \left[\frac{R_i}{R_o + R_i - (j/\omega C)}\right]A_1 v = \frac{[R_i/(R_o + R_i)]A_1 v_i}{1 - [j/\omega(R_o + R_i)C]} \qquad (13.2)$$

The amplitude of the voltage at the input to the second transistor is reduced from the Thevenin output voltage of the first stage $v_{\text{Th}} = A_1 v_i$ by a factor

$$\frac{R_i/(R_o + R_i)}{\{1 + [1/\omega(R_o + R_i)C]^2\}^{1/2}} \tag{13.3}$$

If $\omega \ll 1/(R_o + R_i)C$, then the voltage at the input to the second transistor is very small. When ω is large enough that $1/\omega(R_o + R_i)C$ is negligible, however, then the ratio of the voltage at the input to the second stage to the Thevenin output voltage from the first stage is simply $R_i/(R_o + R_i) = 1/[1 + (R_o/R_i)]$. As pointed out earlier, this is called the *midband approximation*. In the midband approximation the voltage at the input to the second transistor is

$$v = \frac{1}{[1 + (R_o/R_i)]} A_1 v_i = \frac{-8.8 v_i}{1 + \frac{500}{680}} = -5.1 v_i \tag{13.4}$$

The gain of the second stage is

$$A_2 = -\frac{R_C}{r_{tr}} = -\frac{500}{7} = -71 \tag{13.5}$$

where R_C and r_{tr} are the collector resistor and transresistance of the second transistor. The Thevenin output voltage from the second stage is A_2 times the voltage at the input to the second transistor. In the midband approximation this is $A_2 v = (-71)(-5.1)v_i = 362 v_i$. The Thevenin output resistance of the second stage is 500 ohms, which is simply the collector resistor of the second transistor.

The third stage is simply an emitter follower and so the output of this stage is approximately equal to the voltage at the input to the third transistor. The input resistance to the third stage is simply a (4×10^3)-ohm resistor and (2×10^3)-ohm resistor in parallel. We have ignored the input resistance to the third transistor since it is $\beta(R_E + r_{tr}) \simeq 200 \times 207 \simeq 40 \times 10^3$ ohms, which is much larger than either of the two resistors used to provide the dc bias voltage for the base of the third transistor. The input resistance to the third stage is given by

$$\frac{1}{R_i} = \left(\frac{1}{4} + \frac{1}{2}\right) 10^{-3} \tag{13.6}$$

or $R_i = 1.33 \times 10^3$ ohms. Thus the voltage at the input to the third transistor is

$$\left(\frac{1}{1 + \frac{500}{1330}}\right) 362 v_i = 263 v_i \tag{13.7}$$

in the midband approximation. Since the third-stage emitter follower has a gain of approximately 1, the overall gain of the *RC* coupled amplifier neglecting the negative feedback is approximately 263 near midband.

We now wish to discuss the effect of the negative feedback provided by the 10^3-ohm resistor connected between the emitter of the third stage and the emitter of the first stage. A fraction $\beta = 50/(1000 + 50) = \frac{1}{21}$ of the output signal is fedback to the emitter of the input transistor. That this is true can be seen with the help of Fig. 13.3 where we detail only those components of the amplifier that are important for the feedback. Because the gain of the amplifier is very much reduced by the feedback, it follows that v_{be}, the incremental base to emitter voltage of the first stage, must always be very small. In other words, the feedback holds $v_f \simeq v_i$. This is equivalent to the virtual ground approximation that we discussed in the preceding chapter. From $v_{be} \simeq 0$ or $v_f \simeq v_i$, we conclude that the incremental feedback current i through the 1000-ohm resistor must also flow through the 50-ohm resistor. There is no parallel path to ground through T_1. This is entirely a result of the feedback. The point becomes clearer if we remember from Chapter 11 that the output impedance of an emitter follower amplifier is the parallel combination of r_{tr} and the emitter resistor. Thus in Fig. 13.3 if v_o were not linked to v_i through the amplifier and feedback network but was rather just an independent voltage, then v_{be} would not be clamped near zero and v_o would see r_{tr} (7 ohms) in parallel with the 50 ohms.

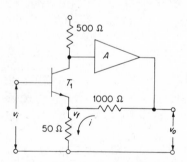

Figure 13.3 A schematic of the circuit of Fig. 13.1 that emphasizes the feedback network.

We may now calculate the amplification with feedback from $1/\beta$ or from the relations $v_i \cong v_f \cong 50i$ and $v_o = 1050i$. We find $A = v_o/v_i = 1/\beta = 21$. It should be noted that the feedback is pure voltage feedback. The current i, which is proportional to v_o, does not appear in the input to the amplifier. The input impedance is increased, not decreased, as a result of the feedback.

As we have already discussed, the gain of an *RC* coupled amplifier falls off at low frequencies because of the coupling capacitors used between stages. At higher frequencies the gain of an *RC* coupled amplifier will fall off due to the stray capacities, especially the stray base to collector capacity and the Miller effect.

Equation 13.3 tells us that $\omega = 1/(R_o + R_i)C \simeq 2000$ rad/sec is a measure of the low-frequency cutoff without feedback. The negative feedback will considerably extend both the low-frequency and high-frequency response.

One final comment is that the 10-microfarad capacitor and the 0.5-millihenry inductor separating the +15-volt supply voltage from the first stage are necessary because without them there may be sufficient gain and positive feedback (see Chapter 16 for a discussion of positive feedback) for the amplifier to oscillate. Because the power supply has some internal resistance, the output signal modulates the output of the 10-volt supply. Unless filtered out this modulation reaches the input and for some frequency range will be in phase with the input signal.

13.2 The Stability of Negative Feedback

We have discussed in Chapter 12 how negative feedback can be used to make the amplification of an amplifier nearly independent of the properties of the amplifier without feedback and almost completely dependent on the parameters of the feedback network only. We also showed in Chapter 12 that the voltage gain of a single transistor amplifier varies as $A_0(\omega) = A_0/[1 + (j\omega/\omega_1)]$. Both $A_0(\omega)$ and A_0 are gains without feedback. A_0 is the midband gain and ω_1 is a characteristic frequency determined by the intrinsic capacitances of the transistor and the Miller effect. For $\omega \gg \omega_1$, the amplification approaches $A_0(\omega) = -j\omega_1 A_0/\omega$ and the input and output signal voltages are related by $v_o = (-j\omega_1 A_0/\omega)v_i$. At midband the corresponding relation is $v_o = A_0 v_i$. Thus at midband the input and output voltages are either in phase or 180° out of phase depending on whether A_0 is positive or negative. As the frequency increases, the phase difference between input and output changes, compared to the phase difference at midband, by an amount that approaches 90° at very high frequencies. The factor $-j$ tells us that this phase change is a lag of the output with respect to the input.

For an amplifier that has two stages, the gain is $A_1(\omega)A_2(\omega) = A_1 A_2/[1 + (j\omega/\omega_1)][1 + (j\omega/\omega_2)]$ and the phase of the output at very high frequencies is shifted by nearly 180° with respect to the output at midband. If the amplifier has three stages, then the phase of the output at very high frequencies with respect to the output at midband can reach and exceed 180°; in fact, it can almost reach 270° at very high frequencies. Now these results have important implications for the behavior of amplifiers using negative feedback. For negative feedback the signal that is fedback from the output must be 180° out of phase with the input. This is assured by requiring that βA_0 be negative in the fundamental equation $A_F = A_0/(1 - \beta A_0)$ (Eq. 12.4). But we have seen that additional phase shifts between input and output are introduced as the frequency increases from midband ($\omega \ll \omega_1$) to the high-frequency limit $\omega \gg \omega_1$. When the additional phase shift approaches $180°$, the feedback signal is in phase with and reinforces the input signal. The feedback becomes positive. As shown in Chapter 16, a circuit with positive feedback will oscillate if the $\beta A_0(\omega)$ of the circuit is greater than 1. Consequently, one must assure that the $\beta A_0(\omega)$ of an amplifier falls below 1 at a lower frequency than the frequency at which the phase shift reaches $180°$. A process called *phase compensation* is used to accomplish this. Phase compensation is discussed with respect to operational amplifiers in Chapter 15.

Referring again to the equation $A_F = A_0/(1 - \beta A_0)$, we note that an increase in frequency tends to reduce A_F because A_0 is decreased, but on the other hand A_F tends to increase because the feedback shifts from negative toward positive feedback. This often results in the gain of an amplifier increasing somewhat above $A_F = -1/\beta$ at some high frequency. As the frequency increases still further, the gain then falls because A_0 becomes very small. This peaking of the gain occurs even though the amplifier gain falls below 1 before the phase shift reaches $180°$ so that oscillation cannot occur. This peaking of the gain is illustrated in Fig. 13.4.

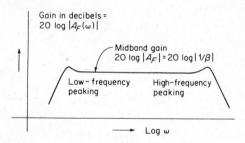

Figure 13.4 A logarithmic plot of gain versus frequency for an amplifier with negative feedback. Gain peaking results from a phase shift of the feedback signal.

One last comment on feedback stability may be in order. For an *RC* coupled amplifier the gain of each stage of the amplifier has the form $A_0(\omega) = A_0/[1 + (j\omega_1/\omega)]$ at low frequencies where A_0 is the midband gain of the individual stage (see Eq. 13.2 and 13.3). Thus for an *RC* coupled amplifier the phase of the output at very low frequencies shifts with respect to the midband output. For an *RC* coupled amplifier with three or more stages the phase shift can reach $180°$. Thus the $\beta A_0(\omega)$ of an *RC* coupled amplifier with negative feedback must fall below 1 at a frequency higher than the frequency at which the phase becomes $180°$ in order to avoid oscillation at some low frequency. Peaking of the gain may occur at a low frequency for an *RC* coupled amplifier with negative feedback; that is, the gain of the amplifier at some low frequencies may be greater than the midband value $A_F = -1/\beta$. Low-frequency peaking is also shown in Fig. 13.4.

13.3 Tuned Amplifiers

The RC coupled amplifier that we discussed is suitable as a broadband amplifier. Of course this means that the signal and the noise to be amplified is that associated with the broadband of frequencies for which the circuit has a large gain. Often this is not desirable and instead one wishes to amplify a narrow band of frequencies centered about a particular frequency. For example, the intermediate frequency amplifier in a superheterodyne receiver is used to amplify only a narrow band of frequencies centered about the difference frequency of the carrier and the local oscillator. Typical intermediate frequencies are from 400×10^3 to 30×10^6 Hz.

We have already learned how to use a several-stage amplifier to obtain more gain than is available from a single stage. Thus what we would like to know is how to build a single stage of a tuned amplifier. A schematic diagram of such a circuit is shown in Fig. 13.5. An amplifier with two transistors arranged as shown is called a cascode amplifier. The circuit can be understood as follows. The lower FET (T_1) acts as a common source amplifier.

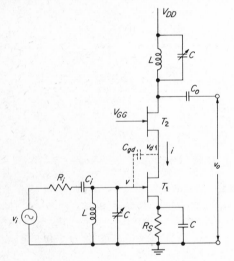

Figure 13.5 A tuned cascode amplifier.

The upper FET (T_2) acts as load for T_1. The dc operating point of the cascode amplifier is determined by the resistor R_S and by the dc supply voltages V_{DD} and V_{GG} as shown. The input and output capacitors C_i and C_o are assumed to be ac shorts and are used to couple one stage to the next. The capacitor C is also assumed to be an ac short. The incremental drain to source current in T_1 is given by $i = g_{fs} v$ where v is the incremental voltage at the gate of T_1. The parallel LC network in the gate circuit of T_1 (that is in the input circuit) determines the frequency at which the amplifier is tuned. The resonant frequency of the network is given by $\omega_0{}^2 = 1/LC$. The LC network at resonance has an impedance that is real and very large. Off resonance the impedance of a parallel LC circuit is complex and small in magnitude. The signal v at the gate of T_1 is given by $v = [Z_{LC}/(R_i + Z_{LC})]v_i$ where Z_{LC} is the impedance of the parallel LC circuit and R_i is the internal resistance of the source v_i. Near resonance, $Z_{LC} \gg R_i$ and $v \simeq v_i$; away from resonance, $Z_{LC} \ll R_i$ and $v \simeq 0$. Consequently the circuit has an appreciable gain only near $\omega = \omega_0$, that is within the frequency range from $\omega_0 - \omega_0/2Q$ to $\omega_0 + \omega_0/2Q$ where Q is the quality factor of the circuit. In the same way the parallel LC circuit in the drain circuit of T_2, that is in the output circuit, tunes the output voltage. Let us now assume that v_i is at the resonant frequency of the two LC circuits. In this case the incremental drain to source current in T_1 is $i = g_{fs}v_i$. This same incremental current must flow in T_2 so that $i = -g_{fs}v_{d1}$ where v_{d1} is the incremental voltage at the drain of T_1 and where we have assumed that both T_1 and T_2 are identical and so have the same transconductance g_{fs}. The output voltage is given by $v_o = -Z_{LC}i = -Z_{LC}g_{fs}v_i$ where Z_{LC} is the impedance of the tuned LC network in the output circuit. The gain of the amplifier is $v_o/v_i = -g_{fs}Z_{LC}$. We have assumed that the LC network in the input and the LC network in the output are identical and that each has the impedance Z_{LC}.

We now wish to point out one of the advantages of the cascode circuit. Let us consider the effect of the gate to drain capacity C_{gd} of the lower FET, T_1. The equations $i = g_{fs}v_i$ and $i = -g_{fs}v_{d1}$ imply that v_i and v_{d1} are equal in magnitude but of course they are opposite in sign. The current that flows through the stray capacity C_{gd} is given by $i_1 = j\omega C_{gd}(v_i - v_{d1}) = j\omega C_{gd}(2v_i)$. Without the cascode arrangement of the two FET's the Miller effect would have caused a current $i_1 = j\omega C_{gd}(1 - A_0)v_i$ to flow through C_{gd}. The cascode arrangement reduces the Miller effect increase in the effective capacity from gate to drain from $(1 - A_0)C_{gd}$ to $2C_{gd}$.

Tuned amplifiers, which do not use the cascode circuit, are sometimes constructed so that a current that is exactly equal in amplitude but opposite in phase to the current through the stray gate to drain capacity is fedback from the output to the input. This process, which is called *neutralization*, is another way to "beat" the Miller effect.

If several identical stages of amplification are used, the bandwidth of the amplifier decreases from the value ω_0/Q for one stage. The reason for this is that the gain of the entire amplifier is the product of the gains of each stage. Thus if each stage has a gain of A_s at $\omega_0{}^2 = 1/LC$, then the gain of the overall amplifier is $A_s{}^N$ at ω_0 where N is the number of stages. Now consider the frequency where the gain of an individual stage is $A_s/2$. At this frequency the gain of the amplifier is $(A_s/2)^N = (1/2^N)A_s{}^N$. Thus the bandwidth of the amplifier can become smaller than is desirable if several stages are used. There are two common methods of overcoming this problem. The simplest method is to tune the *LC* resonant circuits of successive stages to slightly different frequencies but such that the bandwidth of one resonant circuit overlaps the others. Now the gain of one stage will be high at one frequency and gain of the next stage will be low at this frequency, whereas at a slightly different frequency the situation will be reversed. By the use of this technique, which is called *stagger tuning*, the bandwidth can be spread to a useful value but with some loss of gain. The other method of increasing the bandwidth is to use a mutual inductor to couple one stage to the next. This is illustrated in Fig. 13.6.

Figure 13.6 Interstage coupling using a mutual inductance.

Only the essential ac components are shown. The collector tank circuit of one stage is coupled by the mutual inductance *M* to the base tank circuit of the next stage. It will be recalled that the resonant frequency of a tank circuit coupled by a transformer to an identical tank circuit is split into two frequencies given by $\omega_0/\sqrt{1 \pm k}$ where $k = M/L$. If one selects $k \sim 1/Q$, then it can be recalled that the secondary current will have the following appearance (Fig. 13.7). Thus by the use of a mutual inductance for coupling one stage to the next one can broaden the bandwidth of the amplifier. This discussion of the coupling of one stage of a tuned amplifier to the next stage has neglected completely the very important effect of the loading of the output of one stage by the input of the next stage so that our statements are only intended to acquaint the reader with some of the ideas and terminology. If the reader wishes actually to build or use a tuned amplifier, he should follow this subject further by studying a more detailed treatment.

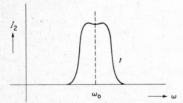

Figure 13.7 The frequency response of an amplifier using the mutual inductance coupling shown in Fig. 13.6.

13.4 Power Amplifiers

Power amplifiers are amplifiers that are used to supply large amounts of power to a load. Typical values of the power supplied by a power amplifier range from about 1 watt to many kilowatts. Power amplifiers are used for driving a loudspeaker system at audio frequencies, for driving a radio trans-mitter antenna at radio frequencies, and for a variety of other functions.

Power amplifiers are classified by the portion of input signal during which current flows in the output circuit. The various classes of amplifiers are discussed and illustrated below.

In class A operation (Fig. 13.8) the current flows in the output circuit during the entire period of the input signal. The amplifier circuits we pre-viously considered have all been class A. Distortion of the output current occurs as shown because the output current has a nonlinear dependence on the input voltage. For most applications one must use considerable care to eliminate distortion.

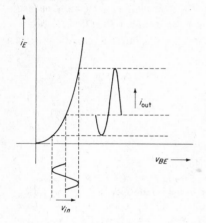

Figure 13.8 The input and output signals for a class A amplifier.

Let us consider a simple class A power amplifier for audio fre-quencies. The circuit for this amplifier is shown in Fig. 13.9. The first thing to note about the power amplifier shown is that the load R_L is coupled to the collector of the power transistor by a transformer. This is often a useful thing to do because the dc bias current I_C in the transistor is large. If the resistance R_L were inserted into the collector circuit without the transformer, then the power $I_C^2 R_L$ would be wasted in the load resistor. The power dissipated in the transformer coil due to the dc current can be small. The transformer permits us to separate the dc and ac signals in the collector circuit. Only the ac signals reach the load. The 1/1 ratio for the transformer is merely an arbitrary ratio selected to simplify the discussion and this ratio could be adjusted as desired for impedance matching.

The Q point or operating point for the amplifier is located in Fig. 13.10. Suppose I_C and V_{CE} are the dc values of i_C and v_{CE}, respectively, at the Q point. For ac signals the current i_C will vary about I_C on a dynamic load line of slope $-1/R_L$. In order to have both the maximum swing in voltage and current about the Q point, the slope of the dynamic load line should be such as to run from $v_{CE} = 0$, $i_C = 2I_C$ to $v_{CE} = 2V_{CE}$, $i_C = 0$. When this condition is satisfied, then it is possible to deliver the power

$$P = \frac{V_{CE}}{\sqrt{2}} \frac{I_C}{\sqrt{2}} = \frac{V_{CE} I_C}{2} \qquad (13.8)$$

to the load. The power dissipated in the amplifier is $P' = V_{CE} I_C$. It should be noted that V_{CE} is essentially the power supply voltage V_{CC} since the dc load line has a very large slope as shown in Fig. 13.10. Therefore the maximum possible efficiency of a class A power amplifier is $e = P/P' = \frac{1}{2}$. In practice a good amplifier can actually be used with efficiencies of 40% or more. In order to obtain efficiencies as high as 40%, it is usually necessary to use a current drive or negative feedback instead a voltage drive. This is because transistors are more linear in their response to a current drive than to a voltage drive.

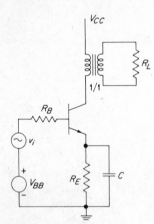

Figure 13.9 Schematic of a class A power amplifier with a transformer load.

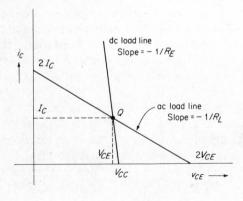

Figure 13.10 The operating point Q and load lines of the amplifier of Fig. 13.9.

In practice the load R_L is not often such that the slope $-1/R_L$ provides the correct dynamic load line. In these cases the transformer is used to transform the impedance R_L so that the power amplifier can operate along a satisfactory dynamic load line.

Class A power amplifiers are used for a variety of purposes. One of the most common uses for a class A power amplifier is as a driver stage for a class B output stage.

In class B operation the transistor is biased near cutoff so that current flows in the output circuit for one-half of the input signal as shown in Fig. 13.11. Obviously an output of this type is highly nonlinear. In order to obtain a more linear output a push-pull amplifier is usually used. In this amplifier two transistors are each run with a class B bias. The circuit is such that one transistor conducts on one-half of a cycle and the other transistor conducts on the other half cycle. The outputs of the two individual transistors are combined to provide an output signal that is nearly linear with respect to the input signal. An example of a class B push-pull amplifier for audio frequencies is shown in Fig. 13.12.

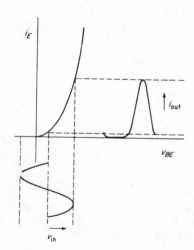

Figure 13.11 The input and output signals in class B operation of a power amplifier.

For the amplifier shown the two out of phase signals from two terminals of the input transformer are fed to the bases of two transistors T_1 and T_2. Each of these transistors is operating in class B. The output signal from the transistors is combined and coupled to R_L by the output transformer.

The maximum efficiency of a class B amplifier can be estimated as follows. The quiescent point is such that $I_C = 0$. Therefore the quiescent voltage V_{CE} is the full supply voltage V_{CC}. The power delivered to the amplifier by the power supply is $P' = \langle V_{CC} i_C \rangle = (2/\pi)(i_{C_{max}} V_{CC})$. Now the maximum power that can be supplied to the load is

$$P = \frac{V_{CC}}{\sqrt{2}} \frac{i_{C_{max}}}{\sqrt{2}} \qquad (13.9)$$

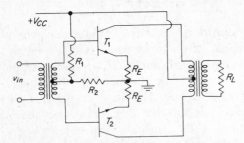

Figure 13.12 A class B push-pull power amplifier.

Therefore the maximum possible efficiency is $e = P/P' = \pi/4 = 78\%$. In order to operate with this maximum efficiency the voltage v_{CE} across an individual transistor must swing from V_{CC} to zero during the on part of the ac cycle.

The dc bias of the transistors in a class B push-pull amplifier is very important if distortion is to be minimized. At first one might think that the bias would be best set with the quiescent voltage $V_{BE} = 0$. From a composite plot of v_{BE} versus i_C for the two transistors in the push-pull amplifier, however, it is easily seen that this results in serious crossover distortion. This is shown in Fig. 13.13. The crossover distortion can be avoided by biasing both transistors so that they are carrying a small steady state current at the quiescent point. This is done by selecting R_1 and R_2 to be such that V_{BE} is just slightly greater than 0.6 volt for silicon transistors. The near elimination of crossover distortion by biasing the two transistors so that they are slightly turned on is shown in Fig. 13.14.

The emitter resistors R_E in the class B amplifier must not be bypassed with a capacitor. The class B amplifier must have a bias very near zero. If a large capacitor is used in parallel with R_E, then because of the rectifying action of the base-emitter junction the capacitor will charge up and alter the dc bias.

In a power amplifier it is important that care be used to assure thermal stability. This is particularly important because of the large currents and power dissipation in a power amplifier. Silicon transistors with their relatively large forbidden energy gap are used because they are less sensitive to temperature changes than germanium transistors. Often other measures are also taken to assure thermal stability. Although the techniques for assuring the thermal stability will not be discussed here, Prob. 13.9, 13.10, 13.11, and 13.12 are used to bring out the important features of bias circuits that are stable against thermal runaway.

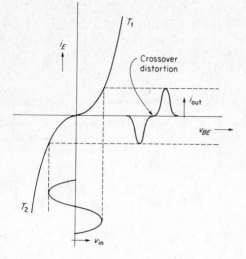

Figure 13.13 Crossover distortion in the output of a class B push-pull power amplifier.

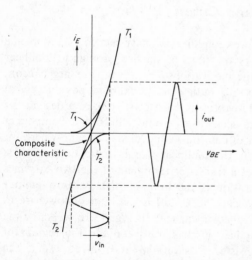

Figure 13.14 The input and output of a class B power amplifier biased to minimize crossover distortion.

In class C operation current flows in the output circuit for less than half the period of the input signal. This is shown in Fig. 13.15.

The output of class C amplifiers is extremely nonlinear. Consequently they are normally used when the output signal can be filtered so that the signal after the filter contains only the fundamental of the input signal frequency. Class C amplifiers are often used at radio frequencies.

The major advantage of a class C amplifier is that very high efficiencies may be obtained (for example, as high as 90 %).

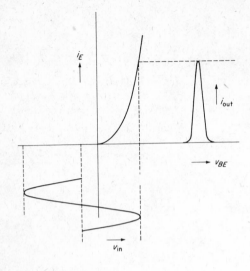

Figure 13.15 Input and output signals in class C operation.

13.5   Saturation and Cutoff

In our discussion of power amplifiers we have met for the first time the deliberate use of transistors or tubes in regions of highly nonlinear response. In power amplifiers the motivations for such use are largely economic. A broadcast station wants to radiate the maximum possible signal for the smallest possible capital investment and electric bill, or a designer of mass-produced audio amplifiers seeks the minimum bulk and component cost. In later chapters we shall discuss pulse and digital circuits where for quite different reasons one exploits the nonlinear characteristics of transistors.

The useful properties of a transistor terminate at the two extremes of saturation and cutoff. Saturation is the limit of high base to emitter voltage and maximum collector current. Cutoff is the limit of low base to emitter voltage and minimum collector current. In Fig. 13.16 we show the collector characteristics of a typical bipolar transistor and the regions of saturation and cutoff. A transistor is saturated at a given v_{CE} if the collector current can be increased only by increasing v_{CE} but not by increasing the base-emitter voltage. At saturation the collector-emitter voltage v_{CE} may be as low as 0.2 or 0.3 volt. This is less than v_{BE}, the base-emitter voltage, and both collector and emitter are forward biased with respect to the base. The concentration of minority carriers in the base is very high, i.e., much higher than when the transistor is in the active region. They must be removed to drive the transistor out of saturation. This requires a time which somewhat limits the usefulness of the transistor in those high speed pulse and switching circuits which drive the transistor into saturation.

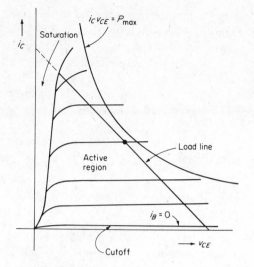

Figure 13.16 Collector characteristics showing the saturation and cutoff regions of a bipolar transistor. The load line drawn makes almost maximum use of the power dissipation capabilities of the transistor, which are represented by the curve $i_C v_{CE} = P_{\max}$.

By cutoff is meant the region where i_B is zero or less. Collector currents are then very small. Consider a bipolar transistor operating at zero base current. A small forward emitter bias will inject minority carriers into the base where some recombine and the remainder are expected to fall across the base collector junction. The recombination current must be provided to the base, however, and it can come only from the reverse bias current from collector to base or from leakage currents. If we ignore the latter and assume a reverse bias current I_{CO} from collector to base, then the emitter takes on a very small forward bias and provides a current $(\beta + 1)I_{CO}$ to the base. The same total current flows from base to collector. Of this current βI_{CO} is the flow of minority carriers from base to collector and I_{CO} is the reverse bias current (minority carriers of the collector flowing into the base and neutralizing the recombination current). The subscript O in the notation I_{CO} is used because I_{CO} is the back bias current that will flow from the collector to the base when the emitter is an open circuit (see Sec. 10.4).

It is helpful to think of a transistor as a variable resistance in series with its fixed collector and emitter resistors. The most useful match of circuit parameters and transistor characteristics is achieved if the transistor at saturation is a very low resistance compared to R_C and R_E and at cutoff is a very high resistance compared to R_C and R_E. In this case the maximum output voltage swing is available and it is obvious that the saturated and cutoff states are relatively stable in the sense that large changes in the transistor resistance have only a small effect on the output voltage. This stability is very important in digital circuitry and will be discussed in Chapter 17.

We have drawn in Fig. 13.16 a load line that provides an effective match between circuit parameters and transistor characteristics. Obviously other, less useful, load lines might be largely in the saturation region (too small R_C and R_E) or largely in the cutoff region (too large R_C and R_E).

The hyperbola labeled *maximum power dissipation* is the curve $i_C v_{CE} = P_{max}$ where P_{max} is the maximum rated power dissipation of the transistor. In amplifier design the load line must be inside the hyperbola but efficiency and economics demand that it be close to the hyperbola. In pulse use, excursions outside the hyperbola are permitted if the time spent outside the hyperbola is not too great.

Field effect transistors show similar saturation and cutoff behavior but the internal physics is somewhat different.

13.6   Some Emitter Follower and Source Follower Circuits

Follower circuits are very widely used to match a high impedance source to a low impedance load. We have discussed in Chapter 11 some single transistor versions of follower circuits and have shown in Fig. 11.19 a source follower using two FET's which has excellent stability against thermal drift, a gain very close to 1, and almost exact matching of the dc level of the input and output. The advantages of the circuit of Fig. 11.19 are greatest when slowly varying signals must be amplified and dc coupling is necessary.

The ac properties of follower circuits are also sometimes greatly improved, however, by the use of more than one transistor. We discuss a number of examples in this section. We develop the ideas with the help of Prob. 13.13 through 13.18. If the reader wishes, the material of this section and the associated problems can be omitted without loss of continuity·

The emitter follower circuit of Fig. 11.16 is a good example of a circuit that can be improved for certain applications by the use of a multitransistor circuit. For example, if the load that the emitter follower circuit drives is a small resistance, say, perhaps 50 ohms, then the input resistance of the emitter follower will be only 5000 ohms (we ignore the base bias resistors) for a transistor with $\beta = 100$. In many applications this would be too low an input impedance to be satisfactory. This can be overcome by using one emitter follower to drive another emitter follower. This is illustrated in Prob. 13.13, 13.14, and 13.15.

Another problem with the emitter follower of Fig. 11.16 is that the speed of its response when driving a capacitive load is different depending on the polarity of the input signal to the emitter follower. Let us suppose that the resistor R_E of Fig. 11.16 has a capacitor C_L connected in parallel. When a positive step voltage signal is applied to the input of the emitter follower, the capacitor tends to hold the emitter voltage constant. The charge on the capacitor is increased as current flows through the transistor and the capacitor charges up with a time constant $r_{tr} C_L$ so that the emitter follows the base.

On the other hand if a negative step voltage is applied to the input of the emitter follower, the transistor is cut off until the capacitor discharges through R_E. The time constant for this is $R_E C_L$. Thus the emitter follows the base but with a short time constant $r_{tr} C_L$ for a positive signal and a long time constant $R_E C_L$ for a negative signal. This difficulty can be overcome by using two transistors in an emitter follower circuit so that one turns on for positive signals and the other turns on for negative signals. This is illustrated in Prob. 13.16, 13.17, and 13.18.

Problems

13.1   For the circuit of Fig. 13.1, assume that the input is driven by a voltage source with negligible internal resistance. Find the low frequency for which $v_b/v_i = 1/\sqrt{2}$ where v_b is the actual incremental voltage at the base of the first transistor and v_i is the output voltage from the voltage source.

13.2   For the circuit of Fig. 13.1 if the 1000-ohm negative feedback resistor were absent, what would be the input resistance and the output resistance of the circuit?

13.3   What is the input resistance and the output resistance of the circuit of Fig. 13.1 with the 1000-ohm negative feedback resistor in the circuit?

13.4   Calculate the voltage gain, the input impedance and the output impedance of the cascode amplifier shown in Fig. 13.17. Discuss how the Miller effect is nearly eliminated by this circuit.

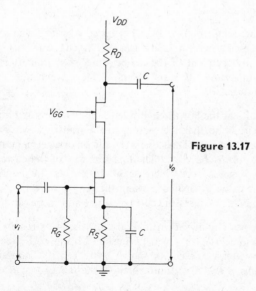

Figure 13.17

13.5   For the circuit of Fig. 13.5, what would be reasonable values of L and C if the cascode amplifier is to be tuned for 10^6 Hz?

13.6   The class A amplifier of Fig. 13.9 carries a peak current $i_c = 100$ milliamp and has a supply voltage V_{CC} of 10 volts. What is the theoretical maximum power that the amplifier can deliver? What is the quiescent current I_C?

13.7   If the class B push-pull amplifier of Fig. 13.12 carries a peak current $i_c = 100$ milliamp and has a supply voltage $V_{CC} = 10$ volts, what is the maximum power the amplifier can deliver? What is the quiescent current I_C in each transistor?

13.8   Sketch as a function of the time both the input voltage signal and the output current for a class C amplifier that might have an efficiency of 90%.

13.9   From Chapter 10 we recall that $I_C = \alpha I_E + I_{co}$ and $I_C = \beta I_B + (\beta + 1)I_{co}$. If we wish to know the rate of change of I_C with the temperature T, then one term that is important is $\partial I_{co}/\partial T$. From Eq. 10.10 we know that $I_{co} = j_0 A e^{-\Delta E/kT}$ where $\Delta E \simeq E_g$. Show that $\partial I_{co}/\partial T \simeq (E_g/kT^2)(I_{co})$. For a silicon pn junction, $E_g = 1.1$ eV and a typical value of $I_{co} = 10^{-11}$ amp. What is $\partial I_{co}/\partial T$ for a back-biased silicon pn junction at room temperature? For a germanium pn junction, $E_g = 0.7$ eV and a typical value of $I_{co} = 10^{-6}$ amp. What is $\partial I_{co}/\partial T$ for a back-biased germanium junction?

13.10   Consider the power amplifier circuit of Fig. 13.9. Suppose R_B is zero. Determine the quiescent emitter current I_E in terms of V_{BB}, V_{BE}, and R_E. If V_{BE} does not change with the temperature, show that I_E is independent of the temperature. If α is also independent of the temperature, show that $\partial I_C/\partial T = \partial I_{co}/\partial T$. It is often a good approximation that V_{BE} and α are independent of the temperature since they usually change slowly with the temperature compared to I_{co}.

13.11   Again consider the circuit of Fig. 13.9. Let us now assume that $R_E = 0$. Determine the quiescent current I_B in terms of V_{BB}, R_B, and V_{BE}. If V_{BE} is assumed to be independent of the temperature, show that I_B is independent of the temperature. If α and therefore β are independent of the temperature, show that $\partial I_C/\partial T = (\beta + 1)I_{co}$.

13.12   Based on the results of the last three problems, discuss carefully how one would design a circuit so that the current in the transistor will not run away if the temperature rises. Does one use a silicon or a germanium transistor? Should R_E or R_B be large? Thermal runaway of the current in a transistor occurs when a small increase in the current in a transistor causes more heat to be dissipated causing the temperature of the transistor to increase, which causes the current to rise still more, and so on.

13.13   The emitter follower circuit shown in Fig. 13.18 is an extremely useful circuit. There are some uses for which it is unsuited, however. For example, suppose the transistor in the circuit shown has a $\beta = 100$, and suppose that $R_L = 15$ ohms $\ll R_E$. Then unfortunately the input resistance of the emitter follower depends on R_L and is too low for many applications. What is the input resistance of the emitter follower? What change is produced in the input resistance if R_L is lowered to 10 ohms? If R_L is raised to 150 ohms?

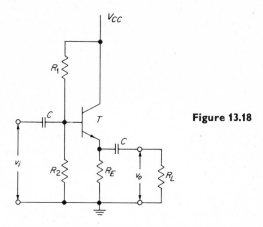

Figure 13.18

13.14 This problem and the next problem show two ways of using multitransistor circuits to overcome the difficulty that the input impedance of an emitter follower depends on the load resistance and the input impedance of the emitter follower may be too low if the load resistance is low. For the circuit shown in Fig. 13.19, calculate the voltage gain, the current gain, the input impedance, and the output impedance. You may assume that both transistors are identical and have a forward current gain β.

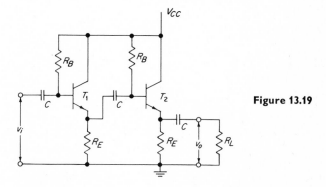

Figure 13.19

13.15 The circuit in Fig. 13.20 is called a *Darlington pair* or a *β multiplier*. Calculate the voltage gain, the current gain, the input impedance, and the output impedance of this circuit in the midband approximation. Sometimes the two transistors in this circuit are formed as an integrated circuit.

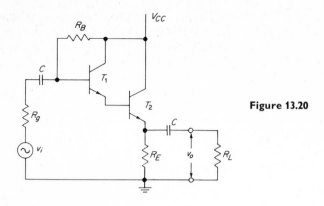

Figure 13.20

13.16   There is still another problem from which followers suffer. Consider the source follower circuit shown in Fig. 13.21 for which the load is a capacitor' C_L. Show that if a positive step voltage is applied to the gate, the FET turns on strongly until the capacitor is charged to its final voltage with a time constant that is approximately C_L/g_{fs}. On the other hand if a negative step voltage is applied to the gate, the FET turns off until the capacitor is discharged, which occurs with a time constant $R_S C_L$. Thus the follower circuit responds very rapidly to a positive voltage signal but only slowly to a negative voltage signal.

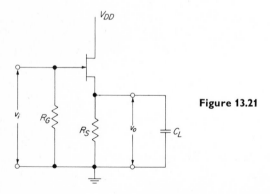

Figure 13.21

13.17   The next two problems show multitransistor follower circuits that can respond rapidly to a pulse of either polarity even when driving a capacitive load. The circuit shown in Fig. 13.22 is called a *White source follower*.

The voltage source V_{GG} is chosen so that the upper FET is biased identically to the lower FET. For this circuit, show that if a positive step voltage
is applied to the input, then the upper FET turns on strongly until the load
capacitor is charged, which occurs with a time constant C_L/g_{fs}. Show also
that if a negative step voltage is applied to the input, then the lower FET
turns on strongly until the load capacitor is discharged, which also occurs
with a time constant C_L/g_{fs}. In addition, find the voltage gain, the input
impedance, and the output impedance of this circuit. What are reasonable
values for R_D, R_G, R_S, V_{GG}, and V_{DD}?

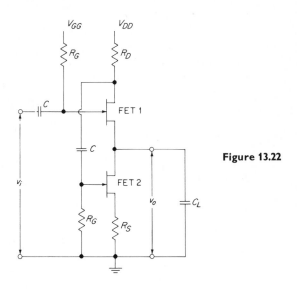

Figure 13.22

13.18   Another follower circuit that will respond rapidly to negative pulses even
when driving a capacitive load is the complementary symmetry source
follower shown in Fig. 13.23. The two FET's are different, one is an
n channel FET and the other is a p channel FET. You may assume for
simplicity that the characteristic curves of the p channel FET are identical
to those of the n channel FET except of course that one must reverse the
signs of all the voltages and currents. The two gate bias supplies V_{GG1}
and V_{GG2} are such that both FET's are carrying the same quiescent current
and have the same quiescent drain to source voltage. Find the voltage
gain, the input impedance, and the output impedance of the complementary
symmetry source follower. Show that if a positive step voltage is applied
to the input, then the upper FET turns on strongly until the capacitor C_L
is charged to its steady state value, which occurs with a time constant that

is C_L/g_{fs}. Show also that if a negative step voltage is applied to the input, then the lower FET turns on strongly until the capacitor C_L is discharged to its steady state value which occurs with a time constant that is C_L/g_{fs}.

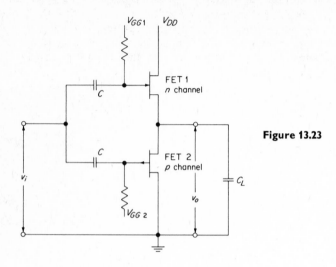

Figure 13.23

14/DC Amplifiers

If very low-frequency signals are to be amplified, coupling capacitors cannot be used at the input and output and between the stages of an amplifier. There must be direct resistive couplings throughout. The amplifier can be thought of as a dc network whose currents and voltages are determined by the dc voltage across the input terminals. The input signal can vary as slowly as one wishes, hence the name dc amplifier.

Direct coupled amplifiers involve a number of special problems and design considerations, which we shall discuss in this chapter.

14.1 Interstage Coupling

Consider an *RC* coupled amplifier. Figure 14.1(a) shows such an amplifier and gives the dc voltages at various points in the circuit. Silicon *npn* transistors are used. It is obvious that the capacitor used for coupling one stage to the next supports a voltage difference of $5.2 - 2.2 = 3.0$ volts. Therefore the capacitor could be replaced by a 3.0-volt battery without affecting the operating conditions as in Fig. 14.1(b). Now let us suppose that the voltage at the first base changed by 0.1 volt from 2.2 to 2.3 volts. The result would be that the voltage at the collector of the first transistor would drop to 4.9 volts, the base of the second transistor would drop to 1.9 volts, and the collector of the second transistor would rise to 6.1 volts. This change would persist as long as the base of the first transistor remained at 2.3 volts. This change would not fall off in a short time as would have happened in an *RC* coupled ac amplifier. Thus the direct coupled amplifier will amplify dc voltage changes.

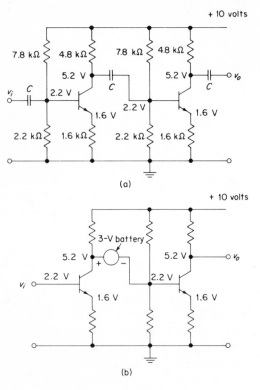

Figure 14.1   (a) An *RC* coupled amplifier. (b) The amplifier of Fig. 14.1(a) using a 3.0-volt battery instead of a capacitor to provide coupling between stages and to preserve the proper biases.

Using batteries for interstage coupling is usually not desirable in a dc amplifier. Three commonly used methods of interstage dc coupling are shown in Fig. 14.2. Leapfrogging is a suitable method if only a small number of stages are used. Unsatisfactorily high supply voltages become necessary, however, if more than a few stages are involved. The method of resistor dividing involves using two power supplies and because of the voltage divider some gain is lost; however, this method is commonly used. The *npn-pnp* complementary coupling works well and is also a commonly used method of interstage coupling in dc amplifiers.

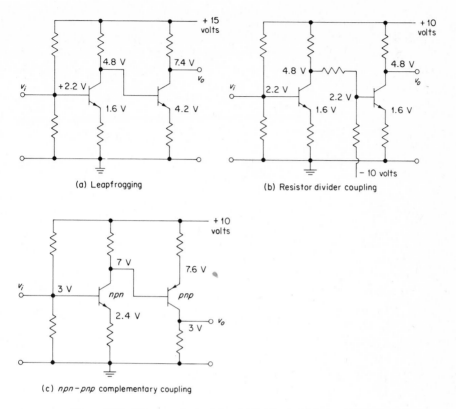

(a) Leapfrogging

(b) Resistor divider coupling

(c) *npn–pnp* complementary coupling

Figure 14.2   Three methods of dc coupling in multistage amplifiers.

Having discussed methods that will provide a dc coupling, the reader might think that he is actually ready to put a dc amplifier together. If the reader were to construct an amplifier such as one of those in Fig. 14.2, he would find to his dismay that even with a perfectly constant input voltage the output drifts in voltage. The output voltage change at constant input is usually expressed in terms of the zero offset voltage at the input. The zero offset voltage is the dc voltage required at the input in order to restore the output to its original value. Thus if a dc amplifier with a gain of 100 for incremental signals drifts by 10^{-2} volt at the output, then the zero offset voltage at the input needed to cancel this output is 10^{-4} volt.

The source of drift in most transistor dc amplifiers can be traced to changes in temperature. As the temperature changes, the emitter current changes, the reverse leakage current from the collector to base changes, the value of β changes, and the transresistance changes. Because some of these quantities change rapidly with temperature (that is, in an exponential manner), the output voltage of a high gain dc amplifier with a fixed input voltage can drift rapidly as the temperature changes. In order to construct a useful dc amplifier it is common to employ a balanced circuit arrangement that automatically cancels many of the drifts introduced by a temperature change. By the use of balanced circuits it is possible to construct a dc amplifier that has an input offset drift due to temperature change as low as 1 microvolt/°C and has a long-term drift as low as a few microvolts per week. By controlling the temperature of the amplifier the dc drift can be made very small; for example, it is easily possible to use amplifiers that drift by less than 0.2×10^{-6} volt equivalent input offset over a day.

Probably the most satisfactory balanced dc amplifier circuit is the difference (or differential) amplifier, which we discuss in the next section.

14.2   Difference Amplifiers

A simple difference amplifier is shown in Fig. 14.3. The circuit has two inputs, 1 and 2, as indicated. The output can be taken either as a double ended signal observed between the two output terminals indicated as 3 and 4 or as a single ended signal from either of the output terminals to ground.

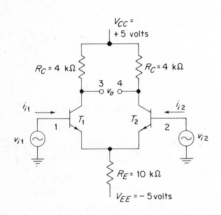

Figure 14.3 The circuit for a simple difference amplifier. In the absence of incremental signals both inputs are at ground potential.

In order to analyze this circuit, let us assume that the two transistors are identical and in the quiescent condition carry a dc current $I_{C1} = I_{E1} = I_{C2} = I_{E2}$. Under these conditions the transresistances of the two transistors are the same $r_{tr} = (26/I_{E1} + r_{\text{ohmic}}) = (26/I_{E2} + r_{\text{ohmic}})$. Now if i_{e1} and i_{e2} are the incremental currents in transistors 1 and 2, respectively, produced by incremental input signals v_{i1} and v_{i2}, we can write

$$v_{i1} = i_{e1}(r_{tr1} + R_E) + i_{e2} R_E$$

and   (14.1)

$$v_{i2} = i_{e2}(r_{tr2} + R_E) + i_{e1} R_E$$

Since $r_{tr1} = r_{tr2}$, we find that $v_{i1} - v_{i2} = (i_{e1} - i_{e2})r_{tr}$. If one uses the voltage between terminals 3 and 4 as the output, then $v_o = (i_{c1} - i_{c2})R_C$. Since $i_{c1} = i_{e1}$ and $i_{c2} = i_{e2}$, we find that $v_o = (-R_C/r_{tr})(v_{i1} - v_{i2})$. The output signal is seen to be proportional to the difference in the signals at the two inputs and the gain of the difference amplifier is $-R_C/r_{tr}$. For the circuit shown in the quiescent state the base of each transistor is at ground potential. If the transistors are made of silicon, then the quiescent voltage at the emitter is about -0.6 volt and thus $I_{E1} + I_{E2} = 4.4 \times 10^{-4}$ amp and $I_{E1} = I_{E2} = 2.2 \times 10^{-4}$ amp. Consequently $r_{tr} = (26/0.22) + 4 = 122$ ohms. Therefore the gain of the amplifier is about $A = (-4 \times 10^3)/122 = -33$.

Often it is convenient to use a single ended output such as the signal from terminal 3 only rather than the double ended output that was taken between terminals 3 and 4. The single ended output has both a dc and an incremental component. We are interested in the incremental component of the single ended output. For the single ended output from terminal 3 the incremental output voltage can be written $v_o = -i_{c1}R_C$. Combining this expression for v_o with Eq. 14.1 we obtain

$$v_o = -i_{c1}R_C = -i_{e1}R_C = -\frac{R_C}{2}\left(\frac{v_{i1} - v_{i2}}{r_{tr}} + \frac{v_{i1} + v_{i2}}{r_{tr} + 2R_E}\right) \qquad (14.2)$$

This shows that the single ended output contains a signal proportional to the difference of the two input voltages and a signal proportional to the sum of the two input voltages. The output signals proportional to the difference of the two inputs and to the sum of the two inputs are called the *differential mode signal* and the *common mode signal*, respectively. The gain of the differential mode is $-R_C/2r_{tr} = -17$, and the gain of the common mode is $-R_C/2(r_{tr} + 2R_E) = -\frac{1}{10}$. It is seen that the gain of the differential mode is 170 times as great as for the common mode. The ratio of these two gains expressed in decibels is called the *common mode rejection ratio*. For the circuit of Fig. 14.3 the common mode rejection ratio is 20 log 170 = 45 dB.

We see that the resistor R_E provides a large negative feedback for the common mode signal only. In the purely differential mode, equal and opposite signal currents flow through R_E and no feedback voltage is developed. The larger R_E, the better is the common mode rejection ratio, but large values of R_E demand inconveniently large supply voltages. This difficulty can be circumvented by using a transistor as the common emitter resistor as shown in Fig. 14.4. Connected as shown, the transistor has a conveniently low dc quiescent resistance (V_{CE}/I_C) but a very high incremental resistance (v_{ce}/i_c). The use is very similar to that of the lower FET in the symmetric source follower circuit of Fig. 11.19. Sometimes a transistor in such an application is referred to as a constant current sink because of its high incremental resistance. The transistor T_3 might easily have an incremental resistance of 1 megohm. Thus if it replaces $R_E = 10^4$ ohms in the circuit of

Fig. 14.3, the common mode rejection ratio increases to 65 dB. For multi-stage amplifiers with more complicated negative feedback circuits, common mode rejection ratios as high as 100 dB may be obtained.

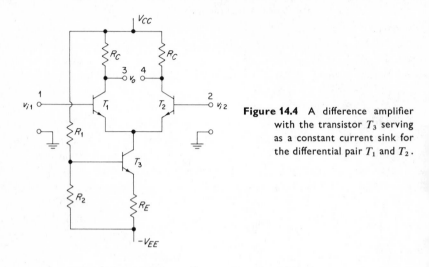

Figure 14.4 A difference amplifier with the transistor T_3 serving as a constant current sink for the differential pair T_1 and T_2.

Our discussion of differential and common mode amplification has assumed single ended output. In a good difference amplifier the second output is swinging in phase with the first on the common mode input and 180° out of phase on the differential input. Thus if we take the double ended output rather than the single, the common mode rejection is complete provided only that the amplifier is symmetric. This is not a useful way to design and operate a difference amplifier, however. If the common mode amplification (for single ended output) is too large, the common mode input signal will drive both transistors to new operating points and the differential amplification will depend on the common mode signal. In addition, identity of the two transistors is demanded over a large range of operating conditions. A good difference amplifier must have a low common mode amplification as measured with single ended output. Low common mode amplification is accomplished by using negative feedback for the common mode only. The differential mode amplification remains large since no negative feedback is used in the differential mode.

The difference amplifier shown works very well. Because the total current i_E in either transistor varies exponentially with the difference of the input voltages, however, the circuit shown is linear only for very small input voltages. Linearity can be improved at the expense of gain by putting a resistor R in each emitter circuit as shown in Fig. 14.5. These resistors introduce negative feedback into the differential mode.

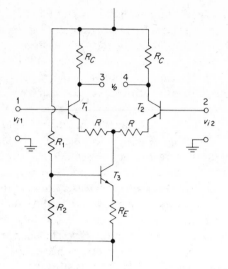

Figure 14.5 A differential amplifier with two resistors each of value R in the emitters of the transistors T_1 and T_2. This differential amplifier is more linear but has a lower amplification than the circuit of Fig. 14.4.

Earlier we pointed out that balanced systems tend to cancel drifts due to changes in the temperature. We should consider this statement more carefully. Let us first consider an unbalanced single-stage amplifier as shown in Fig. 14.6. If the temperature increases, the emitter current increases. As the emitter current increases, however, the voltage across R_E increases and hence the base to emitter voltage decreases since the base voltage is independent of the temperature. As the base to emitter voltage decreases, this tends to reduce the emitter current. Thus the resistor R_E produces a negative feedback that tends to stabilize the current through the transistor. If $R_E \gg r_{tr}$, the feedback produces a change in the base to emitter voltage that largely compensates for the effects of the change in temperature. The changes in temperature ordinarily encountered will change the emitter potential and therefore the base to emitter potential by a few hundredths of a volt. Near room temperature the base to emitter potential decreases by 2.0 millivolts for a 1°C increase in temperature.

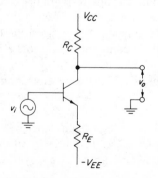

Figure 14.6 A single transistor amplifier.

Let us denote this temperature induced change in the emitter potential by $\Delta v(T) = \Delta i_E R_E$. Assuming $\Delta i_E \simeq \Delta i_C$, the corresponding change in the output potential is $-\Delta v(T) R_C / R_E$. If we divide by the amplification $A = -R_C / R_E$, we find that the input zero offset that would compensate for the output drift is just $\Delta v(T)$. For many applications, however, input signals will be much less than 10^{-2} volt and the temperature drift of an unbalanced amplifier cannot be tolerated.

Now let us compare these results to the temperature drifts found in the difference amplifier of Fig. 14.3 using single ended output. The first important thing to note is that a temperature change is essentially a common mode effect in a difference amplifier. If, for instance, both transistors T_1 and T_2 increase in temperature, then both i_{E1} and i_{E2} will increase, both emitters will rise in potential, and the potentials of both collectors will fall. Assuming similar transistors are used in the circuits of Fig. 14.3 and 14.6, the increase in the emitter potentials of T_1 and T_2 will be given by $\Delta v(T)$. The change in the collector potentials is $-\Delta v(T) R_C / 2 R_E$. The factor $\frac{1}{2}$ occurs because in the common mode R_E carries the emitter currents of both transistors.

We can now see one of the reasons for the superiority of the difference amplifier. In the single transistor amplifier R_C / R_E represents the useful signal amplification and must be appreciably greater than 1. In the difference amplifier $R_C / 2 R_E$ is proportional to the common mode amplification and may be deliberately made much less than 1. Thus the temperature induced drift in the output voltage is very much less for a difference amplifier than for a single transistor amplifier.

The comparison is more striking if we express the results in terms of input zero offsets. For the single transistor amplifier the input zero offset is $\Delta v(T)$. For the difference amplifier with single ended output it is the output voltage change $-\Delta v(T) R_C / 2 R_E$ divided by the differential mode amplification $-R_C / 2 r_{tr}$ or $\Delta v(T) r_{tr} / R_E$. The factor r_{tr} / R_E is essentially the reciprocal of half the common mode rejection ratio. As we have mentioned in the earlier discussion, r_{tr} / R_E might be 10^{-3} or less.

The difference amplifier of Fig. 14.3 or 14.4 does an excellent job of canceling dc drifts provided the two differential transistors are selected to have identical characteristics and are located in precisely the same environment. This can be most easily accomplished by producing the entire difference amplifier as an integrated circuit so that all the components are located on the same small silicon chip (see Chapter 15 for a discussion of integrated circuits). The same purpose can also be accomplished fairly easily by mounting two selected identical transistors on the same heat sink so that their temperature is always the same.

Before completing our discussion of the difference amplifier we shall discuss the input impedance of this type of amplifier. The differential mode and common mode input impedances of a difference amplifier are arbitrarily defined as $2(v_{i1} - v_{i2})/(i_{i1} - i_{i2})$ for the differential mode and $(v_{i1} + v_{i2})/$ $2(i_{i1} + i_{i2})$ for the common mode where v_{i1} and v_{i2} are the incremental input voltages at inputs 1 and 2 and i_{i1} and i_{i2} are the incremental currents flowing into inputs 1 and 2. The differential mode input impedance of the amplifier of Fig. 14.3 is $2(\beta + 1)r_{tr}$ where β and r_{tr} are the forward current gain in the common emitter configuration and the transresistance, respectively, of either of the identical transistors. The common mode input resistance is given by $(\beta + 1)(r_{tr} + 2R_E)/2 \simeq (\beta + 1)R_E$. If $\beta = 100$ for each of the transistors, then the differential mode input resistance is equal to 24×10^3 ohms and the common mode input resistance is 10^6 ohms. For the differential mode the input current flows in one input terminal and out the other input terminal and does not go to a real ground connection. Because of this the equivalent circuit of a differential amplifier is often drawn as shown in Fig. 14.7. The input resistance $R_{i, \text{diff}}$ is connected between the two inputs and is not connected to ground. The equivalent circuit shown in Fig. 14.7 assumes that the output of the amplifier is single ended. If one wishes to include $R_{i, \text{comm}}$ in the equivalent circuit, then a resistor $(\beta + 1)R_E \simeq R_{i, \text{comm}}$ should be connected between the center of $R_{i, \text{diff}}$ and ground.

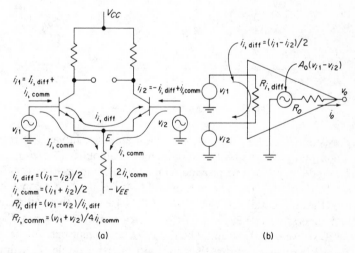

(a)   (b)

Figure 14.7 (a) A differential amplifier showing how the differential current flows in one input and out the other. Thus the differential input impedance lies between the two inputs and does not go to a real ground. The point E is sometimes called a virtual ground. Since no incremental differential current flows through R_E, the point E is at the same potential for incremental signals as the power supply $-V_{EE}$, which is an incremental ground for the differential mode. (b) A circuit equivalent to the circuit of Fig. 14.7(a). The circuit has a differential input and a single ended output. Note that by definition $i_{i,\text{diff}} = (i_{i1} - i_{i2})/2$ and $i_{i,\text{comm}} = (i_{i1} + i_{i2})/2$ so that $i_{i1} = i_{i,\text{diff}} + i_{i,\text{comm}}$ and $i_{i2} = -i_{i,\text{diff}} + i_{i,\text{comm}}$. Also by definition $R_{i,\text{diff}} = (v_{i1} - v_{i2})/i_{i,\text{diff}}$ and $R_{i,\text{comm}} = (v_{i1} + v_{i2})/4i_{i,\text{comm}}$.

14.3 A Multistage DC Amplifier

The differential amplifier forms the basis of many multistage dc amplifiers. An example of a multistage dc amplifier is shown in Fig. 14.8.

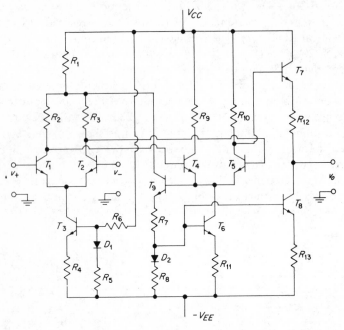

Figure 14.8 A three-stage dc amplifier.

The transistors T_1 and T_2 form a differential amplifier input. The transistor T_3 is a constant current sink for the differential amplifier formed by T_1 and T_2. The diode D_1 is used to provide temperature compensation for the base to emitter junction of T_3. The purpose of T_3 is to maintain a constant total current to the differential pair T_1 and T_2. If the base to emitter voltage is fixed and the temperature increases, then the current through transistor T_3 would increase if D_1 were not present. The diode D_1 is a small part of the voltage drop from V_{CC} to $-V_{EE}$ so that the current through the diode D_1 is almost independent of the temperature and has the value $(V_{CC} + V_{EE})/(R_6 + R_5)$. Thus as the temperature changes the voltage drop across the diode must change by a small amount so that the current through the diode remains constant. If the diode and the base to emitter junction of the transistor T_3 are both made of silicon and have identical current-voltage characteristics, then as the temperature changes the base to emitter junction of T_3 changes in voltage just as D_1 does and consequently the current through T_3 is maintained as relatively constant. The transistors T_4 and T_5 form a differential amplifier that acts as the second stage in this dc amplifier. This second-stage differential amplifier is run with a push-pull input that is taken from the double ended output of the first-stage differential amplifier. The transistor T_6 is a constant current sink for the second-stage differential amplifier and the diode D_2 provides temperature compensation for T_6. The transistor T_9 provides negative feedback for the common mode only. If there is a common mode signal in the second stage, then the voltage at the common emitters for T_4 and T_5 will vary in voltage because the total current through the differential pair T_4 and T_5 will vary. With a differential mode signal the total current to the differential pair will not change. Suppose there is a common mode signal so that the voltage at the base of transistor T_9 increases. Then the transistor T_9 turns on and draws more collector current. This increases the voltage drop across R_1 and consequently the voltage at the input to both T_4 and T_5 is decreased, tending to drop the voltage at the common emitters of T_4 and T_5. Thus the feedback is negative for a common mode signal. There is no feedback for the differential mode. The transistor T_7 acts as an emitter follower output stage. The transistor T_8 acts as large passive resistance in the emitter circuit of T_7 for incremental signals. The output stage with T_7 and T_8 enables the output to be shifted down in voltage to the same quiescent dc level as the input. This is possible since T_8 has a much lower resistance for large dc signals (V_{CE}/I_C) than it has for incremental signals (v_{ce}/i_c). The power supplies V_{CC} and $-V_{EE}$ are such that both the input and the output are at ground potential. The connection from the emitter of T_9 to the bases of T_6 and T_8 provides a feedback that helps cancel drifts in the voltages provided by the power supplies V_{CC} or $-V_{EE}$. Suppose the voltage V_{CC} decreases by a small amount. The voltage at the common emitters of T_4 and T_5 tends to decrease. This causes the voltage at the base and hence also at the emitter of T_9 to decrease. This causes the current carried by T_6 and hence by T_4 and T_5 to decrease. The decrease in the current carried by T_4 and T_5 causes the collector of T_5 to rise in voltage. This causes the emitter of T_7 to rise in voltage. In addition, when the voltage at the emitter of T_9 tends to decrease, the current carried by the transistor T_8 is decreased and so the collector of T_8 increases in voltage. Thus the feedback tends to keep the output voltage constant even when the supply voltages change.

A dc amplifier such as is shown in Fig. 14.8 can be produced as an integrated circuit on a single chip of silicon. Because the entire circuit can be put on a single chip of silicon, the transistors T_1 and T_2 and the transistors T_4 and T_5 are kept at identical temperatures and the effects of thermal drifts can be made small.

The circuit of Fig. 14.8 was based on the circuit of an integrated circuit amplifier manufactured by RCA (model CA3008A). The actual circuit diagram of the RCA circuit is given in Fig. 14.9. It is similar to the circuit of Fig. 14.8. The circuit of Fig. 14.8 was simplified slightly, however, to make the discussion more straightforward. The primary difference in the circuits is that the output stage of the actual RCA model 3008A amplifier has three transistors instead of the two transistors in the output stage of the circuit of Fig. 14.8. The three-transistor output stage of the model 3008A amplifier not only shifts the dc level and provides a low output resistance but it also provides some voltage gain which the two-transistor output stage does not do. The RCA model 3008A amplifier has a voltage gain of 10^3 and a common mode rejection ratio of 90 dB.

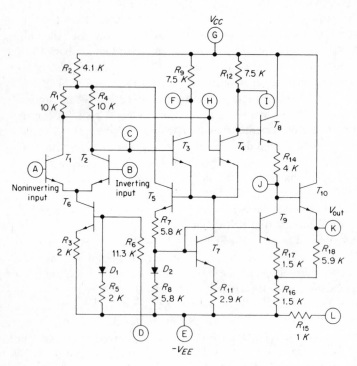

Figure 14.9 The schematic diagram for the RCA model CA3008A integrated circuit dc amplifier. The main difference between this diagram and the diagram of Fig. 14.8 is that the model CA3008A amplifier has three transistors T_8, T_9, and T_{10} in the output stage, whereas the circuit diagram of Fig. 14.8 has only two transistors T_7 and T_8 in its output stage. Note there is a connection between two wires only where there is a black dot. (*Schematic diagram courtesy RCA Corporation.*)

More complicated dc amplifiers with gains as great as 10^5 or 10^6 are commonly used. A dc amplifier such as the one just discussed is called an *operational amplifier*. The applications of operational amplifiers are discussed in the next chapter.

14.4 Mixed AC–DC Amplifiers

Many amplifiers use a mixture of ac and dc techniques in an attempt to exploit the particular advantages of each. For instance, the accurate measurement of a small slowly varying signal may be difficult using a dc coupled amplifier because of the inherent drift. This can be avoided by chopping the dc signal, that is, converting it into an ac signal whose amplitude is proportional to the dc signal. The ac signal is amplified and finally rectified. Large and relatively drift-free amplification of the original dc signal can be achieved.

On the other hand an experiment may directly present an ac signal whose amplitude and perhaps phase are the parameters of interest. An ac amplifier will be used, but the final presentation of the information will be after rectification and averaging in a circuit with a relatively long time constant. Among the advantages are the reduction of the effective bandwidth of the amplifier and therefore the reduction of noise.

Figure 14.10(a) shows the basic elements of a chopper amplifier. An electrically driven switch alternately connects the input signal and then grounds the input capacitor. The switch should spend equal times in each position and this time should be large compared to the switching time.

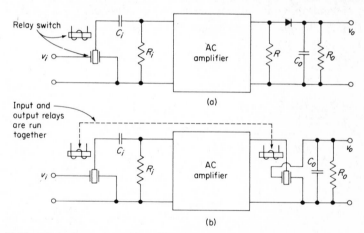

Figure 14.10 (a) A chopper amplifier. The input and output time constants $R_i C_i$ and $R_o C_o$ must be long compared to the chopper period. (b) A synchronous chopper amplifier. The input and output relays must connect to the input and output at the same time and must ground at the same time; that is, the input and output relays are such that the input and output of the amplifier are grounded simultaneously and when v_i is connected to the input of the amplifier the output of the amplifier must be connected to the top of C_o. The synchronous chopper amplifier preserves the polarity of the input signal. Both ac amplifiers contain an interior blocking capacitor in the output (not shown). The resistor R in Fig. 14.10(a) is necessary in order to discharge the blocking capacitor when the diode is not conducting.

Electromechanical switches are useful up to frequencies of a few hundred hertz. Transistor choppers can be used to much higher frequencies although the ratio of off resistance to on resistance is not as great as with mechanical switches.

Rectification of the output can be achieved in a variety of ways. We illustrate the use of a diode just as shown earlier in Fig. 11.6. With this output there is no memory of the sign of the input. As drawn, the output is necessarily positive. Instead of the diode one may arrange that the signal that actuates the input switch also connects the output to the load at the same time that the input is connected to the amplifier. On the other throw both are disconnected. This is called *synchronous rectification* and the output has the same sign as the input (assuming that the amplifier does not invert). This is shown in Fig. 14.10(b).

Let us examine in greater detail another circuit which is used for synchronous rectification and which is called a *phase-sensitive detector circuit*. It is widely used to extract amplitude and phase information about an ac signal of a given frequency that may be mixed with signals of other frequencies and with noise. An example of a phase-sensitive detector is given in Fig. 14.11. The signal of interest is $\varepsilon_1 = |\varepsilon_1| e^{j(\omega t + \phi)}$. The input to the transformer $2\varepsilon_2 = |2\varepsilon_2| \varepsilon^{j\omega t}$ is a reference signal that can be considered to be of fixed amplitude and phase during any particular sampling of ε_1. Under the proper operating conditions a dc signal appears across R_L whose magnitude is proportional to $2|\varepsilon_1| \cos \phi$.

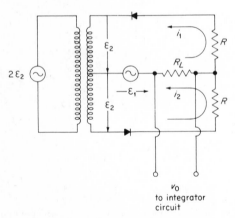

Figure 14.11 A phase-sensitive detector circuit.

v_O
to integrator circuit

To see this we first note that if $\varepsilon_1 = 0$, a current ε_2/R flows through the diodes and the resistances R but because of the symmetry no current flows through R_L. The current ε_2/R flows for only half of each cycle, however. Considering just ε_2 the circuit behaves as a half-wave rectifier. If $\varepsilon_2 = 0$ and $\varepsilon_1 \neq 0$, then the current $\varepsilon_1/(R + R_L)$ flows through R_L and no rectification occurs. During one-half of the cycle the current returns through the upper rectifier and during the other half cycle it returns through the lower rectifier. If both voltage signals are present, we find a much more complicated behavior. We apply Kirchhoff's second law to the upper and lower loops in Fig. 14.11 and, ignoring the voltage drop across the diodes, we obtain

$$
(i_1 - i_2)R_L + i_1 R = \varepsilon_1 + \varepsilon_2
$$
$$
-(i_1 - i_2)R_L + i_2 R = -\varepsilon_1 + \varepsilon_2
$$
(14.3)

Subtracting the second equation from the first we obtain for the current through R_L

$$
i_1 - i_2 = \frac{2\varepsilon_1}{R + 2R_L} = \frac{2|\varepsilon_1| e^{j(\omega t + \phi)}}{R + 2R_L}
$$
(14.4)

It is well to pause for a moment since our discussion has pitfalls for the unwary circuit analyst. In introducing the circuit we remarked that with $\varepsilon_2 = 0$ the source ε_1 sends a current $\varepsilon_1/(R + R_L)$ through R_L and that with $\varepsilon_1 = 0$ the source ε_2 sends no current through R_L. With both ε_1 and ε_2 present, Eq. 14.3 tell us that the current through R_L indeed depends only on ε_1 but is given by $\varepsilon_1/[(R/2) + R_L]$, not $\varepsilon_1/(R + R_L)$. This apparent failure of the superposition principle results from the presence of the highly nonlinear diodes. The source ε_1 alone ($\varepsilon_2 = 0$) can turn on only one diode at a time. The current i_1 or i_2 can flow, but not both, and ε_1 sees the load resistance $R + R_L$. If ε_2 is nonzero, however, both diodes may be conducting. The source ε_1 then sees the two resistances R in parallel and the effective load resistance becomes $R/2 + R_L$.

In the actual use of the circuit of Fig. 14.11 as a phase-sensitive detector the reference voltage ε_2 is required to be large compared to ε_1. For one-half the cycle the diodes are forward biased by ε_2 and currents proportional to ε_1 and ε_2 can flow. Equations 14.3 correctly describe this half cycle. During the other half cycle, however, both diodes are reverse biased by ε_2 and no currents at all can flow. Thus the reference voltage ε_2 and the diodes act as a synchronous rectifier for the output of ε_1. This action is made clear in Fig. 14.12 where we draw the reference voltage ε_2 and the current through R_L. We see that the phase difference between ε_2 and ε_1 determines the relative amount of plus and minus current through R_L during the diode "on" cycle. The average value of v_o, the output voltage across R_L, may be found by integrating $2|\varepsilon_1| e^{j(\omega t + \phi)}$ between $t = -\pi/2\omega$ and $\pi/2\omega$ and dividing by the total period $2\pi/\omega$. The result is $v_o = [(2|\varepsilon_1| \cos \phi)/\pi][R_L/(R + 2R_L)]$.

In Fig. 14.13 we give a phasor diagram of Eq. 14.3 and 14.4 that makes obvious the final result that the rectified output current is proportional to $(2|\varepsilon_1| \cos \phi)/(R + 2R_L)$. This is just the component in phase with ε_2. The other component $(2|\varepsilon_1| \sin \phi)/(R + 2R_L)$ averages to zero during the "on" half cycle.

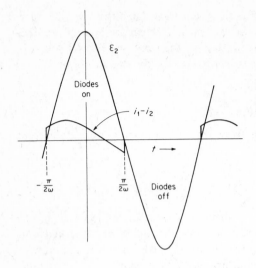

Figure 14.12 The time dependence of ε_2 and also $i_1 - i_2$, the current through R_L. The curves are drawn for $\phi = 45°$.

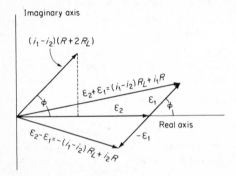

Figure 14.13 The phasor diagram of the voltages of the circuit shown in Fig. 14.11. The component of $i_1 - i_2$ that is in phase with ε_2 appears in the rectified output. It is proportional to $|\varepsilon_1| \cos \phi$. The component of $i_1 - i_2$ that is 90° out of phase with ε_2 averages to zero during the "on" half cycle of the diodes.

The question that one might ask is, "Why should one ever use phase-sensitive rectification when the signal ε_1 could have been converted into a dc signal by a simple half-wave rectifier"? The answer to this question has several parts.

First, the phase-sensitive technique permits one to rectify only the signal at frequency ω. If ε_1 contains several frequencies, that is, if $\varepsilon_1 = |\varepsilon_1| e^{j(\omega t + \phi)} + \sum_n v_n e^{j\omega_n t}$ where $\omega_n \neq \omega$, then the average value of the output voltage v_o is still given by $(2/\pi)|\varepsilon_1| \cos \phi [R_L/(R + 2R_L)]$. This comes about because the current $i_1 - i_2$ is proportional to the component of the phasor ε_1 along ε_2. If ε_1 and ε_2 have different frequencies, then ε_1 rotates about ε_2 and the average output voltage is zero. Therefore only that component of the signal at frequency ω is converted into dc.

Second, one must consider the effect of the noise on the output signal. Noise voltages are produced by various sources. For example, thermal fluctuations in a resistor produce a time average square voltage $\langle v_{noise}^2 \rangle = [4kTR'(\Delta\omega/2\pi)]$ where k is Boltzmann's constant, T is the absolute temperature, R' is the resistance, and $\Delta\omega$ is the bandwidth over which the noise is detected. In electronic circuits there are other random fluctuations producing noise voltages that may have a different magnitude or spectrum than the thermal noise in a resistor. It is characteristice, however, that the Fourier spectrum of a noise voltage extends over a wide range of frequencies. Obviously only the Fourier components of the noise voltage with frequencies near ω will contribute to the phase-sensitive rectified signal. If one takes the output voltage v_o and averages this signal for a time τ, then only the frequencies about ω such that $\Delta\omega\tau \lesssim 1$ will contribute significantly to the output voltage. Thus the output time constant τ determines the bandwidth of noise voltages that appear in the output signal ($\Delta\omega \sim 1/\tau$). Suppose the voltage ε_1 has a noise voltage superimposed on it. The time average of the output voltage can be shown to be $v_o = v_{signal} + v_{noise} \propto (2/\pi)|\varepsilon_1| \cos \phi + (4kTR'/\tau/2)^{1/2}$ where we have assumed that the noise voltage is due to thermal fluctuations in a resistor R' in the voltage source producing ε_1. It should be noted that the signal to noise ratio increases as $\tau^{1/2}$. Whatever the source of the noise, the integration of the output will increase the signal to noise ratio as $\tau^{1/2}$

A third reason for using phase-sensitive detection can be understood by examining the shape of the current-voltage characteristic for a diode used as a half-wave rectifier. For large signals the diode will have a small resistance. In this case noise voltages that are superimposed on the signal will simply appear in the output voltage added to the signal. If the signal is very small or zero, however, the diode characteristic will cause the noise to appear as a dc voltage in the output. Consequently the noise on top of a large signal can be averaged to zero, but the noise on a small signal cannot be averaged to zero. Thus a half-wave rectifier can be used for large voltage signals but not small ones. If $|\varepsilon_2|$ is large as we assumed, however, then the phase-sensitive rectification does not have any problems because of small signals even though ε_1 may be small since the diodes are always used such that the forward resistance of the diode is negligible compared to the resistance of the load.

Finally, of course, phase-sensitive detection preserves the information on the sign of the input which a simple half-wave rectifier does not.

It should be pointed out that there are many other phase-sensitive detector circuits. The particular circuit one chooses will depend on the frequency to be rectified, signal level, signal to noise requirements, and various other considerations.

There remains an important point that must be made clear. A phase-sensitive detector will almost never be used if ε_1 and ε_2 represent phase-independent voltage sources. Frequency drift between the two will usually quickly average the rectified output to zero. An exception might be the determination of the frequency stability of an oscillator by comparing it to an independent standard oscillator of very great frequency stability. A phase-sensitive detector would provide an output whose frequency is the frequency difference of the two oscillators (beat frequency). The stability of the beat frequency is a very sensitive test of how constant is the frequency of the oscillator.

In the usual application the signal ε_1 is derived from the signal ε_2 and phase stability is assured. An example of such an application of a phase-sensitive detector circuit, also called a *lock-in detector*, follows. The voltage ε_2 might be used to chop mechanically or optically a weak light beam and simultaneously drive the detector circuit. The signal ε_1 would be the output of a photocell on which the beam impinges. The signal from the light beam can be separated from noise and from other signals due to stray light. Here the advantages of relatively drift-free and noise-free amplification are sought rather than phase information.

On the other hand if ε_2 is used to drive some resonant system and ε_1 is made proportional to current flowing in the system, then the detector output times ε_2 is proportional to the power absorbed by the resonant system. This comes about because only the component of ε_1, which is in phase with ε_2, appears in the rectified output.

14.5   Voltage Regulators

Most instrumentation circuits demand highly stable dc power supplies of low internal resistance. The design of such supplies offers interesting examples of the use of dc amplifiers and applications of the principles of negative feedback. Previously we have discussed how to construct a rectifier circuit and to filter the output of this circuit in order to produce a dc voltage power supply. A filter circuit can reduce the ac component of the output of a power supply to less than 1% of the dc output. If one wants the ac component of the output of a power supply reduced to a value much less than 1% (say, 10^{-5} of the dc output), however, then it is necessary to provide electronic regulation of the output voltage. The voltage input to a voltage regulator is the output of a rectifier with a good filter so that the ac voltage is already a small fraction of the dc voltage.

A very simple voltage regulator circuit is shown in Fig. 14.14. In this circuit the transistor T_3 forms a dc amplifier. The Zener diode fixes the voltage of the emitter of T_3 (R_4 is used to carry the current necessary for the Zener diode). The collector current of the transistor T_3 should be small so that the current in the Zener diode does not vary too much. The resistors R_2 and R_3 sample the output voltage, and the voltage sample is used to drive the base of T_3. The voltage at the base of T_3 is approximately $[R_3/(R_2 + R_3)]v_O$ provided the base current of T_3 is small compared to $v_O/(R_2 + R_3)$. The voltage at the base of T_3 is about 0.6 volt greater than the reference voltage from the Zener diode if T_3 is a silicon transistor. The resistor R_1 allows current to flow to the collector of T_3 and to the base of T_2. The combination of transistors T_1 and T_2 is used as one super transistor. The combination is called a Darlington pair or a β *multiplier* because the β of the pair is equal to the product of the β's of the two individual transistors. This can be seen from the fact that the emitter current of T_2 is the base current of T_1. Thus the total current gain from the base of T_2 to the emitter of T_1 is the product of the separate β's.

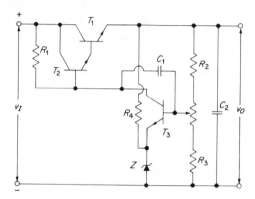

Figure 14.14 A simple voltage regulator circuit.

Now let us suppose that the output voltage tries to go up slightly. The base to emitter voltage v_{BE} of the transistor T_3 goes up, and the collector current in T_3 goes up. When this happens, the current into the base of T_2 decreases, and the voltage across T_1 increases (that is, when the current through T_1 decreases, T_1 moves along its load line increasing the voltage across T_1). Consequently the output voltage is pulled back down again. The Darlington pair is used because the collector current of T_3 is small and changes in the collector current of T_3 may not be sufficient to drive the base of T_1 alone. Another way to look at this is to note than when v_{BE} of the transistor T_3 increases, the collector current in T_3 increases. This means that the voltage at the collector of T_3 drops. The voltage at the collector of T_3 is the voltage at the base of T_2 and this is almost the same as the voltage at the emitter of T_2 and consequently at the emitter of T_1. Therefore if the output voltage starts to go up, the feedback is negative, and the voltage swing is limited to a very small value. One final point about the circuit is the use of capacitors C_1 and C_2. The capacitor C_1 is used to reduce the gain of the dc amplifier at high frequencies. This is necessary in order to prevent the system from oscillating at a high frequency. Because of C_1 the negative feedback is not very effective for very high frequencies and so the capacitor C_2 is used to provide a low impedance path for high-frequency fluctuations at the output. Problems 14.14 and 14.15 illustrate how one calculates numerically the output voltage and how well the voltage regulator of Fig. 14.14 can eliminate, from the output voltage, fluctuations in the input voltage. The result obtained is that $v_o/v_i \times -1/\beta A_v$ where v_o and v_i are, respectively, the fluctuations in the output and input, A_v is the gain of the dc amplifier T_3, and βv_o is the voltage at the base of T_3. The higher A_v, the better the voltage regulation.

The previous voltage regulator circuit has several bad features: (1) Temperature fluctuations cause the properties of T_3 to change and so the dc voltage comparison is not temperature compensated. (2) The resistor R_1 permits part of the ripple voltage present at the input to drive the base of T_2. (3) The overall gain of the dc amplifier is low (resulting from the use of only one transistor T_3 for dc voltage amplification) and so the regulation is not so good as can be achieved with higher dc gain (see Prob. 14.15). A circuit that provides a solution to the problems listed above is shown in Fig. 14.15.

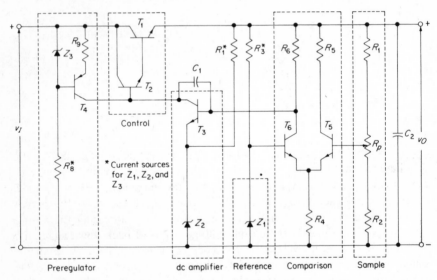

Figure 14.15 A practical voltage regulator circuit providing protection against thermal drift and ripple. (*Schematic diagram from Transistor Circuit Design by Engineering Staff of Texas Instruments Incorporated. Copyright 1963 by Texas Instruments Incorporated. Used by permission of McGraw Hill Book Company.*)

The symbols Z_1, Z_2, and Z_3 represent Zener diodes and resistors R_3, R_1, and R_8, respectively, carry the current for these Zener diodes. In this circuit the difference amplifier formed by transistors T_6 and T_5 compares a fraction $R_5 v_o/(R_4 + R_5)$ of the output voltage with the reference voltage provided by the Zener diode Z_1. The transistor T_3 is a simple dc amplifier stage just as in the previous circuit. The input to T_3 is proportional to the difference between $[R_5/(R_4 + R_5)]v_o$ and the reference voltage provided by Z_3. The transistors T_1 and T_2 form a β multiplier just as in the previous circuit. The transistor T_4

offers a very high resistance to fluctuating voltages. Ripple on the input voltage does not cause the current in T_4 to vary because it does not change the base to emitter voltage v_{BE} for T_4 (note that a *pnp* transistor must be used for this application). The transistor T_4 is called a *preregulator* because it eliminates input ripple that would otherwise get into the regulator circuit. It also serves as a very large collector resistor for T_3 and thus increases the amplification provided by T_3. The constant current from T_4 partly passes through T_2 and partly through T_3. Now suppose that the output voltage tries to go up. Then the base of T_3 goes up in voltage and T_3 carries more current. Since T_4 provides a fixed current, the transistor T_2 carries less current and consequently the voltage drop across T_1 increases. Therefore the output voltage is forced back down, that is, the feedback is in a negative sense. This circuit overcomes the bad features of the previous circuit because the difference amplifier is temperature compensated and adds dc gain to the circuit and because the preregulator prevents ripple from getting into the regulator circuit.

Both circuits provide regulation against output voltage changes caused by changes in output current. Thus the output impedance of the voltage regulator is reduced.

Problems

14.1   Show that the voltage gain is $-g_{fs}r_{os}/2$, and calculate the output impedance for the circuit shown in Fig. 14.16. Discuss how the symmetry of the circuit helps eliminate the dc drift of the output voltage due to changes in the temperature provided both the upper and the lower FET's are carefully matched and provided the two FET's are maintained at the same temperature.

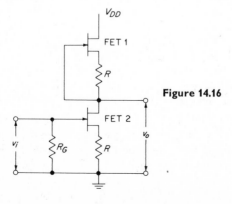

Figure 14.16

14.2   The differential mode input impedance of a difference amplifier is defined as $R_{i, \text{ diff}} = 2(v_{i1} - v_{i2})/(i_{b1} - i_{b2})$ where v_{i1} and v_{i2} are the input signals to the two transistors in the difference amplifier and i_{b1} and i_{b2} are the input base currents to the two transistors. The common mode input impedance of a differential amplifier is defined as $R_{i, \text{ comm}} = (v_{i1} + v_{i2})/2(i_{b1} + i_{b2})$. For the difference amplifier circuit of Fig. 14.3, show that, as stated in the text, $R_{i, \text{ diff}} = 2(\beta + 1)r_{tr}$ and $R_{i, \text{ comm}} = (\beta + 1)(r_{tr} + 2R_E)/2$.

14.3   Calculate the output impedance of the difference amplifier circuit of Fig. 14.3 when used with a double ended output and when used with a single ended output.

14.4   Calculate the differential mode gain for the circuit of Fig. 14.5. Assume all three transistors are identical.

14.5   Suppose that the two bipolar transistors of Fig. 14.3 were replaced by FET's for which $g_{fs} = 4 \times 10^3$ micromhos and $r_{os} = 10^5$ ohms, and suppose that the bias supplies are adjusted to be suitable for FET operation. What is the differential mode voltage gain? Is the voltage gain higher when FET's or when bipolar transistors are used?

14.6   Although a difference amplifier using FET's has a lower differential mode voltage gain than one using bipolar transistors, it is somewhat more linear. By examining a typical set of FET characteristic curves (such as given in Chapter 11), show that a bipolar transistor difference amplifier such as the one in Fig. 14.5 with emitter resistors chosen so that the gain is reduced to be the same as for an FET difference amplifier is more linear than the FET difference amplifier. What advantages might an FET difference amplifier have?

14.7   Show that the lower constant current sink transistor in the difference amplifier of Fig. 14.4 acts like a large passive resistance for incremental signals. Show that the magnitude of the passive resistance is

$$\frac{v_{ce}}{i_c} = \frac{[1/h_{fe} + R_E/(h_{ie} + R_E)]}{[h_{oe}/h_{fe} - h_{re}(h_{ie} + R_E)]} \simeq \beta r_c$$

provided R_E is sufficiently large.
Note. In Chapter 11 we define $r_c = 1/h_{oe}$.

14.8   Using the results of Prob. 14.7, calculate the common mode rejection ratio, the differential mode input resistance, and the common mode input resistance for the circuit of Fig. 14.4.

14.9 For the circuit shown in Fig. 14.17, calculate the differential mode gain and the common mode gain. (*Hint.* This problem is most easily treated by using the general circuit theorems of Chapter 2.)

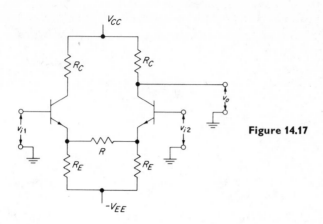

Figure 14.17

14.10 A phase-sensitive detector has an input signal that is derived from a photocell whose input contains light chopped at 124 Hz. The reference voltage is of course also at 124 Hz. Unfortunately the lights in the room flicker at 120 Hz. The output of the photocell contains a 120-Hz signal that is introduced into the input to the phase-sensitive detector. How long must the output of the phase-sensitive detector be integrated so that only the information at 124 Hz is retained in the ultimate output?

14.11 Sketch an *RC* network that will integrate the output of a phase-sensitive detector. The *RC* network should have a time constant of 1 sec. Sketch an *RC* network that will integrate the output of a phase-sensitive detector twice, that is, a network that integrates the output of the phase-sensitive detector and then integrates the output of the first integrator. What advantages would integrating twice have over a single integration but with a longer time constant?

14.12 The circuit shown in Fig. 14.18 is a commonly used phase-sensitive detector circuit. Discuss the operation of this circuit, and show that the output is indeed a phase-sensitive rectification of the input. The input voltage is v_1 and the reference voltage is v_2.

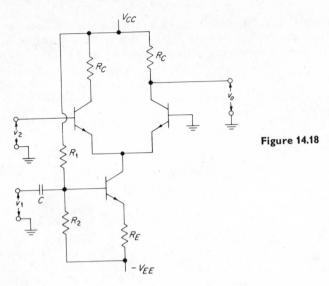

Figure 14.18

14.13   The input signal to a phase-sensitive detector has a frequency of 100 Hz and is 10 microvolts in magnitude. The equivalent rms input noise to the phase-sensitive detector is white and has a magnitude of 3 microvolts rms when the bandwidth is 10 Hz. What output integration time constant is needed so that the signal to noise ratio in the output of the phase-sensitive detector is 100?

14.14   Consider the voltage regulator of Fig. 14.14. Suppose that the resistor chain connected to the base of the transistor T_3 divides the output voltage so that the voltage at the base of T_3 is βv_o. If $\beta = \frac{1}{4}$ and if the Zener diode has a reverse breakdown voltage of 5.4 volts, what is v_o?

14.15   Let us estimate how well regulated the output voltage v_o is for the voltage regulator of Fig. 14.14. As in Prob. 14.14 the voltage at the base of transistor T_3 is βv_o. Let us assume that the Darlington pair T_1 and T_2 have such a high current gain that the input current to the pair is negligible. Let us denote incremental changes in the input and output voltages by v_i and v_o. Show that the voltage at the collector of T_3 is $v_i + \beta A_v v_o = v_i - \beta(R_1/r_{tr})v_o$ where $A_v = -R_1/r_{tr}$ is the voltage gain of T_3. The Darlington pair T_1 and T_2 simply act as an emitter follower. Use this to show that $v_o = v_i + \beta A_v v_o$ or $v_o/v_i = 1/(1 - \beta A_v) \simeq -1/\beta A_v$.

14.16   Suppose for the circuit of Fig. 14.14 that $R_1 = 10^3$ ohms, $\beta = \frac{1}{4}$, and the transistor T_3 carries 10 milliamp. What is v_o/v_i?

14.17   Suppose that for a voltage regulator the overall gain of the dc amplifier is 10^3, and suppose that the input to the dc amplifier is $(\frac{1}{4})v_o$. If the total input voltage has an rms value of the ac ripple that is 0.1 volt, what will be the magnitude of the ac ripple on the output voltage?

15/Operational Amplifiers

The name *operational amplifier* is applied to a large and important class of commercially available, high gain, dc coupled amplifiers. The name derives from their use in analog computers. With the proper feedback all the common mathematical operations, addition, multiplication, integration, and so on, can be performed on time-varying voltage signals. This by no means exhausts the applications of operational amplifiers, however. Aside from a few specialized fields, for instance, the handling of high power or very high frequencies, almost all linear electronics can be performed by operational amplifiers.

15.1 Integrated Circuit Operational Amplifiers

Operational amplifiers come in a variety of forms ranging from a simple difference amplifier such as previously discussed to a multistage chopper stabilized amplifier. Perhaps the most common form of the operational amplifier is an integrated circuit containing the entire operational amplifier on a single silicon chip. Integrated circuits will be discussed in Sec. 15.6. The input stage of the amplifier is a two-transistor difference amplifier with another transistor used as a constant current sink in the common emitter of the differential pair. The input stage is followed by a second differential amplifier stage that also introduces additional negative feedback for the common mode. A single ended output is taken from the second stage and is amplified in the final stage. The output stage also shifts the dc level of the output so that zero volts at the input corresponds to an output of zero volts. A three-stage dc amplifier of this general type was discussed in Sec. 14.3 and is shown in Fig. 14.9.

Leads provide access to the integrated circuit at 10 or 12 different points but for most of our discussion we can represent the operational amplifier schematically as in Fig. 15.1. Both v_+ and v_- of the differential input and the output v_o are referenced to the same ground. We omit the leads to the power supply. Both a positive and negative supply are needed in order to permit the common grounding of input and output and the direct proportionality of input and output voltage.

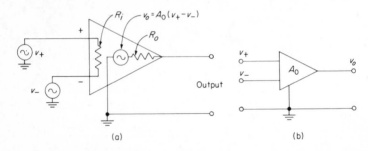

(a)   (b)

Figure 15.1 (a) A schematic diagram of an operational amplifier with a differential input. Both the input voltages v_+ and v_- and the output voltage v_o are measured with respect to ground. The coupling is dc and the output voltage is zero if the input voltage $v_i = v_+ - v_-$ is zero. The input resistance is R_i, the output resistance is R_o, and the amplification without feedback is A_o so that $v_o = A_o(v_+ - v_-)$. (b) A more compact representation of the amplifier of (a).

The output voltage is given by $v_o = A_o(v_+ - v_-)$. Single chip operational amplifiers with voltage gains from 10^3 to more than 10^5 are available. The output voltage will saturate at something a little less than the supply voltage. Maximum v_o of the order of ± 10 volts are common.

Because operational amplifiers are almost always used in external circuits that provide large negative feedbacks, there is usually no attempt to increase the linearity of the amplifier by the provision, internally, of negative feedback for the differential mode. A plot of v_o versus $v_i = (v_+ - v_-)$ is shown in Fig. 15.2 for an amplifier with a voltage gain of 10^5 and a maximum output voltage of 10 volts.

The general comments of the previous chapter on input offset drifts and the advantages of the differential amplifier and of integrated circuit construction apply, of course, to the operational amplifier.

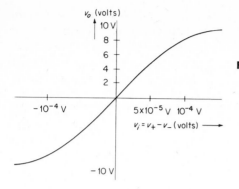

Figure 15.2 The output versus the input for an operational amplifier having a small signal amplification $A_o = 10^5$ and a saturated output voltage of ± 10 volts.

15.2 Some Simple Applications of Operational Amplifiers

All the various forms of negative feedback that we discussed in Chapter 12 find application with operational amplifiers. In Fig. 15.3 we illustrate the use of operational voltage feedback. The v_+ input is grounded and the voltage to be amplified is applied to v_-. Thus v_i and v_o have opposite signs and S is a virtual ground as is required. The input current to the amplifier itself is very small and therefore almost the same current flows through R_1 and R_2. To an excellent approximation we have $v_i = i_i R_1$, $v_o = -i_i R_2$, and $A = v_o/v_i = -R_2/R_1$. The resistances R_1 and R_2 will be chosen so that $R_2/R_1 = |A| \ll |A_o|$; therefore the amplification with feedback is almost independent of the properties of the operational amplifier. In an analog computer such a circuit would be used to multiply the input voltage by the factor $-R_2/R_1$. Figure 15.2 shows that the amplifier without feedback has only a limited region of linearity (constant amplification). With feedback, however, almost the full saturation output voltage is available before linearity is lost.

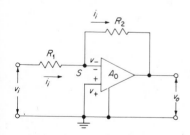

Figure 15.3 An operational amplifier with operational voltage feedback so that $v_o = -(R_2/R_1)v_i$.

A circuit for summing two voltages is shown in Fig. 15.4. Assuming S to be a virtual ground (in this application it is called the *summing point*), we have $v_o = -R_2(i_1 + i_2) = -(R_2/R_1)(v_1 + v_2)$. Thus the output voltage is the sum of the input voltages times an adjustable constant. Clearly more than two voltages can be summed by increasing the number of inputs. Current sources (rather than voltage sources) can be summed by applying them directly to the point S.

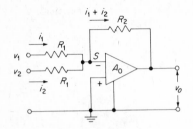

Figure 15.4   A circuit for summing two voltages. The output voltage is given by $v_o = -(R_2/R_1)$ $(v_1 + v_2)$.

We illustrate in Fig. 15.5 an important circuit that furnishes the integral of a time-varying input signal $v_i(t)$. For this circuit $v_o = -q/C$ where $q = \int_0^t i_i \, dt$ is the charge accumulating on the capacitor in the time t. Substituting $i_i = v_i/R$ we have finally

$$v_o(t) = -\frac{1}{RC}\int_0^t v_i(t) \, dt$$

If several inputs are connected through separate resistors each of magnitude R, then the output voltage will be the negative of the sum of the integrals of each of the input voltages divided by RC.

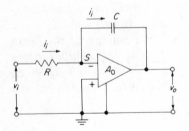

Figure 15.5   An integrator circuit. The output voltage is given by $v_o = (-1/RC)\int_0^t v_i(t)\,dt$.

It is also possible to differentiate a voltage signal using an operational amplifier. Differentiation presents serious difficulties, however, that we shall discuss later when we consider some of the limitations of operational amplifiers. Fortunately most analog computing can be done without differentiation. Let us consider as a simple example the differential equation $dy/dt = -y/\tau$. The solution is $y = y_0 e^{-t/\tau}$. The circuit that provides the solution is shown in Fig. 15.6. A battery placed across the condenser imposes initial conditions on the solution and determines the scale factor between the variable y and the output voltage. Opening the switch in series with the battery defines $t = 0$. After the switch is opened, the charge on the condenser $q = Cv_o$

decays exponentially from its initial value $q = CV_0$ to $q = 0$ through the resistance R and traces out the solution $y(t)$ at the output terminal. We see that $RC = \tau$ really defines the unit of time. It is not necessary and usually not desirable that this be real time. Thus if $\tau = 10$ sec for some process, we don't need to choose $RC = 10$ sec in our circuit. The circuit can be run at any convenient speed and the necessary scaling done later.

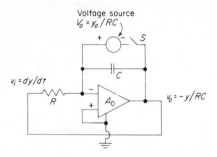

Figure 15.6 An analog computer solving the equation $dy/dt = -y/\tau$. The battery imposes the initial condition that $y(0) = y_0 = RCV_0$.

In Fig. 15.7 we show a circuit that solves a more difficult differential equation whose solution is not immediately obvious. We write the equation in the form $y'''(t/RC) + 3y'(t/RC) + 5y(t/RC) = 0$ where a prime indicates differentiation with respect to the time. This makes clear that an RC time constant of the circuit is defining the unit of time and also eliminates the factor $1/RC$ from the integrals. The equation is solved by starting with the voltage y''' and integrating until we have formed successively $-y''$, y' and $-(\frac{5}{3})y$. The output $-(\frac{5}{3})y$ is multiplied by -1 to give $+(\frac{5}{3})y$ and is added to y'. This sum is multiplied by -3 to give $-3y' - 5y$. Since $-3y' - 5y$ equals y''', these are equated electronically by connecting the voltage $-3y' - 5y$ to the original input voltage y'''. The function $y(t/RC)$ can be obtained by a voltage divider from $(\frac{5}{3})y$. The solution must be started with voltages across each of the condensers that give the desired initial conditions.

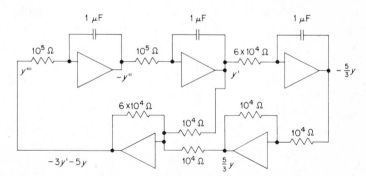

Figure 15.7 A schematic diagram of an analog computer circuit for solution of the differential equation $y''' + 3y' + 5y = 0$. The v_- inputs are used, and all v_+ inputs are grounded. The v_+ inputs and the ground leads are not shown.

The circuit constants chosen in Fig. 15.7 imply that the time unit for the problem is $RC = 10^5$ ohms $\times$ 1 microfarad $= 10^{-1}$ sec. For the last integration the time constant is 0.06 sec; thus the output is $-(0.1/0.06)y = -(\frac{5}{3})y$.

A driving force $F(t)$ can be put into a differential equation easily. For example, for the differential equation $y''' + 3y' + 5y = F(t)$ the analog computer is the same as before except that the voltage $-F(t)/3$ must be summed together with $+y' + (\frac{5}{3})y$ so that the output of the final operational amplifier is $-3y' - 5y + F(t)$. This quantity is then equated to y'''. The use of a driving force $F(t)$ means that one must generate a voltage equal to $-F(t)/3$ by some technique.

Analog computers are a very convenient and inexpensive way of solving complicated differential equations; however, analog computers are not capable of the high precision that can be obtained using a digital computer. In the solution of a second- or third-order differential equation an analog computer can achieve at best an accuracy of a few tenths of 1%. A digital computer is limited only by the time available for the computation.

15.3   The High-Frequency Response of Operational Amplifiers

Before discussing further applications of operational amplifiers we must point out certain limitations of these devices. By far the most important limitation is the unavoidable loss of amplification at high frequencies. The gain $A(\omega)$ in a typical case might fall from 10^5 at dc ($\omega = 0$) to unity at $\omega = 10^8$ rad/sec. Some of the reasons for this were discussed in Sec. 12.6 on the high-frequency response of a transistor. For the reasons discussed in Chapter 12 (mostly the Miller effect) the gain of a single-stage dc amplifier as a function of the frequency has the general form $A(\omega) = -A_0/(1 + j\omega RC)$ $= -A_0/(1 + j\omega/\omega_1)$, where ω_1 is the frequency parameter that limits the frequency response of the amplifier and $-A_0$ is the dc amplification. At $\omega = \omega_1$ we have $A(\omega_1) = -A_0/\sqrt{2}$ [down 3 dB] and a phase of $\pi/4$ between $A(\omega_1)$ and $A(0)$. For $\omega \gg \omega_1$ the amplification $A(\omega) \to 0$ and the phase shift $\to \pi/2$.

For an amplifier with three stages the gain of the entire amplifier can be written as

$$A(\omega) = \frac{-A_0}{[1 + (j\omega/\omega_1)][1 + (j\omega/\omega_2)][1 + (j\omega/\omega_3)]} \tag{15.1}$$

where ω_1, ω_2, and ω_3 are the frequency parameters for the three stages of the amplifier. A typical integrated circuit operational amplifier might have three stages and the gain would then have this form. Although we have not yet discussed oscillators, we shall find in later discussions that an amplifier will oscillate if the amplifier is such that the signal fed back to the input is in phase with the input signal at $\omega = \omega'$ and if the product $\beta A(\omega')$ is greater than 1. For a single-stage amplifier the phase is π at $\omega = 0$ and goes to $\pi/2$ at high frequencies. Obviously the phase can never be zero so the amplifier cannot oscillate. For a three-stage amplifier, however, the phase of the feedback signal can go through zero since a phase shift of almost $\pi/2$ per stage is possible. It is sufficient to assure that for the frequency at which the circuit has a feedback signal that is in phase with the input signal the gain of the amplifier is less than 1. This process was mentioned in Sec. 13.2 on feedback stability. This process of assuring feedback stability is called *phase compensation*. Phase compensation can be accomplished in several ways. We shall discuss only one method since the others are easily understood once the basic idea is grasped. The idea of phase compensation is to roll the gain off with only a single RC network (so that the maximum phase shift by the RC network is $\pi/2$) until the gain is less than 1. A system such as this cannot oscillate since the total phase shift cannot reach π while the gain is greater than 1. Suppose we connect an RC network across the input to the operational amplifier as shown in Fig. 15.8. Let us assume that the internal impedance of the input to the amplifier is much greater than $[R + (1/j\omega C)]$.

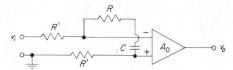

Figure 15.8 An operational amplifier with phase compensation at the input.

Further, let us assume that the amplification $A(\omega)$ without feedback and without the RC input network and using the negative input (with the positive input grounded) is given by Eq. 15.1. The amplification $A'(\omega)$ with the RC input network is

$$A'(\omega) = \frac{A(\omega)(v_- - v_+)}{v_i} = A(\omega)\frac{(R + 1/j\omega C)}{[R + 2R' + (1/j\omega C)]}$$

$$= A(\omega)\frac{(j\omega RC + 1)}{[j\omega(R + 2R')C + 1]}$$

(15.2)

Substituting for $A(\omega)$ we find

$$A'(\omega) = \frac{-A_0(1 + j\omega RC)}{[1 + j\omega(R + 2R')C][1 + j(\omega/\omega_1)][1 + j(\omega/\omega_2)][1 + j(\omega/\omega_3)]}$$

Plots of log $A(\omega)$ (upper curve) and log $A'(\omega)$ (lower curve) versus log ω are shown in Fig. 15.9. Let us consider first the plot of log $A(\omega)$. We assume $\omega_1 < \omega_2 < \omega_3$. These parameters are usually called *corner frequencies* because at each the slope of log $A(\omega)$ versus log ω becomes more negative. It is customary to show a sharp corner although, of course, the slope really changes smoothly. For $\omega_1 < \omega < \omega_2$ a tenfold increase in ω produces a tenfold (20 dB) decrease in $A(\omega)$. Thus we speak of a 20-dB/decade roll-off of the gain. Between ω_2 and ω_3 the roll-off is about 40 dB/decade and beyond ω_3, 60 dB/decade. In theory a roll-off of 40 dB/decade is safe because a phase shift of π is approached only asymptotically. In practice a lower roll-off rate must be used. At frequencies higher than ω_3 a phase shift of π is soon reached and with feedback the amplifier would be expected to oscillate.

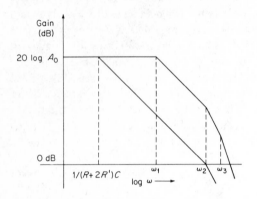

Figure 15.9 The gain in decibels (dB) versus the frequency plotted for an uncompensated three-stage amplifier (upper curve) and plotted for the same amplifier but with phase compensation as shown in Fig. 15.8 (lower curve). The frequency is plotted logarithmically.

The lower curve is essentially Eq. 15.2 but RC is so chosen that $RC = 1/\omega_1$. The term $[1 + (j\omega/\omega_1)]$ in the denominator is then canceled and we have

$$A'(\omega) = \frac{-A_0}{[1 + j\omega(R + 2R')C][1 + (j\omega/\omega_2)][1 + (j\omega/\omega_3)]} \qquad (15.3)$$

Considered in the complex plane the corner frequencies introduce poles ($\omega = -j\omega_1$) in the gain function. The elimination of a pole by cancellation with a similar term in the numerator (the zero) is referred to as *pole zero cancellation*. This has been achieved by introducing a still lower frequency pole at $\omega = -j/(R + 2R')C$ but this result is most advantageous. By starting the gain roll-off at a lower frequency one can reduce the gain to 0 dB without exceeding the safe roll-off rate of 20 dB/decade. R' has been chosen so the gain goes to 0 dB at $\omega = \omega_2$. Note that the slope does not change as $\omega = \omega_1$ is crossed. The pole at ω_1 has been canceled.

There are other methods of phase compensation. For example, many integrated circuit operational amplifiers provide connections between the second and third stages of the amplifier so that the gain can be rolled off at that point.

Both the RC network at the input and the connection so that one can roll off the gain between the second and third stages of the operational amplifier enable phase compensation to be achieved by changing the open loop gain of the amplifier. Some commercially available operational amplifiers are already internally compensated. In addition, some operational amplifiers (such as the Fairchild model 741) use only two stages of voltage amplification. If only two stages are used, then phase compensation is easily accomplished but special designs are required to achieve high gain.

Integrating and differentiating circuits provide examples of feedback networks that themselves introduce a phase shift and a frequency dependence of the feedback factor and therefore of the amplification with feedback. If we interchange the resistance and capacitor of an integrating circuit (Fig. 15.5), we obtain the differentiating circuit of Fig. 15.10. We have $v_i = q/C$ or $dv_i/dt = i_i/C$ and $v_o = -Ri_i$. Eliminating i_i gives $v_o = -RC\,dv_i/dt$.

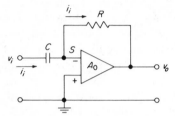

Figure 15.10 A differentiating circuit. The output is given by $v_o = -RC\,dv_i/dt$.

Let us consider the response of these circuits to a sinusoidal input $i_i = |i_i| e^{j\omega t}$ For the integrating circuit, $v_i = i_i R$ and $v_0 = -i_i/j\omega C$. Therefore $A = v_0/v_i = -R/j\omega C$. For the differentiator, $A = v_0/v_i = -j\omega C/R$. Thus the integrator feedback network rolls off the gain at 20 dB/decade starting at zero frequency. Relatively little high-frequency performance is demanded of the operational amplifier itself. At zero frequency the gain becomes very large. It is limited by the gain of the operational amplifier. This means that small dc offset potentials can be troublesome since they will be continuously integrated until the output saturates. This problem is avoided by shunting the capacitor with a high resistance that reduces the dc gain but does not interfere with the integration of frequencies of interest.

The problem with the differentiator is just the opposite. Its amplification must increase linearly with frequency to beyond those input frequencies that one seeks to differentiate. Put another way, a fast change at the input requires a fast change at the output to keep S, the virtual ground, at zero potential. Thus differentiation demands good high-frequency performance of the operational amplifier and this is not consistent with phase compensation. In addition to this, differentiators, because of the large amplification at high frequencies, are susceptible to high-frequency noise. For these and other reasons they are not widely used.

In general, because the stabilization of negative feedback necessitates phase compensation and consequently reduced high-frequency gain, the response of an operational amplifier to transients is not instantaneous. A useful measure of the high-frequency response of an amplifier is its response to an input step function voltage. The output does not immediately assume the new dc output voltage but rather starts toward it at an almost constant rate of change of voltage called the *slewing rate*. Typical slewing rates for operational amplifiers are in the range 1–10 volts/microsec.

It should be obvious from our previous remarks that since $A(\omega)$ falls as ω increases, the gain at high frequencies of an amplifier using operational feedback will not be given by the simplified formulas that assume S is a virtual ground, that is, that $(v_+ - v_-) \simeq 0$. If one is interested in the true gain at high frequencies, then this must be calculated using feedback theory and without assuming that $A_0 \gg 1$.

In addition to gain-bandwidth limitations there are other limitations of operational amplifiers. One of the most serious limitations concerns input zero offset. Generally there are constant dc input zero offset voltages and small incremental drifts of the input zero offset superimposed. The dc input zero offset voltage can be canceled out by applying a suitable dc voltage to the positive terminal of the amplifier. This voltage is called the *input voltage offset*. There is also the problem of drift of the input zero offset. As discussed under the topic of dc amplifiers this can be made small by the use of a differential amplifier input stage held at a constant temperature. The zero offset often arises because either the β's or the base resistances of the two transistors in the input differential amplifier are not identical. Obviously the zero offset may also be corrected by putting a small dc current into the input of the operational amplifier. This current is called the *input current offset*.

15.4   Some Additional Uses of Operational Amplifiers

Although most of our discussion has concerned an operational amplifier with operational voltage feedback, all the various feedback modes can be used. Figure 15.11 shows a circuit that provides pure voltage feedback. Since v_- and v_+ must be at almost the same potential, we have $v_o/v_- = v_o/v_+ = v_o/v_i = A = (R_i + R_f)/R_i$. The input impedance is high, in fact, the full input impedance of the operational amplifier itself times A_0/A. The output has the same sign as the input. This is in contrast to operational

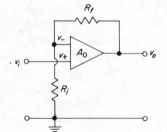

Figure 15.11 A noninverting voltage feedback amplifier. The gain is given by $A = v_o/v_i = (R_i + R_f)/R_i$.

voltage feedback for which the input impedance is low and the output is inverted.

Using a combination of operational voltage feedback and voltage feedback, a differential amplifier can be built. Of course the operational amplifier is already a differential amplifier. What we mean is that a relation between input and output voltages of the form $v_o = A(v_2 - v_1)$ can be arranged where A is much less than A_0, the amplification of the operational amplifier without feedback, and where A is therefore almost independent of A_0. Figure 15.12 shows such a circuit. The amplification is $A = R_f/R_i$. To see this, note that if $v_2 = 0$, we have $v_o = -(R_f/R_i)v_1$. If $v_1 = 0$, we have $v_2 R_f/(R_i + R_f) = v_o R_i/(R_i + R_f)$ or $v_o = (R_f/R_i)v_2$. These last equations follow at once if we remember that v_+ and v_- must be at almost the same potential. Since we now have the outputs for v_1 and v_2 as separate sources, superposition gives $v_o = (R_f/R_i)(v_2 - v_1)$ if both are present.

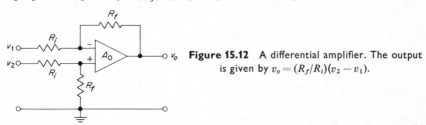

Figure 15.12 A differential amplifier. The output is given by $v_o = (R_f/R_i)(v_2 - v_1)$.

It is well to note that this circuit is not really symmetric in the v_1 and v_2 inputs. The v_2 input sees the constant input impedance $R_i + R_f$. The input impedance seen by v_1 depends on the value of v_2, however. Only if $v_2 = 0$ does it have the expected value R_i. One also requires that the internal resistances of the voltage sources v_1 and v_2 be small compared to R_i. For instance, changes in v_2 can drive appreciable currents through v_1.

It is also possible using operational amplifiers to perform nonlinear electronic functions. For example, it is possible to form logarithms of voltage functions. A circuit for doing this is shown in Fig. 15.13.

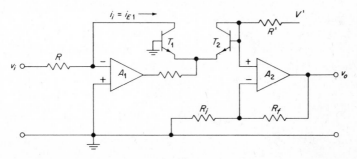

Figure 15.13 A logarithmic amplifier. The output is given by $v_o = -(kT/e)[(R_i + R_f)/R_i]$ ln $[(R'/R)(v_i/V')]$.

In the circuit shown, a nonlinear element, a transistor T_1, is inserted into the feedback network of the first operational amplifier. The base-emitter *pn* junction of a transistor has very well-understood properties. In particular the total emitter current of transistor T_1 is given by $i_{E1} = i_o e^{ev_{BE1}/kT}$ or $v_{BE1} = (kT/e) \ln (i_{E1}/i_0)$. This expression is accurate as long as the ohmic drop across the base to emitter of the transistor is negligible, that is, as long as $r_d = kT/eI_E \gg r_{ohmic}$ where I_E is the quiescent dc value of i_E. The current i_{E1} through the transistor T_1 is forced to be equal to $i_i = v_i/R = i_{E1}$ by the first operational amplifier.

The second transistor T_2 operates such that $v_{BE2} = (kT/e) \ln (i_{E2}/i_0)$. Now if we select the transistors T_1 and T_2 such that the back bias current i_0 is the same for the two, then we find that $v_{BE1} - v_{BE2} = (kT/e) \ln (i_{E1}/i_{E2})$. Since the emitters of T_1 and T_2 are at the same voltage and since the base of transistor T_1 is at groundp otential, it follows that $v_{B2} = (-kT/e) \ln (i_{E1}/i_{E2})$. The current i_{E2} in T_2 is given by $i_{E2} \simeq V'/R'$. Combining the previous results we obtain

$$v_{B2} \simeq -\frac{kT}{e} \ln \left[\left(\frac{R'}{R} \right) \left(\frac{v_i}{V'} \right) \right]$$

The second operational amplifier acts as a noninverting amplifier of gain $(R_i + R_f)/R_i$ so that $v_o = -(kT/e)[(R_i + R_f)/R_i] \ln [(R'/R)(v_i/V')]$. For satisfactory operation it is necessary to set i_{E2} to be relatively small. For example, if i_{E2} is set so that $i_{E2} = V'/R' = 2 \times 10^{-5}$ amp, then the ohmic resistance in the base-emitter junctions will be negligible. If $i_{E2} = 2 \times 10^{-5}$ amp, then $v_o = 0$ when $i_{E1} = v_i/R = 2 \times 10^{-5}$ amp. Thus if one wants $v_o = 0$ if $v_i = 1$ volt, then $R = 5 \times 10^4$ ohms. Obviously this system operates satisfactorily only if $v_i > 0$. A major problem with a circuit of this sort is that the output voltage is very dependent on the temperature. It is imperative that T_1 and T_2 be maintained at the same temperature and that the temperature be constant.

It is also possible to form antilogarithms using operational amplifiers. A circuit for doing this is shown in Fig. 15.14. The transistors T_1 and T_2 are a matched pair. The current in T_2 is set by varying the resistor R. If R is set at 10^4 ohms, then the current in T_2 is about 5×10^{-5} amp ($v_{BE2} \simeq 0.5$ volt). Now the first operational amplifier holds the negative input at the same voltage as the positive input. The positive input is at $-v_{BE2}$. The current in T_1 is given by $i_1 = i_0 e^{e(|v_{BE2}| + v_i)/kT}$. The output from the first operational amplifier is

$$v_1 = -|v_{BE2}| - R_f i_0 e^{e(|v_{BE2}| + v_i)/kT}$$

The second operational amplifier is simply a difference amplifier of gain 3 so that

$$v_o = (3R_f i_0 e^{e|v_{BE2}|/kT}) e^{ev_i/kT}$$

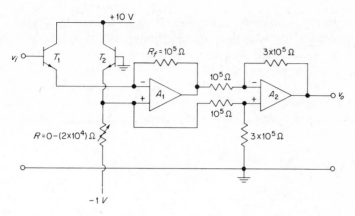

Figure 15.14 An exponential amplifier. The output is given by $v_o = \{3R_f i_0 \exp[(e|v_{BE2}|/kT]\}$ $\exp(ev_i/kT)$.

It is seen that the output voltage is the antilog of the input voltage. As with the logarithmic amplifier the current in the individual transistors should be kept small so that $kT/eI_E \gg r_{ohmic}$ and the temperature of the matched pair of transistors should be carefully regulated.

Commercial logarithmic and antilog amplifiers are available in compact packages. These are exceedingly useful as function generators. For example, suppose one wants to form the function $v_1{}^2 v_2$ where v_1 and v_2 are voltages that are functions of the time. A circuit for forming this function is shown in Fig. 15.15. Circuits of this type are very important. For example, circuits of this type enable one to use analog computers to solve problems involving nonlinear differential equations. In addition, circuits of this type enable one to perform nonlinear analysis of signals and subsequently use the analyzed signals with a digital computer. Hybrid circuits using analog techniques for nonlinear analysis of data followed by computations using digital computers are important at the present time and are certain to become extremely important in the future.

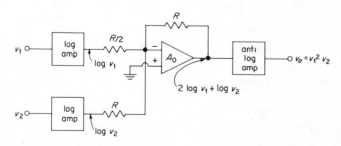

Figure 15.15 A circuit for performing the nonlinear operation $v_o(t) = v_1{}^2(t)v_2(t)$.

Operational amplifiers can also be used to produce a very narrow band amplifier. This is done by using a circuit with a frequency-dependent response in the feedback network. An example of such a circuit is shown in Fig. 15.16. The circuit shown will have a very high gain at the resonant frequency ω_0 of the twin T and will have a much lower gain away from the resonant frequency of the twin T. The twin T was discussed in Chapter 6. It is an RC coupled circuit that presents a very high impedance at its resonant frequency.

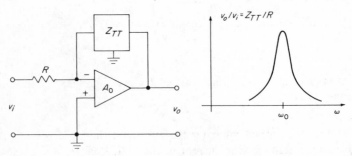

Figure 15.16 A narrow band amplifier using a twin T as the feedback impedance. The impedance Z_{TT} of the twin T is high and sharply peaked at its resonant frequency ω_0.

Another use of an operational amplifier is as a voltage comparator. This use is illustrated in Fig. 15.17. For this circuit, since there is no feedback, the output voltage will simply saturate at, say, ± 10 volts depending on whether v_1 is greater than or less than v_2.

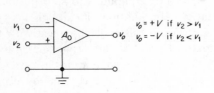

Figure 15.17 A voltage comparator. In a range of $v_2 - v_1$ of the order of V/A_0 the output voltage v_o jumps from $-V$ to $+V$ where V is the saturation output voltage of the operational amplifier.

15.5 Linear Gates

Gates are widely used to control the passage of a signal from one part of a circuit to another. They are essentially on-off devices. When the gate is closed, passage of the signal is completely blocked. When the gate is open, the signal is freely passed. A gate is said to be linear if during the open period the signal is passed without distortion. The opening and closing of the gate is under independent control although it may be linked to the signal to be blocked or passed as in the sweep trigger circuit of an oscilloscope.

An operational amplifier may be used to provide a linear gate. An example of such a circuit is shown in Fig. 15.18. In this circuit the operational amplifier is connected as a difference amplifier. Connected to the negative input of the difference amplifier is an FET. When the gate of the FET has a small bias, then the FET acts like a resistor (that is, it has an ohmic behavior) with perhaps 100-ohm resistance and any signal at the input v_{in} is amplified by the amplifier (which has a gain of 1). When the gate of the FET has a strong positive bias, however, the FET is cut off and has a drain to source resistance of several megohms. The signal v_i is now amplified but with a gain of approximately zero. The purpose of the capacitor C is to cancel out any part of the gating signal that is passed capacitively to the minus input of the difference amplifier. The capacitor C is simply adjusted until this is accomplished.

Typical waveforms for v_i, v_o, and v_g are shown in Fig. 15.19.

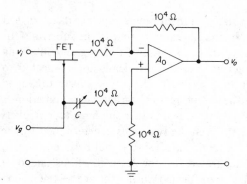

Figure 15.18 A linear gate circuit. A gating pulse v_g sets the amplification $A = v_o/v_i$ at either one or zero.

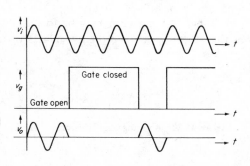

Figure 15.19 Graphs of v_i, v_g, and v_o as functions of the time for the linear gate circuit of Fig. 15.18.

15.6 The Structure of Integrated Circuits

One of the most fascinating and useful achievements in electronics has been the development and mass production of inexpensive integrated circuits. An integrated circuit is an entire circuit including transistors, diodes, resistors, and capacitors all fabricated on a single chip of silicon. Integrated circuits enable the laboratory worker to use entire complex circuits with only a minimum of wiring to connect the circuit. The circuits are very small and consequently are useful in devices such as computers where space is an important consideration. In addition, because integrated circuits are entirely contained on a single small chip of silicon, all the transistors and other elements in the circuit can be maintained at the same temperature. This leads to a great simplification whenever one wishes either to stabilize the temperature of a given circuit or when one wishes to compensate for changes in the temperature. Of course the most important advantage of integrated circuits is that integration permits the economic production of very complicated circuits.

When we described the operation of a bipolar transistor, we described the structure of the transistor; however, we did not emphasize the manufacturing processes. We shall follow a similar plan in describing the nature of integrated circuits.

Most integrated circuits use a chip of single crystal silicon doped to be a p-type semiconductor as the initial material (that is, the substrate) for an integrated circuit. In order to make a resistor, an n-type region containing a p-type region is formed in the p-type chip and two leads are connected to the inner p-type region as shown in Fig. 15.20. The p-type substrate is held at a negative voltage relative to the n-type region so that the pn junction is back biased. This isolates the n-type region from other similar n-type pockets in the p-type chip. The resistance between the two leads is determined by the size, shape, and conductivity of the semiconductor material between the two resistor leads. The leads to the integrated resistor are metal films laid over the insulating oxide layer that has been formed on the surface of the silicon chip. In order to form an integrated transistor an additional n-type region is formed as shown in Fig. 15.21. An integrated capacitor may be formed by using the insulating oxide layer as the dielectric between the two plates of a capacitor as shown in Fig. 15.22.

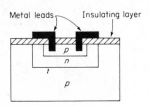

Figure 15.20   An integrated resistor.

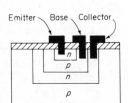

Figure 15.21   An integrated bipolar transistor.

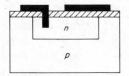

Figure 15.22 An integrated capacitor. The oxide layer is the dielectric between one lead and the n-type conducting material.

One can also form integrated diodes. In addition to the methods of production indicated, still other methods may be used to produce circuit elements; for example, a back-biased pn junction may be used as a capacitor. Complicated integrated circuits are formed by producing the necessary components (transistors, resistors, and so on) on a silicon chip and interconnecting these as desired to produce the entire electronic circuit on a single chip of silicon. Thus, for example, a simple emitter follower circuit is shown in Fig. 15.23. Also indicated on the figure is how this circuit might be formed as an integrated circuit.

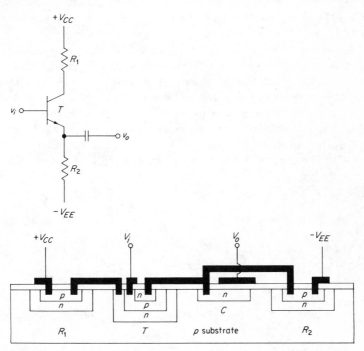

Figure 15.23 A simple emitter follower amplifier and its construction as an integrated circuit. The connection between C and R_2 is on the oxide layer but goes around v_o. The most negative potential $- V_{EE}$ is connected to the p-type substrate.

It should be noted that the p-type substrate is held at the most negative voltage in the circuit so that all the n-type pockets are isolated from one another due to the back-biased pn junctions formed.

It is possible to produce very complicated circuits as integrated circuits. For example, operational amplifiers and logic gates for digital computers are very often made as integrated circuits. It is now possible even to produce integrated circuits containing both bipolar transistors and FET's.

The use of various circuit elements in integrated circuits is not the same as in discrete component circuits. For example, there is no way to produce an inductor in an integrated circuit so that inductors are never used. In an integrated circuit the area on a silicon chip required to produce a 10^{-11} farad capacitor is about three times as large as to produce a transistor and the area required to produce a 10^3-ohm resistor is about two times as large as to produce a transistor. As a result the cost of an integrated circuit containing capacitors is large compared to the cost of a similar circuit containing no capacitors. Consequently capacitors are used in integrated circuits only if they are completely unavoidable. Most integrated circuits are dc coupled (that is, use no capacitors). Because of these considerations the integrated circuit of Fig. 15.23 would not normally be constructed containing a capacitor but would be dc coupled. The capacitor in Fig. 15.23 was shown only to illustrate how a capacitor is formed in an integrated circuit.

Because the area used by transistors in an integrated circuit is small, and because two transistors produced on the same small silicon chip can be nearly identical in their properties and can be maintained at nearly the same temperature, transistors are often used to set dc bias conditions in integrated circuits. These transistors have nothing to do with the incremental properties of the circuit. An example of the use of transistors in dc biasing is shown in Fig. 15.24. The current in T_1 is approximately V/R_1 (this ignores the voltage drop across the forward-biased base-emitter junction of the transistor T_1).

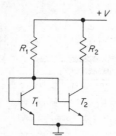

Figure 15.24 The transistor T_1 provides the base bias for T_2.

Since the transistor T_2 is identical with T_1 (that is, has the same β, and so on) and since the base to emitter voltage for T_2 is the same as for T_1, the collector current in T_2 is also V/R_1. Circuits such as this are almost never seen in discrete component circuitry because of the cost of finding identical transistors and the cost of the additional wiring and components. If one wishes to understand the detailed dc and incremental properties of various integrated circuits, it is necessary simply to learn which circuits are commonly used as bias circuits. These parts of the circuits can, of course, then be ignored in the incremental analysis of an integrated circuit. Figure 15.25 shows a photograph of an integrated circuit amplifier.

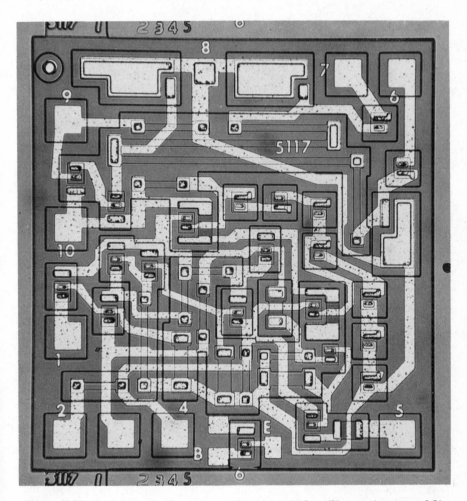

Figure 15.25 A photograph of an integrated circuit amplifier. (*Photograph courtesy RCA Corporation.*)

Problems

15.1   Suppose one is given voltage sources v_1, v_2, and v_3. Using high gain operational amplifiers, form the following:

(a) $v_1 + v_2 + v_3$
(b) $5v_1 + 3v_2 + 4v_3$
(c) $5v_1 + 3v_2 - 4v_3$

15.2   Given voltage sources v_1, v_2, and v_3, use high gain operational amplifiers to form the following functions:

(a) $\int v_1 \, dt + \int v_2 \, dt + \int v_3 \, dt$
(b) $5 \int v_1 \, dt + 3 \int v_2 \, dt + 4 \int v_3 \, dt$
(c) $5 \int v_1 \, dt + 3 \int v_2 \, dt - 4 \int v_3 \, dt$
(d) $5 \int v_1 \, dt + 3 \, v_2 - 4v_3$

15.3   Given the voltage sources v_1 and v_2, use high gain operational amplifiers to form $5 - 3 \int v_1 \, dt + 4v_2$.

15.4   Sketch the circuits of analog computers that will solve the differential equations $dy/dt + y = 0$ and $dy/dt - y = 0$. The initial condition in both cases should be $y = 3$. Over what time intervals will the output be $y(t)$, and over what time intervals will the output simply be saturated at the maximum output of the operational amplifier (assumed to be ± 10 volt) for the two equations?

15.5   Sketch the circuit diagram of an analog computer for solving each of the following differential equations:

(a) $d^2y/dt^2 + a\,(dy/dt) + by = 0$
(b) $d^2y/dt^2 + 6\,(dy/dt) + 4y = 0$
(c) $d^2y/dt^2 + 4y = 0$

You may ignore the initial conditions for the solutions.

15.6   Sketch the circuit for an analog computer that will solve for the steady state solution to the differential equation $d^2y/dt^2 + 4\,(dy/dt) - 6y = 5 \sin 10^3 t$. Approximately how long is required after the analog computer is started for the transients to damp out leaving only the steady state solution?

15.7   Sketch the circuit diagram for an analog computer that can solve the following differential equations:

(a) $d^3y/dt^3 + 6\,(d^2y/dt^2) + 9\,(dy/dt) + 3y = 0$
(b) $d^4y/dt^4 + 4\,(dy/dt) + 7y = 0$

You may ignore the initial conditions.

15.8   For the circuit given in Fig. 15.26 show that

$$v_o = -\left[\frac{R_2}{R_1} v_i + \left(\int v_i \frac{dt}{R_1 C_2}\right)\right].$$

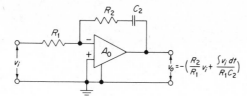

Figure 15.26

15.9   Find v_o in terms of v_i for the circuit shown in Fig. 15.27.

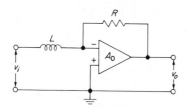

Figure 15.27

15.10   A three-stage operational amplifier has a gain given by

$$A(\omega) = \frac{10^5}{[1 + (j\omega/\omega_1)][1 + (j\omega/\omega_2)][1 + (j\omega/\omega_3)]}$$

where $\omega_1 = 10^5$ rad/sec, $\omega_2 = 10^7$ rad/sec, and $\omega_3 = 10^8$ rad/sec. Find the appropriate values of R and C for the phase compensation network of Fig. 15.8 so that the gain of the amplifier rolls off as shown in Fig. 15.9.

15.11   Draw a differential amplifier circuit using a high gain operational amplifier with negative feedback. The output voltage should be $v_o = 10(v_1 - v_2)$ where v_1 and v_2 are the two input voltages.

15.12   Use an operational amplifier to design a noninverting amplifier with a total voltage gain of 25.

15.13   Use logarithmic and exponential amplifiers to make a circuit that will form v_i^2 and $v_i^{1/2}$.

15.14   For the circuit shown in Fig. 15.28, find v_o in terms of v_i. Plot both the magnitude of v_o and the phase of v_o versus frequency ω for a constant amplitude but variable frequency input signal $v_i = Ve^{j\omega t}$.

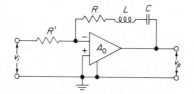

Figure 15.28

15.15   There exist circuits that can multiply two analog numbers x and y together to give either $+xy$ or $-xy$. A circuit that can perform this multiplication for either sign of x and for either sign of y and give the result including the correct sign is called a *four-quadrant multiplier* (the four quadrants refer to the possible sign combinations for the input signals). Let us consider an amplifier that takes two inputs x and y and forms $-xy$ as an element in the feedback circuit for an operational amplifier. What is v_o in terms of v_i for the circuit as given in Fig. 15.29?

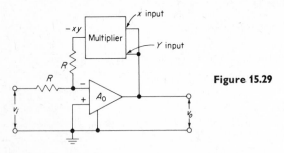

Figure 15.29

15.16   For the circuit shown in Fig. 15.30, find v_o in terms of v_i.

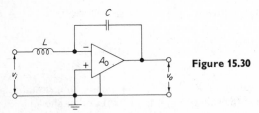

Figure 15.30

15.17   Using a high gain differential input operational amplifier with negative feedback, design a follower circuit that has an input resistance greater than 10^6 ohms and an output resistance of 50 ohms. You may assume that the operational amplifier has an open loop gain of $A_0 = 3 \times 10^4$, an intrinsic input resistance of $R_i = 150 \times 10^3$ ohms, and an intrinsic output resistance of $R_o = 5000$ ohms.

15.18   Show that the time constant for the decay of a charge from the capacitor in the integrator circuit of Fig. 15.5 is $A_0 RC$ when the input is grounded. Thus the integrator circuit using an operational amplifier holds the integral for a much longer time than the simple RC integrator of Fig. 6.27.

16/Nonlinear Electronic Circuits

There are many electronic circuits for which the output signal is not proportional to the input signal. These circuits are called *nonlinear electronic circuits*. An example of a nonlinear circuit is an amplifier with distortion. For the amplifiers we previously studied, negative feedback was commonly employed to reduce distortion and consequently make the circuit more nearly linear. There are many useful circuits, however, that deliberately employ nonlinear elements. An example of such a circuit is the voltage comparator or the logarithmic amplifier previously discussed. Nonlinear circuits usually employ positive feedback. When positive feedback is used, the output of an amplifier is fed back to the input in phase with the input signal so that the input is reenforced. This leads to the rapid increase of both the input and output signals. In fact, as we shall show, the output of a circuit using strong positive feedback will increase exponentially until nonlinear effects limit the increase of the output signal.

In this chapter we shall discuss oscillators, multivibrators, and nonlinear devices used to control power such as SCR's (silicon controlled rectifiers). In the next chapter we shall discuss the nonlinear circuits used in logic operations such as digital computation.

16.1   Oscillators

An oscillator is a circuit that will provide a continuous time-varying output signal even in the absence of an applied input signal. All oscillators function by using positive feedback.

In order to understand better how an oscillator works, let us consider the particular circuit shown in Fig. 16.1.

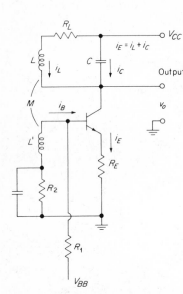

Figure 16.1   An oscillator circuit employing a mutual inductance M for positive feedback.

The oscillator in Fig. 16.1 works as follows. The mutual inductance M provides positive feedback from the collector circuit of the transistor to the base circuit. If a small noise voltage occurs at, say, the base, this signal is amplified by the transistor. The amplified signal at the collector is fed back positively to the base. The circuit begins to run away. Sizable amplification by the transistor occurs only if the noise signal that initiates the process is at the resonant frequency of the collector LC tank circuit. The magnitude of the output signal will continue to increase until the energy dissipated in R_L in the collector tank circuit is as large as can be supplied by the transistor. At this point the circuit simply oscillates with a steady ac amplitude at the frequency $\omega \simeq 1/\sqrt{LC}$.

With these introductory comments in mind, let us analyze the circuit of Fig. 16.1 in detail. Since an oscillator is an inherently nonlinear device, it might be thought that using linear amplifier analysis on the circuit would not be a sensible procedure. This is correct in the steady state condition; however, a linear analysis can show us how the oscillations begin to build up in an oscillator circuit. For the circuit shown, let us assume that initially the oscillator has a total emitter current i_E being drawn (i_E includes both the quiescent and the incremental currents in the emitter). If we further assume that the transistor is not saturated, then the collector current is nearly the same as i_E and the base current i_B is very small. The capacitor across R_2 is assumed to be very large so that its impedance for an ac signal is zero. The voltage drive applied to the base of the transistor is $M\ di_L/dt$ where i_L is the current in the inductor L in the collector circuit and M is the mutual inductance between the inductor in the collector circuit and the coil in the base circuit. We neglect the term $L'\ di_B/dt$ since i_B is usually very small. For small signals we can write $M\ di_L/dt = v_{be} + R_E i_e = (r_{tr} + R_E)i_e$ where i_e is the small signal incremental ac component of the emitter circuit. The quantity $M\ di_L/dt$ is an incremental voltage even though i_L is a total current. For this reason the right-hand side of the equation must contain only the incremental variables v_{be} and i_e. In addition, we can write $i_E = i_L + i_C$ where i_C is the current flowing through the capacitor in the collector circuit. Note that i_C is not the collector current. We have assumed that the collector current is equal to i_E. Kirchhoff's voltage law applied to the parallel LC network in the collector circuit gives us the equation

$$L \frac{di_L}{dt} + R_L i_L = \frac{Q_C}{C} \tag{16.1}$$

where R_L is the resistance of the inductor L and where Q_C is the charge stored on the capacitor C. This equation assumes that all the resistance is in the inductive leg and that the capacitive leg has zero resistance, and further this equation assumes that the base current is very small so that $M\ di_B/dt$, the emf induced in L by the mutual inductance, is zero. By differentiating Eq. 16.1 we find

$$L \frac{d^2 i_L}{dt^2} + R_L \frac{di_L}{dt} = \frac{i_C}{C} \tag{16.2}$$

We have treated i_L and i_C as total quantities and, in fact, in Eq. 16.1 the terms $R_L i_L$ and Q_C/C contain large dc contributions. As a result of the differentiation, however, the terms in Eq. 16.2 are all incremental. The current i_C is both total and incremental because there can be no dc current through a capacitor. Thus the equation $i_C = i_E - i_L$ holds for both the total currents i_E and i_L and also for their incremental components. The dc parts of i_E and i_L are equal. Using also $M\ di_L/dt = (r_{tr} + R_E)i_e$ we eliminate i_C from Eq. 16.2 and obtain Eq. 16.3 below. It must be remembered, however, that the i_L in Eq. 16.3 is now only the incremental part of the current through L. The solution we shall obtain for i_L also refers only to the incremental part.

$$L \frac{d^2 i_L}{dt^2} + R_L \frac{di_L}{dt} = \frac{M}{C(r_{tr} + R_E)} \frac{di_L}{dt} - \frac{i_L}{C}$$

which may be rewritten as   (16.3)

$$LC \frac{d^2 i_L}{dt^2} + \left(R_L C - \frac{M}{r_{tr} + R_E} \right) \frac{di_L}{dt} + i_L = 0$$

As a trial solution to this equation, let us try $i_L = i_0 e^{\alpha t}$. This leads to the equation

$$LC\alpha^2 + \left(R_L C - \frac{M}{r_{tr} + R_E} \right) \alpha + 1 = 0 \qquad (16.4)$$

The solution for α is

$$\alpha = \frac{[M/(r_{tr} + R_E) - R_L C] \pm \sqrt{[M/(r_{tr} + R_E) - R_L C]^2 - 4LC}}{2LC} \quad (16.5)$$

The solutions are classified by whether $[M/(r_{tr} + R_E) - R_L C]^2$ is greater than, equal to, or less than $4LC$. We shall consider only the case where $4LC$ is greater than $[M/(r_{tr} + R_E) - R_L C]^2$. Only for this case do we obtain oscillations. The solution for i_L is

$$i_L = i_0 e^{[(1/2LC)M/(r_t + R_E) - R_L C]t} \cos(\omega t + \phi) \qquad (16.6)$$

where

$$\omega = \sqrt{\frac{1}{LC} - \frac{1}{4L^2 C^2} \left(\frac{M}{r_{tr} + R_E} - R_L C \right)^2}$$

If $M/(r_{tr} + R_E) - R_L C \geq 0$, then the current i_L will increase exponentially with the time. Physically the oscillations will die out if M is such that the feedback is negative or if the feedback is positive but is so weak that it cannot make up for the energy dissipated in R_L (see Prob. 16.1). The oscillations will increase, however, if the feedback is positive and large enough to compensate for the losses in R_L.

As i_L increases, i_e and i_b must also increase; however, i_L cannot increase forever. The question we now pose is the following: What limits the increase of i_L? The answer to this question is that the linear approximation that enabled us to write $v_{be} = r_{tr} i_e$ cannot be used as the variations in i_E about the quiescent point become large. Obviously as the variations in i_E about the quiescent point become large, the transistor will saturate as i_E increases and the transistor will reach cutoff as i_E decreases. In fact, if one introduces a value of r_{tr} averaged over the swing of i_E, then as variations in i_E about the quiescent point become large the average value of r_{tr} over the cycle increases; that is, the gain of the amplifier decreases as the variation in i_E about the quiescent point becomes large. This occurs because as one drives near $i_E = 0$, the value of r_{tr} becomes very large; whereas r_{tr} is limited to a lower value of r_{ohmic} as i_E increases. The value of i_L will increase (but not exponentially except in the linear region) until r_{tr} has some average value $\langle r_{tr} \rangle$ such that $M/(\langle r_{tr} \rangle + R_E) = R_L C$. When this condition is satisfied, the level of oscillation will be constant, and $i_L = i_0 \cos(\omega t + \phi)$ where $\omega = 1/\sqrt{LC}$. A sketch of i_L versus time is shown in Fig. 16.2.

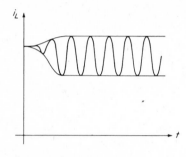

Figure 16.2 The total current in the inductor L of Fig. 16.1 as a function of the time.

It is interesting that Eq. 16.3, whose solution gives the exponentially increasing oscillation, also says that if the incremental current i_L and its first and second derivatives are zero at any time, then i_L will remain zero. We have already mentioned that very small noise voltages and transients that contain some Fourier components near $\omega = 1/\sqrt{LC}$ are sufficient to start the positive feedback and the buildup of the oscillations. Usually preventing unwanted oscillations in high gain amplifier circuits is more of a problem than starting an oscillator.

It is instructive to look at this circuit from a somewhat different viewpoint. Suppose an emf source $\varepsilon_b(t) = v_i e^{j\omega t}$ is inserted in the base circuit as shown in Fig. 16.3. For this circuit $(r_{tr} + R_E)i_e = j\omega M i_L + v_i e^{j\omega t}$. Using Kirchhoff's voltage law on the plate circuit we find that $j\omega L i_L + R_L i_L = i_C/j\omega C$. Using these two equations plus the relationship that $i_L + i_C = i_e$ it follows that

$$i_L = \frac{v_i e^{j\omega t}/(r_{tr} + R_E)}{(1 - \omega^2 LC) + j\omega[R_L C - M/(r_{tr} + R_E)]} \tag{16.7}$$

Consequently it follows that the output voltage v_o is given by

$$v_o = (j\omega L + R_L)i_L$$

$$= -\left(\frac{j\omega L + R_L}{r_{tr} + R_E}\right) \frac{v_i e^{j\omega t}}{\{(1 - \omega^2 LC) + j\omega[R_L C - M/(r_{tr} + R_E)]\}} \tag{16.8}$$

From our previous discussion of feedback we recall that $v_o = A_0 \varepsilon_b/(1 - \beta A_0)$ where A_0 is the amplification without feedback. By comparison with the two equations we find that

$$1 - \beta A_0 = \left[1 - \omega^2 LC + j\omega\left(R_L C - \frac{M}{r_{tr} + R_E}\right)\right] \frac{Z_{\parallel}}{j\omega L + R_L} \qquad (16.9)$$

where $Z_{\parallel}$ is the impedance of the parallel LC network in the collector circuit and where we have used the result $A_0 = -Z_{\parallel}/(r_{tr} + R_E)$.

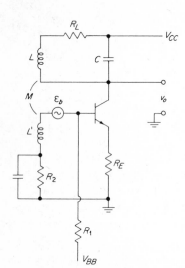

Figure 16.3 The circuit of Fig. 16.1 redrawn as an amplifier for the signal source ε_b.

Previously we found that in the steady state the average value of the transresistance was such that $R_L C = M/(\langle r_{tr}\rangle + R_E)$ and that $\omega = 1/\sqrt{LC}$. Thus we see that, viewing the circuit as an amplifier with feedback, the condition for steady state oscillation is $1 - \beta A_0 = 0$ or $\beta A_0 = +1$. This is called the *Barkhausen criterion* for oscillation. It must be emphasized that an oscillator does *not* need to have $\beta A_0 = +1$ initially for oscillations. Instead it is merely necessary for the oscillator to be such that as the small signals build up due to positive feedback, the amplifier saturates in such a way that the condition $\beta A_0 = +1$ is reached in the steady state.

The Barkhausen criterion, while useful, does not enable us to look at the small signal gain and the feedback network and determine whether a given circuit will oscillate. A theorem due to Nyquist, which we shall not attempt to prove, states that if one plots the imaginary part of βA_0 (Im βA_0) against the real part of βA_0 (Re βA_0) with the frequency as a parameter running from $-\infty$ to $+\infty$, then a circuit will oscillate if the plot is a closed curve enclosing $+1$ or if the plot runs from some point at infinity in the complex plane to another point at infinity such that it circles $+1$ in a clockwise sense. Several plots are shown in Fig. 16.4 to illustrate Nyquist's theorem. The first Nyquist plot shown is for the oscillator circuit we previously analyzed. Obviously the Nyquist theorem gives us a method of examining the small signal parameters of a circuit and determining whether the circuit will oscillate. Of course a circuit that satisfies the Nyquist condition for oscillation will not operate anywhere on the small signal curve but will instead change in a nonlinear manner until the circuit satisfies the condition that $\beta A_0 = +1$ when the oscillator is operating in the steady state. It should now be obvious why phase compensation is used with operational amplifiers to prevent the phase of the feedback from reaching 180° until $\beta A_0(\omega)$ has fallen to less than 1. Unless phase compensation is used, the operational amplifier circuit will oscillate.

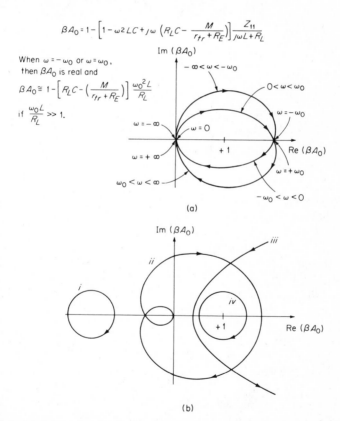

Figure 16.4 (a) A Nyquist plot for the circuit of Fig. 16.3. The circuit will oscillate. (b) Nyquist plots for other circuits. The circuits corresponding to plots (i) and (iii) will not oscillate, whereas the circuits corresponding to (ii) and (iv) will oscillate. The arrows on the curves show the direction of the algebraic increase of ω.

Oscillators using tuned LC tank circuits operate from around 100 kHz up to about 800 MHz. There are many different methods used to develop the positive feedback other than the use of a mutual inductance. Two of the various possibilities are shown in Fig. 16.5.

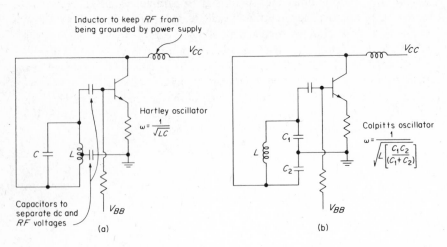

Figure 16.5   Two oscillator circuits. (a) A Hartley oscillator. (b) A Colpitts oscillator. In each case the oscillator output is taken between the collector and ground.

In an oscillator with a tuned LC tank circuit the frequency stability is determined primarily by the Q of the tank circuit. Typical values of the Q for a radio-frequency tank circuit are 100–200. The stability of an LC oscillator with a tank circuit having a Q of 200 can be of the order of 1 in 10^4 or 10^5 parts for short times. For some purposes this is not sufficient stability.

Another disadvantage of oscillators that use a tuned LC tank circuit as the frequency determining element is that the frequency will vary if either the capacitor or inductor in the tank circuit is altered. As the temperature, pressure, and so on, change the physical properties and hence the values of either the capacitor or the inductor, then the frequency will change. Crystal oscillators are often used when an oscillator with a very stable frequency is desired. In a crystal oscillator a piezoelectric crystal such as quartz is used in the feedback circuit. The crystal is carefully cut so that the mechanical resonant frequency of the crystal has a desired value. The piezoelectric property of the crystal causes a potential difference to be produced across the crystal when it is strained. Thus the mechanical resonant frequency of the crystal is converted into an ac voltage. The mechanical resonance of a good quartz crystal has a very high Q; for example, a Q of 10^5 is possible. With a very high Q element such as the piezoelectric crystal in the feedback circuit the frequency stability of an oscillator can be as good as 1 in 10^7 parts or even

1 in 10^8 parts if the temperature of the crystal is carefully controlled. Crystal oscillators are used for frequencies of about 10^5 to 10^7 Hz.

For low frequencies (from 1 Hz to 2×10^5 Hz) it is common to use RC phase shift oscillators. These oscillators use an RC network to produce the phase shift necessary for the positive feedback. No tuned LC tank circuit is used in the feedback network of an oscillator of this type. One of the most commonly used RC phase shift oscillators is the Wien bridge oscillator. A Wien bridge oscillator using an operational amplifier is shown in Fig. 16.6.

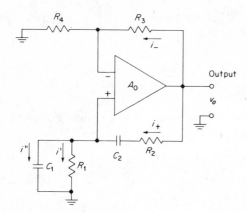

Figure 16.6 A Wien bridge oscillator circuit. The circuit employs an RC network to produce the phase shift necessary for positive feedback.

The circuit is very similar to that of the Wien bridge discussed in Chapter 6. For the Wien bridge oscillator shown, when $\omega = \sqrt{1/R_1 R_2 C_1 C_2}$, then the output is fed back to the plus input with zero phase shift providing the positive feedback necessary for oscillation. The proof of this is left as a problem. The resistors R_3 and R_4 provide negative feedback that stabilizes the gain at a desired value so that steady state sinusoidal oscillations occur. In order to better understand the Wien bridge oscillator, let us consider the linear equations that determine how the oscillations build up in such a circuit. Upon examining the circuit the following relations are obvious:

$$v_o = A_0(v_+ - v_-) = A_0(i' R_1 - i_- R_4)$$

$$v_o = i_-(R_3 + R_4) = i_+ R_2 + \frac{q_+}{C_2} + i' R_1$$

$$i' R_1 = \frac{q''}{C_1}$$

(16.10)

and

$$i_+ = i' + i''$$

where A_0 is the small signal gain of the amplifier without feedback which is normally very large. The quantities q_+ and q'' are the incremental parts of the charges on C_2 and C_1, respectively. Upon differentiating the equations containing q_+ and q'', combining these equations with the remaining equations, and making the simplifying assumption that $A_0 \gg 1$, the following equation for i_- can be derived:

$$\frac{1}{2}\frac{d^2 i_-}{dt^2} + B\frac{di_-}{dt} + Ci_- = 0$$

where

$$B = \frac{1}{2}\left[\frac{R_1(C_1 + C_2) + R_2 C_2}{R_1 C_1 R_2 C_2} - \frac{R_3 + R_4}{R_2 R_4 C_1}\right] \tag{16.11}$$

and

$$C = \frac{1}{2R_1 C_1 R_2 C_2}$$

To determine i_- we substitute the trial solution $i_- = i_0 e^{\alpha t}$ into Eq. 16.11. We find that α must satisfy the relation

$$\tfrac{1}{2}\alpha^2 + B\alpha + C = 0$$

or $\hspace{12cm}$ (16.12)

$$\alpha = -B \pm (B^2 - 2C)^{1/2}$$

If the current i_- is to be an exponentially increasing oscillation, we must have $B^2 - 2C < 0$ for oscillation and $B < 0$ for exponential growth. The oscillation stabilizes at a constant amplitude when nonlinear effects make $B = 0$. The frequency of oscillation is then given by $\omega^2 = 2C$. From the expressions for B and C given in Eq. 16.11 we can write the conditions for stable oscillation as follows:

$$\frac{R_3}{R_4} = \frac{R_2}{R_1} + \frac{C_1}{C_2} \qquad \text{(from } B = 0\text{)}$$

and $\hspace{12cm}$ (16.13)

$$\omega = \sqrt{2C} = \sqrt{\frac{1}{R_1 C_1 R_2 C_2}}$$

These are the same as the balance conditions for a Wien bridge given in Chapter 6. See Eq. 6.26 and Fig. 6.22 and note that the labeling of the components is different in Chapter 6. Thus the Wien bridge oscillator can be thought of as a balanced Wien bridge being driven at its balance frequency. The balance is maintained because the high gain operational amplifier will keep the differential input $v_+ - v_- \simeq 0$.

Once i_- is known, the output voltage v_o follows from

$$v_o = i_-(R_3 + R_4) \tag{16.14}$$

The resistance R_4 is usually chosen as the nonlinear element that assures the buildup of oscillations and finally the stabilization of the oscillation at a constant amplitude. For small i_- and v_o the resistance R_4 must be small. We see from Eq. 16.11 that for sufficiently small R_4 the parameter B will be negative; therefore oscillations will build up. As i_- and v_o increase, R_4 increases until finally $B = 0$ and the oscillation stabilizes. A small tungsten filament light bulb can be used for R_4 because the resistance of the filament rises rapidly when its temperature increases due to an increase in i_-. Another method of changing R_4 is to use an FET as R_4 and let the bias level for the gate be derived from the output signal by rectification. In this way the channel width and hence the resistance R_4 is controlled by the output voltage so that the oscillation level is stabilized.

If a Wien bridge oscillator is to be used as a variable oscillator, it is necessary to have both C_1 and C_2 variable (or R_1 and R_2). In order that the oscillation level be constant it is necessary that C_1 and C_2 track so that C_1/C_2 remains constant.

There are many other types of useful RC phase shift oscillators; however, the mode of operation is similar to the Wien bridge oscillator. The RC positive feedback network produces a phase shift that is large enough that it is possible to satisfy the criterion $1 - \beta A_0 = 0$.

The oscillators we have discussed so far have been designed to produce a sinusoidal waveform. In the oscillator circuit using a tuned LC tank the sinusoidal waveform occurs in spite of the nonlinear behavior of the amplifier because the LC tank circuit acts as a filter and rejects harmonics of the fundamental frequency. For the RC phase shift oscillator, negative feedback is used to keep the loop gain near 1 so that the nonlinearity does not have to clip much to limit the amplitude. This means that only the fundamental of the oscillator frequency occurs strongly in the output since the amplifier is essentially linear. The nonlinearity needed for stability is in the negative feedback loop.

There is a class of oscillators in which the output is extremely nonsinusoidal. These oscillators are called *heavily biased relaxation oscillators*. In the next section we shall use various types of multivibrators as illustrations of relaxation oscillators.

16.2 Multivibrators

A multivibrator is a circuit that must be in one of two possible states. The transition from one state to the other is assumed to be very short compared to the time spent in either state. A multivibrator can be constructed by coupling two amplifiers together using strong positive feedback. All multivibrators have positive feedback and a loop gain in excess of 1. The particular circuit configuration is usually not critical. Multivibrators are classified as bistable, monostable, and astable. The bistable multivibrator is such that either of the two possible states of the circuit are stable, that is, a bistable multivibrator will remain in either of its two states until an external signal causes the state to be changed. A bistable multivibrator is also called a *flip-flop*. A monostable multivibrator has only one permanently stable state so that after a signal causes the multivibrator to go from its stable state to an unstable state, then the circuit will later return to its stable state automatically. An astable multivibrator is one that has no stable state and the circuit changes from one state to the other repeatedly in a free running manner.

In order to understand multivibrator circuits better, let us discuss the bistable multivibrator in detail. We shall then give a somewhat briefer discussion of the other classes of multivibrators. Figure 16.7 shows a bistable multivibrator. The inputs to the multivibrator are labeled R and S and the outputs are labeled Q and $\bar{Q}$. The reader should note that in this section Q is not a charge. The bistable multivibrator shown operates with one transistor saturated and the other transistor cut off. Let us assume initially that T_1 is saturated and T_2 is cut off. With this condition prevailing we shall take $v_{CE1} \simeq 0.2$ volt for transistor T_1 since the collector to emitter voltage for a saturated transistor is small and a reasonable value for a high gain transistor is about 0.2 volt. The dc base voltage of transistor T_2 is therefore $v_{B2} = 0.2R_B/(R_B + R)$ volt. If we arbitrarily take $R_B = 5 \times 10^3$ ohms and $R = 20 \times 10^3$ ohms, then $v_{B2} = 0.2 \times \frac{1}{5} = 0.04$ volt. Since the emitter of both T_1 and T_2 is grounded, the base to emitter voltage of T_2 is $v_{BE2} = 0.04$ volt. Thus if T_1 is saturated, then T_2 must be cut off. If T_2 were turned on, then $V_{BE2} \simeq 0.6$ volt assuming T_1 and T_2 are silicon transistors. Since T_1 is saturated, the base to emitter voltage for T_1 is approximately 0.6 volt. Since T_2 is cut off and therefore $i_{C2} = 0$, the Thevenin equivalent parameters for the circuit viewed between the base of T_1 and ground are given by $\varepsilon_{Th} = 10R_B/(R_C + R_B + R)$ and $R_{Th} = R_B(R + R_C)/(R_C + R_B + R)$. If we take $R_C = 2 \times 10^3$ ohms, we find that $\varepsilon_{Th} = 1.85$ volts and $R_{Th} = 4.1 \times 10^3$ ohms. The base current to transistor T_1 is found from $v_{BE1} = 0.6$ volt $= 1.85$ volts $- 4.1 \times 10^3 i_{B1}$, which leads to $i_{B1} \simeq 3 \times 10^{-4}$ amp. The collector current of T_1 can be calculated from 10 volts $- R_C i_{C1} = 0.2$ volt, which leads to $i_{C1} \simeq 4.9 \times 10^{-3}$ amp. The ratio $i_{C1}/i_{B1} = 16$. The criterion for saturation in a transistor is simply that $\beta i_B > i_C$ so that a transistor with $\beta = 50$ easily satisfies the saturation criterion for the circuit shown.

What we have shown with the previous discussion is that the circuit is in fact actually stable when one transistor is saturated and the other is cut off. Finally, what is the value of the collector voltage for the transistor that is cut off? This may be calculated from $v_{C2} = (10 - 0.6)R/(R + R_C) = 8.5$ volts. Thus the output voltages are $Q = 0.2$ volt, $\bar{Q} = 8.5$ volts. We also point out the obvious fact that while we have shown that the circuit is stable with T_2 cut off and T_1 saturated, the reverse situation with T_1 cut off and T_2 saturated, that is, the state with $Q = 8.5$ volts and $\bar{Q} = 0.2$ volt, is also stable as seen from the symmetry of the circuit.

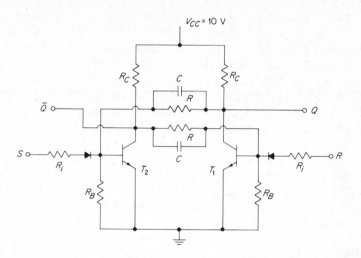

Figure 16.7 A bistable multivibrator.

Let us now ask what happens when a short positive pulse is applied to the S (for set) input. When the voltage at the base of transistor T_2 reaches a value near 0.6 volt, then T_2 begins to turn on. As T_2 turns on, the collector of T_2 begins to fall in voltage. As the collector of T_2 falls in voltage, the base of T_1 begins to drop in voltage. This process is aided by the capacitor C, which presents a negligible impedance to the base of T_1 for the fast falling voltage of the collector of T_2. As the base of T_1 begins to decrease in voltage, the transistor T_1 moves out of saturation and the collector voltage of T_1 increases. As the collector of T_1 increases in voltage, the base of T_2 is pulled still higher in voltage. Strong positive feedback is clearly occurring and the system runs rapidly to the state where T_2 is saturated and T_1 is cut off so that now $Q = 8.5$ volts and $\bar{Q} = 0.2$ volt. The process in which positive feedback produces the rapid change of state is called *regeneration*. Further positive pulses at the S input have no effect on the state of the flop-flop. The flip-flop can now be restored to its original state by applying a positive pulse to the R (for reset) input. The diodes in the S and R inputs are used to prevent triggering on negative pulses. Some flip-flops are designed to trigger on a negative pulse which turns off the saturated transistor.

The flip-flop that we described is a very simple one. Other more complicated flip-flops will be discussed in the next chapter, which is on digital circuits.

Flip-flops have many uses such as storing numbers in binary form. For this use the value of Q, which can be either 0.2 or 8.5 volts for the circuit we discussed, is used to represent the binary numbers 0 and 1, respectively.

If a series of pulses is applied, each pulse being properly shaped and steered to the input of the off transistor, then the output at, say, Q has only half as many pulses as the input. Thus Q completes a cycle for every two inputs. Flip-flops can be used in a circuit so as to provide a division by any power of 2 (see Chapter 17 for a discussion of this process). This forms the basis for using flip-flops in scaling units. Flip-flops are widely used in decade-scaling units, binary scalers, registers, and many other devices. These uses will be described in the next chapter, which discusses digital circuits, and also in the final chapter, which discusses common electronic laboratory instruments.

Before leaving the bistable flip-flop we mention why this circuit was treated along with oscillators. The answer is that after the circuit is triggered, the output of the flip-flop increases rapidly to its final value and remains at this value even in the absence of any input signal after the trigger signal because of the positive feedback and because the loop gain is greater than 1. The two transistors simply act as inverting amplifiers and the positive feedback circuit assures the rapid transition from one stable state to another. The similarities of an oscillator and a flip-flop become clearer if we imagine a flip-flop operating with completely symmetric currents and voltages. Neither transistor is cut off and Q and $\bar{Q}$ are at the same potential. The flip-flop is now in the same unstable condition as an oscillator just before oscillations start their exponential growth. Noise voltages or a small transient will tip the flip-flop one way or the other and the positive feedback will drive it into one of its two stable configurations.

A monostable multivibrator has only one stable state and is often called a *univibrator* or *trigger pair*. A univibrator is shown in Fig. 16.8. The univibrator operates in the absence of a trigger signal with T_2 saturated and T_1 cut off. The transistor T_2 is held in a saturated state by the resistor R_{B2}. If we assume $v_{CE} \simeq v_{BE} \simeq 0$, then $R_{B2} i_B \simeq R_C i_C$. Since $\beta i_B > i_C$ for saturation, we see that $\beta R_C > R_{B2}$ assures that the transistor is saturated. The transistor T_1 is held in cutoff by the resistor R_{B1}, which goes to ground. The collector of T_2 is at 0.2 volt so the base of T_1 is at a voltage $0.2R_{B1}/(R_{B1} + R)$ volt, which is less than 0.6 volt so that T_1 is indeed cut off.

Now suppose that a positive pulse v_i is put into the input of the univibrator. The transistor T_1 begins to turn on when the base to emitter voltage reaches approximately 0.6 volt. As T_1 turns on, the collector voltage of T_1 drops and hence the base voltage of T_2 drops, and the transistor T_2 is driven toward cutoff. Positive feedback acts to accelerate the process and soon the transistor T_2 is completely cut off and T_1 is saturated. However, this situation does not persist indefinitely as it would for a bistable flip-flop. The capacitor C' will charge up with a time constant $\tau = R_{B2} C'$. The base to emitter voltage of T_2 will increase as this occurs. When the base to emitter voltage of T_2 reaches approximately 0.6 volt, the univibrator will switch back to the stable state where T_2 is saturated and T_1 is cut off. Voltages as a function of the time at various points in the circuit are shown in Fig. 16.9.

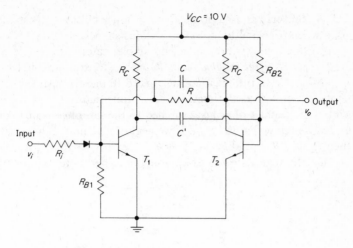

Figure 16.8 A monostable multivibrator.

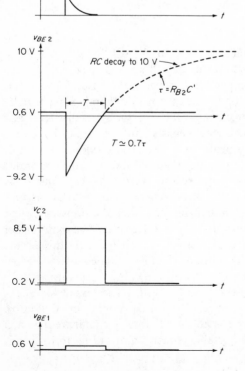

Figure 16.9 Typical voltages as a function of the time in a monostable multivibrator circuit.

Univibrators are used primarily to produce pulses of constant height and length when triggered by pulses of various shapes.

An astable multivibrator has no stable states. An astable multivibrator is also called a *free running multivibrator*. A free running multivibrator is shown in Fig. 16.10. In the circuit of Fig. 16.10 one transistor is saturated and the other is cut off at all times. Let us assume that transistor T_1 has just switched to cutoff and T_2 has just switched to the saturated state. For T_2 to be saturated we know that $\beta i_B > i_C$. If we neglect v_{BE} and v_{CE} in the saturated state, then $R_B i_B \simeq R_C i_C$ and $i_B = (R_C/R_B)i_C$. Thus $\beta R_C > R_B$ to assure saturation. In the saturated state $v_{CE} \simeq 0.2$ volt and when cutoff $v_{CE} \simeq 10$ volts.

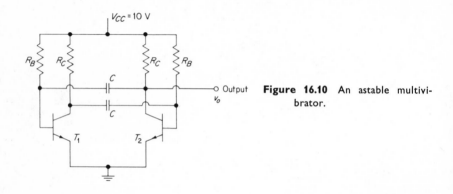

Figure 16.10 An astable multivibrator.

Thus the base of T_1 switched rapidly from about 0.6 volt to $\simeq -10 + 0.2 + 0.6 \simeq -9.2$ volts when T_1 switched to the cut off state. The base of T_1 now begins to rise in voltage as the capacitor C between the base of T_1 and the collector of T_2 charges up. The time constant for this charging is $\tau \simeq R_B C$. When the base of T_1 reaches approximately 0.6 volt, the system switches and T_1 saturates while T_2 is cut off. This state persists until the base of T_2 returns to about 0.6 volt and the system again switches. The astable multivibrator will run indefinitely first with one transistor cut off and the other saturated and then with the reverse situation and so on. We can easily estimate the time T that the multivibrator remains in one of its two states. We shall assume that the base is switched from $+0.6$ to -10 volts when the transistor is cut off (actually we know it switches from $+0.6$ to -9.2 volts). We shall further assume that the transistor turns on when $v_B = 0$ volt (actually this occurs when $v_B \simeq 0.6$ volt). With these assumptions we can calculate the time T from $(1 - e^{-T/R_B C}) = \frac{1}{2}$, which leads to $T = R_B C \ln 2 \simeq 0.7 R_B C$. In Fig. 16.9 we wrote $T \simeq 0.7 R_B C$ for the monostable multivibrator. The reason for this is identical to the argument we just gave for the free running multivibrator. The voltages at various points in the astable multivibrator circuit are shown in Fig. 16.11. Although the particular astable multivibrator of Fig. 16.10 spends the same time in each of of its two states, this is not generally necessary and the relative time spent in each of the two states can be changed by making the transistor bias resistors different or using different coupling capacitors. The astable multivibrator that we discussed may not be self-starting when it is turned on. For example, if both transistors are saturated, the circuit is stable and not oscillatory. Some astable multivibrators contain circuitry that assures self-starting but we shall not discuss these circuits.

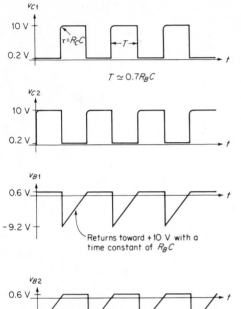

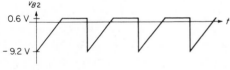

Figure 16.11 Typical voltages as a function of the time in an astable multivibrator circuit.

There are many uses for an astable multivibrator. One common use is to divide the frequency of a regular series of pulses by a fixed number. This can be accomplished in the following way. Suppose we have a series of pulses. The period of the pulses is just slightly less than $\frac{1}{3}T$, that is, just less than one-sixth the period of the multivibrator. We now apply the pulses to the bases of the transistor T_1 and T_2. Figure 16.12 shows the voltages at the two bases of the transistors and v_o. It can be seen that the pulse that comes along just before the multivibrator would have changed its state naturally, causes the multivibrator to alter its state slightly early, and synchronizes the multivibrator with the pulses. The output of the multivibrator has one-sixth as many pulses as the input to the bases of the transistors in the multivibrator. Division by a factor of 10 is common.

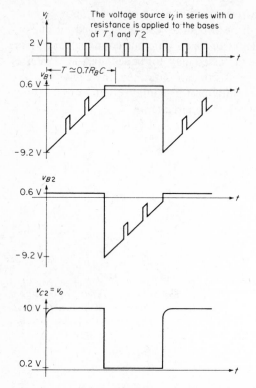

Figure 16.12 Typical voltages as a function of the time in a synchronized astable multivibrator circuit.

We might comment before leaving multivibrators that all multivibrators (bistable, monostable, and astable) satisfy the Nyquist criterion. This illustrates the fact that the Nyquist criterion merely indicates that a circuit is unstable because of positive feedback. The Nyquist criterion does not imply that the output of an oscillator will have a sinusoidal time variation.

There is one final circuit that we wish to discuss in this section, the Schmitt trigger. Figure 16.13 shows the basic circuit of a Schmitt trigger. For the Schmitt trigger circuit if $v_{B1} = 0$, then transistor T_1 is cut off and T_2 is saturated. The voltage across R_E, that is, at the emitters of the two transistors, is given by $v_{E2} = (10 - 0.2)R_E/(R_E + R_{C2})$ volt. Now suppose that the input voltage is raised until $v_i \simeq v_{E2} + 0.6$ volt. Then transistor T_1 begins to turn on. The positive feedback will drive the system until T_1 is saturated and T_2 is cut off. The output voltage v_o will increase from $(v_{E2} + 0.2)$ to 10 volts. The voltage at the emitter of the transistors is now given by $v_{E1} = (10 - 0.2)R_E/(R_E + R_{C1})$ volt. Now if $R_{C1} > R_{C2}$, then $v_{E2} > v_{E1}$. As v_i is now lowered, the circuit will switch back to the state where T_1 is cut off and T_2 is saturated when $v_i \simeq (v_{E1} + 0.6)$ volt. The fact that the switching occurs at different voltage levels as v_i goes up and as v_i comes down is called *hysteresis*. Schmitt triggers are commonly used to discriminate against voltage signals less than a given size. This is illustrated in Fig. 16.14.

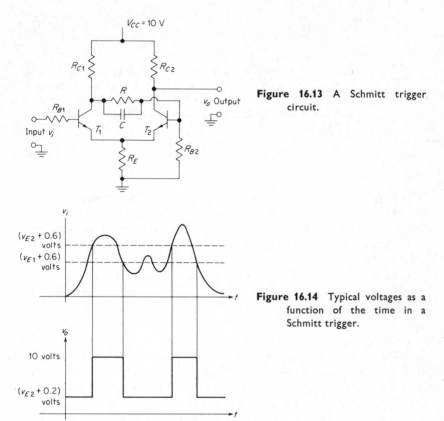

Figure 16.13 A Schmitt trigger circuit.

Figure 16.14 Typical voltages as a function of the time in a Schmitt trigger.

Although the circuit we have discussed uses transistors that are driven into saturation, it is possible to construct a Schmitt trigger where the transistors are not driven to saturation. As we have discussed in Chapter 13, saturation should be avoided if very high speed operation is desired.

16.3 The Silicon Control Rectifier (SCR) and Similar Devices

The silicon control rectifier (SCR) is a four-layer three-terminal device. A schematic diagram of an SCR is shown in Fig. 16.15. Also shown in the figure is the circuit symbol for an SCR.

For the SCR shown in Fig. 16.15 there are two stable states, one with very small current through the SCR and the other with the SCR conducting a large current from the anode to the cathode. The anode is maintained at a positive potential with respect to the cathode. In the nonconducting state the pn junctions J_1 and J_3 of Fig. 16.15 are zero biased and junction J_2 is back biased. Now let us suppose that some minority carriers, electrons, are introduced into layer 3 from layer 4 of Fig. 16.15 by making the gate more positive.

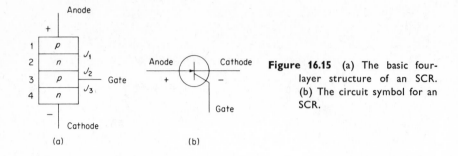

Figure 16.15 (a) The basic four-layer structure of an SCR. (b) The circuit symbol for an SCR.

This slightly forward biases junction J_3. Some of the minority carriers will drift to junction J_2. The minority carriers that drift to the junction J_2 can simply fall over into layer 2 since the back-biased junction does not present a barrier to the flow of minority carriers. When the electrons enter layer 2, they are now majority carriers. These electrons in layer 2 lower the potential of layer 2 so that junction 1 is forward biased. This leads to a flow of current from layer 1 into layer 2. Some of this current consists of holes which are minority carriers in layer 2. These minority carriers in layer 2 can drift to junction J_2 and fall over into layer 3 where the holes are majority carriers. These holes will raise the potential of layer 3 so that the forward bias of the junction J_3 is further increased. This action leads to more current flowing from layer 3 to layer 4. Some of this current will be due to electrons injected into layer 3 from layer 4. These electrons can now repeat the cycle just described. It is clear that a great deal of positive feedback exists and the SCR will run away into a high current mode. The initiation of the current is usually accomplished by making the gate more positive than the cathode. If the anode to cathode voltage is large enough, however, this process can be started by an avalanche breakdown at junction J_2 which introduces minority carriers in layers 2 and 3. The two processes are not independent. The more positive the gate, the lower is the anode to cathode voltage at which the SCR goes into the conducting state. Because of the intrinsic positive feedback the SCR will remain on even after the initial gate signal is no longer present.

The operation of an SCR may also be understood by considering the SCR to be equivalent to two transistors, one a *pnp* transistor and the other an *npn* transistor, as shown in Fig. 16.16.

The SCR is normally run with the anode held at a positive voltage with respect to the cathode. Suppose the gate is held at a voltage such that the base to emitter junction of the transistor T_2 is zero biased or negatively biased. Under this condition no current (or only a small back bias current) can be conducted from the anode to the cathode of the SCR. Now suppose that the gate is made slightly more positive. In other words a current is injected into the base of the *npn* transistor T_2 in our model of the SCR. This current surge is amplified by transistor T_2. The collector current in T_2, which is β_2 times larger than the input current to the gate, is the base current for the transistor T_1. The collector current in transistor T_1 is β_1 times the base current in T_1 and so is $\beta_1\beta_2$ times as large as the input base current to T_2. The collector of T_1 now feeds current into the base of T_2. It should be clear that there exists a strong positive feedback in the pair of transistors which represent the SCR. Consequently after a small input current is injected into the gate, the SCR will rapidly run away into a state where the SCR is carrying a large current which is mostly determined by the external resistance in the circuit. The SCR will not return to the off state when the input to the gate is removed. The input signal to the gate merely serves the purpose of providing an initial current that will enable the instability inherent in the positively fed back system to grow. Thus the initial gate current serves the same purpose as the initial noise voltage in an oscillator, as previously discussed. One might ask why the SCR does not trigger on the small back bias current that flows in the SCR when the gate is negatively biased. The answer can be understood on the model of the SCR as two transistors. When a silicon transistor is near cutoff, the value of β may be very small, less than 1. In the off state both β_1 and β_2 of transistors T_1 and T_2, respectively, are less than 1. The result of this is that the overall current gain $\beta_1\beta_2$ is not large enough to cause the system to run away. When the gate current is large enough so that $\beta_1\beta_2$ is equal to or greater than 1, then the system will run away to the on state. The critical minimum value of the gate current i_{gt} necessary to cause the SCR to run away is called the *gate trigger current*. Let us ask why it is true that β is less than 1 for small values of the collector current for a silicon transistor.

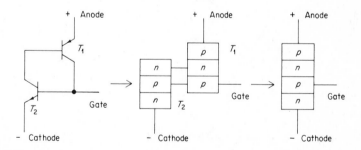

Figure 16.16 An SCR represented as two transistors, one an *npn* transistor and the other a *pnp* transistor.

The result that the β of a silicon transistor can be less than 1 at low currents is due to the fact that at low collector currents a large fraction of the current is due to recombination in the barrier. This is not the case for large collector currents and so when enough minority carriers are injected into layer 3, then transistors T_1 and T_2 operate in a mode where $\beta_1\beta_2 = 1$ and the on state will persist. Germanium devices with $\beta_1\beta_2 < 1$ cannot be built (at least are not built), so all four-layer devices are made of silicon. In Prob. 16.16 and 16.17 the reader is asked to show that the SCR will regeneratively run away if $\beta_1\beta_2 \geq 1$.

Once the SCR is turned on, the gate no longer controls the forward flow of current through the SCR. Consequently in order to return the SCR to a nonconducting state one must either reverse the anode to cathode voltage or interrupt the current through the SCR with an external switch.

Suppose that the gate current to an SCR is held at zero. Now let the total anode to cathode voltage v_{AC} increase. At small values of the anode to cathode voltage almost no current flows in the SCR. As the voltage increases, however, eventually the anode to cathode voltage will be large enough that a Zener-type breakdown will occur in the back-biased junction. As soon as the avalanche current starts to flow, the inherent positive feedback takes over and the SCR regenerates to its normal conducting state.

Let us now ask what happens when the anode to cathode voltage is reversed. The current through an SCR with a negative anode to cathode voltage is very small until the Zener-type breakdown voltage occurs. The characteristics of the SCR are *not* the same for negative anode to cathode voltages as for positive anode to cathode voltages because there is no positive feedback mechanism in the reverse direction.

The current-voltage characteristics for a typical SCR are shown in Fig. 16.17.

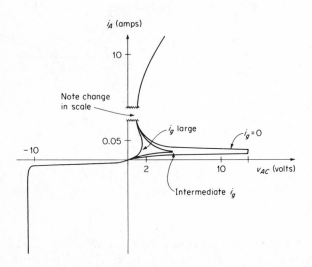

Figure 16.17 The anode current i_A versus anode to cathode voltage v_{AC} for various values of the gate current i_g for an SCR.

Some SCR's that are manufactured can carry forward currents larger than 1000 amp. Currents as large as 1000 amp are possible because the β's involved do not need to be too much greater than 1 so that design parameters can differ substantially from those of a transistor. These design changes make high breakdown voltages and high current capabilities possible.

Let us discuss some of the uses for the SCR. A circuit that illustrates a typical application of an SCR is shown in Fig. 16.18. In the circuit shown the SCR is used to control the power delivered by an ac voltage v to the load resistor R_L. The SCR is cut off entirely on the half cycle of v when the anode to cathode voltage on the SCR is negative. The SCR remains cutoff during the first part of the half cycle during which the anode to cathode voltage of the SCR is positive. When the gate current reaches the critical value i_{gt}, however, the SCR will trigger into its conducting state. The critical value of the gate current i_{gt} is specified by the manufacturer of the SCR. In order to set the trigger at some particular point in the cycle of the ac voltage we vary R_T. If we ignore the voltage drop across the two forward-biased pn junctions, one in the diode and one in the SCR (~ 1.2 volts), then the voltage v_T at which the SCR triggers is given by $v_T = i_{gt} R_T$. The diode is inserted into the circuit in order to block the negative voltage off the gate and prevent breakdown in the reverse-biased gate to cathode pn junction on the half cycle of v when the anode to cathode voltage of the SCR is negative. The ac voltage v and the current in the load resistor are shown in Fig. 16.19. The average power delivered to the load resistor R_L may be calculated as follows:

$$P = \frac{1}{T}\int_0^T i_L^2(t) R_L \, dt = \frac{1}{\omega T}\int_\alpha^\pi i_0 \sin^2 \omega t R_L \, d(\omega t)$$

$$= \frac{1}{2\pi} i_0^2 R_L \left(\frac{\pi - \alpha}{2} + \frac{\sin 2\alpha}{4}\right)$$

(16.15)

where $T = 2\pi/\omega$ and $\alpha = \sin^{-1}(R_T i_{gt}/v_0)$. From the previous figure and from this equation it is clear that the power can be varied from $i_0^2 R_L/4$ to $i_0^2 R_L/8$ by the circuit shown, that is, the phase of the trigger can be varied from $0°$ to $90°$.

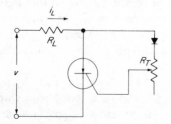

i_L

R_L

v

R_T

Figure 16.18 A circuit using an SCR to regulate the power to a load resistor R_L.

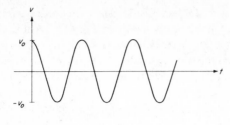

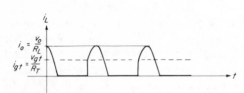

Figure 16.19 The total load current i_L through the SCR in Fig. 16.18 as a function of the time. Also shown is the ac voltage v across the input of the circuit as a function of the time.

The circuit discussed is a particularly simple one. It is possible to devise more complicated circuits that produce full-wave power control and that will trigger at any phase of the voltage from $0°$ to $360°$.

The most common use for an SCR is just the one illustrated. It provides a method of varying the power to a given load without having to vary the input voltage v. SCR's are used to enable one to dim lights continuously or to vary the power delivered to an ac motor. A common laboratory circuit using SCR's is a circuit to stabilize the temperature of an oven. The temperature is sensed and the power delivered to the heater element is decreased or increased depending on whether the temperature of the oven is hotter or colder than is desired.

We shall briefly discuss four other SCR-like devices. The first of these is the light activated silicon control rectifier (LASCR). In an ordinary SCR the trigger current is injected into the circuit at the gate. In an LASCR the trigger is produced by light that falls on the base region of transistor T_2 in the model of an SCR shown in Fig. 16.16. The light can excite electrons directly from the valence band of the silicon into the conduction band. The electrons in the conduction band of the base region of the *npn* transistor are minority carriers and can serve to trigger the four-layer SCR structure into its conducting state. Obviously an LASCR must have a window so that light can get directly to the silicon material. In order to be effective the light must have a minimum frequency given by $hf = eV_{gap} = 1.1$ eV where eV_{gap} is the band gap in silicon. This implies that the wavelength of the light must be shorter than about 11,000Å.

Another useful SCR-type device is the triac. A triac is actually an integrated circuit containing a pair of SCR's put onto a single silicon chip back to back. The advantage to a triac is that its current-voltage characteristics are symmetric with respect to the sign of the anode to cathode potential difference. The current-voltage characteristics of a triac look just like the SCR characteristics shown in Fig. 16.17 in the forward direction (positive values of v_{AC}). In the backward direction (negative values of v_{AC}) however, the current-voltage characteristics of the triac are not the same as for a back-biased SCR. The current-voltage characteristics of a triac have an odd symmetry about $v = 0$; that is, $i(v) = -i(-v)$. In other words the current-voltage characteristics of a triac are unchanged if one reverses the sign of all voltages and currents.

The silicon control switch (SCS) is a four-layer device that has a connection to all four layers; in other words, the SCS has two gates. In a two-transistor model of an SCS the bases of both transistors are used as gates. In the SCS either gate can trigger the circuit into its conducting state; that is, either gate can be used to control when the device is turned on.

There is still another very useful four-layer device. This is the gate turn off switch (GTO). The GTO is constructed so that a small reverse gate current can turn the device off. None of the devices such as SCR, SCS, or LASCR are affected by the gate once the device is in its conducting state. The GTO is useful in circuits where it is advantageous to turn current flow off with a low power control signal rather than reversing the polarity or interrupting the current flow.

Figure 16.20 shows the circuit symbols for the SCR, LASCR, SCS, the triac, and GTO.

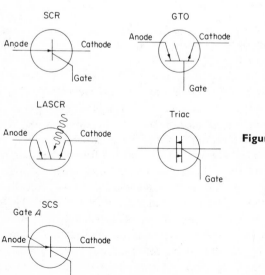

Figure 16.20 The circuit symbols for an SCR, LASCR, SCS, GTO, and triac.

16.4　The Unijunction Transistor

The unijunction transistor has a number of important applications in nonlinear circuits. It is a three-terminal device containing a single *pn* junction. In this respect the unijunction transistor is similar to the FET. In a unijunction transistor, however, the *pn* junction is usually forward biased rather than reverse biased. This, and certain structural differences, leads to very different characteristics for the two devices.

In Fig. 16.21 we show schematically the structure of a unijunction transistor, the standard symbol, and an equivalent circuit that illustrates its mode of operation. A pellet of lightly doped *n*-type silicon has ohmic contacts made at opposite ends. These contacts are called the base 1 (B_1) and base 2 (B_2) terminals. Between the two bases, at the side of the pellet, is a small region of *p*-type material. The electrical contact to this region is called the *emitter (E) terminal*.

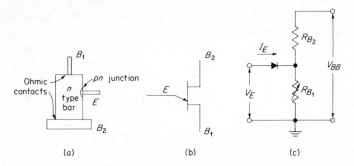

Figure 16.21　(a) Schematic showing the structure of a unijunction transistor. (b) The circuit symbol for an unijunction transistor. (c) An equivalent circuit for a unijunction transistor.

Let us assume that B_1 is grounded and a positive voltage V_{BB} is applied to B_2. If the emitter is open circuited so that no emitter current can flow, the silicon pellet is simply a resistor. Typically the resistance between the two bases might be a few thousand ohms. The open circuited emitter assumes a potential somewhere between ground and V_{BB} for whose value we write ηV_{BB}. The fraction η of the voltage between B_1 and B_2 that appears between the emitter and B_1 is called the *intrinsic standoff ratio*. Typically η might be between 0.5 and 0.8. In the equivalent circuit we show the *pn* junction as a diode and divide the total resistance of the silicon bar into R_{B_1} between the emitter and base 1 and R_{B_2} between the emitter and base 2. From the previous definition we see that we can write $\eta = R_{B_1}/(R_{B_1} + R_{B_2})$.

Note that the resistance R_{B_1} has been drawn as a variable resistance. This is the essential point in the operation of the unijunction. To see how the resistance R_{B_1} is varied, let us assume that a voltage $V_E > \eta V_{BB}$ is applied to the emitter. The *pn* junction is now forward biased and positive holes (minority carriers) are injected into the *n*-type silicon bar. The positive holes can move toward the base B_1, which is negative with respect to the emitter, but they cannot move toward the base B_2. Because of the increase in the number of current carriers, the resistance R_{B_1} decreases. The emitter is more heavily doped than the bar itself; therefore, the positive holes injected may greatly exceed the original concentration of majority current carriers (electrons) in the *n*-type bar. Thus as the forward bias on the *pn* junction is increased, the resistance R_{B_1} may decrease from a few thousand to a few tens of ohms.

In Fig. 16.22(a) we plot the emitter characteristics of a unijunction. We see that there is a region of negative resistance where an increase in I_E, the emitter current, is accompanied by a decrease in V_E, the emitter to base 1 potential. The negative resistance region is a result of the rapid decrease in R_{B_1} as I_E increases. When R_{B_1} decreases, the potential with respect to base 1 of the *n* side of the *pn* junction decreases, Therefore V_E, the potential of the *p* side of the junction, must also decrease since the forward potential drop across a highly conducting silicon *pn* junction will not be much greater than 0.6 volt prior to saturation. The minimum in V_E denotes the onset of saturation.

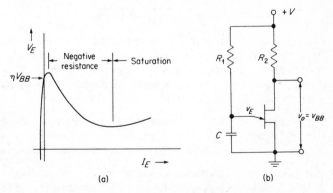

Figure 16.22   (a) The emitter characteristics for a unijunction transistor. The voltage V_{BB} of Fig. 16.21(c) is held at some fixed value. (b) A relaxation oscillator using a unijunction transistor.

In Fig. 16.22(b) we show a simple relaxation oscillator circuit using a unijunction transistor. Let us suppose that the capacitor C is originally almost uncharged. The emitter voltage v_E is then close to ground and the *pn* junction is heavily back biased since the *n* side of the junction is at the potential ηv_{BB}. The current through R_1 charges the capacitor and in a time of the order of $R_1 C$ the voltage v_E increases to slightly above v_{BB}. The *pn* junction is now forward biased and minority carriers rapidly reduce the value of R_{B_1}. The capacitor discharges through R_{B_1} with the time constant $R_{B_1} C$ that can be very much less than $R_1 C$. When the discharge is complete, the minority carriers disappear, R_{B_1} increases rapidly, the pn junction is again back biased, and the cycle starts over again. The potential v_E executes a sawtooth wave whose period is approximately $R_1 C$.

During most of this wave the base to base current through the unijunction is constant and therefore v_o is constant. During the capacitor discharge, however, the current through the unijunction and R_2 increases and v_o drops. Thus the output v_o gives a sequence of short negative pulses whose separation is approximately $R_1 C$.

Problems

16.1 Consider the circuit of Fig. 16.1. Show that the power input to the parallel LC circuit in the collector by incremental signals is given in magnitude by

$$P_{\text{in}} = \left(\frac{M \, di_L/dt}{r_{tr} + R_E} \right)^2 Q^2 R_L$$

$$= \frac{M^2 \omega_0{}^2 Q^2 R_L}{(r_{tr} + R_E)^2} i_L{}^2$$

where Q is the quality factor of the resonant LCR circuit. Show also that the power dissipated in the LC circuit is $P_{\text{dis}} = i_L{}^2 R_L$. The condition that oscillations will build up rather than damp out in the oscillator is that $P_{\text{in}} \geq P_{\text{dis}}$. Show that this is the same as to the condition $M/(r_{tr} + R_E) > R_L C$, which was deduced in Sec. 16.1.

16.2 For the circuit of Fig. 16.1, assume that $i_E = 10$ milliamp, $R_E = 10$ ohms, $L = 10^{-6}$ henry, $C = 10^{-8}$ farad, and the Q of the resonant circuit is 100. What is the minimum value of M necessary to sustain oscillations?

16.3 Discuss qualitatively the feedback loop in the Hartley oscillator, and show that the feedback is positive (not negative). Show that the feedback ratio for this oscillator is $\beta = -fL/(1 - f)L$ where f is the fraction of L between the base and ground.

16.4 Discuss qualitatively the feedback loop in the Colpitts oscillator, and show that the feedback is positive (not negative). Show that the feedback ratio is $\beta = -C_1/C_2$ for this oscillator.

16.5 A quartz crystal cut with parallel faces vibrates with a particular resonant frequency that depends only on the dimensions of the quartz crystal. The crystal can have a very high Q. If metal film electrodes are placed on two opposite parallel sides, then because of the piezoelectric property of the crystal (see Sec. 16.1) the crystal will act just like a series resonant circuit but with a capacity C_s, due to the metal electrodes in parallel with the resonant circuit as shown in Fig. 16.23. Calculate the impedance of the crystal as a function of the frequency. Find the series resonant frequency f_s where $|Z|$ is a minimum, and find the parallel resonant frequency f_p where $|Z|$ is a maximum (not $\omega = 0$). If $C_s \gg C$, what is the frequency separation $f_p - f_s$? What is the phase shift at the series and parallel resonant frequencies?

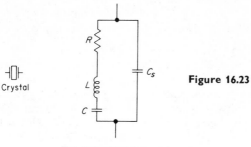

Figure 16.23

Circuit equivalent to crystal

16.6 A crystal controlled oscillator is shown in Fig. 16.24. The values of the quantities in the equivalent circuit for the crystal that is used as a series resonant circuit are $L = 100$ henry, $C = 4 \times 10^{-14}$ farad, $R = 600$ ohms, and $C_s = 5 \times 10^{-12}$ farad. What is the Q of the equivalent circuit for the crystal? What should the value of R_1 be if the negative feedback is to just be equal to the positive feedback? Discuss how R_1 should vary as a function of the output voltage if stability of the oscillator amplitude is to be achieved. Assume R_1 is the only nonlinear element in the circuit.

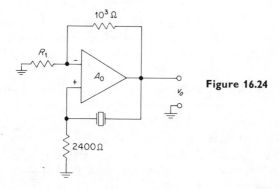

Figure 16.24

16.7 For the Wien bridge oscillator of Fig. 16.6, select values of R_1, R_2, R_3, R_4, C_1, and C_2 such that stable operation is possible.

16.8 Show that when $\omega^2 = 1/R_1 R_2 C_1 C_2$, then the voltage at the plus input v_+ is in phase with the output voltage v_o for the Wien bridge oscillator circuit of Fig. 16.6. What is the feedback ratio β of the positive feedback network when $\omega^2 = 1/R_1 R_2 C_1 C_2$?

16.9   For the circuit shown in Fig. 16.25, find the frequency at which the feedback becomes positive so that oscillations may occur.

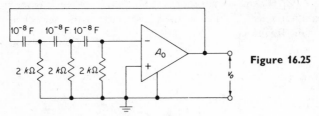

Figure 16.25

16.10   Figure 16.8 shows a monostable multivibrator. The β of both transistors is 100. Suppose we want to generate an output pulse that is 10^{-3} sec wide for each input pulse. Suppose that $R_{B1} = 5 \times 10^3$ ohms, $R = 3 \times 10^4$ ohms, $R_C = 3 \times 10^3$ ohms, and $C' = C = 10^{-8}$ farad. Select satisfactory values for the remaining resistors in the circuit. Show clearly that your choice of resistors will lead to a stable state where T_2 is saturated and T_1 is cut off, and show clearly that the output pulse will be 10^{-3} sec long.

16.11   For the astable multivibrator shown in Fig. 16.10, select appropriate values for R_B, R_C, and C so that the multivibrator will run with a period of 10^{-3} sec. You may assume $\beta = 100$ for both transistors. Clearly show that for your choice of resistors the circuit will have one transistor saturated and the other cut off except when switching from one state to the other occurs.

16.12   Draw the circuit with appropriate values for all the components of a bistable flip-flop that will have an off state where 0 volt $< Q < 0.3$ volt and an on state with 4.2 volts $< Q < 5.0$ volts.

16.13   Design a Schmitt trigger with appropriate values for the components, that is, the resistors and the capacitors. The Schmitt trigger should change its state at 5.0 volts for increasing voltage and near 4 volts as the voltage decreases. You may assume the transistors in the circuit have $\beta = 100$.

16.14   Calculate the minimum wavelength for a photon that can excite an electron from the top of the valence band to the bottom of conduction band in silicon ($E_g = 1.1$ eV). Repeat this calculation for germanium ($E_g = 0.7$ eV).

16.15   In the circuit of Fig. 16.18, assume that the input voltage is the 110 volts line voltage and that $R_L = 100$ ohms. The gate trigger current for the SCR is 25 milliamp. If one wishes to dissipate 40 watts in the load resistor, how large should R_T be?

16.16   Consider the two-transistor model of the SCR shown in Fig. 16.16. Assume that the transistors have the forward current gains β_1 and β_2. Use the relations $I_{c1} = \beta_1 I_{B1} + (\beta_1 + 1)I_{co1}$ and $I_{c2} = \beta_2 I_{B2} + (\beta_2 + 1) I_{co2}$ to show that the anode current is given by $I_A = I_{c1} + I_{c2} =$

$(1 + \beta_1)(1 + \beta_2)I_{co1}I_{co2}/(1 - \beta_1\beta_2)$. This shows that positive feedback causes the SCR to run away regeneratively into its conducting state when $\beta_1\beta_2 = 1$. (*Hint.* You should use the fact that $I_{B1} = I_{C2}$ and $I_{B2} = I_{C1}$. Assume that no gate current flows.)

16.17   Use the results of Problem 16.16 to show that positive feedback produces regeneration in the SCR when $\alpha_1 + \alpha_2 = 1$.

17/Nonlinear Circuits—Digital Circuits

One of the most important uses of nonlinear electronic circuits is to perform logical operations. The outstanding example of this use of electronic logic circuits is the digital computer. We shall discuss digital circuits including the use of these circuits to carry out logical operations. Our discussion will focus primarily on how these circuits are used to carry out the arithmetic operations necessary in a digital computer. However, the student who masters the ideas associated with the use of digital circuits will find numerous common laboratory applications for the logic operations these circuits can perform.

17.1 Binary Numbers

In a digital computer, electronic circuits are used to perform arithmetic operations such as addition or multiplication. The circuits that perform arithmetic operations are called *digital* or *logic circuits.*

In a digital computer, calculations are almost always done using a binary number system. The reason for this will be obvious once we understand how to express a number in a binary system. Let us consider the decimal number 1523. This number may be expressed using the base 10 as

$$1523 = (1 \times 10^3) + (5 \times 10^2) + (2 \times 10^1) + (3 \times 10^0)$$

In the same way a binary number is expressed in the base 2 in terms of powers of 2. For example, in a binary system

$$1011 = (1 \times 2^3) + (0 \times 2^2) + (1 \times 2^1) + (1 \times 2^0)$$

Thus the binary number 1011 has a decimal equivalent of 11. In a similar way the decimal number 1523 has a binary equivalent that is

$$\begin{aligned}10111110011 = &(1 \times 2^{10}) + (0 \times 2^9) + (1 \times 2^8) + (1 \times 2^7) + (1 \times 2^6) \\ &+ (1 \times 2^5) + (1 \times 2^4) + (0 \times 2^3) + (0 \times 2^2) + (1 \times 2^1) \\ &+ (1 \times 2^0)\end{aligned}$$

Only two numbers, 0 and 1, occur as the number multiplying a power of 2 in a binary number. As a consequence of this a circuit that is used in the expression of a binary number need have only two states, which we shall represent as 0 and 1. If we were to attempt to use a circuit to represent a decimal number, the circuit would have to have 10 possible states. It is much easier to construct a circuit that has 2 well-defined easily identifiable states than one that has 10 states. It is for this reason that digital computers use the binary number system. An individual digit, a 0 or 1, of a binary number is sometimes called a *bit*.

There are other number systems that are sometimes used in conjunction with digital computing systems. One is the octal system in which numbers are written in the base 8. The decimal number 17 has an octal equivalent that is $21 = (2 \times 8^1) + (1 \times 8^0)$. The octal system is convenient because $2^3 = 8$ and, therefore, conversion from a binary number to an octal number is very simple. This results in a digit of an octal number being represented by three digits in the binary equivalent. For example, the binary number 101001111

Binary	101	001	111
Octal	5	1	7

has an octal equivalent which is 517. The corresponding decimal number is equal to $(5 \times 8^2) + (1 \times 8^1) + (7 \times 8^0) = 335$.

Another common number system is the hexadecimal system. In this system numbers are written in the base 16. The 16 basic hexadecimal numbers followed by their decimal equivalents are the following; 0–0, 1–1, 2–2, 3–3, 4–4, 5–5, 6–6, 7–7, 8–8, 9–9, A–10, B–11, C–12, D–13, E–14, F–15. Thus the hexadecimal number C6 has a decimal equivalent which is given by $(12 \times 16^1) + (6 \times 16^0) = 198$ and a binary equivalent which is 11000110. Note that the first four digits in the binary equivalent of C6 are simply the binary equivalent of C and the last four digits in the binary equivalent of C6 are the binary equivalent of 6. In what follows we shall always separate the digits of a binary number into groups of four. This makes it easier to recognize at a glance the positions of the various powers of 2 and facilitates translation into the hexadecimal system.

In some computers a binary coded decimal (BCD) system is used for the input or output. In this system each digit of a decimal number is coded in binary individually. Thus the decimal number 975 has a binary coded decimal equivalent which is

Binary coded decimal	1001	0111	0101
Decimal	9	7	5

Of course in this case one must not read the entire sequence of 12 binary digits as a binary number.

Let us now consider the rules for addition and multiplication in a binary system. The four possible sums of single digits in a binary system are

$$
\begin{array}{lcccc}
A & 0 & 1 & 0 & 1 \\
B & 0 & 0 & 1 & 1 \\
\hline
\text{Sum} & 0 & 1 & 1 & 10
\end{array}
\tag{17.1}
$$

The sum of two arbitrary numbers can be computed using these rules. For example,

Problem: Find A plus B.

		Solution	Decimal Equivalent	
$A = 10\ 1101$		10 1101	45	
$B = 11\ 0101$		11 0101	53	(17.2)
	Carry	111 101	0	
	Sum	110 0010	98	

The rules for multiplication are also easily understood. The four possible products of single-digit numbers in a binary system are

$$
\begin{array}{lcccc}
A & 0 & 1 & 0 & 1 \\
B & 0 & 0 & 1 & 1 \\
\hline
\text{Product} & 0 & 0 & 0 & 1
\end{array}
\tag{17.3}
$$

Using these rules it is possible to multiply any two binary numbers together. For example,

	Binary Product	Decimal Equivalent	
	1011	11	
	110	6	
	0000		
	1 011		(17.4)
	10 11		
Carry	111 100		
Product	100 0010	66	

The subtraction of one binary number from another is a straight-forward process entirely analogous to what is done in the decimal system. It is merely necessary to pay attention to the mechanics of borrowing from the next higher power of 2 when that is necessary. A computer is more efficient, however, if the variety of its basis arithmetic operations can be kept to a minimum. Thus it is customary to use a notation for negative numbers that permits treating subtraction as a problem in addition.

Let us assume for simplicity that a computer is working with 8-bit binary numbers. The left-hand bit, which we underline for clarity, tells whether the number is positive or negative. A 0 sign bit stands for a positive number, a 1 sign bit for a negative number. Thus $\underline{0}010\ 1100$ is the decimal number $+44$. Because the left-hand bit is used to give the sign, the largest number we can express with 8 bits is $2^7 - 1$. Now a negative number is expressed as the 1's complement of the corresponding positive number. To obtain the 1's complement we write the positive number, for example $\underline{0}010\ 1100\ (+44)$, and then change all 0's to 1's and all 1's to 0's. This gives $\underline{1}101\ 0011\ (-44)$. The procedure automatically changes the sign bit to a 1, which indicates a negative number. The sign bit 1 is also a warning that the remaining bits are not to be treated as the digits of an ordinary binary number. Thus $\underline{1}101\ 0011$ is *not* -83.

With these definitions we can treat subtraction in the same way as addition. If we wish to subtract 44 from 70, we simply add 70 and -44 where -44 is expressed in the 1's complement notation. We obtain

$$
\begin{array}{lr}
\underline{0}100\ 0110 & 70 \\
\underline{1}001\ 0011 & -44 \\
\hline
1\ \underline{0}001\ 1001 & \\
\end{array}
$$

End around carry $\qquad\qquad 1 \hookleftarrow$

$$
\begin{array}{lr}
\hline
\underline{0}001\ 1010 & 26
\end{array}
$$

(17.5)

Note that the sign bits are treated exactly as number bits in the addition. This automatically gives the correct information on the sign of the difference. A 1 that appears to the left of the sign bit must be added in the 2^0 position to obtain the right answer. This is called an *end around carry*. Now let us add 20 and -44. We obtain

$$
\begin{array}{lr}
\underline{0}001\ 0100 & 20 \\
\underline{1}101\ 0011 & -44 \\
\hline
\underline{1}110\ 0111 & -24
\end{array}
$$

(17.6)

In this case there is no end around carry. The sign bit 1 tells us that the sum is negative and that the sum is expressed as a 1's complement. If we complement the sum, we obtain $\underline{0}001\ 1000$ or $+24$ as expected. From the definition the sum of any number and its 1's complement is $\underline{1}111\ 1111$. But this sum must be zero; therefore, $\underline{1}111$ 1111 is zero expressed in the 1's complement notation for a negative number. This can be checked by adding $\underline{1}111$ 1111 to any positive number and noting that the sum is just the original number. Since $\underline{0}000$ 0000 is also zero, it is seen that the 1's complement scheme has two expressions for zero. It must be noted that the machinery works only for the addition of a positive and a negative number. If two negative numbers are to be added, they must be added as positive numbers. The information on the sign is carried along separately.

Computer subtraction is also done using the 2's complement of a number. This is just the 1's complement plus 1. We shall not go into the details of this system.

The decimal equivalent of 1's complement subtraction is 9's complement subtraction. It will make clearer what we have been doing. Let us write $+273$ as $\underline{0}$ 0273. The $\underline{0}$ says that the number is positive. We have limited ourselves to numbers no larger than $10^4 - 1$. The 9's complement of $\underline{0}$ 0273 is $\underline{9}$ 9726 (-273). The $\underline{9}$ now indicates that the number is negative and expressed as a 9's complement. Any number plus its 9's complement is $\underline{9}$ 999 (zero). Now let us demonstrate the addition of positive and negative numbers

$$
\begin{array}{cccc}
\underline{0}\ 0400 & +400 & \underline{0}\ 0200 & +200 \\
\underline{9}\ 9726 & -273 & \underline{9}\ 9726 & -273 \\
\hline
1\underline{0}\ 0126 & & \underline{9}\ 9926 & -\ 73 \\
\end{array}
$$

End around carry $\qquad 1\!\hookleftarrow$

$$
\underline{0}\ 0127 \qquad +127 \qquad\qquad\qquad (17.7)
$$

That $\underline{9}$ 999 is zero is clear from $\underline{9}$ 999 $+\ \underline{0}$ 0400 $= 1\underline{0}$ 0399 $= \underline{0}$ 0400.

17.2 Logic Gates

It is necessary before proceeding further to discuss the types of logic circuits used in computers. The first circuit we shall discuss is the AND gate. Suppose that this circuit is to have two inputs A and B. Suppose further that the signal applied to either A or B can have only two values, represented by 0 or 1. An AND circuit is such that the output of the AND gate is given by

$$
\begin{array}{ll}
0 & \text{if } A = 0 \text{ and } B = 0 \\
0 & \text{if } A = 1 \text{ and } B = 0 \\
0 & \text{if } A = 0 \text{ and } B = 1 \\
1 & \text{if } A = 1 \text{ and } B = 1
\end{array}
\qquad (17.8)
$$

These relationships indicate that the output is a 1 if and only if both A and B are simultaneously 1. The output of an AND gate is usually denoted as $A \cdot B$. There are many ways to construct an AND gate. A simple method is as shown in Fig. 17.1. In the circuit shown if the signal at either A or B is at ground potential (denoted as state 0), then the output is held at about $+0.6$ volt (the drop across a forward-biased silicon diode). When both A and B are raised above ground by some voltage v' (denoted as state 1), however, then the output voltage becomes $(v' + 0.6)$ volt. It is clear that if we let the output voltage $v_o = 0.6$ volt be state 0 and $v_o = (v' + 0.6)$ volt be state 1 for the output, then the circuit shown is an AND gate and that $v_o = A \cdot B$. Of course it is obvious that multiple inputs can be used instead of just two inputs. For the multiple input case, $v_o = A \cdot B \cdot C \cdots$ and the output is 1 if and only if all the inputs are 1. It is important to note that if v_o is to have only two possible values, then the signals applied to A and B can have only two possible states.

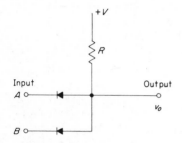

Figure 17.1 An AND gate.

In general, it is necessary for a logic gate to be able to detect any voltage signal higher than a certain voltage as a logical 1 and any voltage below some other particular voltage as a logical 0. In addition, the circuit should have an output that is easily recognized by a subsequent logic circuit as either 1 or 0. Figure 17.2 illustrates the fact that a logic gate must have a logical 1 output that is higher in voltage than the minimum input signal detected as a logical 1 and the logic gate must have a logical 0 output that is lower in voltage than the maximum signal detected as a logical 0. In order to assure that this occurs it is necessary to amplify the signals in a logic gate.

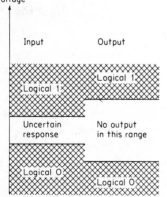

Figure 17.2 The voltage levels required in a logic circuit.

An AND gate followed by an inverting amplifier is shown in Fig. 17.3. If either input A or input B is at ground potential, then the base of the transistor is at about -0.6 volt. This follows from the fact that point S is at $+0.6$ volt and from the fact that the voltage drop across two forward-biased diodes in series is 1.2 volts. The transistor is turned completely off. When the transistor is off, the diode connected to the 4.4 volt-power supply holds the output voltage at 5.0 volts. This diode is called a *clamping diode*. When both inputs A and B go up in voltage by more than about $+1.2$ volts, then the transistor will turn on and saturate and the output of the logic gate will drop to near ground potential (0.2 volt). When the transistor is saturated, the clamping diode is back biased and no current flows in it. The logical signal at the output of this gate is

$$
\begin{array}{lll}
1 & \text{if } A = 0 \text{ and } B = 0 & \\
1 & \text{if } A = 1 \text{ and } B = 0 & \\
1 & \text{if } A = 0 \text{ and } B = 1 & (17.9) \\
0 & \text{if } A = 1 \text{ and } B = 1 &
\end{array}
$$

Thus the output of this gate is the inverse or negation of the output of an AND gate (in the sense that 0 is the inverse of 1 and 1 is the inverse of 0). A gate such as this called a NAND gate (NOT–AND). The output of a NAND gate is often written as $v_o = \overline{A \cdot B}$ where the bar indicates the act of inversion.

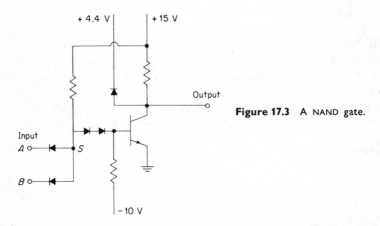

Figure 17.3   A NAND gate.

The gates we have discussed or will discuss use positive logic where the logical 1 is represented by a higher voltage than the logical 0. There are gates that use negative logic where the logical 1 is represented by a lower voltage than the logical 0. We shall not discuss negative logic gates.

We shall utilize NAND gates extensively in the logical electronics to be described. This primary use of NAND gates is an arbitrary choice and not a necessity although it does represent the commercial availability of such gates. NAND gates may be purchased as integrated circuits. For example, it is common to find four NAND gates with two inputs each or to find two NAND gates with four inputs each formed on a single silicon chip costing about 50 cents. Some typical voltages for an integrated NAND gate such as the one shown in Fig. 17.5 are $+5$ volts supply voltage, $+3.5$ volts for the logical 1 output, $+2.0$ volts for the lowest input voltage recognized as a logical 1, 0.8 volt for the highest input voltage recognized as a logical 0, and 0.2 volt for the normal logical 0 output.

There are other types of gates. For example, an OR gate can be constructed as is shown in Fig. 17.4. As long as A and B are both at ground potential (logical 0 at both inputs), then the output is about -0.6 volt (a logical 0 at the output). If either A or B rise above 0 (to a logical 1), however, then the output rises (to a logical 1). Thus we find that the output of an OR gate is given by

$$
\begin{array}{ll}
0 & \text{if } A = 0 \text{ and } B = 0 \\
1 & \text{if } A = 1 \text{ and } B = 0 \\
1 & \text{if } A = 0 \text{ and } B = 1 \\
1 & \text{if } A = 1 \text{ and } B = 1
\end{array}
\qquad (17.10)
$$

This is usually written as $v_o = A + B$.

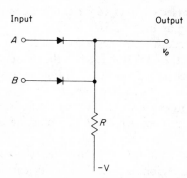

Figure 17.4 An OR gate.

As before it is usually necessary to amplify the output of an OR gate before it is a really useful circuit. Since the amplification inverts the output of the OR gate, the resulting circuit is called a NOR gate (NOT–OR). The logical output of a NOR gate is given by

$$
\begin{array}{ll}
1 & \text{if } A = 0 \text{ and } B = 0 \\
0 & \text{if } A = 1 \text{ and } B = 0 \\
0 & \text{if } A = 0 \text{ and } B = 1 \\
0 & \text{if } A = 1 \text{ and } B = 1
\end{array}
\qquad (17.11)
$$

This is often represented as $v_o = \overline{A + B}$.

The logic gate shown in Fig. 17.3 employs diodes and transistors and consequently is called DTL (Diode Transistor Logic). The response of such a logic gate to an applied signal is not infinitely fast. The limitations of the response time will be discussed next.

Let us first discuss the time delays in the transition when the output of the NAND gate of Fig. 17.3 goes from a logical 1 to a logical 0. When the output is a logical 1, the transistor is cut off. Before the transistor can begin to carry current, one must supply enough charge to the capacities of all the pn junctions in both the diodes and the transistor so that these junctions have the proper biases for the output to be in a logical 0 state. In an integrated circuit the capacity of a diode pn junction is usually larger than the capacity of a pn junction in a transistor as discussed shortly. Thus the time required to supply charge to a diode junction is usually larger than the time to supply charge to a transistor junction. The time delay to supply charge to a single junction is short but the time delay necessary to supply charge to all the diode and transistor junctions while still short is significant. After the base to emitter junction is forward biased, there is a very short time delay as the first minority carriers diffuse across the base region of the transistor. Even after the first current carriers reach the collector, the current must build up to the saturation level. In order for the current in the transistor to build up to saturation level one must supply the charge that is stored as minority carriers in the base region of the transistor (see Chapter 12 for a discussion of charge stored in the base region). As long as the transistor is in the active region, that is, the transistor is not saturated, this time delay is relatively short. There are also stored minority carriers in the depletion layer of the diodes and time delays are required to supply this charge. Thus the total time necessary after the input changes for the output of the NAND gate to go to a logical 0 from a logical 1 (that is, for the transistor to go from cutoff to the start of saturation) is a relatively short time. Now let us consider the situation as the output of the NAND gate goes from a logical 0 to a logical 1. The transistor is saturated when the output is a logical 0. When a transistor is saturated, there is a great deal of charge stored in the base region as excess minority carriers. Base neutrality is maintained by a compensating excess of majority carriers. The excess stored charge in the base is much greater in the saturated state than when the transistor is in the active region. In order for the transistor to come out of the saturated state a relatively long time delay is needed to remove the excess charge stored in the base. Because of the stored charge diodes also go into and come out of their conducting states in different times. After the transistor leaves the saturated state, there is a short time delay before the transistor is cut off. The time delays other than the time to remove the excess stored charge from the base of the saturated transistor are comparable to the time to go from the cutoff state to the start of saturation. Thus the total time delay for the output of the NAND gate of Fig. 17.3 to go from a logical 0 to a logical 1 is longer by perhaps a factor of 2 than the time to go from a logical

1 to a logical 0, mostly because of the time delay required to remove excess base charge from the transistor in the saturated state. For simp icity in what follows, however, we shall merely speak of an average propagation delay time, and we shall ignore the difference in the delay depending on whether the output changes from 1 to 0 or from 0 to 1. The reader should keep in mind, however, that such differences do exist.

As a final note on delay times one should realize that there is a stray capacitance at the output of the circuit that must be charged as the state of the gate is changed. This charging requires a short time delay and also makes the total time delay depend somewhat on the number of circuits the output is driving since the stray capacity will depend on the number of circuits being driven. This effect is called the time delay due to *fan out*.

It is important to recognize that a major time delay in DTL logic gates is the time required to remove stored charge from the base of the saturated transistor in the circuit and that this time is greater than the time for minority carriers to diffuse across the base region of the transistor or the time to change the charge stored at a *pn* junction. For a typical DTL gate the total average propagation delay time τ is about 25 to 50×10^{-9} sec.

Integrated circuits such as are used for logic gates have many interesting properties. One of these is the fact that transistors can be made physically smaller than diodes. Consequently logic gates are often made using only transistors. As a result of the small size of the transistors the charge stored as minority carriers in the base region of these transistors when saturated is small and the time to deliver this charge can be relatively short. In addition there are no physically large diodes that cause time delays as there are in DTL circuits. Gates using only transistors in an integrated circuit are called TTL (transistor-transistor-logic). A typical response time for a TTL gate is $(5-15) \times 10^{-9}$ sec.

A typical TTL gate is shown in Fig. 17.5. The input transistor T_1 has two emitter junctions. The total collector current of this transistor is simply equal to the sum of the two emitter currents minus the base current.

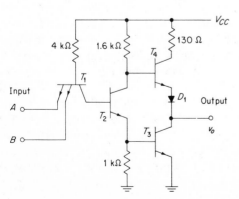

Figure 17.5 A TTL NAND gate. (*This schematic diagram courtesy Texas Instruments Incorporated.*)

This transistor behaves just like two separate transistors that have their collectors wired together and their bases wired together. As long as either input A or input B or both inputs are held at a logical 0 (that is, near ground potential), then T_1 is saturated and the collector of T_1 is low in voltage. The transistor T_2 is turned off. The base of transistor T_3 is therefore also low so that transistor T_3 is cut off. Transistor T_4 is turned on to saturation. Thus the output voltage v_o is high. If both A and B go up in potential to a logical 1, so that v_{BE} is negative then the collector of the transistor T_1 goes up in voltage, transistor T_2 turns on, and transistor T_3 saturates. The transistor T_4 turns off. This means that the output voltage drops to about 0.2 volt. Thus this logic gate performs the NAND logic function. Let us suppose that the output of this NAND gate is used to drive the input of the next NAND gate. When the output is in the logical 0 state, the transistor T_3 "sinks" the current drawn from the input of the next gate. When the output of the gate is in the logical 1 state, the transistor T_4 provides current to the input of the next gate. A given TTL gate can only drive a limited number of other gates because of the current that the output must either sink or provide. For a typical TTL gate this number is about 10 and this is called the *fan out limitation* of the gate.

One may wonder why the diode D_1 in the TTL gate does not cause this gate to be a DTL gate. The answer is that this diode is used only to provide a voltage drop so that when T_3 is saturated, T_4 is cut off and is not required to have an extremely high back bias resistance. Consequently it is made simply as a transistor with the base and the collector wired together. Thus it is small and does not limit the speed of the circuit.

As we have already mentioned, typical delay times for TTL gates are $(5–15) \times 10^{-9}$ sec. It is possible to reduce this time to about 3×10^{-9} sec by the use of Schottky diodes. When a transistor goes into saturation, the base to emitter voltage is about 0.7 volt and the collector to emitter voltage is about 0.2 volt. Thus the base to collector junction is forward biased by about 0.5 volt. A Schottky diode is a metal semiconductor diode that is placed across the base to collector, and the Schottky diode is forward biased when the base is positive with respect to the collector. This diode conducts strongly at about 0.25 volt. Thus the collector to base voltage is clamped at 0.25 volt when the transistor is saturated. Since the collector can't go as far below the base as it can when the Schottky diode is not present, the charge stored in the base is less with the Schottky diode present than when it is absent. Thus the time delay to remove the stored charge from the base region is reduced by the Schottky diode. When the transistor is off, the collector to base junction is back biased, and the Schottky diode is also back biased.

In Fig. 17.6(a) we show another TTL NAND gate. The output resistor R_L of this gate is external to the integrated circuit so that the gate is called an *open collector gate*. One very useful feature of the circuit of Fig. 17.6(a) is that several NAND gates can use the same output resistor R_L as is shown in

Fig. 17.6(b). If any one of the NAND gates is a logical 0, then the current through R_L to the output transistor of that gate is sufficient to cause the common output of all the gates to be a logical 0. This is called the wired OR function. Thus the output of the circuit shown in Fig. 17.6(b) is $\overline{A \cdot B} + \overline{C \cdot D}$. The wired OR gates have some disadvantages. When the output of this gate is used to drive the input of another, then if the output is a logical 0, the transistor T_3 of Fig. 17.6(a) can "sink" the current drawn from the input of the next gate. When the output of the gate is a logical 1, however, then the current into the input of the next gate must be drawn through the large resistor R_L. This causes some problems with fan-out limitations and causes the gate to be somewhat slower than the TTL gate of Fig. 17.5.

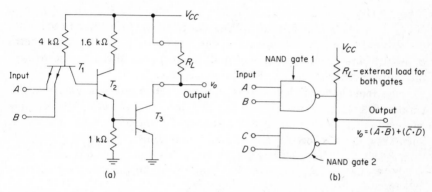

(a)   (b)

Figure 17.6 (a) A TTL NAND gate with an external connection for the resistor R_L so that the wired OR connection can be used. (b) Two TTL NAND gates with an output wired OR connection. Only one resistor R_L is used for the two gates. If either gate has an input such that its output alone would be a logical 0, then the current drawn through R_L is sufficient to pull the output of both gates to 0. The output is $\overline{A \cdot B} + \overline{C \cdot D}$. The symbol for the NAND gate is a standard one that we shall repeatedly use since the use of such a symbol saves drawing out each circuit. The symbol is used for any NAND gate. and is not restricted to representing TTL NAND gates. (*The schematic diagram of* (a) *courtesy Texas Instruments Incorporated.*)

The TTL NAND gate of Fig. 17.5 also shows another difficulty of TTL logic gates. Suppose that the output of this gate is a logical 1. Thus transistor T_4 is saturated and T_3 is cut off. Now suppose because the input changes that the circuit starts toward a logical 0. The transistor T_3 can go into saturation faster than T_4 can come out of saturation. Thus for a short time both T_3 and T_4 are in saturation and a large current can be drawn from the power supply V_{CC}. If the same supply is used for power to many logic gates, then unless the output resistance of the power supply is very low this transient current pulse may be somehow presented to the other gates as a stray pickup input voltage signal that will cause unwanted changes of logic gates.

When logic gates have delay times as short as 3×10^{-9} sec, even short wires act like transmission lines since the velocity of light is about 1 ft/nanosec (10^{-9} sec). One must be extremely careful with these very fast gates to allow for propagation times on the wires in a circuit and to eliminate problems with reflected pulses.

Another problem with very fast logic circuits of the type we are discussing is that the pulses that occur in circuits are very short. The high frequencies that are involved in these pulses mean that capacitive and inductive effects are very important, and cross talk from one circuit to another can be a very serious problem. One should always be alert for this type of problem.

For various reasons that we shall not discuss, the input to a TTL logic gate such as is shown in Fig. 17.5 sometimes tries to go negative in voltage. Often such logic gates will have a diode clamp between each input and ground. The diode is such that if the input goes more than a few tenths of a volt below ground, the diode conducts strongly and clamps the voltage, preventing the input from going further negative.

Some modern logic gates use inputs that are essentially several emitter followers with their emitters connected together. This common emitter is then used to drive the emitter of an amplifier. The parts of the circuit essential for speed are operated without going into a saturated region. As a result these logic gates are very fast. This type of gate is called ECL (emitter coupled logic). A typical response time for an ECL gate is about 1×10^{-9} sec.

A typical ECL gate is shown in Fig. 17.7. The input signals would be about -0.7 volt for a logical 1 and -1.7 for a logical 0. In the ECL circuit shown if the inputs to transistors T_1 to T_4 are logical 0's, then T_1 to T_4 are cut off and transistor T_5 is carrying a current I. The current I is determined by the base to emitter voltage of T_5. The transistor T_6 merely serves to provide a constant reference voltage of about -1.2 volts at the base of T_5. The base of T_6 is held at a constant voltage by the resistors R_1 and R_2 and by the two diodes. The two diodes provide some temperature compensation as follows. The emitter of T_5 is two pn junction drops below the base of T_6. Thus the voltage across R_E and the voltage across R_2 must be identical. The voltage across R_2 is just $(5 \cdot R_2)/(R_1 + R_2)$ independent of the temperature (ignoring the drop across the two diodes) and so the current in the transistor T_5 is constant independent of the temperature. In other words as the temperature of the circuit rises, the current in the transistor would like to increase but the voltage drop across the two diodes changes in just such a way as to minimize this tendency. The current I in the transistor T_5 is selected to be small enough that T_5 is not saturated. For example, the collector voltage of T_5 is probably about -1.0 volt when T_5 is carrying the current I. Now if a logical 1 is supplied to any one of the four inputs, that transistor turns on and conducts the current I. The transistor T_5 is cut off in this state. The resistor R_E is large enough that the current in the circuit is approximately constant. The current is simply switched from transistor T_5 to one of the input transistors by a logical 1 at one or more of the inputs. When the transistor T_5 is cut off, the voltage at the collector of T_5 goes to 0 volt. The OR output is one *pn* junction below the collector of T_5 in voltage. Consequently the logic voltages at the OR output are -0.7 volt for a logical 1 and -1.7 volts for a logical 0.

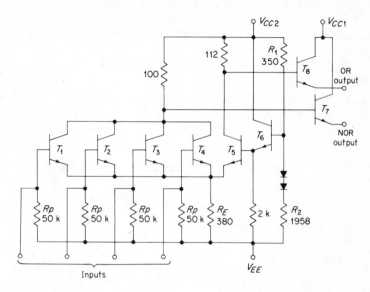

Figure 17.7 An ECL circuit. This ECL circuit has both an OR and a NOR output. For this circuit typical supply voltages are $V_{EE} = -5$ volts and $V_{CC1} = V_{CC2} = 0$ volt. (*This schematic diagram courtesy Motorola Semiconductor Products Incorporated.*)

Obviously the outputs must be loaded in order for the circuit to function. The base resistors R_p are selected to be such that when the inputs are disconnected, then the transistors T_1 to T_4 are cut off. In addition to being extremely fast (delay times of about 10^{-9} sec), which arises from the non-saturated operation, the ECL gates have several other advantages. One of these advantages is that the ECL gates generate both OR and NOR signals simultaneously. With the DTL and TTL gates previously described one must use an additional gate to generate the inverse (or complementary) signal. This would mean that the OR and NOR signals from DTL circuitry would not be simultaneous in time but instead one signal would lag behind the other by the delay time of one gate.

There is one more type of logic gate that should be mentioned, the MOSFET gate. The MOSFET gates have the advantage that it is possible, because of technical advantages in the manufacturing process, to produce really large-scale integration of circuits containing MOSFET gates. This means that it is possible to produce integrated circuits on a single chip of silicon containing hundreds or even thousands of logic gates. Thus it is possible to produce whole sections of a computer as a single integrated circuit. It may also be possible in the near future to produce large-scale integration of logic circuits using bipolar transistors.

We stress that the speed of logic gates is very important since this determines how fast a computer can make calculations. For example, if the delay in a given gate is 15×10^{-9} sec and a particular calculation involves a sequence of 10^6 logical operations, then the minimum time possible for such a calculation is 1.5×10^{-2} sec.

Before we complete our discussion of logic gates we should point out that if one uses different types of gates, for example, DTL and TTL gates, in the same circuit, one must be sure to translate the DTL logic voltage levels to those appropriate to the TTL gates and vice versa before one supplies, for example, the TTL gate with a logical 1 from the DTL gate.

17.3   Uses of Logic Gates for Arithmetic Operations

We shall next consider how to utilize logic gates. Although the particular type of logic gate used will determine how fast the resulting circuits work and what voltages represent a logical 0 or 1, the various methods of combining digital logic circuits to perform various functions are not dependent on the particular type of logic gate, that is, whether the logic gate is a DTL or TTL gate. Thus it is convenient to introduce symbols for the logic gates. The symbols used are shown in Fig. 17.8.

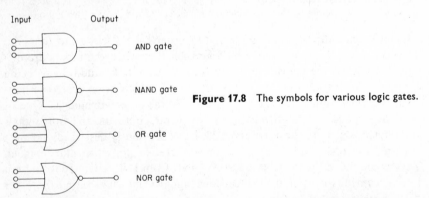

Figure 17.8 The symbols for various logic gates.

The logic gates described can be viewed as a method of performing various algebraic manipulations. Thus a NAND gate takes input signals A and B and forms $\overline{A \cdot B}$. Since we have logic gates that will execute various algebraic manipulations with variables that can take on only two possible values, 0 or 1, we need to know the properties of an algebra (called Boolean algebra) where the functions (and variables) can take on only two possible values. The table below illustrates how one studies the properties of an algebra of this type.

A	$\bar{A}$	B	$\bar{B}$	$A + \bar{A}$	$A \cdot \bar{A}$	$A + B$	$A \cdot B$	$\bar{A} + \bar{B}$	
0	1	0	1	1	0	0	0	1	
0	1	1	0	1	0	1	0	1	
1	0	0	1	1	0	1	0	1	(17.12)
1	0	1	0	1	0	1	1	0	

From this table it is clear that

$$A + \bar{A} = 1$$
$$A \cdot \bar{A} = 0 \qquad (17.13)$$

In addition to these two properties another very interesting but less obvious result can be seen. This is the result that $A \cdot B$ is the inverse of $\bar{A} + \bar{B}$. Therefore we see that

$$A \cdot B = \overline{\bar{A} + \bar{B}} \qquad (17.14)$$

This result is called *De Morgan's theorem* and is particularly useful in understanding logic circuits. The logical symbol 1 is sometimes called *true* and the logical 0 is sometimes called *false*. A table such as

A	B	$A + B$	
0	0	0	
0	1	1	
1	0	1	(17.15)
1	1	1	

is called a *truth table*. Any theorem in Boolean algebra can be proved by simply writing out the corresponding truth table. Some properties of Boolean algebra are the following

Commutative Properties

$$A + B = B + A$$
$$A \cdot B = B \cdot A$$

Associative Properties
$$(A + B) + C = A + (B + C)$$
$$(A \cdot B) \cdot C = A \cdot (B \cdot C)$$

Distributive Properties
$$A \cdot (B + C) = (A \cdot B) + (A \cdot C) \qquad (17.16)$$

Special Properties

$$(A + B) \cdot (A + C) = A + (B \cdot C)$$
$$A + B = \overline{A} \cdot \overline{B} \qquad \text{(De Morgan's theorem)}$$
$$A \cdot A = A$$
$$A + A = A$$
$$A + \overline{A} = 1$$
$$A \cdot \overline{A} = 0$$
$$\overline{\overline{A}} = A$$

In the properties listed for Boolean algebra it should be remembered that + means OR, · means AND, and an overbar or ("‾") indicates inversion.

Previously we said that the logic we shall use will be constructed using primarily NAND gates. As we shall show, using only NAND gates, one can construct all the other logical functions. We could equally well have elected to use a logic based on NOR gates only. Our use of NAND gates is arbitrary but it reflects the varieties of available integrated circuit gates.

When using a multiple input NAND gate, if an input is not used, it must be connected to a logical 1. This is plainly necessary because if it were connected to a logical 0, then the output would always be 1.

We shall now show how AND, OR, and NOR functions can be constructed using only NAND gates. An AND gate constructed from NAND gates is shown in Fig. 17.9. The AND gate is also sometimes called a coincidence circuit since the output is 1 only if both inputs are 1 simultaneously. The first NAND gate obviously forms the NAND function. The second NAND gate simply inverts the output of the first. The second NAND gate has one or more unused inputs connected to a logical 1 or to the used input. The unused inputs are not shown. An OR gate constructed from NAND gates is shown in Fig. 17.10. A NOR gate constructed with NAND gates is the same as the OR gate above except that an additional NAND gate is used to invert the output forming $\overline{A + B}$.

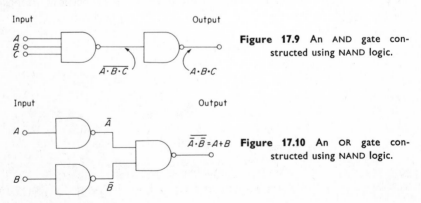

Figure 17.9 An AND gate constructed using NAND logic.

Figure 17.10 An OR gate constructed using NAND logic.

We shall now continue to investigate various logical functions that can be obtained by forming various combinations of logic gates. A very useful function is the EXCLUSIVE OR gate. This is given by $\bar{A} \cdot B + A \cdot \bar{B}$. The symbol for this gate, shown in Fig. 17.11, is often used. A truth table for the EXCLUSIVE OR gate is given below:

$$
\begin{array}{ccccc}
A & B & \bar{A} & \bar{B} & (\bar{A} \cdot B) + (A \cdot \bar{B}) \\
0 & 0 & 1 & 1 & 0 \\
0 & 1 & 1 & 0 & 1 \\
1 & 0 & 0 & 1 & 1 \\
1 & 1 & 0 & 0 & 0 \\
\end{array}
\tag{17.17}
$$

An EXCLUSIVE OR gate can be formed as is shown in Fig. 17.12.

Figure 17.11 The symbol for an EXCLUSIVE OR gate.

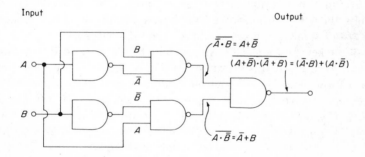

Figure 17.12 An EXCLUSIVE OR gate formed using NAND logic.

Let us now look at the process of adding two binary numbers with digital circuitry. If we have only two bits to add, then the possible combinations are the following:

$$
\begin{array}{lcccc}
A & 0 & 0 & 1 & 1 \\
B & 0 & 1 & 0 & 1 \\
\hline
\text{Sum} & 0 & 1 & 1 & 10 \\
\end{array}
\tag{17.18}
$$

Now the right-hand bit of the sum has the values that an EXCLUSIVE OR gate will provide and the left-hand bit has the values that an AND gate will provide.

Therefore a circuit for adding two bits can be constructed as shown in Fig. 17.13. This circuit is called a *half adder*.

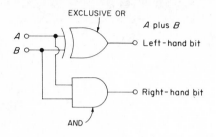

Figure 17.13 A circuit for producing the sum of two single-digit binary numbers. This circuit is called a *half adder*.

Now let us consider the addition of two numbers each containing several bits. As an example, consider the following sum:

$$
\begin{array}{lr}
A & 101\ 1011 \\
B & \underline{110\ 1011} \\
\text{Carry} & \underline{1111\ 011} \\
\text{Sum} & 1100\ 0110
\end{array}
$$

(17.19)

It is clear that in addition of two binary numbers one generally is adding three bits, the Nth bit from A, the Nth bit from B, and the carry bit generated in the $N - 1$ sum. It is necessary to generate the Nth sum bit and a carry bit that is taken to the $(N + 1)$ step. A circuit that will do this is called a *full adder*. Such a circuit is shown in Fig. 17.14. There are integrated circuits that will form sums for two four-bit binary numbers or even two eight-bit binary numbers. These can be combined to calculate sums of still larger numbers. The symbol for a four-bit adder is shown in Fig. 17.15. The adder shown will add two four-bit numbers plus a carry in bit to give a four-bit sum plus a carry out bit. The power supply leads are omitted as they are on all the symbols for logic gates.

Forming a product of two binary bits is very easy. The possible combinations of two-bit binary products are

$$
\begin{array}{lcccc}
A & 0 & 1 & 0 & 1 \\
B & 0 & 0 & 1 & 1 \\
\text{Product} & \overline{0} & \overline{0} & \overline{0} & \overline{1}
\end{array}
$$

(17.20)

Consequently an AND gate can form a product between two binary bits.

Using various combinations of AND gates and full adders it is possible to form any product between two arbitrary binary members.

It is also possible to construct combinations for subtracting although as we have mentioned earlier subtraction is usually accomplished by adding a negative number expressed as a 1's or 2's complement. Digital circuits that perform division are also possible.

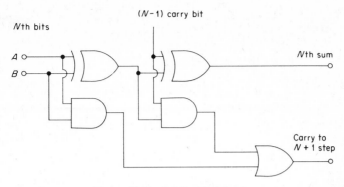

Figure 17.14   A full adder circuit.

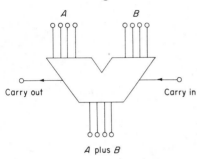

Figure 17.15   The symbol for a four-bit adder.

One often wishes to test two numbers for equality. One can test two bits for equality as follows. Let A be one bit and B be the other bit. If A equals B, then $(A \cdot B) + (\bar{A} \cdot \bar{B}) = 1$ but if A is not equal to B, then $(A \cdot B) + (\bar{A} \cdot \bar{B}) = 0$. A circuit for performing an equality test between two bits is shown in Fig. 17.16. In order to compare two numbers of several digits each it is necessary to compare each of the corresponding pairs of digits for equality and then use the output of all the equality comparisons as the input of an AND gate to assure that every pair of bits was identical. This operation is represented schematically in Fig. 17.17.

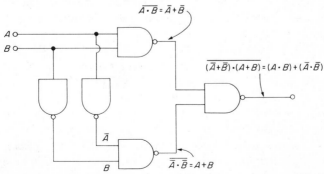

Figure 17.16   The circuit for comparison of the equality of two bits A and B. This circuit is called an EXCLUSIVE NOR circuit.

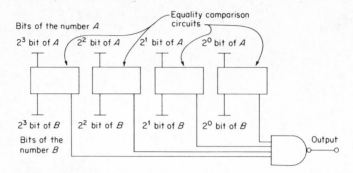

Figure 17.17 A circuit for testing the equality of all the bits in two different numbers A and B.

Inside a modern high speed digital computer most calculations are performed using binary numbers. At the input and output, however, it is sometimes desirable to convert the binary numbers to decimal numbers. This operation is straightforward but requires an enormous number of logic circuits if the numbers are large. To illustrate the method we shall decode only the four two-digit binary numbers into their decimal equivalents. The four two-digit binary numbers and their decimal equivalents are

Binary Number		Decimal Equivalent	
Second bit	First bit		
0	0	0	(17.21)
0	1	1	
1	0	2	
1	1	3	

A circuit for converting from these binary numbers to decimal numbers is shown in Fig. 17.18. Binary to decimal conversion is straightforward but tedious because many NAND gates are required. For numbers greater than 10 it is necessary to include "divide by 10" circuits and other logic.

The decoding of a binary number into a decimal number can also be done by a diode matrix. A diode matrix for decoding a two-bit binary number into a decimal number is shown in Fig. 17.19. By simply following the various connections through it is seen that the output is simply the decimal equivalent of the binary input. The inputs might be about 0.2 volt for a logical 0 and 3.5 volts for a logical 1.

The decoding of a binary into a decimal number is only one example of decoding. Recently decoders have become very popular. These devices also go under the names of *read only memory* (ROM) and *translators*. Another example of a decoder or a read only memory might be a table of sin x. The input might be the number x in binary form and the output would be sin x.

Binary input

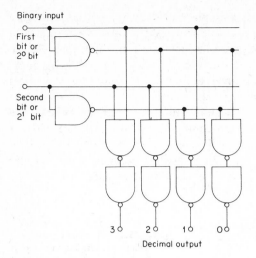

Figure 17.18 A circuit for converting a two-bit binary number into its decimal equivalent. This circuit is called a *binary to decimal decoder*. Note that connections of two wires occur only where there is a black dot and not when two wires simply cross.

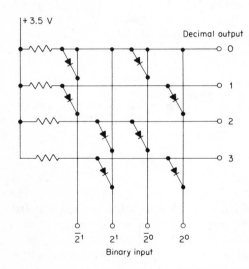

Figure 17.19 A diode matrix for decoding a two-bit binary number into its decimal equivalent. Two wires are connected only where there is a black dot and not where two wires simply cross.

In this read only memory the number x might be the binary equivalent of the 91 integers corresponding to 1° steps from 0° to 90°. The output would be simply the value of sin x for each of the 91 values of x. In the example of the binary to decimal decoder the output was the decimal equivalent of a binary number at the input, whereas in the read only memory for sin x the output is sin x where x is presented at the input. Read only memories are now made as integrated circuits with as many as 8192 bits in the table. A device of this type is called an 8k ROM. The tables of quantities given in read only memories include sin x, cos x, tan x, x^2, x^3, $x^{1/2}$, $x^{1/3}$ and many others.

A decoder can be used in a simple selector switch for data. In order to understand this we first must examine a simple circuit that can transfer a data bit from the input to the output of a circuit on command. Figure 17.20 shows such a circuit. When the command line is at a logical 0, the output is at a logical 0. When the command line is at a logical 1, the output is identical to the input data bit so that the input data bit is transferred to the output upon command. A selector switch is shown in Fig. 17.21. The input to the decoder determines the nature of the command signal and hence whether data bit A or data bit B is transferred. The OR gate simply passes either data bit that is presented to it. Thus depending on what the input to the decoder is the output of, the OR gate can be either data bit A or data bit B, and the system acts as a selector switch.

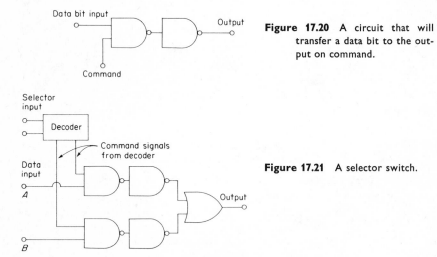

Figure 17.20 A circuit that will transfer a data bit to the output on command.

Figure 17.21 A selector switch.

Thus far we have discussed the use of logic gates to form circuits that carry out arithmetic operations such as addition or multiplication and that can decode from binary numbers to decimal numbers. These circuits provide the basis for digital computations in a computer. There is also a need in computers for storing digital information both before and after computations are performed. There are a variety of ways to store digital information. Circuits for storing digital information will be discussed in the next section of this chapter.

17.4 Flip-Flops

For long-term storage outside a computer, digital information is often recorded on magnetic tape, paper tape, removable magnetic discs, or paper cards. Inside a computer, digital information is often stored using an individual magnetic core for each bit or sometimes with flip-flops. In either type of magnetic memory (tape or core) the direction of magnetization of a ferromagnetic material provides the two possible states (the logical 1 or the logical 0). Digital information is recorded by magnetizing the material and the information is subsequently read out by sampling the magnetization.

For short-term storage and sometimes for the internal long-term memory of digital information in a computer a flip-flop circuit is commonly used. The simplest flip-flop circuit that is used is the set-reset flip-flop (*RS* flip-flop). The circuit for this flip-flop is shown in Fig. 17.22. Normally both R and S are maintained at a logical 1. Suppose that $\bar{Q}$ is a logical 0 and Q is a logical 1. Now suppose a pulse to a logical 0 (that is, a change from the state 1 to 0) comes into R. The result is that $\bar{Q}$ changes to 1 and this change drives Q to 0. A second logical 0 pulse at R will have no effect on the output state of the flip-flop. This change will persist until a pulse comes into S and drives S to a logical 0, which causes the outputs to flip again. Thus an *RS* flip-flop is capable of recording information and then remembering it until an external signal alters the state of the flip-flop. An *RS* flip-flop is an extremely useful circuit because of its simplicity and speed. It has one major drawback, however. If a logical 0 pulse occurs simultaneously at R and S, then both Q and $\bar{Q}$ become 1. This state is ambiguous and the eventual final state is determined by which input remains a logical 0 the longer or by component asymmetries. Therefore when using an *RS* flip-flop, one must assure that logical 0 pulses do not occur at both R and S simultaneously.

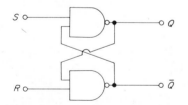

Figure 17.22 An *RS* flip-flop.

It is often useful to synchronize a flip-flop so that the information at the inputs is shifted to the output of the flip-flop when a clock pulse is applied. A clocked *RS* flip-flop is shown in Fig. 17.23. In the quiescent condition, R, S, and the clock input Ck are held at a logical 0. Let us suppose initially that $Q = 1$ and $\bar{Q} = 0$. Now suppose a logical 1 signal is applied to R. As long as Ck remains at 0, no change in the output occurs. When Ck changes to 1, however, then $\bar{Q}$ flips to 1 and Q to 0. Thus the transfer of the input signal to the output of the *RS* flip-flop is allowed only while the Ck is 1.

The next type of flip-flop we shall discuss is the D-type (for data) flip-flop. This flip-flop has built-in protection against both Q and $\bar{Q}$ assuming the same state as can happen with an *RS* flip-flop. A D flip-flop is shown in Fig. 17.24. The flip-flop has only a single data input D. A NAND gate is used to invert the input signal to the R input so that it is impossible to have the same signal applied at both the S and R inputs. Suppose initially $Q = 0$ and $\bar{Q} = 1$, and suppose the clock input Ck is 0 in the quiescent state. Now if Ck is pulsed to 1, then the output remains at $Q = 0$, $\bar{Q} = 1$ if $D = 0$ and the output becomes $Q = 1$, $\bar{Q} = 0$ if $D = 1$. The S reset and R reset leads are used to determine the initial state of Q and $\bar{Q}$. Normally a D flip-flop is used with an external lockout circuit that prevents D from changing for the duration of the clock pulse. The information at D is transferred to the output on the leading edge of the positive going clock pulse.

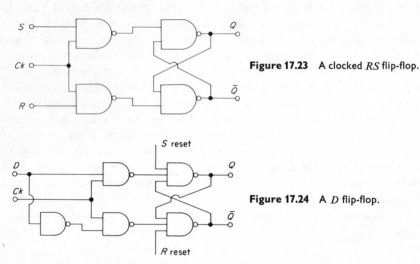

Figure 17.23   A clocked RS flip-flop.

Figure 17.24   A D flip-flop.

The final type of flip-flop we shall discuss is a rather complicated master-slave flip-flop known as a JK flip-flop. A schematic diagram of a master-slave JK flip-flop is shown in Fig. 17.25. Since an analysis of the JK flip-flop is straightforward but tedious, we shall simply state the properties leaving the analysis for a problem. The information at the J and K inputs is transferred to the master flip-flop on the initial positive going edge of the clock pulse and the information in the master flip-flop is transferred to the slave flip-flop on the trailing (negative going) edge of the clock pulse. A truth table for a JK flip-flop is given.

J	K	Q_f	$\bar{Q}_f$
1	1	$\bar{Q}_i$	Q_i
1	0	1	0
0	1	0	1
0	0	Q_i	$\bar{Q}_i$

(17.22)

In the table Q_f is the final value of Q after the clock pulse and Q_i is the initial value of Q before the clock pulse. The symbol Cl in Fig. 17.25 stands for *clear*. Normally the Cl is at 1. The Cl is used to set Q at 0 when desired. A symbol used for the JK flip-flop is shown in Fig. 17.26.

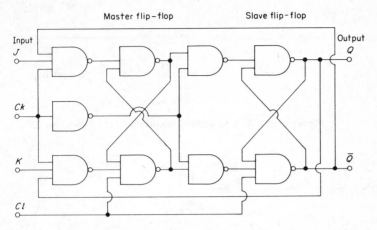

Master flip–flop   Slave flip–flop

Figure 17.25   A "master-slave" JK flip-flop.

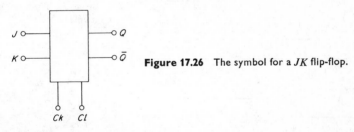

Figure 17.26   The symbol for a JK flip-flop.

All three types of flip-flops discussed are commercially available as integrated circuits on a single silicon chip so that it is not necessary to connect the various NAND gates to form these circuits.

We shall illustrate the use of flip-flops with two very useful circuits. The first of these circuits is a counter or scaler. A counter is a circuit that will count and record individual events. Thus after 200 pulses enter a counter, the output of the counter should read 200. A circuit using JK flip-flops that will count pulses is shown in Fig. 17.27. For the counter shown the clock is used as the input. Both J and K are set to 1 for each flip-flop. On the trailing edge of the first pulse the first flip-flop switches to $Q = 1$ and $\bar{Q} = 0$. The output Q of the first flip-flop serves as the clock pulse of the next flip-flop. On the trailing edge of the second pulse the first flip-flop switches back to $Q = 0$ and this causes the second flip-flop to switch to $Q = 1$. Consequently the outputs of the various flip-flops indicate the number of input pulses as a binary number. The output of the first three flip-flops is indicated in Fig. 17.28.

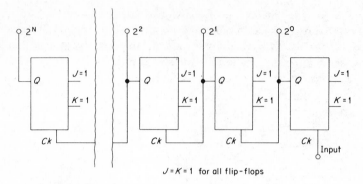

$J = K = 1$ for all flip-flops

Figure 17.27   An up counter constructed using JK flip-flops.

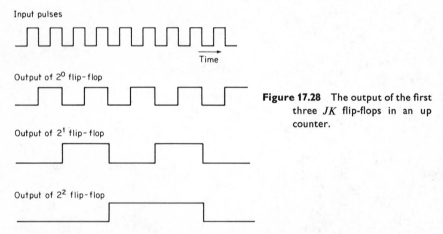

Figure 17.28   The output of the first three JK flip-flops in an up counter.

The counter of Fig. 17.27 is called an *up counter* because each input pulse increases by 1 the number that the counter reads. There is another counter called a *down counter* that decreases the number that the counter reads by 1 for each input pulse. A down counter can be made by using the circuit of Fig. 17.27 except that the $\bar{Q}$ output instead of the Q output is fed to the clock input of the next JK flip-flop (see Prob. 17.12). There are counters that can count either up or down upon command.

The counter shown has one defect. When a pulse is put into the 2^0 counter input there is a time delay τ caused by each JK flip-flop so that the number stored in the counter is not correctly recorded at the output of all the various JK flip-flops until after a total elapsed time $N\tau$ where N is the number of flip-flops. This time is called the *ripple-through time*. There are counters that can take an input pulse and then at a slightly later time all the output voltages change to their new value at the same time. This type of counter is called a *synchronous counter*.

The next circuit using JK flip-flops that we shall discuss is the shift register. A shift register is a series of flip-flops that store information. An input pulse to the shift register will move to the right by one bit the information stored in each flip-flop. A shift register is shown in Fig. 17.29. For the shift register shown the outputs Q and $\bar{Q}$ of one JK flip-flop serve as the inputs J and K for the next flip-flop. The shift signal is supplied simultaneously to the clock inputs of all flip-flops. By remembering that if $J = 1$, $K = 0$, then the output changes to $Q = 1$, $\bar{Q} = 0$ and if $J = 0$, $K = 1$, then the output changes to $Q = 0$, $\bar{Q} = 1$; it is obvious that the trailing edge of every clock pulse shifts the contents of the register one bit to the right. The input to the first flip-flop can be varied by setting J and K for the first flip-flop as desired. Shift registers are useful, for example, in multiplication for shifting the individual products before adding. Shift registers have many other uses. There are bidirectional shift registers that can shift either to the right or to the left, and there are synchronous shift registers.

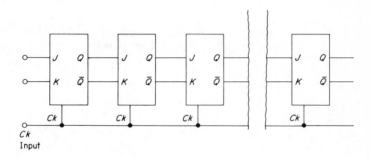

Figure 17.29 A shift register made using JK flip-flops.

There are yet a couple of very simple circuits that are extremely useful that we shall describe. The first of these is a circuit for driving an indicator light from a NAND gate. For the circuit of Fig. 17.30 the lamp is on if the output of gate A is a logical 1 and off if the output of gate A is a logical 0. The second circuit is one for removing the bounce of a contact of a switch. When a mechanical switch is closed, the contact does not settle down to a steady contact instantly because the switch contact bounces. For a fast digital circuit this is very aggravating because each bounce can look like a separate digital pulse. The circuit of Fig. 17.31 is useful for removing this problem. When the switch first goes from position B to position A, the RS flip-flop switches its state, and the output of the flip-flop is altered. Subsequent bounces do not affect the state of the output as long as the switch doesn't bounce so far as to contact position B again. When the switch is thrown back from A to B, the RS flip-flop again changes its state.

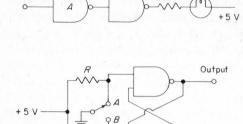

Figure 17.30 A circuit for indicating the output of a NAND gate by a lamp. If the lamp is lighted, then the output of NAND gate A is a logical 1 and if the lamp is off, then the output of NAND gate A is a logical 0.

Figure 17.31 A circuit for eliminating the bounce of a mechanical toggle switch in producing a voltage pulse.

We have now discussed many of the basic circuits used in a computer or in other digital circuits. The organization and uses of a digital computer will be discussed in Chapter 18.

17.5   The Conversion of a Digital Signal into an Analog Signal and Vice Versa

An analog system uses a continuous quantity such as the voltage to represent a given number. For example, a voltage might be used to represent displacement. Thus if 1 volt is used to represent a displacement of 1 meter, then a voltage of 2 volts would represent a displacement of 2 meters and so on. We have already studied how an analog computer can be used to study in a very simple manner the time evolution of a quantity that can be represented as a voltage signal. Because all electrical systems have noise voltages present as well as the desired signal voltage, there is a limit as to how accurately one can determine the value of an analog voltage.

A digital system uses fixed voltage levels to determine a logical 0 or a logical 1. A number is then represented digitally in the base 2. Small noise signals added to the voltages for a logical 1 or a logical 0 do not change the state of the system (for example, from a logical 1 to a logical 0). Therefore digital circuits can be read as accurately as the number of digits permits. Unfortunately a high speed electronic digital computer is a more complicated device than an analog computer (see Chapter 16 for a very brief discussion of the principles involved in an analog computer).

Sometimes it is useful to do some calculations using the precision of a digital computer and then after the digital calculations are complete to convert the results into an analog voltage signal. A digital to analog (D to A) converter is used for this purpose. The inverse operation analog to digital conversion is also a very useful function. We shall study circuits that perform these operations.

A circuit that will perform digital to analog conversion is shown in Fig. 17.32. The digital to analog converter shown works as follows. The summing point S of the operational amplifier is held at ground potential approximately. Therefore on the other side of the two silicon pn junction diodes in the minus input circuit of the operational amplifier the potential is $+1.2$ volts (0.6 volt per diode) when the diodes are forward biased and current is flowing. Now suppose that for the digital system the logical 0 is 0.2 volt and the logical 1 is $+3.5$ volts. Consider the 2^0 input. Suppose a logical 0 is applied to the 2^0 input. The current through the 40 kilohm-resistor will be passed to the 2^0 input terminal. On the other hand if a logical 1 is applied to the 2^0 input, then the current through the 40-kilohm resistor will pass into the summing point of the operational amplifier. Obviously similar considerations will apply to the other inputs. The current through the 40-kilohm resistor in the 2^0 circuit will be $(10 - 1.8)$ volts/(4×10^4) ohms $= (2.1 \times 10^{-4})$ amp. The current in the 2^0 circuit is half as large as the (4.2×10^{-4})-amp current through the 20-kilohm resistor in the 2^1 circuit.

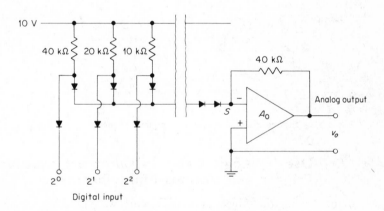

Figure 17.32 A digital to analog converter.

The current in the 10-kilohm resistor in the 2^2 circuit is (8.4×10^{-4}) amp and so on. The output voltage v_o from the operational amplifier is $(-4 \times 10^4) \times (\Sigma i)$ where the term Σi is the sum of the currents that are passed into the summing point of the operational amplifier. Therefore the output voltage v_o from the operational amplifier is proportional to the digital number that is present at the digital inputs. There are other digital to analog conversion circuits, but the circuit shown will do a satisfactory job if extremely high precision is not required.

Once one has a digital to analog converter, it is possible to form an analog to digital converter. A diagram for a very simple analog to digital converter is shown in Fig. 17.33. In the diagram shown the clock puts out regular pulses; the gate passes these pulses to the counter, which records digitally the number of pulses which have entered the counter. The digital to analog (D to A) converter then converts the digital number from the counter into an analog voltage which is compared to the analog voltage input by an operational amplifier which acts as a comparator. The counter is started from zero. When the counter reaches a digital number whose analog equivalent is infinitesimally larger than the analog input voltage, the comparator changes its output. This causes the gate to stop passing the pulses from the clock to the counter. The digital number equivalent to the analog input can then be read off the counter. This particular analog to digital converter is slow because one must start counting from zero each time. There are analog to digital converters that can sidestep this difficulty. They are somewhat more complicated than the analog to digital converter we have discussed, however, and they will not be discussed in this text.

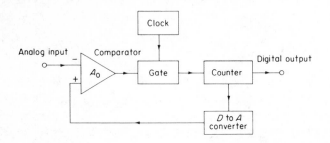

Figure 17.33 An analog to digital converter.

17.6   Uses of Logic Gates to Construct a Schmitt Trigger and Various Multivibrators

Before leaving the subject of digital circuits we should like to discuss how one can use logic circuits to form some common laboratory circuits. It is often useful to have a circuit that is able to produce a logical 1 when the dc input voltage to the circuit exceeds some previously selected voltage. A circuit that performs this function is called a *Schmitt discriminator* or *Schmitt trigger*. We previously (in Chapter 16) discussed the Schmitt discriminator. We shall now show how NAND gates can be used to construct this circuit. A circuit for doing this is shown in Fig. 17.34. For the circuit shown if $v_i = 0$ volt, then the input to the first NAND gate is a logical 0 and the output of the second NAND is a logical 0. The input circuit to the first NAND gate can be redrawn as shown in Fig. 17.35. When $v' = [R_{eq}/(R_1 + R_{eq})]v_i$ exceeds the level where the first NAND gate will switch (~ 2.0 volts, say), then both NAND gates will change their state. The input circuit to the first NAND gate will now appear as shown in Fig. 17.36.

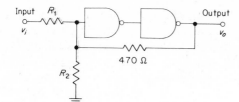

Figure 17.34   A Schmitt trigger constructed using NAND gates.

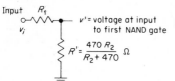

Figure 17.35   The equivalent circuit for the input of the Schmitt trigger of Fig. 17.34 when the output of the trigger is a logical 0 (0 volt).

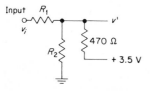

Figure 17.36   The equivalent circuit for the input of the Schmitt trigger of Fig. 17.34 when the output of the trigger is a logical 1 (3.5 volts).

For this diagram we have assumed that the logical 1 of the second gate is $+3.5$ volts. Now if v_i is lowered, then when v' passes the trigger level, the NAND gates will again switch. A plot of the output voltage v_o versus the input voltage v_i for a Schmitt discriminator is shown in Fig. 17.37. Schmitt discriminators are often used to produce pulses that are triggered at a particular phase of a sine wave or in a pulse height analysis to give a pulse if another pulse exceeds a particular voltage in magnitude. We shall also show how to construct monostable and astable multivibrators using NAND gates.

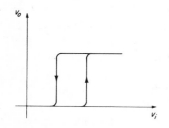

Figure 17.37   The output voltage v_o versus the input voltage v_i for a Schmitt trigger. Note that v_o versus v_i depends on whether v_i is increasing or decreasing; that is, the circuit exhibits hysteresis.

Figure 17.38 shows a monostable multivibrator. In the diagram shown the resistor R is assumed to be small enough that the input to NAND gate B is normally held at a logical 0. This means that the output of NAND gate B is a logical 1 and consequently the upper input to NAND gate A is 1. The input voltage v_i is normally held at a logical 1. Since both inputs to NAND gate A are logical 1's, the output of NAND gate A is a logical 0. Now suppose a short negative signal (a logical 0) is applied at the input. When this happens, both

NAND gates change their state. Now even when the short negative input signal is over, one of the inputs (the upper input) is still held at a logical 0 by the output of NAND gate B. Thus the output of NAND gate A remains at a logical 1. The input of NAND gate B will return to 0 with a time constant $\tau = RC$. When the input to NAND gate B reaches a low enough voltage to be reliably recognized as a logical 0, both NAND gates will change their state again. The input and the output of the circuit are shown in Fig. 17.39.

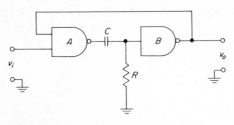

Figure 17.38 A monostable multivibrator constructed using NAND gates. The resistor R must be less than about 500 ohms if a typical TTL NAND gate is used because when the input is a logical 0 a TTL NAND gate draws about -1.6 milliamp.

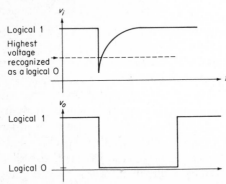

Figure 17.39 The input and output voltages as a function of the time for the monostable multivibrator of Fig. 17.38.

A monostable multivibrator is often used to produce a pulse of standard shape and length from an input pulse. The positive feedback network in the monostable multivibrator is obviously the wire connecting the output of NAND gate B and the input of NAND gate A and the RC network connecting the output of NAND gate A and the input of NAND gate B. The NAND gates contain the amplification necessary in an oscillator. In fact in this circuit the NAND gate simply acts as an inverting amplifier.

In Chapter 16 we discussed circuits using transistors that are connected to form multivibrators. The advantage to using NAND gates is that the NAND gates come as integrated circuits so that the wiring necessary to produce a finished circuit is very simple. In addition, by using the integrated NAND gates it is easy to know how many other gates the multivibrator will drive since the manufacturer of the NAND gates gives this information. On the other hand for a circuit using transistors one must make tedious calculations to determine whether the multivibrator is adequate to drive a desired load.

An astable or free running multivibrator is shown in Fig. 17.40. Let us assume initially that point A is a logical 0, point B is a logical 1, point C is a logical 0, point D is a logical 1, point E is a logical 1, and point F is a logical 0. Because points F and A are at the same potential, no current flows through R_1. In the same way no current flows through R_3. However, current flows out of the output of the first NAND gate at point B and charges the capacitor C_2 with a time constant $R_2 C_2$. When C_2 is charged so that it is recognized as a logical 1, the point D goes to a logical 0 and the capacitor C_3 discharges with a time constant $R_3 C_3$. When C_3 discharges so that it is recognized as a logical 0, point F goes to a logical 1 and the capacitor C_1 is charged with a time constant $R_1 C_1$. This process continues indefinitely. The voltage at point F, which is the output of the circuit, goes from a logical 0 to a logical 1 and back again indefinitely. The period of the entire multivibrator is approximately $2 \times (R_1 C_1 + R_2 C_2 + R_3 C_3)$. The circuit uses the RC time constants to slow up a voltage wave that propagates around the loop. Provided the time constants $R_1 C_1$, $R_2 C_2$, and $R_3 C_3$ are not too different, then it can be shown (see Prob. 17.18) that whatever the initial states of the three NAND gates are in the circuit, the circuit will start running as described above after a short time (approximately the shortest of the times $R_1 C_1$, $R_2 C_2$, or $R_3 C_3$). Thus the oscillator is always self-starting.

As described in the previous chapter an astable multivibrator can be synchronized to run at a submultiple of the frequency of another train of pulses. An input for the synchronizing pulses is also shown in Fig. 17.40.

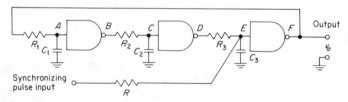

Figure 17.40 A free running multivibrator (oscillator) constructed using NAND gates.

Problems

17.1   What are the binary numbers that are equivalent to the decimal numbers 5, 18, 325, and 1753? What are the octal and hexadecimal numbers equivalent to the decimal numbers above?

17.2   What are the decimal equivalents of the binary numbers 11 0110 1001 and 1 0001 0110? What are the decimal equivalents of the hexadecimal numbers AC35 and 42B? What are the binary equivalents of the two hexadecimal numbers? What are the decimal and binary equivalents of the octal numbers 743 and 542?

17.3   Use NAND logic to form the following:

    (a) $(A \cdot B) + (C \cdot D)$

    (b) $(A + B) \cdot (C + D)$

    (c) $(A \cdot B) + \overline{(C \cdot D)}$

    (d) $(A + B) \cdot \overline{(C + D)}$

17.4   Use NOR logic to form the following:

    (a) $A + B$

    (b) $A \cdot B$

    (c) $\overline{A \cdot B}$

    (d)   EXCLUSIVE OR

17.5   Use NAND logic to form the following functions:

    (a) $\overline{(\overline{A} + B)} \cdot C$

    (b) $\overline{A + B + \overline{C}}$

    (c) $(\overline{A} + B) + C$

    (d) $A + (B \cdot \overline{C})$

17.6   The following is the input and output table for a read only memory (a decoder):

Input	Output
0000	0001
0001	0010
0011	0011
0010	0100
0110	0101
1110	0110
1010	0111
1011	1000
1001	1001
1000	1010

You are to design a diode matrix that will act as the read only memory for this table.

17.7   Design a read only memory (a decoder) using NAND gates that will take the input of the table in Prob. 17.6, and provide the output of that table.

17.8   Seven NAND gates are connected as follows. The output of the first NAND gate acts as the input for the second gate. The output of the second gate acts as the input of the third gate and so on until the output of the seventh gate is brought back as the input for the first gate. The NAND gates form a ring and the circuit is called a *ring oscillator*. If each NAND gate has a propagation delay $\tau = 10^{-8}$ sec, show that the circuit oscillates and that the frequency of oscillation is about 7 MHz Measuring the frequency of a ring oscillator such as the one described is a common way to determine the average propagation delay of a NAND gate. Show that the ring oscillator must have an odd number of gates.

17.9 Write out the binary equivalent for the subtraction of the decimal number 75 from the decimal number 113. Carry out this operation using a 1's complement.

17.10 Given a shift register, a four-bit adder, and various NAND gates, sketch a circuit that will permit one to multiply a pair of two-bit numbers together.

17.11 Actually analyze the K flip-flop shown in Fig. 17.25, and verify the following table:

J	K	Q_f	$\bar{Q}_f$
1	1	$\bar{Q}_i$	Q_i
1	0	1	0
0	1	0	1
0	0	Q_i	$\bar{Q}_i$

where Q_f is the final state of Q after a clock pulse and Q_i is the initial state of Q before the clock pulse.

17.12 Alter the up counter of Fig. 17.27 to produce a down counter. This can be done by using the $\bar{Q}$ output of each flip-flop as the clock for the next flip-flop. The readout of the counter is still from the Q output. Show the voltage levels at the Q outputs as a function of the time for the 2^0, 2^1, 2^2, and 2^3 flip-flops as pulses come into the input. Show that if the down counter output goes below zero that the negative number is represented as a 2's complement.

17.13 Using the two-NAND gate circuit of Fig. 17.20 for the transfer of a data bit on command, sketch the circuit diagram of an up counter, the output of which is shifted to a register synchronously on command. The register is simply a set of JK flip-flops, one for each bit; that is, there is a JK flip-flop for the 2^0 bit, one for the 2^1 bit, and so on.

17.14 In a nuclear physics experiment one wishes to count the number of times the two γ rays resulting from the annihilation of a positron and an electron are recorded in coincidence in two different nuclear particle detectors that we call A and B. Sketch a logic circuit that will produce an output pulse if there are coincident input pulses from the outputs of both nuclear detectors A and B but not otherwise.

17.15 In a nuclear physics experiment one wishes to count the number of events for which there are coincident pulses from nuclear particle detectors A, B, and C and for which there is no pulse from nuclear particle detectors D, E, and F. Use logic circuits to sketch a circuit that will permit one to carry out these operations. This logic circuit should record when A, B, and C are in coincidence and in anticoincidence with D, E, and F.

17.16   Consider the circuit diagram of Fig. 17.38 for a monostable multivibrator using NAND gates. The multivibrator is to put out a pulse 10^{-5} sec in length. What RC product is required for this circuit if the NAND gates used have a normal logical 1 of 3.5 volts and a minimum voltage recognized as a logical 1 of 2.0 volts and if the NAND gates have a normal logical 0 of 0.2 volt and a maximum voltage recognized as a logical 0 of 0.8 volt? If the logic gate has an input current of -1.6 milliamp when the input is a logical 0, what is the maximum value R can have? Based on the above, what would be good choices of R and C?

17.17   Consider the oscillator circuit of Fig. 17.40. Suppose the three NAND gates have logic levels that are the same as those given in Prob. 17.16. Show that the period of the oscillator is $T = A(R_1 C_1 + R_2 C_2 + R_3 C_3)$ where A is a constant. Calculate the numerical value of A. Assume the propagation delay of each of the gates is much less than T.

17.18   Suppose that at time $t = 0$ the oscillator circuit of Fig. 17.40 is in a state such that the input of all three NAND gates are logical 0's and the outputs of all the NAND gates are logical 1's. Show that as described in the text the oscillator will be free running in a short time. Suppose that $R_1 C_1 > R_2 C_2 > R_3 C_3$. How long a time is required before the oscillator is free running? This problem illustrates the fact that no matter what the initial state of the NAND gates that oscillations will begin; that is, the oscillator is self-starting. You may assume that $R_1 C_1$, $R_2 C_2$, and $R_3 C_3$ are not too different and that all three time constants are large compared to the propagation delay of the NAND gates.

17.19   In this chapter we discussed only positive logic. Let us consider how negative logic differs from positive logic. Suppose we take the state where the output voltage is $+5$ volts as the logical 0 and the state where the output voltage is 0.2 volt as the logical 1 for the circuit of Fig. 17.3. Show that when one uses negative logic, the circuit of Fig. 17.3 acts to provide the NOR function, whereas with positive logic the circuit provides the NAND function. In general the logic statements such as De Morgan's theorem are true for either positive or negative logic, but a particular logic gate will provide different logic functions depending on whether positive or negative logic is employed.

17.20   A digital to analog converter similar to the one of Fig. 17.32 has four inputs to 2^3, 2^2, 2^1, and 2^0. Select the resistors and the power supply so that the analog output is such that the decimal number 10 is represented by 10.0 volts.

17.21   A crystal oscillator runs at 10^5 Hz. The output of the oscillator is pulses (not a sinewave). Suppose an up counter such as the one in Fig. 17.27 is used to count the output of the oscillator. How many JK flip-flops are needed if the final flip-flop is to drive a light that is to flash on and off alternately at a rate not to exceed one flash a second.

17.22   Using the various types of logic gates and flip-flops, design a circuit that will take the output of the oscillator described in Prob. 17.21 and will generate a logical 1 for 10^{-4} sec followed by a logical 0 for 3×10^{-4} sec, followed by a logical 1 for 10^{-4} sec, and followed by a logical 0 for 251×10^{-4} sec. This pattern is to be repeated indefinitely.

18/Common Laboratory Instruments

There are a number of laboratory instruments that are used very commonly with electric and electronic instruments. In this section we shall discuss the basic modes of operation of some of these instruments and the relative advantages of the various instruments.

18.1 The Oscilloscope

Aside from the volt-ohm-milliammeter that was discussed in Chapter 3, the oscilloscope is, no doubt, the most commonly used laboratory instrument. The oscilloscope is indispensable as a test instrument for electronic circuitry and for direct measurement of voltages as a function of time in an experimental setup.

An oscilloscope consists of a cathode-ray tube (CRT), the associated circuits to power the CRT, and various circuits to permit the cathode-ray beam to be deflected in either vertical or horizontal directions.

The basic components of a CRT are shown in Fig. 18.1. The CRT has an exterior glass envelope that is evacuated to a high vacuum. The cathode of the CRT usually consists of an oxide material with a low work function. The cathode is heated indirectly by the radiant energy from a filament that is in close proximity to the cathode. The cathode is run at a temperature that is hot enough to boil off electrons. In most CRT's the cathode is almost a planar surface.

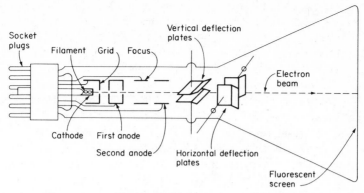

Figure 18.1 A cathode-ray tube as seen in cross section.

The voltage applied between the grid and the cathode determines the electron current drawn from the cathode. The CRT is usually run in a space charge limited region of the current-voltage characteristic. In normal operation the grid is run lower in voltage than the cathode (by about 5–30 volts) just as is done in an ordinary triode electron tube such as the ones discussed in Chapter 8. In a CRT, however, the grid electrode is normally a single plate with a circular hole and not a mesh as is common in vacuum triodes.

The electron beam is accelerated to a final energy that is determined by the voltage difference between the second anode and the cathode. Typically this voltage might be 3–10 kilovolts but might be as high as 20–25 kilovolts. The entire front end of the CRT is covered by a conducting film held at the same voltage as the second anode. Because of this film an electron that has passed the second anode simply travels in a straight line with a constant velocity through the field-free region to the fluorescent screen (assuming no deflecting voltages). This conducting film also prevents the fluorescent screen from charging up. The first anode serves initially to accelerate the electron beam after it passes through the grid, and the first anode also shields the grid to cathode region from the focus voltage; that is, the first anode makes the electron beam intensity independent of the voltage on the focus electrode.

The impact of the electron beam on the fluorescent screen causes it to fluoresce, and a bright spot is visible where the beam strikes the screen.

An electron beam leaving the cathode diverges beyond the grid electrode (the beam divergence, if not corrected by the focus electrode, will cause the beam to strike the fluorescent screen over a broad diffuse region). The focus electrode enables one to reverse the divergence of the electron beam leaving the grid, focusing the electron beam into a fine spot on the fluorescent screen. If one increases the beam intensity, then the mutual electrostatic repulsion of the electrons in the electron beam causes the beam to diverge. The focus electrode cannot correct for the divergence of the beam produced by mutual electrostatic repulsion of the electrons in the beam since the repulsion continues beyond the focusing electrode. A result of these considerations is that one cannot move the electron beam arbitrarily rapidly across the fluorescent screen and still produce a visible image. This occurs because one cannot increase the intensity in a well-focused spot above some particular value assuming the high voltage supply is fixed. Although the total electron current in the beam rises, the space charge causes the beam to diverge so that the beam intensity (that is, current/area) remains relatively constant. In order that a spot on the fluorescent screen be visible a certain minimum number of electrons per unit area must strike the screen during the time Δt that the beam spends covering the width of the electron beam. As the electron beam is moved more rapidly across the fluorescent screen, the number of electrons per unit area striking the screen in the time Δt decreases. One can compensate for this effect by increasing the electron beam intensity. However this works only as long as the space charge in the beam itself doesn't cause the beam to diverge. Increasing the acceleration voltage on the second anode does increase the light emitted by the fluorescent screen so that oscilloscopes with fast horizontal sweeps usually have a high voltage (15–25 kilovolts) supply for acceleration.

The vertical and horizontal deflection plates have connections that permit a voltage difference to be applied between the plates. A voltage difference V between the plates produces an electric field $E = V/d$ between the plates where d is the spacing between the plates. This electric field produces a force on an electron that is passing through the plates $-eE$ so that the electron beam is accelerated in a direction perpendicular to the plates. If v, the original velocity of the electron beam, is large compared to $v_\perp$, the transverse component of the electron's velocity, then the angular deflection of the beam is given by

$$\theta \simeq \tan \theta = \frac{v_\perp}{v} = \frac{(-eE/m)(l/v)}{v} = \frac{-e}{m} \frac{lE}{v^2} = \left(\frac{eV}{\frac{1}{2}mv^2} \right)\left(\frac{l}{2d} \right) \quad (18.1)$$

where m is the mass of the electron and l is the length of the plates parallel to the direction of the incident electron beam. In this expression $-eE/m$ is the acceleration of the electrons perpendicular to their direction of motion and l/v is the time an electron spends between the plates. By applying suitable voltages between the horizontal and vertical deflection plates the electron beam can be deflected to an arbitrary set of coordinates (x, y) on the fluorescent screen.

The voltage difference between the horizontal deflection plates (the x deflection plates) can be produced either by an external voltage or by an internal sweep voltage generated by a circuit in the oscilloscope. If an external voltage is to be used to drive the x deflection plates, then the horizontal display mode should be set to external. With the oscilloscope operating in this mode a voltage that is applied to the horizontal input is suitably amplified and applied to the x deflection plates. The horizontal amplifier is a dc coupled amplifier and the horizontal gain can be attenuated by the horizontal gain control and by the horizontal amplifier in-out switch. The horizontal gain control is accurately calibrated in volts per division deflection on the CRT's fluorescent screen. Thus by observing the deflection produced by a given signal, one can determine the magnitude of the voltage signal. The vertical deflection plates (y deflection) are used in a similar fashion to the x deflection plates (the vertical deflection system will be discussed in detail later). Thus one can observe on an oscilloscope a given external voltage corresponding to deflections in the y direction versus another voltage corresponding to deflections in the x direction.

An example of a use for the external horizontal sweep is in measuring the ratio of two ac frequencies by forming a Lissajous figure. When an ac voltage of frequency $f_1 = nf$ (n is an integer) is applied to the x deflection plates and an ac voltage of frequency $f_2 = mf$ (m is an integer) is applied to the y deflection plates, then a stationary pattern is formed on the fluorescent screen of the oscilloscope. The shape of the pattern depends on the ratio of the frequencies of the two ac voltages and on the relative phase of the two ac voltages. By observing and analyzing the stationary pattern formed one can deduce both the ratio of the frequencies and the relative phase of the two voltages. A typical Lissajous figure that might occur when the frequency of the ac voltage applied to the horizontal (x) deflection plates is 60 Hz and when the frequency of the ac voltage applied to the vertical (y) deflection plates is 30 Hz is shown in Fig. 18.2. The Lissajous figure shown implies that the two signals both pass through zero voltage at the same time.

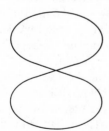

Figure 18.2 A Lissajous figure. The frequency of the ac voltage applied to the horizontal plates is 60 Hz. The frequency of the ac voltage applied to the vertical plates is 30 Hz. The two signals pass through zero voltage at the same time.

In most applications the voltage waveform to be applied to the horizontal deflection plates as a function of time is generated internally (that is, inside the oscilloscope) by a circuit called a *sweep generator*. After suitable amplification of the sweep voltage, by the horizontal amplifier, the voltage is applied across the horizontal deflection plates. A typical form for the sweep voltage is shown in Fig. 18.3. The sweep voltage is a linear function of the time from $t = 0$ to $t = \tau_1$. After the time τ_1 the voltage rapidly falls to zero in a time $\tau - \tau_1$. During the time from 0 to τ_1, the electron beam is swept in the plus x direction with a constant x component of velocity. During the time from τ_1 to τ the electron beam is blanked out so that one does not observe the "flyback" of the beam to its initial position on the x axis.

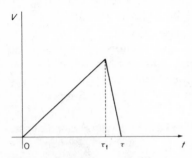

Figure 18.3 A typical sweep voltage applied to the horizontal deflection plates of an oscilloscope.

The sweep voltage starts at $t = 0$ because of a trigger signal. The trigger signal on an oscilloscope can be derived from an external source, from the 60-Hz line, or internally from the signal being observed. In order to appreciate better the importance of the trigger signal, let us consider how one uses an oscilloscope to observe a pulse voltage as a function of the time. Suppose pulses are occurring randomly in time as might be the case if they were produced by a radioactive source. Let us further suppose that the pulses have been amplified and shaped so that the pulses as a function of time are identical in magnitude and width. The incoming pulses themselves are used to trigger the sweep internally. Suppose the pulse has a magnitude of 5 volts. It might be convenient to trigger (that is, start) the horizontal sweep when the pulse amplitude passes 1.0 volt. In Fig. 18.4 we illustrate several possible pulses and the horizontal sweep voltage as a function of the time. The display on the oscilloscope appears to be the same for each pulse so that the display stands still even though the pulses occur at random intervals. The oscilloscope display is also shown in Fig. 18.4.

It is obvious that the internal trigger can be used with periodic repeating voltage waveforms as well as with the nonperiodic pulses just discussed. The external trigger is useful for starting the horizontal sweep in coincidence with some externally controlled event. The 60-Hz ac line trigger is useful if a waveform occurs synchronously with the ac power line. In addition, the horizontal sweep may be allowed to operate in a free running mode; that is, as soon as one sweep is completed, the next sweep begins.

In its most common application an oscilloscope is used to plot an external voltage versus time. In order to do this the time base of the horizontal sweep is adjusted to an appropriate value. The external voltage is amplified by the vertical amplifier and is applied to the vertical deflection plates of the oscilloscope. The vertical amplifier is usually a dc amplifier although it can be used in an ac coupled mode. The gain of the vertical amplifier in a common oscilloscope can be varied so that the beam deflection ranges between about 10^{-3} volt per division and about 10 volts per division. Commonly, although not always, a division on the oscilloscope screen is about 1 cm.

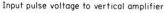

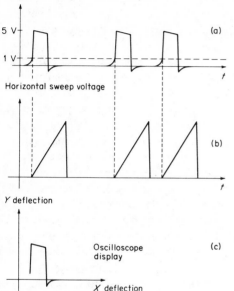

Figure 18.4 (a) A random sequence of pulse voltages. (b) The horizontal sweep of the oscilloscope that is triggered when the input signal reaches a 1-volt level. (c) The stationary display of the pulse on the face of the CRT.

On many oscilloscopes there are two inputs, A and B. If voltages V_A and V_B are applied to the inputs A and B, respectively, then one can present on the oscilloscope screen a trace of either V_A versus time or V_B versus time. One can also use the vertical amplifier as a difference or a sum amplifier and present $V_A - V_B$ or $V_A + V_B$ versus time as the display on the oscilloscope screen. In addition, many oscilloscopes permit one to select an alternate mode in which V_A versus time and V_B versus time are displayed alternately; that is, on one sweep V_A is applied to the vertical deflection plates and on the next sweep V_B is applied to the vertical deflection plates. One final display mode is called *chopped*. In this display the voltage applied to the vertical deflection plates switches rapidly back and forth from V_A to V_B. In this mode of operation both V_A and V_B are displayed simultaneously. The chopped mode of operation is useful, for example, in comparing the frequency or the phase angle between two ac voltages.

The intensity of the cathode-ray beam can be varied by applying a voltage signal to the terminals labeled *CRT grid*. The connection to the grid usually is located at the rear of the oscilloscope. This permits one to vary the intensity of the electron beam trace on the fluorescent screen. This control is often called the Z input. The Z input is used, for example, to put timing markers on the oscilloscope trace.

The focus, astigmatism, and intensity controls are used to adjust properly the electron beam in the CRT. The horizontal and vertical position controls are used to center the display on the fluorescent screen.

There are *dual beam* oscilloscopes. A dual beam oscilloscope has two separate electron guns and two separate vertical and horizontal deflection systems. The dual beam oscilloscope can be used to display two voltages V_A and V_B simultaneously on the fluorescent screen. The display of the dual beam oscilloscope is similar to the chopped display previously discussed. With a dual beam oscilloscope, however, both V_A and V_B are displayed continuously and the vertical deflection plates are not switched between V_A and V_B as in the chopped mode. The dual beam oscilloscope has an especial advantage over chopping when the frequencies of the voltages V_A and V_B are as large or larger than the frequency at which the chopping occurs.

Finally we should like to describe the properties of a standard commercially available oscilloscope, the Tektronix model 7503 oscilloscope. Figure 18.5 shows a Tektronix model 7503 oscilloscope. The accelerating voltage for the CRT in the Tektronix 7503 is 18 kilovolts. Both the vertical and the horizontal deflection systems are very flexible since they can be chosen from various plug-in units. For example, if the time base plug-in unit model number 7B50 is used with the 7503, then one can obtain the following properties. The internal horizontal sweep can be selected in ranges from 5×10^{-9} to 5 sec per division. For use with an external horizontal input the horizontal deflection amplifier can be used from dc to 500 kilohertz and with a sensitivity variable between 90 millivolts per division and 9 volts per division. There are a wide variety of other time base plug-in units that can be used with the Tektronix 7503 oscilloscope. These various time base plug-in units have a number of different and useful properties.

The vertical deflection system for the Tektronix 7503 also utilizes interchangeable plug-in amplifier units. We shall just mention the properties of a few of these plug-in vertical amplifier units. The Tektronix model 7A12 amplifier permits the use of two inputs, channel 1 and channel 2 for voltages V_1 and V_2, respectively. The oscilloscope with a 7A12 amplifier can display V_1 only, V_2 only, $V_1 + V_2$, and $V_1 - V_2$. In addition, the alternate and chopped displays are available. The gain of each channel can be varied from 5×10^{-3} to 5 volts per division. The vertical amplifier may be used from dc to about 75×10^6 Hz. A step voltage applied to the amplifier will emerge as a step with a rise time of 4.7×10^{-9} sec. The Tektronix 7503 oscilloscope can use two 7A12 amplifiers at the same time to permit three or four trace displays.

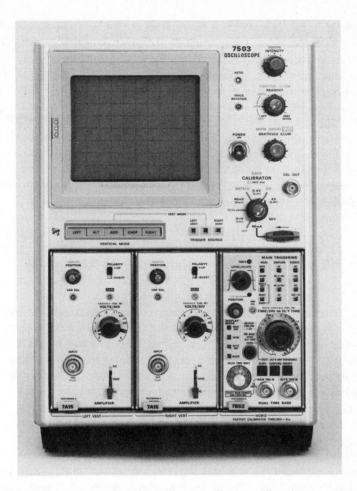

Figure 18.5 A Tektronix model 7503 oscilloscope. The time base plug-in unit is model 7B52 and both vertical channels have 7A15 plug-in amplifiers. (*Photograph courtesy Tektronix, Inc.*)

Another vertical amplifier that can be used with the Tektronix 7503 oscilloscope is the Tektronix 7A22 amplifier. This amplifier also has two inputs A and B. One can display V_A only, V_B only, and $V_A - V_B$. There are no alternate or chopped modes available with this amplifier, however. This vertical amplifier has a very high gain so that deflection sensitivities ranging from 10^{-5} to 10 volts per division may be used. The input amplifier has a drift rate equivalent to 10^{-5} volt/hr. This vertical amplifier may be used from dc to 10^6 Hz. However, it is possible using front panel switches to vary the bandwidth of the amplifier by varying the high-frequency and independently the low-frequency cutoffs, that is, the frequencies at which the amplifier gain has fallen by -3 dB at either the high- or low-frequency ends of the pass band. This is important at high gain sensitivity since the input noise voltage increases with the pass band of the amplifier (see Chapter 12 for a discussion of noise). This amplifier is different from the previous one in that the vertical sensitivity is much higher but this has been achieved at the cost of decreasing the bandwidth.

The oscilloscope is an astoundingly versatile instrument. We have discussed only the most common applications. Space does not permit us to discuss the use of such instruments as sampling oscilloscopes, storage oscilloscopes, and many other uses for this instrument. However, a person interested in electronic instrumentation would do well to study carefully an up-to-date manufacturers' catalog in order to appreciate completely the usefulness of this instrument.

18.2   The Tunnel Diode and Its Use in Very Fast Trigger Circuits

Oscilloscopes and other devices sometimes require extremely fast trigger circuits. A TTL gate with Schottky diodes can have a propagation delay time as short as 3×10^{-9} sec and an ECL gate can have a delay time as short as 10^{-9} sec. If one wishes to have a delay time for a trigger circuit less than 10^{-9} sec, however, then one may need to use a tunnel diode circuit. Tunnel diodes can be extremely fast, and a trigger circuit with a tunnel diode can have a time delay of 10^{-10} sec or even less. Before we can understand a trigger circuit that uses a tunnel diode, we must first understand how a tunnel diode works.

Figure 18.6 shows the band structure of the pn junction that forms the tunnel diode. Both the p and the n regions of the diode are very heavily doped with impurities. The impurities are so dense that they cannot be considered to form isolated energy levels in the forbidden region, but instead the impurity centers are close enough together that the interaction of neighboring impurities with each other causes the impurity energy levels to form a continuous band of allowed energy levels. In the n-type material this band of donor impurity energy levels lies immediately below the conduction band, and in the p-type material the band of acceptor impurity levels lies immediately above the valence band. Because the impurity concentration is high, the depletion layer of the pn junction is very thin, as small as 10 to 100 lattice spacings. With no applied potential the Fermi energy level is the same in both the p- and n-type material and no net current flows. As a small forward bias is applied, the electrons in the donor impurity band in the n-type material can readily tunnel into empty states of the acceptor impurity band in the p-type material. The current increases rapidly as the voltage increases.

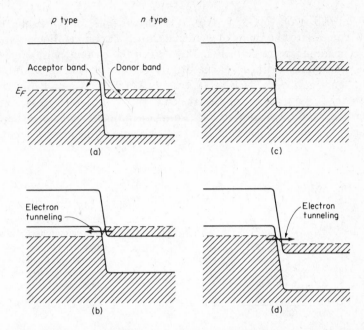

Figure 18.6 (a) The energy band structure of a tunnel diode with no applied bias voltage. Notice the electrons in the donor band in the n-type material and the holes in the acceptor band in the p-type material. (b) A small forward bias voltage is applied to the tunnel diode. The p-type material is at a positive potential with respect to the n-type material. This means that the negatively charged electrons in the n-type material are raised in energy and can tunnel from the n-type material into the p-type material. The current flows from the p- to the n-type material. (c) A larger forward bias is applied to the junction and tunneling is now improbable. (d) A back bias voltage is applied to the junction. Tunneling of electrons from the p- to the n-type material occurs.

With a still further increase in the applied potential, however, the tunneling becomes improbable and the current decreases. At still higher voltages the normal diode action occurs and the current increases as $I = I_0\, e^{eV/kT}$. In the back-biased condition, tunneling is possible at any voltage and the back-biased current is large and increases rapidly for very small back bias voltages. When a tunnel diode is deliberately used for its rapid turn on with a back bias, it is called a *back diode*. The current-voltage characteristic of a tunnel diode is shown in Fig. 18.7. Most tunnel diodes are made of germanium.

   Let us now consider how one might use a tunnel diode in the circuit shown in Fig. 18.8. When the NAND gate has a logical 1 output, the output voltage is about 3.5 volts assuming the gate is a typical TTL circuit. The current I through the tunnel diode is then $I = (3.5 - V_o)/500$ where V_o is the voltage across the tunnel diode. If the voltage across the diode is approximately 0.03 volt, then $I \simeq 7$ milliamp and the diode operates at the quiescent point labeled A on Fig. 18.7. Now if a current pulse of $i_i = 3$ milliamp or larger comes into the input of the circuit, then the tunnel diode will switch to the operating point B on Fig. 18.7. When the current pulse is gone, the diode will operate with a voltage drop $V_o \simeq 0.4$ volt across the diode and with a current $I = (3.5 - V_o)/500 = (3.5 - 0.4)/500 = 6.2$ milliamp through the diode. This operating point is labeled C on Fig. 18.7. The important point is that the circuit does not go back to its operation at point A. The output voltage has switched from 0.03 to 0.4 volt because of the input current pulse.

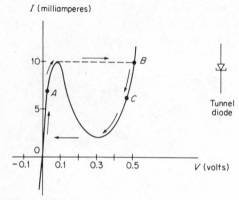

Figure 18.7 The current-voltage characteristic of a tunnel diode. The region where the slope of current-voltage characteristic is negative is called the *negative resistance region*. The circuit symbol for a tunnel diode is also shown.

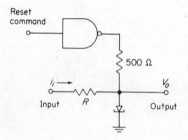

Figure 18.8 A bistable circuit employing a tunnel diode.

In order to return to operation at point A one must reduce the current through the diode and hence also the voltage across the diode to a point near the 0.3-volt minimum in the current-voltage characteristics. A further reduction in the current will now switch the diode as shown in Fig. 18.7 to the low voltage state. In the circuit shown the reset can be accomplished by dropping the output of the NAND gate to a logical 0 (0.2 volt) and then returning the output to a logical 1. This occurs when a logical 1 signal is applied at the input labeled RESET COMMAND.

The primary importance of the type of circuit shown is that the switching from point A to point B on the current-voltage characteristic of Fig. 18.7 is extremely fast; only 10^{-10} sec or less is required for this switching process. Thus the tunnel diode can be used where a very fast trigger circuit or any other very fast pulse circuit is required.

18.3   The Electrometer

One of the more useful laboratory instruments is an electrometer. The basic measurement made by an electrometer is the measurement of a small current. The Keithley 610C is an example of a very versatile electrometer that is commercially available. This instrument is shown in Fig. 18.9. A much simplified version of the basic circuit is shown in Fig. 18.10. The Keithley 610C has a high gain dc amplifier, the output of which appears across a galvanometer movement connected, using a resistor R', as a voltmeter. The input stage of the dc amplifier is a MOSFET so that the current drawn is very small (the input impedance of the MOSFET is about 10^{14} ohms shunted by 22×10^{-12} farad so that the dc current is usually less than about 10^{-14} amp).

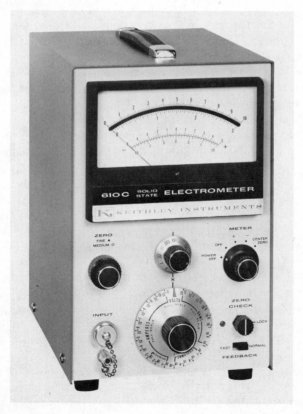

Figure 18.9   A Keithley model 610C electrometer. (*Photograph courtesy Keithley Instruments, Inc.*)

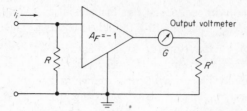

Figure 18.10 The normal mode of measuring a current with a Keithley 610C electrometer. The galvanometer movement G and the series resistor R' form the output voltmeter.

A small current can be measured in the normal mode by passing the current through a large resistor R and using the dc amplifier connected with negative feedback so that its gain is -1 as a voltmeter to measure the voltage across the resistor R. The output voltmeter is calibrated to read the input current directly. The dc amplifier with unit gain is very stable because of the negative feedback that produced the gain of -1. The resistor R in the input of the Keithley 610C electrometer is selected by the range switch and is numerically equal to $1/I$ where I is the full-scale current reading indicated by the range switch. The resistor R varies in steps that are powers of 10 from 10 to 10^{11} ohms. The multiplier switch changes the full-scale sensitivity of the Keithley by changing the resistor R' used to convert the galvanometer movement into a voltmeter. The multiplier switch varies from 10^2 to 10^{-3}. Thus the maximum sensitivity of the Keithley 610C for measuring current is $10^{-11} \times 10^{-3} = 10^{-14}$ amp for full-scale deflection. In the normal mode of operation just described the Keithley 610C is a relatively slow instrument because the input resistor, which may be as large as 10^{11} ohms, is shunted by a capacitor that is about 22×10^{-12} farad.

There is another mode of operation for the Keithley 610C as an ammeter called the *fast mode*. In this mode the resistor R is connected in an operational feedback configuration as shown in Fig. 18.11. In this configuration the input impedance is R/A_0. Therefore if A_0 is large, then the input impedance of the device is small. In this mode the reduced input resistance gives a much smaller RC input time constant. In this configuration the output voltage is $v_o = -i_i R$, so that the current ranges are the same as for the normal operation previously described (the values of R are the same as for the normal mode of operation).

In addition to measuring current the Keithley 610C electrometer can be used for several other measurements. For example, it is possible to use a capacitor in an operational feedback loop with the instrument in the fast mode in order to integrate the input current. The front panel range switch enables one to select capacitors corresponding to full-scale ranges of 10^{-7}, 10^{-8}, 10^{-9}, and 10^{-10} coulomb (with the multiplier switch set to 1). The multiplier switch again enables one to vary these readings by a factor ranging from 10^2 to 10^{-3}.

The Keithley 610C can be used to measure voltages. In this mode of operation the dc amplifier is operated with unity gain. This mode of operation is illustrated in Fig. 18.12. The input impedance of this instrument as a voltmeter is the input impedance of the input MOSFET in the amplifier. This impedance is about 10^{14} ohms shunted by 22×10^{-12} farad as mentioned previously. The high input impedance is very useful as it enables one to measure voltages without severely loading a high impedance source.

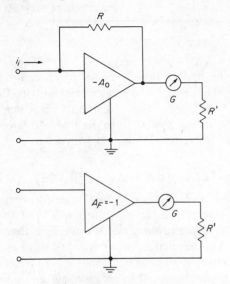

Figure 18.11 The measurement of a current using the fast mode of the Keithley 610C electrometer.

Figure 18.12 The use of a Keithley 610C electrometer used as a voltmeter.

Finally a Keithley 610C may be used as an ohmmeter to measure a resistance. The Keithley 610C can be used in several ways to measure an unknown resistance R_x. One of these methods is called the *fast guarded* method. When operating in this mode, the Keithley is connected internally as shown in Fig. 18.13. In Fig. 18.13 only R_x is external to the instrument. ε is a voltage source of 10^{-1} volt on the 10^{12}-ohm scale, 1.0 volt on the 10^{11}-ohm scale, and 10 volts on the (10^5-10^{10})-ohm scale. The constant current $i = \varepsilon/R$ is driven through R_x. In this expression R is 10^{11} ohms for the $(10^{10}-10^{12})$-ohm scales and then varies in power of 10 from 10^{10} ohms for the 10^9-ohm scale down to 10^6 ohms for the 10^5-ohm scale. The output voltage is $-\varepsilon(R_x/R)$. The output voltmeter is calibrated to read the value of R_x directly. In the normal method of measuring a resistance the same current as in the fast guarded method is put through the unknown resistor and the voltage across the unknown resistor is measured.

If one wishes to measure the current-voltage characteristics of a diode directly, then one can simply connect the diode to the input of the Keithley and set the Keithley to the resistance scale. The current I through the diode is ε/R, which can be selected by varying the resistance scale, and the voltage across the diode is $R_x(\varepsilon/R)$, which can be obtained by reading R_x and multiplying by ε/R.

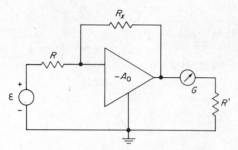

Figure 18.13   A Keithley 610C electrometer used in the *fast guarded* method to measure an unknown resistance R_0. The output voltage is given by $v_o = -(\varepsilon R_x)/R$.

Our discussion has centered around how an electrometer such as the Keithley 610C operates. If one requires a greater current sensitivity than 10^{-14} amp full-scale deflection, then there are other specialized instruments such as the vibrating reed electrometer; however, the Keithley 610C is an excellent example of a useful and interesting laboratory instrument.

18.4   The Vacuum-Tube Voltmeter (VTVM)

The vacuum-tube voltmeter (VTVM) is a device that performs many of the same functions as a volt-ohm-milliammeter (VOM). A VTVM has a dc amplifier and an output that is indicated on a meter movement. Because the input to the VTVM is the grid of the vacuum tube in the first stage of the dc amplifier, the current drawn (either total or incremental) is very small (as discussed in Chapter 8). As a result of the small incremental current drawn, the input impedance of the VTVM can be relatively high, of the order of magnitude of 10^7 ohms, for example. The high input impedance of the VTVM is the primary advantage of a VTVM over a VOM. An additional advantage is that the dc amplifier can have an appreciable gain so that a VTVM may be somewhat more sensitive than a VOM.

An example of a VTVM is the Hewlett Packard 410C multifunction voltmeter shown in Fig. 18.14. This instrument uses a chopper stabilized dc amplifier with a voltage gain of 100. The input impedance of the 410C is 10^7 ohms shunted by about 1.5×10^{-12} farad. The dc voltage ranges are from $\pm 15 \times 10^{-3}$ to ± 1500 volts for full-scale deflection. The 410C also can be used to measure dc currents in the ranges from $\pm 1.5 \times 10^{-6}$ to ± 0.150 amp for full-scale deflection. This is accomplished by measuring the voltage drop the current produces across a resistance.

The 410C may be used as an ac voltmeter. The ac measurements are accomplished by the use of a rectifying probe as shown in Fig. 18.15. In this circuit the capacitor charges to the peak value of the ac signal. Such an instrument does not indicate the rms value of the ac voltage but instead indicates the peak value of the ac voltage (these are not simply related except for a pure sinusoidal ac voltage). The 410C has ac voltage ranges from 0.5 to 300 volts for full-scale deflection. The 410C is useful for ac frequencies up to 700×10^6 Hz.

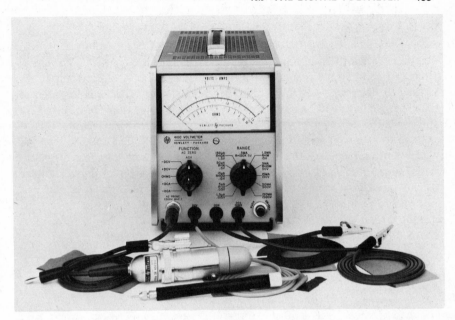

Figure 18.14 A Hewlett-Packard model 410C multifunction voltmeter. (*Photograph courtesy Hewlett-Packard.*)

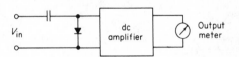

Figure 18.15 Measuring an ac voltage by using a rectifing probe with the Hewlett-Packard model 410C voltmeter.

In addition the 410C can be used to measure resistances with ranges from 10 to 10^7 ohms full-scale deflection.

18.5 The Digital Voltmeter

Normally the reading of a voltage by a voltmeter is indicated by the deflection of the pointer of a galvanometer movement. This analog voltage (the voltage is an analog quantity because it can assume any value) is often converted into a digital quantity by a student who writes down the voltage indicated. Using a student as an analog to digital converter is not the only way to accomplish the conversion. A digital voltmeter (DVM) is an instrument that indicates a voltage as a digital number. The DVM display may be as a discrete set of lighted numerals or the DVM may indicate the voltage either as a decimal number or in a binary coded decimal (BCD) output. A DVM is a very useful device because it helps avoid human error in reading and recording voltages as well as helping eliminate boredom.

There are DVM's that work on a variety of operating principles. We shall describe the operation for three types of DVM. These three types of DVM's are a ramp type, a dual slope type, and a potentiometer type.

The ramp-type DVM works as follows. A linear voltage, called the *ramp*, varies between, say, $+12$ and -12 volts. A comparator is used to determine when the ramp and the unknown voltage are identical. The signal from this comparator is used to gate an oscillator on. Another comparator is used to determine when the ramp crosses 0.0 volt. This signal is used to gate the oscillator off. A counter determines the total number of cycles of the oscillator during the time the oscillator was gated on. This process of voltage to time conversion is illustrated in Fig. 18.16. The total number of cycles is proportional to the input voltage. The measured voltage is displayed with lighted numerals, printed out, or presented as a set of digital voltage signals in BCD. The primary difficulty with such an instrument is that it is not easy to produce a stable linear ramp. Nonlinearities in the ramp voltage are the largest source of error in using a ramp type of DVM; however, this source of error can often be as small as 0.05% or even smaller.

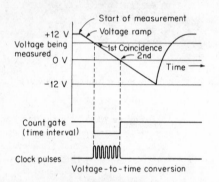

Figure 18.16 The process of voltage to time conversion in a digital voltmeter. (*Diagram courtesy Hewlett-Packard.*)

A typical example of a ramp-type DVM is the Hewlett Packard model 3440 digital voltmeter. This instrument is shown in Fig. 18.17. A block diagram of this instrument is shown in Fig. 18.18. In Fig. 18.18 the box labeled *sample rate* determines how often the voltage is measured and displayed. This rate can be varied between five times per second to once every 5 sec. The 3440 DVM has various plug-in units available to determine the voltage range that can be measured. One of these plug-in units, model 3443A, has full-scale ranges of 99.99 millivolts, 999.9 millivolts, 9.999 volts, 99.99 volts, and 999.9 volts. The correct range for a given voltage can be set manually or there is automatic ranging. A voltage too large for a given range is indicated by the overrange indicator.

A dual slope-type DVM works as follows. In order to measure a given input voltage v_i, a current that is directly proportional to the voltage v_i is produced. This current is integrated for a fixed time interval, perhaps $\frac{1}{60}$ or $\frac{1}{10}$ sec. This is called the *up-slope integration*. The final charge representing the integral of the current is stored on a capacitor. This charge is directly proportional to the average value of the input voltage over the time interval used for integration. This up-slope integration is accomplished simply by using an operational amplifier as an integrator. The input current is v_i/R, and the output voltage is $v_o = - \int (v_i/R) \, dt/C$. The final charge stored on the capacitor is $- \int (v_i/R) \, dt$. The capacitor is now discharged by a precise reference current. This is called the *down-slope discharge*. The time required to discharge the capacitor is accurately measured. This time is proportional to the average value of the input voltage. The time is measured by counting the total number of cycles of a high-frequency crystal oscillator during the time required for discharge. The average value of the input voltage is then presented as a digital number. The dual slope DVM is also used to form the ratio v_1/v_2. In order to do this, v_1 and v_2 are both integrated and the resulting charges are stored on two different capacitors C_1 and C_2. These two charges are proportional to v_1 and v_2, respectively. A current proportional to the charge on C_2 and hence proportional to the average value of v_2 is produced, and this current is used to discharge the capacitor C_1, the charge on C_1 being proportional to v_1. The time necessary to discharge C_1 is proportional to v_1/v_2. The ratio v_1/v_2 is presented as a digital number by counting the number of cycles of an oscillator during the discharge time. The dual slope DVM's are probably the most popular type at the present time.

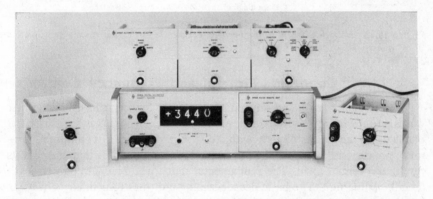

Figure 18.17   A Hewlett-Packard model 3440A ramp-type digital voltmeter. Several plug-in units that can be used with this instrument are shown. (*Photograph courtesy Hewlett-Packard.*)

A potentiometer type of DVM works in a quite different manner. In this type of DVM a potentiometer is balanced automatically usually by having fast relays that switch resistors in the circuit so as to automatically move along a resistor string that is the equivalent of the slide-wire in a potentiometer until the potentiometer is balanced. The final digital readout is determined by the resistors in the resistor string when the potentiometer is balanced. An instrument such as this can be made arbitrarily accurate by using more and more resistors so as to be able to move more nearly continuously along the equivalent of the slide-wire. The main difficulty with a bridge-type DVM is that even using very fast reed-type relays the time required to measure a voltage is appreciable. One other difficulty with an instrument of this type is that small ac fluctuations on the dc voltage may leave the instrument constantly searching for balance but never achieving the balance. To overcome this problem many such instruments have a filter that averages the voltage to be measured.

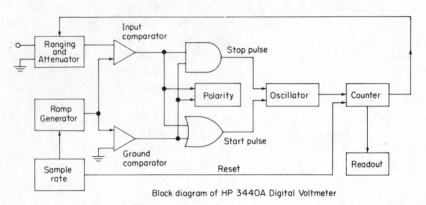

Block diagram of HP 3440A Digital Voltmeter

Figure 18.18 A block diagram of the Hewlett-Packard 3440A digital voltmeter. (*Diagram courtesy Hewlett-Packard.*)

18.6 Counters, Timers, and Frequency Counters

In Chapter 17 we discussed a binary counter (or scaler). This counter was made up of successive JK flip-flops. A binary counter is very useful if one is using the results directly with a computer. For many common laboratory applications, however, it is desirable to have a counter that presents its results in decimal form. The fundamental unit for a decimal counter is the decade-scaling unit. The basic decade-scaling unit uses four binary flip-flops. This would permit one to construct a four-stage binary counter. The outputs of four binary flip-flops have $2^4 = 16$ possible rearrangements and consequently could be used to scale binary numbers whose decimal equivalent would range from 0 to 15. Instead of using all 16 possible states, however, feedback is used to limit the number of states used to 10. With a suitable decoding the decade-scaling unit can count and display decimal numbers from 0 to 9. A simple scheme for constructing a decade-scaling unit is shown in Fig. 18.19. Although Fig. 18.19 appears to be very complicated, it can be understood rather easily. All four flip-flops have both the J and K inputs held at a logical 1. This means that the value of Q changes with each input clock pulse. The scaling unit simply counts in a normal sequence until the binary number 1010 whose decimal equivalent is 10 is reached. NAND gates are used to form the inverse of the output of the 2^0 and the 2^2 flip-flops. We could have also taken the inverse of the output of the 2^0 and the 2^2 flip-flops from the $\bar{Q}$ output rather than by using NAND gates. The output of all the flip-flops (with the two inversions just mentioned) serve as the input to another NAND gate that is used to apply a logical 0 to the clear (Cl) input of each flip-flop when the binary number 1010 (decimal equivalent 10) is reached. This resets all the flip-flops so that $Q = 0$ and the process starts again. The negative pulse that clears the decade-scaling unit is fed to the next decade-scaling unit after the necessary inversion. A decade-scaling unit can be put together using fewer components than were used in Fig. 18.19. The circuit of Fig. 18.19 lends itself to a simple exposition, however, and it was used for this reason.

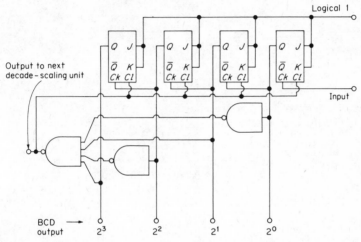

Figure 18.19 A decade-scaling unit constructed using JK flip-flops. The output is an 8421 code.

The decoding of the binary numbers so that a decimal number can be read out from the decade-scaling unit of Fig. 18.19 as a lighted numeral is accomplished in the same way as the binary to decimal conversion that was described in the chapter on digital circuits. It is obvious that a binary coded decimal (BCD) output can be obtained directly from the leads indicated as 2^0, 2^1, 2^2, and 2^3 in Fig. 18.19. In Table 18.1 we show the 16 logical states for the output of the four flip-flops and indicate the decimal number each state represents in the decoded output of the decade-scaling unit.

It is not necessary to use the first 10 of the 16 possible binary states for the four flip-flops as the states representing the 10 numbers from 0 to 9 in a decimal decade. Any 10 of the 16 states could have been used. Table 18.2 gives the binary states used to represent the decimal numbers in some common decade-scaling units. Many other combinations are possible and used.

Table 18.1

Binary	Decimal
0000	0
0001	1
0010	2
0011	3
0100	4
0101	5
0110	6
0111	7
1000	8
1001	9
1010	}
1011	
1100	
1101	} not used
1110	
1111	}

Table 18.2

Decimal	8421	4221	Gray	Old
0	0000	0000	0000	0000
1	0001	0001	0001	0001
2	0010	0010	0011	0010
3	0011	0011	0010	0011
4	0100	1000	0110	0110
5	0101	0111	1110	0111
6	0110	1100	1010	1100
7	0111	1101	1011	1101
8	1000	1110	1001	1110
9	1001	1111	1000	1111

The 8421 code was the code used in the circuit of Fig. 18.19 and is merely the succession of states in ordinary binary counting. The other codes are produced in decade-scaling units by wiring the four flip-flops so that negative feedback produces the desired succession of states. Circuits using the four codes of Table 18.2 to represent the digital numbers 0–9 are in use, as are other codes. The four codes given in Table 18.2 are of two different types. The first two codes are called *weighted* codes. The weights for the codes are given by the numbers above the code. The 8421 code means that if one multiplies the left-hand digit by 8, the next digit by 4, the next digit by 2, and the right-hand digit by 1 and sums these, then the result is the decimal number being represented. The 4221 code is also a weighted code. Weighted codes are particularly useful if one wishes to convert the output from a digital signal to an analog signal. Referring to Chapter 17 where digital to analog conversion was described, it is seen that the precision resistors that determine the current in the digital to analog converter described are in the ratios of one over the weight given the digit in the weighted codes. The decade-scaling unit illustrated in Fig. 18.19 is an 8421 weighted code. The gray code is used in situations where it is important that no ambiguity occur in the switching from one decimal number to the next. To understand this, consider the 8421 code. The switch from the decimal number 1 to 2 requires a switch from the binary state 0001 to the state 0010. If the second digit switches slightly before the first digit, then the binary numbers might be 0001 initially, 0011 for a short time, and 0010 finally. For a short time during the switching transient the binary state was 0011; that is, the state was ambiguous. In the gray code only one digit switches in going from one digital number to the next so that no ambiguity is possible. The code labeled *old* is a code for which the decade-scaling unit is easily wired using clocked set-reset flip-flops. Because of the ease in wiring the negative feedback for this decade-scaling unit, it formerly was often used. Most modern decade-scaling units are produced as integrated circuits, however, and they have a variety of codes. The 8421 code is probably most common. In order to scale numbers larger than 10, several decade-scaling units are used. Each decade-scaling unit divides the input pulse rate by a factor of 10. The first decade-scaling unit registers the 1's digit; the output of the first scaling unit goes to the next scaling unit, which registers the 10's digit; and so on.

A typical integrated circuit decade-scaling unit is the Texas Instruments model SN7490. This integrated unit is not constructed internally the same as the decade-scaling unit of Fig. 18.19, but it is similar in that it has four master-slave flip-flops and a gated reset. The SN7490 uses a 8421 code and has a maximum counting rate of about 18 MHz.

Decade-scaling units such as we have been discussing have a wide variety of uses. In nuclear physics applications, for example, one often has pulses coming at random times. The pulses are to be recorded by a scaler. A scaler consists of several decade-scaling units in series. Usually a nuclear scaler will have a time base (such as may be derived from the 60-Hz ac line), and the scaler can be gated on automatically for a preset time, selected to be suitable for the particular experiment.

It is also common to have a scaler scale a fixed number of counts and record the total lapsed time required for this operation. In this use the scaler is called a *timer*. This use is very important because the error in a measurement of N random counts is $\sqrt{N}$ so that this mode of operation permits preselection of the accuracy of the experiment. This of course is not the case if one preselects the total time for counting.

Other features included with many scalers are integral and/or differential pulse height analyzers. An integral pulse height analyzer is simply a Schmitt trigger (for a description of a Schmitt trigger, see Chapter 16) before the input to the first decade-scaling unit so that one selects only pulses larger than some minimum voltage. The trigger voltage level on the integral discriminator is variable. A differential pulse height discriminator is a device that passes to the first decade-scaling unit only pulses larger than some minimum voltage, say, E, and less than some voltage $E + \Delta E$.

Any scaling unit has some minimum resolving time τ. This means that if more than one pulse arrives within the time τ, only a single pulse is recorded. When pulses arrive randomly, there is always a finite chance that two successive pulses will be separated by less than the time τ. Thus a scaler may not accurately record all the events. Suppose one measures with a scaler N_1 total events in a time T. Then the true counting rate N_{1T} is given by

$$N_{1T} = \frac{N_1/T}{1 - (N_1 \cdot \tau/T)} \tag{18.2}$$

This can easily be seen by noting that the fraction of the time that the scaler is incapable of recording a new event is $N_1 \cdot \tau/T$ and hence the fraction of the time that the scaler is capable of recording a new event is $1 - (N_1 \cdot \tau/T)$. The resolving time of a scaler can be measured as follows. Use two independent sources of random pulses 1 and 2. Measure the total counts N_1 and N_2 recorded independently when source 1 is present alone and when source 2 is present alone, respectively. Each measurement is for a duration T. Also measure the total counts N_3 recorded in the time T when both sources are present simultaneously. We have then

$$N_{1T} = \frac{N_1/T}{1 - (N_1 \cdot \tau/T)}$$

$$N_{2T} = \frac{N_2/T}{1 - (N_2 \cdot \tau/T)} \tag{18.3}$$

and

$$N_{3T} = \frac{N_3/T}{1 - (N_3 \cdot \tau/T)}$$

Since $N_{3T} = N_{1T} + N_{2T}$, we have

$$\frac{N_3/T}{1 - (N_3 \cdot \tau/T)} = \frac{N_1/T}{1 - (N_1 \cdot \tau/T)} + \frac{N_2/T}{1 - (N_2 \cdot \tau/T)} \tag{18.4}$$

Let us assume $\tau/T \ll 1$ and ignore terms involving $(\tau/T)^2$. We then find

$$\tau = \left(\frac{N_1 + N_2 - N_3}{2N_1 N_2} \right) T \tag{18.5}$$

In this discussion we have tacitly assumed that the resolving time τ in an experiment depends only on the decade-scaling units. Actually it is a common situation that the particle or photon detector that initiates the voltage pulse may introduce a major contribution to the total resolving time in a counting experiment.

An example of a scaler-timer is the Hewlett Packard model 5201L shown in Fig. 18.20. This scaler has both preset-count and preset-time modes of operations. Preset-count times are variable from 0.1 to 9999.9 sec in 0.1-sec steps. The resolving time τ of the scaler is 2×10^{-7} sec in the preset-time mode and 10^{-5} sec in the preset-count mode. The scaler has a built-in differential pulse height analyzer. The time base for the scaler is from the 60-Hz ac line. The visual display is with in-line lighted numerals and, of course, in the decimal system. One can also obtain a printed BCD output (8421 code).

Figure 18.20   A Hewlett-Packard model 5201L scaler-timer. (*Photograph courtesy Hewlett-Packard.*)

In a somewhat different function the decade-scaling unit is used in an instrument called a *frequency counter* to measure the frequency of an ac sine wave. A schematic diagram showing how a frequency counter works is shown in Fig. 18.21. The input sine wave is amplified so that its amplitude is large enough to drive the digital circuits. The amplified wave is then shaped into pulses by a Schmitt trigger. These pulses are gated to the decade-scaling units. The gate is controlled by a very stable internal time base that is usually a crystal oscillator. The time base is obtained by dividing the crystal oscillator frequency by successive powers of 10. The division is usually accomplished by a heavily biased relaxation oscillator that is synchronized to to the crystal oscillator frequency (as explained in Chapter 16).

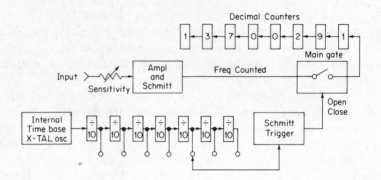

Figure 18.21   A block diagram of the Hewlett-Packard 5360A frequency counter. (*Diagram courtesy Hewlett-Packard.*)

Most commercial frequency counters can be altered to measure the period of a sine wave instead of the frequency. Figure 18.22 shows a block diagram illustrating how a frequency counter can measure the period of a sine wave.

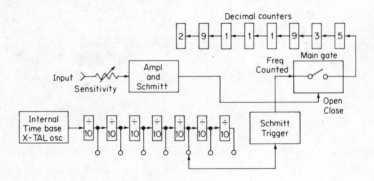

Figure 18.22   A block diagram showing how the Hewlett-Packard 5360A frequency counter can be used to measure the period of a sine wave.

An example of a frequency counter that can be used to measure either frequency or period is the Hewlett Packard model 5360A frequency counter shown in Fig. 18.23. This instrument is very fast, being capable of making measurements of frequencies from 0.01 to 320×10^6 Hz. The crystal oscillator used as the time base is stable to 5 in 10^{10} parts over a 24-hour period. The readout can be made from lighted numerals on the front panel or can be taken as a digital BCD output with an 8421 code.

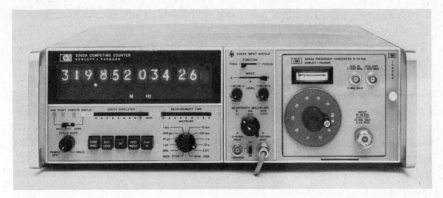

Figure 18.23 A Hewlett-Packard model 5360A frequency counter. (*Photograph courtesy Hewlett-Packard.*)

18.7 The Digital Computer

Digital computers have an extremely wide variety of uses. Scientists, engineers, and others use large, high speed digital computers for calculations or data analysis that would otherwise be exceedingly tedious or even impossible because of the complexity or length of the calculations. For this type of work the bigger and faster the digital computer, the better it can perform the job. Small digital computers also have many uses for a modern scientist or engineer. Small digital computers are widely used in science and engineering to perform on-line data storage and analysis. The term *on-line* means that the computer receives experimental data as electronic signals directly from the instruments used in an experiment. The data are not recorded in a data book and then presented to the computer by the experimenter. The computer itself performs all the data storage, reduction, and analysis, and so on, in an on-line setup. Small on-line digital computers are used by biologists or chemists performing X-ray diffraction experiments on complicated organic or biological crystals, by nuclear physicists carrying out nuclear scattering experiments, and by many other scientists working in a variety of fields. In many areas of research and development the small digital computer is becoming a standard electronic tool just like a VOM or oscilloscope. This trend is sure to continue and even increase. For this reason we describe the organization and the mode of operation for a digital computer.

A major part of a digital computer is the memory. The memory is usually a core memory although other types of memory are possible. One type of memory other than a core memory is an integrated circuit memory.

In an integrated circuit memory a large number of flip-flops are used to store information. These flip-flops are constructed as an integrated circuit with many flip-flops on a single chip of silicon.

A typical core in a core memory is shown in Fig. 18.24. The donut shaped core is made of a very hard magnetic material; that is, the retention field of the material is very large. The hysteresis loop for a core is shown in Fig. 18.25. In Fig. 18.25, B_r is the retention field and I_c is the coercive current for the magnetic material from which the core is made. When the magnetic core is magnetized with magnetization $+B_r$, the system is considered as logic state 1; when magnetized with magnetization $-B_r$, the system is in logic state 0. The field B in the core is produced by a current in a wire running through the core and approximately along the axis of the core. If we assume that the wire is exactly along the axis of the core, then from Ampere's law we have $B = \mu_r \mu_0 I / 2\pi r$ where r is the radial distance from the wire to the point where B is to be determined. At a distance of $\frac{1}{4}$ millimeter from a wire $B = (\mu_r 4 \cdot 10^3) \times \mu_0 I / 2\pi = 8 \times 10^{-4} \mu_r I$ (webers/meter^2). For a typical magnetic material $\mu_r = 1.25 \cdot 10^3$ so that $B = I$ (webers/meter^2). Since 1 weber/meter$^2 = 10^4$ gauss, we have $B = 10^4 \times I$ (gauss). In order to saturate a typical magnetic material, B must be about 3×10^3 gauss so that the current I_0 that will saturate the core must be about $I_0 = 0.3$ amp. Because the hysteresis loop is very square if the magnetization of the core is $B \approx -B_r$, then the current $I_0/2$ will not change the state of the core to $+B_r$ but the current I_0 will change the state of the core to B_r.

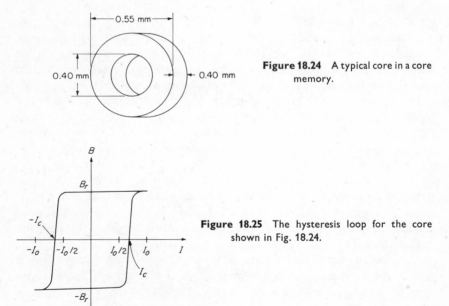

Figure 18.24 A typical core in a core memory.

Figure 18.25 The hysteresis loop for the core shown in Fig. 18.24.

Let us now assume that a wire threading through the core forms a loop. The question we now ask is, How big an emf is induced in the loop when the magnetic core changes its state from $+B_r$ to $-B_r$? This question is answered using Faraday's law of induction, which states that the induced emf $= -d\phi/dt$ where ϕ is the flux through the loop. The core can switch its state in about 2×10^{-7} sec. Therefore,

$$\text{emf} = -\frac{\Delta B A}{t} = -\frac{2 B_r A}{t} = \frac{2 \times 0.3 \times 3 \times 10^{-8}}{2 \times 10^{-7}} = 90 \text{ millivolts}$$

The actual voltage pulses are often smaller than 90 millivolts, being perhaps 40 millivolts.

We have seen that reasonable currents in a wire threading through the core can change the state of the magnetization of the core and that changes in the state of magnetization of the core produce measurable voltages in a circuit composed of a wire loop threaded through the core.

The wire used to thread through the cores is typically about 3×10^{-3} in. in diameter. The currents involved always occur as very short pulses. So thin a wire could not carry a steady current of 0.3 amp.

What we now want to learn is how individual cores are assembled into a useful core memory for a digital computer. The individual cores are assembled on a sheet called the *bit plane*, as is shown in Fig. 18.26. The individual cores are arranged in a square matrix on the bit plane. There is a core at each location (X_i, Y_j). Each location (X_i, Y_j) is called an *address* on the bit plane. There is an X_i driver line wire that runs through each core in any row labeled by X_i. Similarly there is a Y_j driver line wire that runs through each core in any column labeled by Y_j. Let us focus our attention on the particular address (X_1, Y_1) in Fig. 18.26. Suppose that all the cores on the bit plane are initially in the logical state 0. If a current $I_0/2$ is used to drive both the X_1 and the Y_1 lines simultaneously, then the core at location (X_1, Y_j) will have a total current I_0 passing through it and this core will change to the logical state 1. No other core on the bit plane will have its state altered since the maximum current through any other core is $I_0/2$. If the core at (X_1, Y_1) had been in the logical state 1 instead of 0, it would not have been affected by the simultaneous currents $I_0/2$ in both the X_1 and Y_1 lines. By selecting the correct drive, it is possible to write a 1 at any address on the bit plane. In the same way the currents $-I_0/2$ applied simultaneously to both the X_1 and the Y_1 lines will drive the core at (X_1, Y_1) to the logical state 0 if it is initially in the state 1 but will not affect the state of the cores at other addresses. A core memory of this type is called a *coincident current memory*.

In addition to the X and Y drive lines there are two wires each of which thread continuously through every core (that is, through all N^2 cores) on a bit plane one after the other. These two wires are called the *inhibit line* and the *sense line*. These lines are shown in Fig. 18.26. Before discussing the use of these lines we must discuss how the bit planes are organized into the complete core memory. Figure 18.27 shows several bit planes organized into a core memory. In Fig. 18.27 the individual cores are not shown. Of course actually each bit plane contains N^2 cores. There are only four bit planes shown, but in a real memory there are often 16 or more bit planes. A binary number is stored at a given address, for example, (X_i, Y_j), with the 2^0 digit at the (X_i, Y_j) location on the first bit plane, the 2^1 digit at the (X_i, Y_j) location in the second bit plane, and so on until the 2^k digit is stored at the (X_i, Y_j) location in the $k + 1$ bit plane. Obviously the number of bit planes determines the largest number that can be stored at a given address. For a very small computer there might be, for example, 16 bit planes in the core memory. Thus $(2^{16} - 1) = 65,535$ would be the largest number that could be located at a single address.

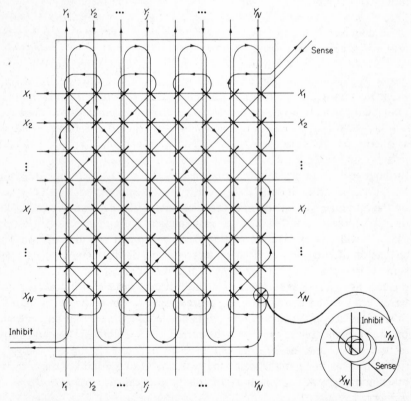

Figure 18.26 Cores arranged on a bit plane. The magnetic cores are represented by the short dark lines at 45° to the horizontal. The insert shows a blow up of how the four wires for the X_N driver line, the Y_N driver line, the sense line, and the inhibit line pass through the hole in the core at the address X_N, Y_N.

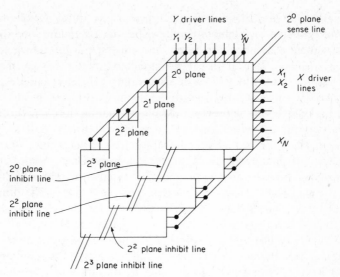

Figure 18.27 The organization of bit planes into a complete core memory as seen from below. Note that all the X_i driver lines for the different bit planes are connected together so that if one drives the X_i driver line, then all the cores threaded by the X_i driver lines in all the different bit planes have the current $I_0/2$ pass through them. Similarly the Y_j driver lines for the different bit planes are connected together. The sense lines for each bit plane are independent although only the 2^0 bit plane sense line shows in the picture. The inhibit lines for each bit plane are independent. Although only four bit planes are shown, there are as many bit planes as there are bits in a word. This would imply that there would be 16 or more bit planes in a core memory stack.

In order to write a digital number into a particular address, say, (X_i, Y_j), the X_i and the Y_j driver lines for all the bit planes are simultaneously pulsed to carry $I_0/2$. The inhibit line of a bit plane is driven with a pulse $-I_0/2$ for each bit that is to be held at zero. Thus if the binary digits 101 are to be stored in the (X_i, Y_j) address of the first three bit planes, then the X_i and Y_j driver lines both carry $I_0/2$ for all the bit planes. The inhibit line for the 2^1 bit plane carries a current $-I_0/2$. The inhibit lines for the 2^0 and 2^2 bit planes carry no current. The process described assumes that all the cores of a given address are in the logic state 0 before the digital number was written into the address.

In order to read the number in the (X_i, Y_j) address, both the X_i and the Y_j drive lines for all bit planes are pulsed to carry currents $-I_0/2$. If a core in a given bit plane was in the logic state 1, then it is flipped to the state 0 and the changing flux will induce a voltage pulse in the sense line for the bit plane. If the core was in the logic state 0, then the pulse induced in the sense line will be small. Since reading a number at a given address destroys the information at the address, it is necessary to make provisions to rewrite the number into the memory if one wishes to save the information. The fact that reading a number at a given address destroys the information at that address is very important. Before a number is written at a given address, the address is always read. Therefore the memory cores in the address are always zero before a new number is written into the address. It was because of this that we assumed in the previous paragraph that all the cores forming a given address were zero before a number was written at that address.

There are two modes of operation for the computer when a number is read out from an address in the core memory. One of these is called *read-modify-write*. In this process the number stored at an address in the memory is read and a new number is written into the address. The other mode of operation is called *read-restore*. In this process the number stored at an address is read and then rewritten into the address. Normally if a computer is in what is called a *fetch cycle*, the computer is wired so as to execute the read-restore operation automatically and the person programming the computer to carry out instructions does not need to include the restore operation in his program.

The number at a given address in the core memory is read out from the core memory into a register called the *memory information register*. A register is simply a series of JK flip-flops that records a binary number.

If a core memory is designed for 16-bit binary numbers, called 16-bit words, each register in the computer will have 16 flip-flops, one for each bit in the 16-bit word. Numbers are read into an address in the core memory from the memory information register.

The core memory address being read is stored in a register called the *memory address register*. All the processes in reading or writing a number into an address are carefully timed. The transfer of information from, say, the core memory to the memory information register occurs at a time determined by internal timing pulses of the computer. The clocking of JK flip-flops was previously discussed in the chapter on digital circuits. The memory cycle time is the time required to complete a read-restore cycle in the core memory. A typical memory cycle time is 10^{-6} sec.

The core memory together with the memory information register and memory address register make up the complete memory of a digital computer.

Next we shall discuss how a computer is organized in order to utilize the core memory, which we have just discussed. Figure 18.28 diagrams a possible organization for a digital computer. In Fig. 18.28 three separate core memories are shown. In a small computer each core memory might have either 4096 or 8192 separate addresses. These are called $4k$ and $8k$ memories, respectively. There actually might be more or fewer core memories than three. There are also two direct memory access stations shown. Again there might be more or fewer of the direct memory access channels. Direct memory access permits one to take information directly out of or put information into any of the core memories without going through the central processing unit. The central processing unit can put information into an address in any of the memories. The central processing unit can put information into one memory, and simultaneously a direct memory access unit can put information into another of the memories thus allowing computation and high speed input or output simultaneously.

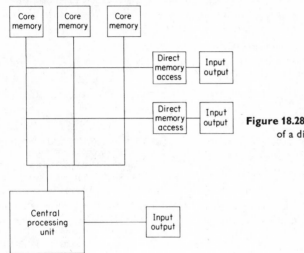

Figure 18.28   The basic organization of a digital computer.

With the rather simplified view of a computer given in Fig. 18.28 in mind, let us consider a somewhat more detailed view of a computer. For simplicity we shall show only one of the several memories and only one of the several direct memory access channels in our detailed picture of the computer. Figure 18.29 shows a fairly detailed block diagram of a computer.

In addition to the information stored in the core memory, information is stored in various registers for short times. In Fig. 18.29 in addition to the boxes labeled *memory information register*, *memory address register*, *memory buffer register*, and *instruction register*, there are registers in the program counter and the accumulator.

Before describing how the computer carries out its instructions we should stress that all the operations of a computer are carefully timed. The clock shown in Fig. 18.30 is a crystal controlled oscillator. A sequence of timing pulses are derived from the oscillator. The time between successive timing pulses is related to the time delay involved in a signal passing through the various logic gates (for example, a NAND gate) used in the computer. The

JK flip-flops used in the computer are driven by the clock pulses, the leading edge of the pulse fixing the state of the input and the transfer of information occurring on the negative trailing edge of the pulse as discussed in Chapter 17. The arrows leaving the box labeled *clock* in Fig. 18.29 are intended to indicate that the clock controls the timing of all the various operations of the computer. The timing in a computer is extremely important because it permits one to look for a pulse at exactly the right time and after the data lines settle. This is an enormous aid in eliminating the problems of distinguishing noise pulses from real data pulses. Without the careful timing of every operation by the clock pulses it probably would not be possible to construct a digital computer.

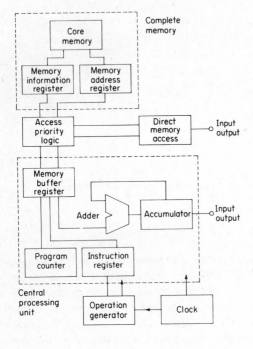

Figure 18.29   A block diagram of a small digital computer. This block diagram is based to some extent on the construction of the PDP8 computer manufactured by Digital Equipment Corporation. Other computers will have some but not all features in common with the diagram shown.

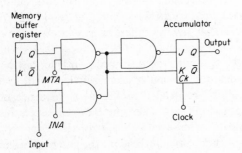

Figure 18.30   A schematic diagram showing how information can be transferred to the accumulator from either the output of the memory buffer register or from an external input.

Perhaps the timing of processes in a computer may be somewhat clearer if we discuss how a particular transfer of information might be made in a computer. Let us consider how one bit of a binary number might be put into the accumulator register from either the memory buffer register or from an external input. In Fig. 18.30 the circuit that will accomplish this transfer is shown. In Fig. 18.30 let us suppose that we wish to transfer a bit of a number from the memory buffer register to the accumulator. The bit to be transferred is at the Q output of one of the memory buffer register's JK flip-flops. Let us assume that regularly spaced timing pulses from the clock in Fig. 18.29 are called $ET1$, $ET2$, $ET3$, ..., and so on. On the trailing edge of the pulse $ET1$ the terminal MTA receives a positive pulse from the operation generator. This produces an input voltage pulse at the input of the JK flip-flop in the accumulator. The pulse $ET2$ is used to clock the accumulator's JK flip-flop so that the signal at the input to this flip-flop is transferred to the output on the trailing edge of $ET2$. At the trailing edge of $ET2$ the signal at MTA is also cut off. All these steps are controlled by the operation generator. If an external signal was to have been put into the accumulator, the operation generator should have produced a pulse at INA rather than at MTA. The sequence of events described in transferring a signal from the memory buffer register to the accumulator are shown in Fig. 18.31.

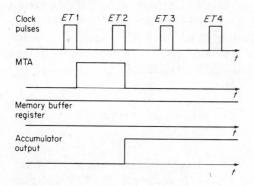

Figure 18.31 The voltages at various points in the circuit of Fig. 18.30 as a function of the time. Note how all transfers of information in the computer are controlled by the timing pulses $ET1$, $ET2$, and so on. The memory buffer register output is a logical 1 for all times shown.

Each direct memory access channel shown in Fig. 18.28 has its own separate memory buffer register. The input and output devices associated with the direct memory access are usually very fast, being, for example, a rapidly rotating magnetic disc or drum.

In order to use a digital computer, a set of instructions, coded in digital form, and whatever data are required are put into the computer. The input to the computer might be through the use of a teletype, paper tape, magnetic tape, or punched cards. This information goes into a section called the *accumulator* or into a direct memory access channel. Although it may not be obvious from the drawing, a digital number or word stored in the accumulator can be transferred to the memory buffer register and then into the core memory. The instructions are put into the memory in the order they are to be carried out so that the first instruction might be address 0, the second instruction at address 1, and so on. Actually the first instruction may not be at the first address, but whatever its address the second instruction is at the next address and so on. The data is stored at a known set of addresses. The computer is now ready to carry out the computations given on the instructions. The instructions are called a *program*. A computer in which the instructions are stored in the memory of the computer is called a *stored program computer*. There is another type of computer in which the machine is wired so as to carry out a particular instruction when a button is pushed. This type of machine is called a *wired program computer*. An example of a wired program machine is a desk calculator that calculates e^x when x is given to the machine and when the e^x button on the front panel is pushed. The great power of a stored program machine occurs because one can write any instructions one wishes and because the instructions can be altered depending on the results of the calculations. These features will be discussed in more detail later.

We should like to discuss the instruction words and data words that are stored in the core memory. In order to do this, let us assume that the core contains words that are 16 bits long. Since $2^{16} = 65,536$, the largest core memory that can be addressed by a 16-bit word contains 65,536 words. A common type of instruction word is shown in Fig. 18.32. In Fig. 18.32 the first 4 bits of the word are the operation code (that is, these bits determine the operation to be carried out) such as add to the accumulator the number whose address is given. The remaining 12 bits are used to give an address.

Figure 18.32   An instruction word.

Since there are only 12 bits, the address given can be only one of $2^{12} = 4096$ addresses, not any one of the 65,536 possible addresses. The 4096 addresses are often used as indirect addresses where the complete address of the number to be operated on is given in the word at the indirect address. A common situation would be for each of the core memory stacks to contain 8192 addresses. In a situation like this, indirect addresses are often required to locate the correct stack and the correct address in the stack. Sometimes several successive indirect addresses are used. We might comment that the 4-bit instruction can only code $2^4 = 16$ instructions. Since this is a small number, it is necessary to devise methods of giving more complicated instructions to the computer. We shall not discuss these methods.

A data word is shown in Fig. 18.33. As seen in Fig. 18.33, the first bit of a data number is used to determine the sign of the word and the remaining 15 bits determine the number. Since $(2^{15} - 1) = 32,767$, no number larger than 32,767 can be recorded in a single address. For numbers larger than 32,767 two or more addresses are used to store one number. The use of several addresses to store a number is called *multiple precision*.

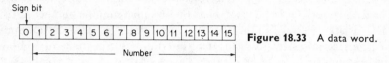

Figure 18.33   A data word.

In order to begin computation the program counter of the computer is set to indicate the first instruction. As various instructions are carried out, the program counter is incremented by 1 (that is, 1 is added to the number in the program counter) to indicate the subsequent instruction. The program counter is set to the first instruction either by an external signal, that is, by pushing a button, or by a signal generated inside the computer after all the instructions and data are stored in the core memory.

The program counter having been set to the address of the first instruction, the computations may begin. The address of the first instruction is entered into the memory address register. The function of the memory address register is to record the address in the core memory that is currently being read or written. The memory address register incremented by 1 is returned to the program counter so that after the execution of the first instruction, the program counter will contain the address of the second instruction. The contents of the first instruction are read and transferred to the memory information register. The first instruction is subsequently transferred from the memory information register to the memory buffer register and rewritten into the address of instruction 1. The first instruction is then transferred to the instruction register that in turn starts the operation generator. This process is called *fetching instruction 1*. The operation generator controls the sequence of events in the entire computer. The arrow leaving the operation generator (Fig. 18.29) indicates that the operation generator controls all the components in the computer. Suppose that the first instruction was to add the contents of address 1000 to the accumulator. The operation generator controls automatically the sequence of events that leads to fetching the contents of address 1000 and executing the first instruction. The events that automatically occur inside the computer would be to set the memory address register to address 1000, read the contents of address 1000 into the memory information register, transfer the contents of the memory information register to the memory buffer register, and rewrite the contents of address 1000 back into the core memory. The data from address 1000 that is in the memory buffer register is put into the adder, added to 0, and put into the accumulator (it has been assumed that the accumulator had been cleared previously). With the first instruction accomplished the computer fetches the second instruction. Suppose that the second instruction is to add the contents of address 1001 to the number in the accumulator, which now contains the contents of address 1000. The process carried out by the computer is the same as for the first step except that after the contents of address 1001 are put into the memory buffer register, both that number and the number in the accumulator are presented to the adder and the total is put into the accumulator. If the second instruction had been to multiply the contents of address 1001 by the number in the accumulator, this would have been accomplished by taking individual products, shifting in the accumulator, and adding in the adder with the final multiplication result stored in the accumulator. It is obvious that if the result of the multiplication is a number with more than 16 digits that additional registers are needed to record the result. The accumulator and the adder together with control logic form the arithmetic unit for the digital computer.

After all the instructions are carried out, the final result is presented to the output, perhaps as a result printed by a teletype machine.

Thus far we have described how a computer is organized and how it carries out the sequence of instructions, one after the other. We should, however, point out that sometimes the operations may involve more than just proceeding from the first instruction through all the other instructions in the order of their occurrence. For example, if one has to carry out the same sequence of instructions many times, one would code the program to repeat a particular sequence of instruction with each set of data points. There are also *jump* instructions or *conditional jump* instructions. An example of a conditional jump instruction might be for the computer program to jump from the present instruction to an instruction several steps removed from the present instruction if some condition is satisfied, for example, if a particular number is less than or equal to another number. In these and other ways the person programming the computer (that is, writing the instructions for the computer) can give alternate instructions to the computer depending on the results of the data analysis or calculation. The fact that the flow of instructions can be modified depending on the results of the previous calculations is very important in making a stored program computer into an extremely useful machine.

As mentioned earlier, small digital computers are being used more and more often in scientific laboratories. An example of a small digital computer is the PDP11 computer that is manufactured by the Digital Equipment Corporation. The PDP11 computer is shown in Fig. 18.34 and 18.35.

Before concluding our discussion of digital computers we should like to illustrate the use of a small digital computer in a scientific laboratory. As a first example, let us discuss the use of a computer for multichannel data storage. A common measurement in a nuclear physics laboratory is the measurement of the pulse height coming from a particle detector that puts out voltage pulses between, say, 0–64 volts depending on the amount of energy a particle loses in a detector. The physicist might want to divide the pulses into 256 channels each of width $\frac{1}{4}$ volt. The number of pulses between $0-\frac{1}{4}$ volt would be recorded in the first channel, the number of pulses between $\frac{1}{4}-\frac{1}{2}$ volt in the second channel and so on. A possible way to do this is the following. First one selects a separate address in the core memory for each of the 256 channels, that is, for each $\frac{1}{4}$-volt spacing from 0–64 volts. In order to analyze the voltage pulses it is convenient to utilize a voltage to time converter. There are several steps in performing this conversion. A voltage pulse that comes from a nuclear detector after pulse shaping might be about 10^{-6} sec in duration. One can construct a fairly simple circuit to produce a voltage which has the same peak value as the pulse but which lasts for a longer time.

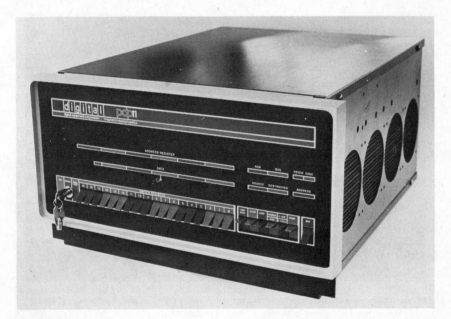

Figure 18.34 A PDP11 digital computer. (*Photograph used by permission of Digital Equipment Corporation, Maynard, Massachusetts.*)

It is now possible to use a circuit similar to the circuit that carries out the voltage to time conversion in a ramp-type digital voltmeter to produce a signal that is a binary digital number proportional to the amplitude of the voltage pulse. It is a simple matter to program the computer to add 1 to the number stored at the address in the core memory that corresponds to the particular pulse height every time the binary number equivalent of that pulse height is presented to the computer. This discussion illustrates two terms that are used with computers, *hardware* and *software*. Hardware is apparatus used in association with the computer and the computer itself. In this case the apparatus for particle detection, pulse shaping and amplification, the formation of a steady voltage of the same magnitude as the peak of the pulse, the pulse height to time conversion, and finally the formation of a binary signal proportional to the time (that is, to the pulse height) are all examples of hardware. The software is the computer program. In this case the software was the program that added 1 to the address corresponding to binary number representing the pulse height. In the use described the computer is used to sort and store multichannel data. Later the computer could be used to analyze the data to yield results that are easily related to theoretical considerations.

Figure 18.35 A dual processer system utilizing two PDP11 digital computers. Note that the digital computer occupies a small fraction of the space occupied by hardware such as teletype printers, tape decks, etc. The cost of this accessory hardware is often as much or more than the cost of the basic computer. (*Photograph used by permission of Digital Equipment Corporation, Maynard, Massachusetts.*)

As a second example of the use of a small digital computer in a scientific laboratory we shall discuss the possible control of an experiment on X-ray diffraction from, for example, a biological crystal by an on-line computer. Suppose a crystallographer wishes to measure the intensity of X rays diffracted from a sample for scattering angles ranging from 1–90° in a particular plane. If the time required to accumulate 1000 counts is to be measured at 1° intervals, then 90 addresses in the core memory are set aside to store the time required for each measurement. The X-ray detector may be moved from one position to the next by a stepping motor.

The voltage pulse from the X-ray detector having been suitably amplified and shaped is presented to the computer. The computer is programmed so that for every pulse entering the computer 1 is added to the number stored at an address we shall call CTR. After a pulse comes into the computer and 1 is added to the number stored at the address CTR, the computer tests to see if the new number (that is, 1 plus the old number) stored at CTR has reached the binary equivalent of 1000 yet. This might be done by subtracting 1000 from the number at CTR and testing to see if the result is 0. If the number stored at CTR has not reached 1000, then the process is repeated when the next pulse comes from the X-ray detector. If the new number at CTR is 1000, then a program is activated to do the following. The time required to record 1000 counts is stored at the appropriate address in the core memory. This time is determined by a separate part of the program that causes the time, as determined by the crystal oscillator clock in the computer, to be stored at an address we call TIME. The number at the address TIME is taken and put into the address corresponding to the angle being measured. The program next clears, that is, resets, to zero the addresses CTR and TIME and puts out a signal that causes the stepping motor to drive the X-ray detector through 1° to the next position where the X-ray intensity is to be measured. The program also increments by 1 the address at which the next time interval will be stored. The entire procedure is repeated until the time required to obtain 1000 counts at each of 90 positions is determined and stored in the computer's core memory. In this application almost everything is done by the software, that is, by the program. The computer acts as a scaler, a timer, and a multichannel data storage device.

The two examples we have presented provide a strong contrast in that the first example uses a large amount of hardware and a simple program. The second example uses almost no hardware and does almost everything with software.

We should like to stress that digital computers are used by scientists in many ways. Complicated tedious arithmetic calculations are often performed. In addition, small digital computers are being used more and more in an on-line manner to replace scalers, timers, and multichannel data storage equipment and to control the sequency of entire laboratory experiments. These uses of small digital computers seem certain to expand greatly in the future and almost all scientists and engineers will need to understand the functions that small digital computers can perform.

Answers to Problems

Chap. 1

Prob. 1.1 $\tau = 2.45 \times 10^{-14}$ sec
Prob. 1.2 $Q = 4.5 \times 10^5$ coulombs
Prob. 1.3 $\rho = 2.55 \times 10^{13}$ ohm-meters
Prob. 1.5 $R_{eq} = 8.5$ ohms; $I = 0.6$ amp; $V_L = 5.1$ volts
Prob. 1.6 $R_{eq} = 1.5$ ohms; $I = 2$ amp; $I_6 = 0.5$ amp; $I_2 = 1.5$ amp; $V_L = 3$ volts
Prob. 1.7 $V = 3.16$ volts
Prob. 1.8 $1/R = 25.9$ mhos
Prob. 1.9 $R_{eq} = 10$ ohms
Prob. 1.10 $R_{eq} = 2/3$ ohms. The resistance is not changed by an electrical contact at the center of the circuit.
Prob. 1.11 One independent current law equation; two independent voltage law equations.
Prob. 1.12 $R' = R/\sqrt{3}$; $V_R = \varepsilon/(2 + \sqrt{3})$
Prob. 1.13 $\varepsilon_{Th} = \varepsilon/2$; $r_{Th} = 3R'/2$
Prob. 1.15 $P = \varepsilon^2/160$ watts
Prob. 1.17 $I_3 = 1.2$ amp; $V = 3.6$ volts
Prob. 1.18 $\varepsilon_{Th} = I_0 r$; $r_{Th} = r$; $I_s = I_0$
Prob. 1.19 $P = 72$ watts
Prob. 1.23 $\varepsilon_{Th} = 101$ volts; $r_{Th} = 1.01$ ohms
Prob. 1.25 $\varepsilon_{Th} = 3.75$ volts; $r_{Th} = 0.25$ ohms
Prob. 1.26 $V = -10$ volts; $I_1 = 3$ amp; $I_2 = -1$ amp; $I_3 = 2$ amp

Chap. 2

Prob. 2.1   6 nodes; 15 branches; 5 independent current law equations; 10 independent voltage law equations.
Prob. 2.2   7 independent current law equations; 6 independent voltage law equations. $I = 0.12$ amp
Prob. 2.3 $I = 0.04$ amp
Prob. 2.4   33.3 volts (edges which connect with the body diagonal); 16.67 volts (edges which do not connect with body diagonal).
Prob. 2.5 $I_1 = 0.04\varepsilon_1 + 0.04\varepsilon_2 = I_{11} + I_{12}$;
$I_{11} = 0.4$ amp ($\varepsilon_1 = 10$ volts);
$I_{12} = 0.2$ amp ($\varepsilon_2 = 5$ volts)

Prob. 2.6   Square, $R_{eq} = R/2$; triangular $R_{eq} = R/3$; hexagonal, $R_{eq} = 2R/3$

Prob. 2.7 $\varepsilon_{Th} = 4$ volts; $r_{Th} = 160$ ohms; $R = r_{Th} = 160$ ohms; $P = 2.45$ watts

Prob. 2.8 $I_N = 0.025$ amp; $r_N = r_{Th} = 160$ ohms

Prob. 2.9 $\varepsilon_{Th} = \varepsilon R_2/(R_1 + R_2)$; $r_{Th} = R_1 R_2/(R_1 + R_2)$

Prob. 2.10 $\varepsilon_{Th} = (\varepsilon_1 R_2 - \varepsilon_2 R_1)/(R_1 + R_2)$; $r_{Th} = R_1 R_2/(R_1 + R_2)$

Prob. 2.11 $I_1 = I_{11} + I_{12} = \dfrac{\varepsilon_1 R_3 (R_2 + R_4) + \varepsilon_2 R_2 (R_3 + R_4)}{R_1 (R_2 R_3 + R_2 R_4 + R_3 R_4) + R_2 R_3 R_4}$

Prob. 2.12 $I = 0.05$ amp in both cases

Prob. 2.13 $I_1 = 0.057$ amp

Prob. 2.14 $R_A = 14.3$ ohms; $R_B = 28.6$ ohms; $R_C = 57.2$ ohms

Prob. 2.15 $R_1 = 175$ ohms; $R_2 = 700$ ohms; $R_3 = 350$ ohms

Prob. 2.16 $I = 1.62$ amp

Prob. 2.17 $R_{11} = R_{22} = 133.3$ ohms; $R_{12} = R_{21} = -400$ ohms
             $g_{11} = g_{22} = 7.5 \times 10^{-3}$ mhos; $g_{12} = g_{21} = -2.5 \times 10^{-3}$ mhos
             $r_{11} = r_{22} = 150$ ohms; $r_{12} = r_{21} = 50$ ohms
             $h_{11} = 133.3$ ohms; $h_{22} = 6.67 \times 10^{-3}$ mhos
             $h_{12} = -h_{21} = 1/3$

Prob. 2.18 $R_{11} = R_{22} = 26.67$ ohms; $R_{12} = R_{21} = -44.4$ ohms
             $g_{11} = g_{22} = 37.5 \times 10^{-3}$ mhos; $g_{12} = g_{21} = -22.5 \times 10^{-3}$ mhos
             $r_{11} = r_{22} = 41.6$ ohms; $r_{12} = r_{21} = 25$ ohms
             $h_{11} = 26.67$ ohms; $h_{12} = -h_{21} = 0.6$
             $h_{22} = 24 \times 10^{-3}$ mhos

Prob. 2.19 $R_{eq} = [R_1 (R_1 + 2R_2)]^{1/2}$;

$$V_n = \varepsilon \left[\frac{R_2}{R_1 + R_2 + R_{eq}} \right]^n$$

Prob. 2.20 $R_{1\pi} = R_1 (R_1 + 2R_2)/R_2$;
             $R_{2\pi} = R_{3\pi} = R_1 + 2R_2$
             where $R_{1\pi}$, $R_{2\pi}$, and $R_{3\pi}$ are three resistors in the Δ connection of
             Fig. 2.11.

Prob. 2.21   The network does not have a Thevenin equivalent circuit.

Prob. 2.22   At either connection between the resistor and the tungsten lamp the
current that leaves one circuit enters the other. Thus, if one characteristic curve is plotted as positive V versus positive I the other must be
plotted as positive V versus negative I. In this case the only intersection
of the characteristic curves is at the origin.

Prob. 2.23   Within the accuracy of the VTVM ($\pm 0.5\%$) and for the currents
drawn the network is linear and $\varepsilon_{Th} = 300$ volts; $r_{Th} = 10$ ohms.

Chap. 3

Prob. 3.1   For maximum current sensitivity use the maximum number of turns
of the finest wire. For maximum voltage sensitivity use a single turn of
the heaviest wire. The angular displacement divided by the power

dissipated in the coil is independent of the choice of the number of turns. However the angular displacement is not proportional to the power dissipated. Thus the power sensitivity is not a constant.

Prob. 3.2 $f = 1$; $I_g = 0.97 \times 10^{-7}$ amp
$f = 0.1$, $I_g = 0.97 \times 10^{-8}$ amp
$f = 0.01$, $I_g = 0.97 \times 10^{-9}$ amp
$f = 0.001$, $I_g = 0.97 \times 10^{-10}$ amp
$f = 0.0001$, $I_g = 0.97 \times 10^{-11}$ amp

Prob. 3.3   Use a single turn of wire in the shape of a square.

Prob. 3.4 $V = 2.5$ volts, $R = 4.5 \times 10^4$ ohms
$V = 10.0$ volts; $R = 1.95 \times 10^5$ ohms
$V = 250$ volts, $R = 5 \times 10^6$ ohms

Prob. 3.5 10^{-2} amp, $R = 25$ ohms
1 amp, $R = 0.25$ ohms

Prob. 3.6 $\varepsilon_{\text{Th}} = 5 \times 10^{-3}$ volts; $r_{\text{Th}} = 1000$ ohms, $I_g = 2.5 \times 10^{-6}$ amp

Prob. 3.8 $R = 2.01 \times 10^5$ ohms

Prob. 3.9   (a) No; (b) Yes

Prob. 3.10 $r_g = 1000$ ohms

Prob. 3.11 $R_1 R_4 = R_2 R_3$, the balance condition is unchanged

Prob. 3.12 $R_4 = 500$ ohms; $\varepsilon_0 = 5$ volts

Prob. 3.13 $R_P = 1500$ ohms; $R = 1500$ ohms

Prob. 3.14 $\varepsilon_{\text{Th}} = 0.25$ volts; $r_{\text{Th}} = 229$ ohms
If Δ, the unbalance voltage, is very much less than $\varepsilon_x = 0.25$ volts then $r_g \simeq 0$ gives maximum power transfer through AB. Note that the circuit to the left of AB is connected to an active network on the right. Thus the relationship $R_{\text{LOAD}} = r_{\text{Th}}$ for maximum power transfer to a passive circuit does not hold.

Prob. 3.17   The full scale meter current ($R_x = 0$) is 50 microamp.

Prob. 3.19   A current of 6 amp is drawn for a short time until the meter coil or hopefully a fuse burns out. Unless the fuse is very fast an angular impulse large enough to destroy the needle and coil is imparted.

Chap. 4

Prob. 4.1 $d = 4.4 \times 10^{-5}$ meters

Prob. 4.2 $C = 2\omega \ell \, \varepsilon_r \varepsilon_0 / d$

Prob. 4.3 $C = 7.1 \times 10^{-8}$ farads; $V_{\text{max}} = 39$ volts

Prob. 4.5 $L_{\text{min}} = 3.3 \times 10^{-3}$ henries;
$L_{\text{max}} = 1.6$ henries
These answers ignore the spreading of the field at the ends of the coil.

Prob. 4.6 $R = 106$ ohms

Prob. 4.9 $I = 10^{-3} e^{-t/10^{-2}}$ amp; $V_R = 10 e^{-t/10^{-2}}$ volts,
$V_C = 10(1 - e^{-t/10^{-2}})$ volts;
$Q = 10^{-5}(1 - e^{-t/10^{-2}})$ coulombs

Prob. 4.10   In most experimental situations an exponential process is completed in less than ten time constants ($\tau = 10^{-2}$ sec). $I = 10^{-4} e^{-t/10^{-2}}$ amp

Prob. 4.11 $I_C = 5 \times 10^{-4} e^{-t/5 \times 10^{-4}}$ amp

Prob. 4.12 $I_L = (1 - 0.5e^{-t/2 \times 10^{-7}})10^{-3}$ amp;
$I_\varepsilon = (1 - 0.25e^{-t/2 \times 10^{-7}})10^{-3}$ amp

Prob. 4.14 $I_1 = (5.0e^{-t/\tau_L} - 1.65e^{-t/\tau_s})10^{-4}$ amp;
$I_2 = (0.27e^{-t/\tau_L} + 3.06e^{-t/\tau_s})10^{-4}$ amp
where $\tau_L = 1.03 \times 10^{-1}$ secs; $\tau_s = 7.3 \times 10^{-3}$ secs
This problem and also the next lead to a system of two simultaneous
first order differential equations. Their solution is instructive but tedious.
Because $RC_2 \ll RC_1$ an approximate solution for I_2 can be obtained
by assuming that C_1 remains uncharged while C_2 charges up. I_1 can
be approximated by assuming that I_2 is very small while C_1 charges up.

Prob. 4.15 $I_1 = (3.33 - 3.06e^{-t/\tau_L} - 0.27e^{-t/\tau_s})10^{-6}$ amp
$I_2 = (3.33 + 1.65e^{-t/\tau_L} - 5.0e^{-t/\tau_s})10^{-6}$ amp
where $\tau_L = 6.85 \times 10^{-9}$ sec; $\tau_s = 4.84 \times 10^{-10}$ sec
Useful approximations to these results can be obtained quickly by
methods similar to those suggested in Prob. 4.14.

Prob. 4.16   At $t = 0$: $I = 0$; $V_L = \varepsilon$; $V_C = V_R = 0$
At $t = \infty$: $I = 0$; $V_C = \varepsilon$; $V_L = V_R = 0$

Prob. 4.17   At $t = 0$: $I_{R_1} = I_C = \varepsilon/R_1$; $I_{R_2} = I_L = 0$
$V_{R_1} = \varepsilon$; $V_{R_2} = V_L = V_C = 0$
At $t = \infty$: $I_{R_1} = I_L = \varepsilon/R_1$; $I_{R_2} = I_C = 0$
$V_{R_1} = \varepsilon$; $V_{R_2} = V_L = V_C = 0$

Prob. 4.18 $V_i = \varepsilon_0 e^{-t/RC}/(1 + e^{-t/RC})$;
$V_f = \varepsilon_0/(1 + e^{-t/RC})$
on charge $V = \varepsilon_0[1 - e^{-t/RC}/(1 + e^{-t/RC})]$
on discharge $V = \varepsilon_0 e^{-t/RC}/(1 + e^{-t/RC})$

Prob. 4.20 $I = 0.2$ amp

Chap. 5

Prob. 5.1   If $f = 10^4/2\pi$ Hz, $Z = 50(3 - j)$ ohms
If $f = 10^6/2\pi$ Hz, $Z = (100 - j)$ ohms
The circuit is capacitive

Prob. 5.2 $Z = 50(1 + j)$ ohms. The circuit is inductive.

Prob. 5.3 $I_L = 0.14$ amp; $I_R = 0.10$ amp; I_R lags the applied emf by 90°.

Prob. 5.4 $Q = \dfrac{\varepsilon_0 C}{[(\omega RC)^2 + 1]^{1/2}} [\cos(\omega t - \phi) - e^{-t/RC}\cos\phi]$

$I = \dfrac{\varepsilon_0 C}{[(\omega RC)^2 + 1]^{1/2}} \left[-\omega \sin(\omega t - \phi) + \dfrac{\cos\phi}{RC} e^{-t/RC} \right]$

where $\phi = \tan^{-1}(\omega RC)$

Prob. 5.5 $I_\varepsilon = 1.41 \times 10^{-2}$ amp; $I_R = 10^{-2}$ amp; $I_C = 10^{-2}$ amp

Prob. 5.6 $\omega_0 = 2.24 \times 10^7$ rad/sec; $Q = 8.94$; V_L and V_C differ in phase by 180°;
$V_L > V_C$ for all $\omega > \omega_0$; $Z = 50$ ohms for $\omega = \omega_0$.

Prob. 5.7 $I = 0.2$ amp; $V_R = 10$ volts;
$V_L = V_C = 89.4$ volts.
All quantities are peak values.

Prob. 5.8   For $\omega = \omega_0/2$: $I = 1.49 \times 10^{-2}$ amp;
$$V_R = 0.74 \text{ volts};$$
$$V_L = 3.32 \text{ volts}; V_C = 13.3 \text{ volts}$$
For $\omega = 2\omega_0$: $I = 1.49 \times 10^{-2}$ amp;
$$V_R = 0.74 \text{ volts};$$
$$V_L = 13.3 \text{ volts}; V_C = 3.32 \text{ volts}$$
All quantities are peak values.

Prob. 5.9 $Z = 100(3 + 19j)$ ohms

Prob. 5.14 $I_S = 10^{-2}$ amp

Prob. 5.15 $I_P = 0.2$ amp; $I_S = 2$ amp; $N_P/N_S = 10$; $P_P = 32$ watts, this is the power dissipated as heat in the primary circuit. The voltage driving the primary circuit is in phase with the primary current.

Prob. 5.16   For $f = 10^4/2\pi$ Hz: $Z = 100(1 + j)$ ohms; $I = 3.54 \times 10^{-2}$ amp; $P_R = 0.125$ watts; The current lags the applied voltage by $45°$. For $f = 10^6/2\pi$ Hz: $Z = 100(1 + 100j)$ ohms; $I = 5 \times 10^{-4}$ amp; $P_R = 2.5 \times 10^{-5}$ watts; the current lags the voltage by $\phi = \tan^{-1} 100$ (almost $90°$).

The applied voltage is 5 volts rms. The currents given are rms values.

Prob. 5.17 $I_P = 0.399$ amp; $I_S = 0.133$ amp

Prob. 5.18 $$I_1 = \frac{(R_L + j\omega L_2)\varepsilon_0\, e^{j\omega t}}{(R + j\omega L_1)(R_L + j\omega L_2) + \omega^2 M^2}$$

$$I_2 = \frac{-j\omega M \varepsilon_0\, e^{j\omega t}}{(R + j\omega L_1)(R_L + j\omega L_2) + \omega^2 M^2}$$

As $\omega \to 0$ inductive reactances become very small compared to resistances. When the inductive reactances are negligible I_1 and I_2 must assume the limiting values given. If $k^2 = 1$ the relationship $I_P/I_S = N_S/N_P$ is valid regardless of the resistances in the primary and secondary circuit. We include in these resistances the resistances of the windings themselves. The relationship $V_P/V_S = N_P/N_S$ is valid if by V_P and V_S we mean the voltages measured across the primary and secondary terminals minus the IR drops in the primary and secondary windings respectively. For a well designed transformer used within its design ratings the IR drops across the windings are small compared to V_P and V_S.

Prob. 5.19 $k^2 = 1$; $N_P/N_S = 1/3$

Prob. 5.20   Energy stored $= 10^{-3}$ joules; $\langle P \rangle = 1$ watt

Prob. 5.22 $$I = \frac{\varepsilon_0\, e^{j\omega t}}{R - \dfrac{j}{\omega(N_2{}^2/N_1{}^2)C}}$$

Chap. 6

Prob. 6.1 $Z_{\text{Load}} = 10^3(1 - j)$ ohms; $R = 10^3$ ohms in series with $C = 10^{-9}$ farads; $P_{\text{max}} = 2.5 \times 10^{-2}$ watts

Prob. 6.2 $\omega < \omega_0$, circuit is inductive;
$\omega > \omega_0$, circuit is capacitive

Prob. 6.3 $\omega_0 = 10^9$ rad/sec; $Z_{res} = 5 \times 10^3$ ohms; $Q = 50$; $\langle P \rangle = 2.5 \times 10^{-3}$ watts
(at resonance); $P_{max} = 5 \times 10^{-3}$ watts (at resonance)

Prob. 6.4 At any frequency the source impedance seen by the resistor is zero
and the average power dissipated in the resistor is $P_R = \varepsilon_0^2/2R$
$\omega \ll \omega_0$: $Z_L = j\omega L$; $I = \varepsilon_0 e^{j\omega t}/j\omega L$
$\omega = \omega_0$: $Z_L = R$; $I = \varepsilon_0 e^{j\omega t}/R$
$\omega \gg \omega_0$: $Z_L = -j/\omega C$; $I = j\omega C\varepsilon_0 e^{j\omega t}$

Prob. 6.5 $Z = \dfrac{10^5}{1 + 10^5 j(\omega/10^6 - 10^6/\omega)}$ ohms; $\omega_0 = 10^6$ rad/sec; $Z_{res} = 10^5$ ohms;

$$I_{res} = 10^{-5} \text{ amp}; \langle P \rangle_{res} = 10^{-5} \text{ watts}$$

Prob. 6.8 Replace the resistor of problem 6.6 by an inductor and the capacitor
of problem 6.6 by a resistor. Make $L/R = RC = 10^{-5}$ sec.

Prob. 6.9 $\omega_0 = 1/RC = 6.25 \times 10^2$ rad/sec

Prob. 6.11 $R_1 R_4 = R_2 R_3$; $L_4 = C_1 R_2 R_3$; $R_4 = 10$ ohms; $L_4 = 10^{-4}$ henries

Prob. 6.12 $R_1 R_4 = R_2 R_3$; $L_4 R_1 = CR_2(RR_1 + RR_3 + R_1 R_3)$

Prob. 6.13 $R_1 R_4 = R_2 R_3$; $R_2(L_3 + M) = R_4(L_1 - M)$

Prob. 6.15 $R = 1.88 \times 10^6$ ohms; $C = 2.5 \times 10^{-11}$ farads

Prob. 6.16 $I_R = 3 \times 10^{-7}$ amp; $V_C = 1.1 \times 10^{-2}$ volts

Prob. 6.17 At very high frequencies the attenuation, α_{dB}, increases by 40 decibels
per decade increase in ω. One may write

$$V_{out} = \frac{1}{R_1 C_1 R_2 C_2} \int dt \int V_{in}\, dt$$

At any frequency:

$$\frac{V_{out}}{V_{in}} = \frac{1}{[(1 - \omega^2 R_1 C_1 R_2 C_2)^2 + \omega^2(R_1 C_1 + R_2 C_2 + R_1 C_2)^2]^{1/2}}$$

Prob. 6.20 $\omega < 1/R_1 C_1$: α_{dB} decreases by 40 decibels per decade increase in ω.
$1/R_1 C_1 < \omega < 1/R_2 C_2$: α_{dB} decreases by 20 decibels per decade increase
in ω.
$\omega > 1/R_2 C_2$: α_{dB} is zero decibels independent of ω;

$$V_{out} = R_1 C_1 R_2 C_2\, d^2 V_{in}/dt^2$$

At any frequency:

$$\frac{V_{out}}{V_{in}} = \frac{\omega^2 R_1 C_1 R_2 C_2}{[(1 - \omega^2 R_1 C_1 R_2 C_2)^2 + \omega^2(R_1 C_1 + R_2 C_2 + R_1 C_2)^2]^{1/2}}$$

Chap. 7

Prob. 7.1 (a) $I_{rms} = I_0$; $\langle P \rangle = I_0^2 R$
(b) and (c) $I_{rms} = I_0/\sqrt{3}$; $\langle P \rangle = I_0^2 R/3$

Prob. 7.2 $V(t) = \left\{ 50 + \sum\limits_{m=1}^{\infty} \left[(-1)^{m-1} \dfrac{200}{(2m-1)\pi} \right] \cos\left[(2m-1)120\pi t \right] \right\}$ volts

Note that the index m runs through all positive integers. However it enters the argument of the cosine in the form $(2m-1)$. Thus only odd multiples of the fundamental frequency $\omega_1 = 120\pi$ rad/sec occur in the expansion.

Prob. 7.3 $V(t) = \left\{ 5 - \sum\limits_{m=1}^{\infty} \left[\dfrac{10}{\pi m} \right] \sin(20\pi mt) \right\}$ volts

The voltage $V(t)$ is neither even nor odd. It is expressed as the sum of the constant term, which is even, and a sum of sine terms which is odd. It is easily seen from the figure that if one subtracts from $V(t)$ its average value (5 volts) the resulting wave form is odd.

Prob. 7.4 $V_{\text{out}} = \sum\limits_{m=1}^{\infty} \dfrac{(-1)^{m-1}200R \cos\left[(2m-1)120\pi t + \phi_m \right]}{(2m-1)\pi\{ R^2 + [1/(2m-1)120\pi C]^2 \}^{1/2}}$ volts

where $\phi_m = \tan^{-1}\{ 1/[(2m-1)120\pi RC] \}$

Prob. 7.5 $I(t) = \left\{ \dfrac{5}{R} - \sum\limits_{m=1}^{\infty} \dfrac{10 \sin(20\pi mt - \phi_m)}{\pi m[R^2 + (20\pi mL)^2]^{1/2}} \right\}$ amp

where $\phi_m = \tan^{-1}(20\pi mL/R)$

Prob. 7.7 $\overline{V}(\omega) = V_0 \tau \sqrt{\pi} e^{-(\omega\tau/2)^2}$

Prob. 7.8 $\overline{V}(\omega) = e^{-j\omega 3\tau/2} 4V_0 \cos\omega\tau \dfrac{\sin(\omega\tau/2)}{\omega}$

Prob. 7.9 The Fourier series contains the frequencies $(100n)$ Hz where n is an integer. The Fourier transform of a single burst of oscillations is given by

$$|\overline{V}(\omega)| = 2V_0 \left| \dfrac{\sin(\omega - \omega_0)\tau/2}{\omega - \omega_0} \right|.$$

The magnitude of the Fourier components are obtained by evaluating $|\overline{V}(\omega)|$ at $\omega_n = 2\pi(100n)$ rad/sec.

Prob. 7.10 The amplifier should have constant gain from 0 to 10^7 Hz.

Prob. 7.12 $v = 2.1 \times 10^8$ met/sec

Prob. 7.13 $R = Z_0 = 50.4$ ohms

Prob. 7.14 $v = 1.33 \times 10^8$ met/sec

Prob. 7.15 $I = 0.133$ amp; Energy $= 1.33 \times 10^{-5}$ joules

Prob. 7.16 At the moment of superposition all of the energy is in the magnetic field.

Prob. 7.17 $V_{\text{ref}} = -5$ volts; $V_{\text{Trans}} = 5$ volts (in each of the other three lines). All pulses have a length of 10 microsec.

Prob. 7.20 Moving back toward A is a reflected pulse of magnitude -5 volts followed after a time $\tau = l/v$ by a pulse of magnitude 5 volts. Both pulses are 10^{-8} sec in length.

Prob. 7.24 $L_0 \Delta X = 5 \times 10^{-5}$ henries per section;
$C_0 \Delta X = 2 \times 10^{-10}$ farads per section
An ordinary delay line with air as the dielectric must be 3000 met. long to provide a delay of 10 microsec between the input and the output pulse.

Chap. 8

Prob. 8.1 $I = 1.891 T^2 e^{-5.22 \times 10^4/T}$
For $I = 10^{-6}$ amp, $T = 1775°K$
For $I = 10^{-3}$ amp, $T = 2270°K$

Prob. 8.2 For $I = 10^{-3}$ amp, $V = 232$ volts;
For $V = 100$ volts, $I = 2.83 \times 10^{-4}$ amp
$I = 2 \times 10^{-4}$ amp if plate and cathode areas are each doubled.

Prob. 8.3 $V_{PC} = 300 - 10^5 I_P$;
The quiescent point is $I_P = 1.5$ milliamp
$V_{PC} = 150$ volts

Prob. 8.4 $I_P = 1.6$ milliamp; $V_{PC} = 140$ volts; $\Delta v_P/\Delta v_G \simeq -50$

Prob. 8.5 No; No

Prob. 8.6 $g_m = 1.7 \times 10^{-3}$ mhos; $r_p = 58 \times 10^3$ ohms; $\mu = 100$; $A = -63$

Prob. 8.7 The quiescent point is $I_P = 10.3$ milliamp, $V_{PC} = 195$ volts.
$g_m = 2.7 \times 10^{-3}$ mhos; $r_p = 7500$ ohms; $\mu = 20$

Prob. 8.8 $A = 0.94$

Prob. 8.9 For Prob. 8.6: $R_{in} = 10^5$ ohms; $R_{out} = 3.7 \times 10^4$ ohms
For Prob. 8.7: $R_{in} = 10^5$ ohms; $R_{out} = 350$ ohms

Prob. 8.10 $g_m = 4 \times 10^3$ mhos; $r_p = 6.4 \times 10^5$ ohms; $\mu = 2.6 \times 10^3$

Prob. 8.11 $R_P = 3.13 \times 10^4$ ohms

Prob. 8.12 $A = -78$; if R_C is bypassed $A = -122$

Prob. 8.13 $C = 1.6 \times 10^{-9}$ farads

Prob. 8.14 $R_{in} = 10^5$ ohms; $R_{out} = 3.04 \times 10^4$ ohms; $R_L = 3.04 \times 10^4$ ohms

Chap. 9

Prob. 9.1 $\lambda = 10^{-9}$ met $= 10$ Å

Prob. 9.2 $\Delta E = 7.5 \times 10^{-8}$ electron volts

Prob. 9.3 $v = 7.25 \times 10^5$ met/sec

Prob. 9.4 $d = 10^{-21}$ met

Prob. 9.5 $\lambda_n = 5/n$ Å, n an integer

Prob. 9.6 $\lambda_1 = 5$ Å
$k_1 = 2\pi/\lambda_1 = 2\pi/5$ Å^{-1}

Prob. 9.11 $d = 1.1 \times 10^{-10}$ met

Prob. 9.12 $V = 1.28 \times 10^{14}$ volts

Prob. 9.15 $f = 6.5 \times 10^{-10}$. This is the probability that a single spin state is occupied.

Chap. 10

Prob. 10.1 In that part of the depletion layer closest to the metal the concentration of carriers is roughly 10^{-16} of the concentration in the bulk material.

Prob. 10.2 $D = 6.3 \times 10^{-8}$ meters $= 630$ Å

Prob. 10.6 $n_n = 4.5 \times 10^5$ electrons/cm^3;
$n_p = 2.25 \times 10^7$ holes/cm^3

Prob. 10.7 $N = 2.4 \times 10^{16}$ pentavalent impurities/cm^3

Prob. 10.8 An ac voltage of amplitude 2 volts (a total swing of 4 volts) will give a total variation of one percent in ω.

Prob. 10.9 $\alpha = 0.995$ gives $\beta = 199$;
$\beta = 50$ gives $\alpha = 0.98$

Prob. 10.10 $V = -5.5$ volts for breakdown. The diode is silicon.

Prob. 10.11 Forward current gain $= \beta = 200$; $\alpha = 0.995$

Prob. 10.12 Forward current gain $= \alpha = 0.95$; $\beta = 20$

Prob. 10.13 $i_C = 0.99 \times 10^{-13} e^{v_{BE}/0.026}$ amp
The expression is accurate only for forward bias.

Chap. 11

Prob. 11.1 $N_S/N_P = 0.488$

Prob. 11.3 $v_T = 1$ volt, $i = 4 \times 10^{-3}$ amp;
$v_T = 1.5$ volts, $i = 8.5 \times 10^{-3}$ amp
$v_T = 2.0$ volts, $i = 13.5 \times 10^{-3}$ amp

Prob. 11.4 $I_B = 1.3 \times 10^{-5}$ amp; $I_C \simeq I_E = 4 \times 10^{-3}$ amp; $V_B = 3$ volts;
$V_C = 7.8$ volts; $V_E = 2.4$ volts; $r_{tr} = 10$ ohms; $A = -3$;
$R_i = 2400$ ohms; $R_o = 1800$ ohms;
The voltage source must be positive.

Prob. 11.6 $I_B = 3.1 \times 10^{-5}$ amp; $I_C \simeq I_E = 4.7 \times 10^{-3}$ amp; $V_B = 3$ volts;
$V_C = 6.5$ volts; $V_E = 2.8$ volts; $r_{tr} = 9$ ohms; $A = -3$;
$R_i = 2400$ ohms; $R_o = 1800$ ohms

Prob. 11.7 $h_{fe} = 250$; $h_{oe} = 1.75 \times 10^{-4}$ mhos

Prob. 11.9 Quiescent point: $I_C = 11.4 \times 10^{-3}$ amp, $V_{CE} = 18.6$ volts;
$I_C \simeq I_E = 11.4 \times 10^{-3}$ amp; $I_B = 64 \times 10^{-6}$ amp; $V_C = 30$ volts;
$V_E = 11.4$ volts; $V_B = 12$ volts; $r_{tr} = 5$ ohms; $A = 0.995$;
$R_i = 3600$ ohms; $R_o = 5$ ohms

Prob. 11.10 Quiescent point: $I_C = 13.4 \times 10^{-3}$ amp, $V_{CE} = 23.2$ volts;
$I_C \simeq I_E = 13.4 \times 10^{-3}$ amp; $I_B = 70 \times 10^{-6}$ amp, $V_C = 36.6$ volts;
$V_E = 13.4$ volts; $V_B = 14$ volts; $r_{tr} = 5$ ohms; $A = -200$;
$R_i \cong 1000$ ohms; $R_o = 1000$ ohms.

Prob. 11.11 The same as the answers to Prob. 11.10 except $A = -18$;
$R_i \simeq 3700$ ohms.

Prob. 11.12 Quiescent point: $I_C = 4 \times 10^{-3}$ amp, $V_{CE} = 20$ volts;
$I_C \simeq I_E = 4 \times 10^{-3}$ amp; $I_B = 12 \times 10^{-6}$ amp; $V_C = 24$ volts;
$V_E = 4$ volts; $V_B = 4.6$ volts; $r_{tr} = 10$ ohms, $A = -13$;
$R_i = 5000$ ohms; $R_o = 4000$ ohms.

Prob. 11.13   The output voltage with the load is reduced to one third of its open circuit output value.

Prob. 11.14 $I_D = 3.8 \times 10^{-3}$ amp; $g_{fs} = 3 \times 10^{-3}$ mhos; $r_{os} > 2.5 \times 10^4$ ohms (This cannot be read accurately from the curves.)

Prob. 11.15   Quiescent point: $I_D = 4 \times 10^{-3}$ amp, $V_{DS} = 6$ volts; The power supply is -10 volts; $I_D = I_S = 4 \times 10^{-3}$ amp; $V_D = -7.5$ volts; $V_G = 0$ volts; $V_S = -1.5$ volts; $g_{fs} = 3 \times 10^{-3}$ mhos; $r_{os} > 2.5 \times 10^4$ ohms; $A = -1.9$; $R_i = 10^6$ ohms; $R_o = 625$ ohms

Prob. 11.16 $A = -0.9$

Prob. 11.17 $A = v_o/v_i = \dfrac{-1.9}{1 - 10^3 j/\omega} = \dfrac{1.9\, e^{-j\phi}}{[1 + (10^3/\omega)^2]^{1/2}}$

where $\phi = \pi - \tan^{-1}(10^3/\omega)$
ω is in radians/sec
as $\omega \to \infty$, v_o and v_i are 180° out of phase
as $\omega \to 0$, v_o lags v_i by 90°

Prob. 11.18   With a 625 ohm load resistor we have $v_o/v_i = -0.95$

Prob. 11.19 $I_D = 4 \times 10^{-3}$ amp; $A = 0.53$; $R_i = 10^6$ ohms; $R_o = 175$ ohms

Chap. 12

Prob. 12.2 $A_F = \dfrac{A_0}{1 - \beta A_0} = \dfrac{A/(1 + j\omega/\omega_0)}{1 - \beta A/(1 + j\omega/\omega_0)} = \dfrac{A/(1 - \beta A)}{1 + j\omega/[\omega_0(1 - \beta A)]}$

With feedback the dc gain is $A/(1 - \beta A)$ and the bandwidth is $\omega_0(1 - \beta A)$. The product of these quantities is $[A/(1 - \beta A)]\omega_0(1 - \beta A) = A\omega_0$. Thus the feedback does not change the gain-bandwidth product.

Prob. 12.3   Without feedback: $\Delta V_{rms} = 4.1 \times 10^{-3}$ volts
With feedback: $\Delta V_{rms} = 4.1 \times 10^{-4}$ volts

Prob. 12.5   With feedback the gain of each circuit is $A_F = 25$. Circuit (a) has the lower distortion, the higher input impedance and the lower output impedance.

Prob. 12.6   Use the $+$ input as the signal input. Connect the $-$ input to the output of the amplifier.
$A_F = 1$; $R_{if} = 10^7$ ohms; $R_{of} = 10^{-3}$ ohms

Prob. 12.9 $A_0 = 2 \times 10^4$

Prob. 12.10 $A = v_o/v_i = \dfrac{1}{1 + R/R_L} \cong 1$ for large R_L; $R_o = R$

Prob. 12.11 $R_{if} = \dfrac{R_i[R_L + R_o + (1 - A_0)R_F]}{R_L + R_o + R_F}$

$R_{if} \simeq \dfrac{-R_i R_F A_0}{R_L + R_o + R_F}$ if A_0 is large

Note that A_0 must be negative since β is taken to be positive.

Prob. 12.12 $\Delta V_{rms} = 4.1 \times 10^{-5}$ volts

Prob. 12.13 $\Delta I_{rms} = 5.7 \times 10^{-8}$ amp

Prob. 12.14 The Johnson noise is $\Delta V_{rms} = 4.1 \times 10^{-6}$ volts. This is essentially the entire noise.

Prob. 12.15 $\omega_\beta = 1.25 \times 10^6$ rad/sec

Prob. 12.17 $\omega = 7.2 \times 10^6$ rad/sec; $f = 1.15 \times 10^6$ Hz.

Prob. 12.19 Without C_F: $A = \dfrac{R_E}{R_E + r_{tr}}$; The input resistance R_i is the equivalent resistance of the bias network $[R + R_1 R_2/(R_1 + R_2)]$ in parallel with $\beta(r_{tr} + R_E)$

With C_F: The capacitor C_F greatly increases the effective resistance of the resistor R for incremental signals since the incremental voltage difference across R is $v_i - Av_i = v_i(1 - A)$ where A is very close to one. The effective value of R is increased to $R/(1 - A)$. The amplification is somewhat reduced because the emitter to ground resistance becomes R_{eq} where $1/R_{eq} = 1/R_E + 1/R_1 + 1/R_2$. We ignore $R/(1 - A)$. However the input resistance is greatly increased. The quantity $R/(1 - A)$ can easily be made large enough so that $R_i \simeq \beta(r_{tr} + R_{eq})$.

Prob. 12.20 $A = \dfrac{j\omega C_{gd} R_D' - A_0}{1 + j\omega R_g[C_{gs} + (A_0 + R_D'/R_g + 1)C_{gd}] - (\omega C_{gs} R_g)(\omega C_{GD} R_D')}$

where $A_0 = g_{fs} R_D'$ and $R_D' = R_D r_{os}/(R_D + r_{os})$

The most important term in reducing the gain at high frequencies is the term $j\omega R_g(A_0 + 1)C_{gd}$ in the denominator (the Miller effect). Thus the amplification is usually written in the simplified form

$$A = \frac{-A_0}{1 + j\omega R_g[C_{gs} + (A_0 + 1)C_{gd}]}$$

Chap. 13

Prob. 13.1 $\omega = 1500$ rad/sec; $f = 239$ Hz

The intrinsic input resistance of the transistor T_1 is very high because of the negative feedback.

Prob. 13.2 $R_i = 1200$ ohms; $R_o = 5$ ohms

Prob. 13.3 $R_i = 1333$ ohms; $R_o = 0.4$ ohms

Prob. 13.4 $A_0 = \dfrac{-g_{fs} R_D}{1 + (R_D + r_{os})/r_{os}(r_{os} g_{fs} + 1)} \simeq -g_{fs} R_D;$

$R_i = R_G$; $R_o = R_D$

The two FET's are assumed to be identical. The Miller effect capacity is reduced from $(1 - A_0)C_{gd}$ to $2C_{gd}$

Prob. 13.5 $f_0 = 10^6$ Hz $= 1/2\pi\sqrt{LC}$;

$C = 10^{-9}$ farads; $L = 2.53 \times 10^{-5}$ henries

Prob. 13.6 P_{max} (to load) $= 1/4$ watt;

$I_C = 0.05$ amp

Prob. 13.7 P_{max} (to load) $= 1/2$ watt;
$I_c = 0$

Prob. 13.9   Silicon: $\partial I_{co}/\partial T = 1.41 \times 10^{-12}$ amp/°K
Germanium: $\partial I_{co}/\partial T = 9 \times 10^{-8}$ amp/°K

Prob. 13.13   If $R_L = 15$ ohms, $R_i = 1500$ ohms; if $R_L = 10$ ohms, $R_i = 1000$ ohms;
if $R_L = 150$ ohms, $R_i = 15,000$ ohms
This assumes that R_1 and R_2 are $\gg \beta R_L$

Prob. 13.14   We assume the input to T_2 does not seriously load the output of T_1.
This is assured since the input resistance to T_2 is never much less
than βr_{tr2} (if $R_L = 0$) while the output resistance of T_1 is r_{tr1} and
$r_{tr1} \simeq r_{tr2}$.
$A_v = [R_E/(R_E + r_{tr})]^2$ if $R_L = \infty$ and $r_{tr1} = r_{tr2} = r_{tr}$
$R_i \cong$ the parallel combination of R_B and $\beta(r_{tr} + R_E)$
$R_o \cong R_E r_{tr}/(R_E + r_{tr})$
$A_i = i_o/i_i = (R_i/r_{tr}) [R_E/(R_E + r_{tr})]$
where i_i is the input current and i_o is the output current which flows
through T_2 and R_L when $R_L = 0$. In the approximation $R_E \gg r_{tr}$ and
$R_B \simeq \beta R_E$ we have $A_i = \beta R_E/2r_{tr}$. The ratio i_o/i_i is lower than
$A_i = \beta R_E/2r_{tr}$ if R_L is greater than zero.

Prob. 13.15   The following results are correct for $\beta \gg 1$ and $R_E \gg r_{tr}$

$$A_v = \frac{R_{eq}}{R_{eq} + R_g} \simeq 1 \text{ if } R_{eq} = \frac{\beta^2 R_E R_B}{R_B + \beta^2 R_E} \gg R_g$$

$$R_i = R_g + \frac{\beta^2 R_E R_B}{R_B + \beta^2 R_E} \quad (\text{if } R_L = \infty)$$

$$R_i = R_g + \frac{\beta^2 r_{tr} R_B}{R_B + \beta^2 r_{tr}} \quad (\text{if } R_L = 0)$$

The input resistance R_i is the resistance seen by the signal generator v_i

$$R_o = r_{tr}$$

$$A_i = i_o/i_i = \frac{\beta^2 R_B r_{tr}}{r_{tr}(R_B + \beta^2 r_{tr})} \simeq \beta^2 \quad \text{if } R_B \gg \beta^2 r_{tr}$$

where i_o is the output current when $R_L = 0$.

Prob. 13.17   With the assumptions $R_G \gg R_S$ or R_D, $r_{os} \gg R_S$ or R_D and $g_{fs} r_{os} \gg 1$
we obtain

$$A_v = \frac{g_{fs} r_{os}[1 + g_{fs}(R_S + R_D)]}{1 + g_{fs} r_{os}[1 + g_{fs}(R_S + R_D)]} \simeq 1$$

$$R_i = R_G$$
$$R_o = 1/g_{fs}[1 + g_{fs} R_D/(1 + g_{fs} R_S)]$$

If the FET's are similar to the 2N3823 discussed in Chapter 11 appropriate values would be

$R_D = R_S = 150$ ohms

$R_G = 10^6$ ohms

$V_{DD} = 20$ volts; $V_{GG} = 9$ volts

Prob. 13.18 $A_v = \dfrac{g_{fs}r_{os}}{1 + g_{fs}r_{os}} \simeq 1;$

$R_i = R_G/2; \ R_o = 1/2g_{fs}$

Chap. 14

Prob. 14.1 $R_o = r_{os}(1 + Rg_{fs})/2$

Prob. 14.3 $R_o = R_C = 4 \times 10^3$ ohms, single ended

$R_o = 2R_C = 8 \times 10^3$ ohms, double ended

Prob. 14.4 $A_{\text{diff}} = -R_C/2(r_{tr} + R)$

Prob. 14.5 $A_{\text{diff}} = -g_{fs}R_C/2$

Since $1/g_{fs} \simeq 250$ ohms is usually larger than r_{tr} the FET difference amplifier will have a lower gain than the bipolar difference amplifier. In order that r_{tr} be less than 250 ohms the current through a bipolar transistor only needs to be greater than 0.1 milliamp.

Prob. 14.6 An FET difference amplifier will have a higher input impedance.

Prob. 14.8 CMRR $= 20 \log [(2\beta r_c)/r_{tr}]$ decibels

$R_{i,\,\text{diff}} = 2(\beta + 1)r_{tr}$

$R_{i,\,\text{comm}} = (\beta + 1)\beta r_c$

Prob. 14.9 $A_{\text{diff}} = -R_C/2[r_{tr} + RR_E/(R + 2R_E)]$

$A_{\text{comm}} = -R_C/2[r_{tr} + RR_E/(R + 2R_E) + 2R_E^2/(R + 2R_E)]$

Prob. 14.10 Integrate for at least 1/4 second.

Prob. 14.13 The time constant should be 90 seconds.

Prob. 14.14 $v_o = 24$ volts

Prob. 14.16 $v_o/v_i \simeq 0.025$

Note that v_o and v_i are incremental quantities.

Prob. 14.17 $v_o = 3 \times 10^{-4}$ volts

Chap. 15

Prob. 15.9 $v_o = (-R/L) \int v_i \, dt$

Prob. 15.10 Possible values of the components are $R = 10$ ohms, $C = 10^{-6}$ farads and $R' = 5 \times 10^3$ ohms. With these values $RC = 10^{-5}$ sec $= 1/\omega_1$ and $(R + 2R')C = 10^{-2}$ sec $= A_0/\omega_2$.

Prob. 15.14 $v_o = -\left[\dfrac{Rv_i}{R'} + \dfrac{L}{R'}\dfrac{dv_i}{dt} + \dfrac{1}{R'C}\int v_i \, dt\right]$

Prob. 15.15 $v_o = -\sqrt{v_i}$, v_i must be a positive voltage

Prob. 15.16 $v_0 = (-1/LC)\int dt \int v_i \, dt$

Chap. 16

Prob. 16.2 $M \geq 1.6 \times 10^{-8}$ henries

Prob. 16.5 $$|Z|^2 = \frac{R^2 + (\omega L - 1/\omega C)^2}{(\omega R C_s)^2 + \left(\omega^2 L C_s - \dfrac{C_s}{C} - 1\right)^2}$$

$\omega_s = 2\pi f_s \cong 1/\sqrt{LC}$ (minimize the numerator)
$\omega_p = 2\pi f_p \cong 1/\sqrt{L[CC_s/(C + C_s)]}$ (minimize the denominator)
$\omega_p - \omega_s = 2\pi(f_p - f_s) = 1/2(C/C_s)1/\sqrt{LC}$

At both ω_p and ω_s the phase shift is very close to zero.
The above results are not exact but are very close to the correct values because the quartz oscillator has a high Q. In typical cases C_s may be of the order of $10^2 C$. Thus ω_p and ω_s are close together but each is well defined, again because of the high Q.

Prob. 16.6 $Q = 8.33 \times 10^4$; $R_1 = 4000$ ohms
$R_1 < 4000$ ohms for low output voltages;
$R_1 = 4000$ ohms at the desired output voltage;
$R_1 > 4000$ ohms for high output voltages

Prob. 16.7 Possible choices are $R_1 = R_2 = 10^3$ ohms, $C_1 = C_2 = 10^{-8}$ farads, $R_3 = 2 \times 10^3$ ohms, $R_4 = 10^3$ ohms. The ratio R_3/R_4 is determined once the ratios R_2/R_1 and C_1/C_2 are chosen. The above choices will give an oscillation frequency $\omega = 10^5$ rad/sec.

Prob. 16.8 At the oscillation frequency $\omega^2 = 1/R_1R_2C_1C_2$ the feedback ratios in the positive feedback network and the negative feedback network are the same.

$$\beta_+ = \frac{1}{1 + R_2/R_1 + C_1/C_2} = \beta_- = \frac{R_4}{R_3 + R_4}$$

Prob. 16.9 $\omega^2 = 1/6R^2C^2$; $\omega = 2 \times 10^4$ rad/sec; $f = 3.2 \times 10^3$ Hz.

Prob. 16.10 $T = 0.7R_{B2}C' = 10^{-3}$ sec.; $R_{B2} = 1.4 \times 10^5$ ohms
Saturation of T_2 is assured since $\beta R_C = 3 \times 10^5 > R_{B2} = 1.4 \times 10^5$

Prob. 16.11 Choose $C = 10^{-8}$ farads, $R_B = 1.4 \times 10^5$ ohms, and $R_C = 5 \times 10^3$ ohms. Saturation is assured since $\beta R_C = 5 \times 10^5 > R_B = 1.4 \times 10^5$

Prob. 16.14 $\lambda = 1.13 \times 10^{-6}$ meters (silicon);
$\lambda = 1.39 \times 10^{-6}$ meters (germanium)

Prob. 16.15 $R_T \simeq 6000$ ohms
First calculate at what time in the positive half cycle the SCR must turn on if the average dissipation in R_L is to be 40 watts. Then calculate the value of R_T which gives 0.025 amp gate current at this time.

Chap. 17

Prob. 17.1

Decimal	Binary	Hexadecimal	Octal
5	0101	5	5
18	0001 0010	12	22
325	0001 0100 0101	145	505
1753	0110 1101 1001	6D9	3331

Prob. 17.2	Decimal	Binary	Hexadecimal	Octal
	873	0011 0110 1001		
	278	0001 0001 0110		
	44085	1010 1100 0011 0101	AC35	
	1067	0100 0010 1011	42B	
	483	0001 1110 0011		743
	354	0001 0110 0010		542

Prob. 17.8 $f = 1/T = 1/2n\tau = 1/(2 \times 7 \times 10^{-8}) \simeq 7 \times 10^6$ Hz

Prob. 17.9

$$113 = \left[\begin{array}{l} \underline{0}110\ 0001 \\ \underline{1}011\ 0100 \\ \hline \end{array} \right.$$

$$\left. {}^{1}\underline{0}010\ 0101 \right]$$

$$\overset{-}{}1 \hookleftarrow$$

$$38 = \quad \underline{0}010\ 0110$$

Prob. 17.16 $RC \simeq 6.8 \times 10^{-6}$ sec
This assumes that the voltage at the input of gate 2 must fall from 3.5 volts to 0.8 volts before the output pulse ends. Possible choices are $R = 100$ ohms, $C = 6.7 \times 10^{-8}$ farads. R cannot exceed 500 ohms since 500 ohms $\times$ 1.6 milliamps equals 0.8 volt and a logical zero may not be reliably recognized. A lower value of R is desirable so that the exponential decay of the voltage is toward a value considerably less than 0.8 volt. This assures that the pulses are reasonably uniform in length.

Prob. 17.17 $A \simeq 2.2$. This assumes that the input voltage level must rise from 0.2 volts to 2.0 volts before the output gate switches from 1 to 0 and that the input must fall from 3.5 volts to 0.8 volts before the output switches from 0 to 1.

Prob. 17.20   One may use a 10 volt power supply, 40×10^3 ohms in the 2^0 digital input, 20×10^3 ohms in the 2^1 input, 10×10^3 ohms in the 2^2 input and 5×10^3 ohms in the 2^3 input. The resistor in the amplifier feedback circuit should be 4.76×10^3 ohms.

Prob. 17.21   At least 17 *JK* flip-flops are needed.

Appendix A/Physical Constants

1. Electronic charge
   $e = 1.6021 \times 10^{-19}$ coulomb
2. Speed of light
   $c = 2.9979 \times 10^8$ meter/sec
3. Mass of the electron
   $m_e = 9.109 \times 10^{-31}$ kilogram
4. Mass of the proton
   $m_p = 1.67252 \times 10^{-27}$ kilogram
5. Avogadro's number
   $N_0 = 6.0225 \times 10^{26}$ molecules/kg-mole
6. Planck's constant
   $h = 6.6256 \times 10^{-34}$ joule-sec
7. Boltzmann's constant
   $k = 1.3805 \times 10^{-23}$ joules/°K
8. Energy conversion
   1 electron volt $= 1.6021 \times 10^{-19}$ joules
9. Length conversion
   1 inch $= 0.0254$ meters
10. Permeability of free space
    $\mu_0 = 4\pi \times 10^{-7}$ newtons/ampere^2
11. Permittivity of free space
    $\varepsilon_0 = 1/(\mu_0 c^2)$ where c is the speed of light
    $\varepsilon_0 = 8.854 \times 10^{-12}$ coulomb^2/newton-meter^2
    $1/(4\pi\varepsilon_0) = 8.987 \times 10^9$ newton-meter^2/coulomb^2

Appendix B/Bibliography

The following is a list of useful books for reference. The list is by no means an exhaustive one.

I. General Physics Textbooks

1. SEARS, F. W. and M. W. ZEMANSKY, *University Physics*, fourth edition. Reading, Mass: Addison-Wesley Publishing Company, 1970.
2. SEMAT, H., *Fundamentals of Physics*, fourth edition. New York: Holt, Rinehart and Winston Inc., 1966.
3. SHORTLEY, G. and D. WILLIAMS, *Elements of Physics*, fifth edition. Englewood Cliffs, New Jersey: Prentice Hall, Inc., 1971.

II. General Books on Electric and Magnetic Fields.

1. FEYNMAN, R. P., R. B. LEIGHTON, and M. SANDS, *The Feynman Lectures on Physics*, Vol. 2. Reading, Mass: Addison-Wesley, Publishing Co., 1964.
2. PANOFSKY, W. K. H. and M. PHILLIPS, *Classical Electricity and Magnetism*. Reading, Mass: Addison-Wesley Publishing Co., 1955.
3. PURCELL, E. M., *Electricity and Magnetism*, Berkeley Physics Course, Vol. 2. New York: McGraw-Hill Book Company, 1963.

III. Textbooks on dc and ac Circuits

1. CLOSE, C. M., *The Analysis of Linear Circuits*. New York: Harcourt Brace Jovanovich, 1966.
2. HAYT, W. H. JR. and T. E. KEMMERLY, *Engineering Circuit Analysis*. New York: McGraw-Hill Book Company, 1962.
3. SCOTT, R. E., *Elements of Linear Circuits*, Reading, Mass: Addison-Wesley Publishing Company, 1965.
4. STOUT, M. B., *Basic Electrical Measurements*, second edition. Englewood Cliffs, New Jersey: Prentice Hall, Inc., 1960.

IV. Electronic Textbooks

1. ALLEY, C. L. and K. W. ATWOOD, *Semiconductor Devices and Circuits*. New York: John Wiley & Sons, Inc., 1971.
2. BENEDICT, R. R., *Electronics for the Scientist*. Englewood Cliffs, New Jersey: Prentice Hall Inc., 1967.

528

3. BROPHY, J. J., *Basic Electronics for Scientists*, second edition. New York: McGraw-Hill Book Co., 1972.

4. DELANEY, C. F. G., *Electronics for the Physicist*. Baltimore, Maryland: Penguin Books, Inc., 1969.

5. DE WAARD, H. and D. LAZARUS, *Modern Electronics*. Reading, Mass: Addison-Wesley Publishing Co. Inc., 1966.

6. THE ENGINEERING STAFF OF TEXAS INSTRUMENTS INCORPORATED, *Transistor Circuit Design*. New York: McGraw-Hill Book Company, 1963.

7. GOSLING, W., W. G. TOWNSEND, and J. WATSON, *Field-Effect Transistors*. New York: Wiley Interscience, 1971.

8. GRAY, P. E. and C. L. SEARLE, *Electronic Principles, Physics, Models and Circuits*. New York: John Wiley and Sons, Inc., 1969.

9. HUNTEN, D. M., *Introduction to Electronics*. New York: Holt, Rinehart and Winston, Inc., 1964.

10. I. C. APPLICATIONS STAFF OF TEXAS INSTRUMENTS INCORPORATED, *Designing with TTL Integrated Circuits*. New York: McGraw-Hill Book Company, 1971.

11. LITTAUER, R. *Pulse Electronics*, New York: McGraw-Hill Book Co., 1965.

12. MALMSTADT, H. V. and C. G. ENKE, *Digital Electronics for Scientists*. New York: W. A. Benjamin Inc., 1969.

13. MALMSTADT, H. V., C. G. ENKE, and E. C. TOREN, *Electronics for Scientists*. New York: W. A. Benjamin Inc., 1963.

14. MILLMAN, J. and H. TAUB, *Pulse and Digital Circuits*. New York: McGraw-Hill Book Co., 1956.

15. SZE, S. M., *Physics of Semiconductor Devices*. New York: John Wiley & Sons, Inc., 1969.

16. TERMAN, F. E. and J. M. STOUT, *Electronic Measurements*, second edition. New York: McGraw-Hill Book Company, 1952.

17. VAN DER ZIEL, A., *Solid State Physical Electronics*, second edition. Englewood Cliffs, New Jersey: Prentice Hall, Inc., 1968.

In addition to these books anyone interested in electronic instrumentation should study carefully the publications and catalogs of the various manufacturers of electronic components and devices.

Index

Ac amplifier, 331
Acceptor level, 238
Accumulator, 503
Active network, 18
Adder, 450
Address (computer), 497
Alpha (α of a transistor), 259
Alpha (α) cutoff frequency, 323
Ammeter, 59
Ampere, 3
Ampere's law, 85
Amplification factor, 211
Amplifier, ac, 331
  cascode, 338
  cathode follower, 222
  chopper, 367
  class A, 341
  class B, 343
  class C, 345
  common base, 261, 303
  common emitter, 261, 277
  difference, 358
  direct coupled, 355
  emitter follower, 348, 350
  linear analysis of, 217, 276

operational, 364, 379
power, 340
push-pull, 343
RC coupled, 331
source follower, 294
tuned, 338
vacuum tube, 213
Analog computer, 382
Analog to digital (A to D) conveter, 461
AND gate, 436
Anode (of a vacuum tube), 205, 208
Anticoincidence circuit (*see also* EXCLUSIVE OR gate), 445
Arithmetic using logic gates, 446
Astable multivibrator, 416, 465
Attenuation in dB, 138
Attenuator, 151
Ayrton shunt, 57

Back diode, 479
Ballistic galvanometer, 58
Band structure of crystals, 230
Bandwidth, 323

Barkhausen criterion for oscillation, 406
Base of bipoler transistor, 260
Base of MOSFET, 264
Base thickness, 261
Battery, 10
BCD (binary coded decimal), 434
Beta (β of a transistor), 262
Beta (β) cutoff frequency, 323
Biasing of transistors, 285
Biasing of vacuum tubes, 214
Binary numbers, 432
  addition, 434
  multiplication, 434
  subtraction using 1's complement, 435
Biot-Savart law, 84
Bipolar transistor, 258
Bistable multivibrator, 411, 454
Bit, 433
Bit plane, 497
Boolean algebra, 447
Bootstrapping, 329
Branch, 11, 28
Branch point, 11, 28

531